ANNUAL REVIEW OF
NEUROSCIENCE

EDITORIAL COMMITTEE (1981)

ANNUAL REVIEW OF NEUROSCIENCE

W. MAXWELL COWAN, *Editor*
Washington University School of Medicine

ZACH W. HALL, *Associate Editor*
University of California School of Medicine

ERIC R. KANDEL, *Associate Editor*
College of Physicians and Surgeons of Columbia University

VOLUME 4

1981

ANNUAL REVIEWS INC.   4139 EL CAMINO WAY   PALO ALTO, CALIFORNIA 94306 USA

ANNUAL REVIEWS INC.
Palo Alto, California, USA

REPRINTS   The conspicuous number aligned in the margin with the title of each article in this volume is a key for use in ordering reprints. Available reprints are priced at the uniform rate of $2.00 each postpaid. The minimum acceptable reprint order is 5 reprints and/or $10.00 prepaid. A quantity discount is available.

International Standard Serial Number: 0147-006X
International Standard Book Number: 0-8243-2404-8

Annual Reviews Inc. and the Editors of its publications assume no responsibility for the statements expressed by the contributors to this Review.

PRINTED AND BOUND IN THE UNITED STATES OF AMERICA

Annual Review of Neuroscience
Volume 4, 1981

CONTENTS

(Note: Titles of chapters in Volumes 1–4 are arranged by category on pages 554–56.)

ANNUAL REVIEWS INC. is a nonprofit corporation established to promote the advancement of the sciences. Beginning in 1932 with the *Annual Review of Biochemistry,* the Company has pursued as its principal function the publication of high quality, reasonably priced *Annual Review* volumes. The volumes are organized by Editors and Editorial Committees who invite qualified authors to contribute critical articles reviewing significant developments within each major discipline. Annual Reviews Inc. is administered by a Board of Directors, whose members serve without compensation.

Publications

Annual Reviews of Anthropology, Astronomy and Astrophysics, Biochemistry, Biophysics and Bioengineering, Earth and Planetary Sciences, Ecology and Systematics, Energy, Entomology, Fluid Mechanics, Genetics, Materials Science, Medicine, Microbiology, Neuroscience, Nuclear and Particle Science, Nutrition, Pharmacology and Toxicology, Physical Chemistry, Physiology, Phytopathology, Plant Physiology, Psychology, Public Health, and *Sociology.*

Special Publications: *History of Entomology* (1973), *The Excitement and Fascination of Science* (1965), *The Excitement and Fascination of Science, Volume Two* (1978), *Annual Reviews Reprints: Cell Membranes, 1975–1977* (published 1978), *Annual Reviews Reprints: Immunology, 1977–1979* (published 1980), and *Intelligence and Affectivity: Their Relationship During Child Development,* by Jean Piaget (1981).

For the convenience of readers, a detachable order form/envelope is bound into the back of this volume.

SOME RELATED ARTICLES IN OTHER *ANNUAL REVIEWS*

From the *Annual Review of Biophysics and Bioengineering,* Volume 9 (1980)

Stimulus-Response Coupling in Gland Cells, B. L. Ginsborg and C. R. House

Modulation of Impulse Conduction Along the Axonal Tree, H. A. Swadlow, J. D. Kocsis, and S. G. Waxman

Nerve Growth Factor: Mechanism of Action, S. Vinores and G. Guroff

The Structure of Proteins Involved in Active Membrane Transport, A. S. Hobbs and R. W. Albers

Special Techniques for the Automatic Computer Reconstruction of Neuronal Structures, I. Sobel, C. Levinthal, and E. R. Macagno

Certain Slow Synaptic Responses: Their Properties and Possible Underlying Mechanisms, J. Kehoe and A. Marty

From the *Annual Review of Pharmacology and Toxicology,* Volume 20 (1980)

Metal Ion Interactions with Opiates, D. B. Chapman and E. L. Way

Radioligand Binding Studies of Adrenergic Receptors: New Insights into Molecular and Physiological Regulation, B. B. Hoffman and R. J. Lefkowitz

The Electrogenic Na^+,K^+-Pump in Smooth Muscle: Physiologic and Pharmacologic Significance, W. W. Fleming

Structure-Activity Relationships of Enkephalin-Like Peptides, J. S. Morley

From the *Annual Review of Physiology,* Volume 43 (1981)

Gap Junctional Communication, E. L. Hertzberg, T. S. Lawrence, and N. B. Gilula

Membrane Charge Movements and Depolarization-Contraction Coupling, M. F. Schneider

Development, Innervation, and Activity-Pattern Induced Changes in Skeletal Muscle, F. Solesz and F. A. Sreter

Regulation of the Cerebral Circulation, H. A. Kontos

Neural Regulation of the Heart Beat, M. N. Levy, P. J. Martin, and S. L. Steuse

Epidermal and Nerve Growth Factors in Mammalian Development, D. Gospodarowicz

"Endorphins" in Pituitary and Other Tissues, H. Imura and Y. Nakai

Anatomical Organization of Central Respiratory Neurons, M. P. Kalia

Roger Sperry

Ann. Rev. Neurosci. 1981. 4:1–15
Copyright © 1981 by Annual Reviews Inc. All rights reserved

CHANGING PRIORITIES   ❖11547

Roger W. Sperry[1]

Division of Biology, California Institute of Technology, Pasadena,
California 91125

INTRODUCTION

Rather than address select aspects of neuroscience, or particular problems with which I happen to have been involved, or to personally reminisce about golden "days that are no more," I plan to respond to the kind invitation to write a prefatory chapter for this volume with a few observations bearing on matters that, while personal in some respects, are also of general ongoing concern to all of us in science and particularly to the future of neuroscience as a discipline.

We used to say that there are two kinds of scientists: those fired-up by a problem and searching for methods to get the answers, and those highly trained in some method who are searching about for some amenable problems. While most of us line up somewhere between these extremes, there is much to be said for, at least in principle, giving preference where possible to problem priorities over methodology. What follows is, above all, problem-oriented and attuned throughout to the query, "What difference does it make?—especially ten, twenty, or more years from now?"

In terms of governmental funding and in other respects, it has become apparent that the overall federal rating for neuroscience is not as high today as it was prior to the Nixon budget reforms of the early 1970s. Nor does it appear that the decline is something temporary from which funding can be expected soon to recover. Nor is it restricted to neuroscience; science in general has been affected with certain exceptions such as cancer-related, energy-related, and other select projects where a major application to current quality-of-life problems is obvious. These changes in relative funding can be assumed to reflect real changes in social priorities and in society's

[1]Preparation of the manuscript was aided by the F. P. Hixon Fund of the California Institute of Technology.

1

collective judgment of the importance of science and what it contributes. We read in *Science* (Sawhill 1979) of the "public disillusionment" and "today's more jaundiced view" of science and that "faith in the beneficence of scientific endeavor and the promise of technology has been steadily eroding."

An underlying cause for these changes can be seen in the new and growing recognition of mounting global, so-called "crisis" problems which science is alleged to have helped to create and which in addition are complicated by social value problems for which science is apparently unable to provide answers. When the quality and even survival of civilized society is potentially in threat, what difference does it make to Congress or the public whether we find some new nerve connections in the brain, some new transmitters or receptors, and so on? Even the ever-strong humanitarian appeal of medical advancements that might eventually save hundreds of thousands of lives does not fully escape the new unspoken perspective of a world already afflicted with population imbalances in the hundreds of millions. The overwhelming priority of the growing demands of today's "global crisis" problems was already perceived in the late 1960s (Platt 1969) to be of sufficient magnitude and urgency to warrant mustering the scientific community in an all-out crash attack with the implication that to continue the practice of "science as usual" is morally indefensible.

Although little has happened in the interim to reduce the specter of global breakdowns, it seems that considerable has happened to discourage public feeling that science and improved technology can be counted on to bring solutions. While the world's production per capita of most major products of the basic biological source systems has already peaked and started on a long downward trend (Brown 1979), world demand continues to rise at the rate of 6 million people per month with predictions of inevitable social turmoil as peoples and nations grow more desperate.

Earlier hopes that science might rise to the occasion with "green revolutions" and other technological answers begin to fade. Science and improved technology, we come to realize, merely make it possible to better maintain more people in better style, for a while, until new limits are reached and the same problems reappear, along with new ones, and all on a greatly magnified scale. Science and advanced technology, whether medical, agricultural, military, energy-related, etc, are now seen, in the long run, to put us in an escalating vicious spiral of technology-population-energy-pollution-etc increases, in which we are now firmly entrapped. There is no reflection here on science or technology per se, of course. As we say, Utopia is tomorrow's level of technology combined with the population levels of the nineteenth century. Remedial suggestions, however, that might in any way involve the highly sensitive, value-laden factor of population controls promptly raise a host of moral issues and value conflicts for which, again, science, it is held, does not provide answers.

Futurists and common logic concur that a substantial change, worldwide, in lifestyle and moral guidelines will soon become an absolute necessity. On a planet of finite resources, the laws and mores of a freely increasing population must eventually be replaced by those of a regulated population, and the sooner this inevitable shift occurs, the better for the residual quality of the biosphere. In brief, it becomes increasingly evident that the prime, urgent need of our times is not for more science and improved technology, medical or otherwise, but for some new ethical policies and moral guidelines to live and govern by.

Once this conclusion is perceived, the common tendency is to by-pass science to look elsewhere for answers. Problems that resolve to basic issues in ethics and morality are traditionally supposed to be beyond science. It is argued that science which describes facts cannot be used to prescribe values. Prevailing doctrine in modern philosophy asserts that it is logically impossible to derive values from a set of scientific facts, or to infer what ought to be from what is. In view of the collective effect of various considerations of this kind, it is hardly surprising that public faith in the promise of science and technology "has been steadily eroding."

A DIFFERENT APPROACH

In what follows I try to defend a position directly counter to the above, which would, in effect, not only restore to science any loss in public favor, but would go further to give science, and the scientific endeavor generally, a new public image and a higher societal role of top priority. On the proposed terms, science becomes the prime hope for escape from the vicious spirals of advancing civilization, but for other reasons. A different approach to the public support and role of science is suggested in which science is upheld, not because it begets improved technology, but because of its unmatched potential for the shaping of ethical values. In the worldview perspectives and truths of science we will find the best key to valid moral guidelines. The arguments are adapted to today's priorities and grow stronger, not weaker, as current global conditions worsen. Even basic "pure" research and the practice of "science as usual" emerge on the proposed terms with a heightened social and moral approval.

The usual appeal to medical, educational, technological, and other direct benefits is by-passed and our bets are placed instead on certain less obvious human value implications that stem from brain research. Particularly relevant are recent changes in concepts relating to the mind of man, the nature of the conscious self, freedom of choice, causal determinacy, and to the fundamental relation of mind to matter and to brain mechanism. Some of man's most enduring concerns are involved, i.e. whether consciousness is mortal or immortal, cosmic or brain-bound, or reincarnate, and the like. It is in terms of the humanistic implications along these and related lines that neuroscience has always had its special interest and greatest meaning. Ideologies, philosophies, religious doctrines, world-models, value systems, and the like will stand or fall depending on the kinds of answers that brain research eventually reveals. It all comes together in the brain.

In brief, recent conceptual developments in the mind-brain sciences are seen to bring changes in worldview perspectives that revise the ultimate criteria and frame of reference for determining human value priorities and resolving value differences. A broad shift of conceptual framework regarding science and human values is involved. Promising prospects can be seen, especially as these changed perspectives apply in those global crisis areas wherein lie today's most serious threats to the quality of life, and where differential outcomes in the resolution of value conflicts tend to have tremendous social consequences. For example, even a slight shift in the delicate balance of value perspectives on the abortion vs right-to-life issue could mean the difference, in itself, of literally many millions of lives, pro or con, in the next few years—with enormous secondary impacts as well—and all compounded on future generations. Similar wide-ranging, quality-of-life consequences result from a shift of values in regard to energy-environmentalist issues, species' rights, and other global concerns.

It is our present contention that a scientific approach to both the theory and the prescription of ethical values is not only feasible, but is by far the best way to go, offering the most promising, perhaps only, visible hope for future generations. The supporting arguments have already been expounded in some detail elsewhere and may be found in the original articles cited and their references (Sperry 1965, 1969, 1972, 1977, 1980). Rather than assume prior knowledge or laboriously restate the reasoning in different words, it is more expedient for present purposes to simply list below some of the principal postulates, propositions, observations, and inferences as excerpted with minor changes from the previous accounts. Because the subject matter ranges somewhat afield from neuroscience, overlap and redundancy are risked rather than the reverse. Some attempt at logical ordering is maintained, but the cross logistics mount rapidly, and a quick grasp of the whole may be found preferable to a logical sequential approach.

COLLECTIVE POSTULATES AND PROPOSITIONS

Subjective Values in Objective Perspective

•In addition to their commonly recognized significance from a personal, religious, or philosophic standpoint, human values can also be viewed objectively in causal control terms as universal determinants in all human decision-making. All decisions boil down to a choice among alternatives of what is most valued, and are determined by the particular value system that prevails.

•Human values, viewed in objective, scientific perspective, stand out as the most strategically powerful causal control force now shaping world events. More than any other causal system with which science now concerns itself, it is variables in human value systems that will determine the future.

•Any given brain will respond differently to the same input and will tend to process the same information into quite diverse behavioral channels depending on its particular system of value priorities. In short, what an individual or a society values, determines very largely what it does.

•As a social problem, human values can be rated above the more tangible global concerns such as those of poverty, pollution, energy, and overpopulation on the ground that these more concrete problems are all man-made and are very largely products of human values. Further, they are not correctable on any long-term basis without effecting adaptive changes in the underlying human values involved.

•The human value factor in biospheric controls stands out as the primary underlying root cause of most of today's difficulties. The more strategic way to remedy mounting adverse global conditions is to go after the social value priorities directly in advance, rather than waiting for man's values to change in response to changing external conditions. Otherwise we are doomed to live always on the margins of intolerability, for it is not until things begin to get intolerable that the voting majority gets around to changing its established values.

•Recent developments in the mind-brain sciences eliminate the traditional dichotomy between science and values and support a revised philosophy in which modern science becomes the most effective and reliable means available for determining valid criteria for moral value and meaning.

Value Theory

•A science of values in the context of decision theory becomes conceivable extending into all branches of behavioral science and forming a skeletal core for the social sciences.

•The seemingly endless complexity of human values is greatly simplified by viewing values in hierarchic structures which are goal dependent, and further, by restricting attention to those areas that are involved in social conflict.

•The innate components of the value structure, which include inherent psychological as well as biological values, can be treated largely as a common invariant denominator of human nature, allowing the focus of attention to be directed to problems of the acquired, cognitive, ideological values where the major sources of value conflict arise.

•On analysis, values are found to be correlates of directed activity. They are always relative to some purpose, goal, or aim, explicit or implicit, and structured in goal-dependent hierarchies. Any concept or belief regarding the purpose and value of life as a whole, once accepted, then logically supersedes and conditions the entire hierarchy of value priorities at subsidiary levels. Values at the ideological plane become ordered and ethical issues judged in accordance with the conceived ultimate purpose of life as a whole. This latter will logically imply at the same time an associated worldview or universe scheme that is consistent.

•Because of the hierarchic structure of values the search for improved ethical guidelines can be narrowed to the search for what ought to be most valued. This in turn leads to problems of the highest determinants of value priorities—the "life goal," "world model" concepts and beliefs that lie at the heart of the problem of moral judgment and logically condition the value structure at all levels.

•Societal values, especially of the kind people disagree on, are always dependent upon, and relative to, some general frame of reference containing the premises, beliefs, presuppositions, etc on which the reasoning about priorities rests. The question may be raised "What makes one reference frame superior or supersedent to another?" and then, "Is there some ultimate frame of reference for values that could logically and rightly be accepted and respected by all countries, cultures, governments, and creeds, and by mankind in general, as the final supreme standard when it comes to judging ethical priorities, resolving value conflicts, and as a guideline for human judgment generally and international decision-making in particular?" The practical need for some such unifying global standard becomes more and more evident for things such as world population control, conserving world resources, protecting the oceans and atmosphere, and for various other modern world problems that increasingly require united effort on a global scale.

•What is needed ideally to make decisions involving value judgments is a consensus on some supreme comprehension and interpretation of the universe and the place and role within it of man and the life experience.

Dependence on Mind-Brain Concepts

•Beliefs concerning the ultimate purpose and meaning of life and the accompanying worldview perspectives that mold beliefs of right and wrong are critically dependent, directly or by implication, on concepts regarding the conscious self and the mind-brain relation and the kinds of life goals and cosmic views which these allow. Directly and indirectly social values depend, for example, on whether consciousness is believed to be mortal, immortal, reincarnate, or cosmic and whether consciousness is conceived to be localized and brain-bound or essentially universal, etc.

•Recent developments in mind-brain theory revise the ultimate criteria and our ultimate frame of reference for determining value priorities. Problems of values, ethics, and morality (questions, i.e., of what is good, right, and ethically true and of what ought to be) become, in these revised terms, something to which science, in the most profound sense, can contribute fundamentally and in which science should be actively and responsibly involved.

•Current concepts of the mind-brain relation involve a direct break with the long-established materialist and behaviorist doctrine that has dominated neuroscience for many decades. Instead of renouncing or ignoring consciousness, the new interpretation gives full recognition to the primacy of inner conscious awareness as a causal reality.

•The phenomena of conscious experience are conceived to play an active, directive role in shaping the flow pattern of cerebral excitation. Instead of being parallelistic and acausal, consciousness in the present scheme becomes an integral part of the brain process itself and an essential and potent constituent of the action. Consciousness is put to work and given a use and a reason for having been evolved in a physical system. Subjective phenomena including values are brought into the causal sequence in human decision-making and behavior generally and thus back into the realm of experimental science from which they had long been excluded.

•The seemingly irreconcilable dichotomies and paradoxes that formerly prevailed with respect to mind vs matter, determinism vs free will, and objective fact vs subjective value become reconciled today in a single comprehensive and unifying view of mind, brain, and man in nature.

•The swing in psychology and neuroscience away from materialism, reductionism, and mechanistic determinism toward a new, monist, mentalist paradigm restores to the scientific image of human nature the dignity, freedom, responsibility, and other humanistic attributes of which it has long been deprived in the materialist-behaviorist approach.

•A nonreductive holistic world model and interpretation of physical reality is supported in which the qualitative pattern properties of all entities are conceived to be just as real and causally potent as those of their components. This preservation of the qualitative value and pluralistic richness of physical reality stands counter to the common tendency to correlate science with reductionism.

Toward the Prescription of Values

•Instead of separating science from values, the current interpretation leads to a stand in which science becomes the best source, method, and authority for determining the ultimate criteria of moral value and those ultimate ethical axioms and guideline beliefs to live and govern by.

•The classic *fact-value* and *is-ought* dichotomies of philosophy logically dissolve in the context of cerebral processing. The operations of the brain are already by nature richly replete with established values and value determinants, both inherent and acquired, with the result that incoming facts regularly interact with and shape values. The resultant value system, along with conceptions of what ought to be, is determined in very large part by the factual input.

•Changing to an ethic based in science would entail in large part a substitution of the natural cosmos of science for the different mythological, intuitive, mystical, or "other-worldly" frames of reference by which man has variously tried to live and find meaning. The aim is not to eliminate value controversy and differences of opinion but only to bring these into a domain set by an agreed-upon frame of reference supported by science— not with the idea that scientific truth is absolute or beyond question, but only with a conviction that it does represent the best and most reliable, credible, and dependable approach to truth available.

' •Once science modifies its traditional materialist-behaviorist stance and begins to accept in theory and to encompass in principle within its causal domain the whole world of inner, conscious, subjective experience (the world of the humanities), then the very nature of science itself is changed. The change is not in the basic methodology or procedures, of course, but in the scope and content of science and in its limitations, in its relation to the humanities and in its role as a cultural, intellectual, and moral force. The kinds of interpretations that science supports, the world picture and attendant value perspectives and priorities, and the concepts of physical reality that derive from science all undergo substantial revisions on these new terms. The change is away from the mechanistic, deterministic, and reductionistic doctrines of pre-1965 science to the more humanistic interpretations of the 1970s. Our current views are more mentalistic, holistic, and subjectivist. They give more freedom in that they reduce the restrictions of mechanistic determinism, and they are more rich in value and meaning.

•Accepting as self-evident the ultimate value of what man generally has held most sacred, namely, the cosmic forces that made, move, and control the universe and created man, and interpreting these in accordance with science, one emerges with a value system that includes a strong reverence for nature promoting the values of the recycle philosophy, population regulation, protecting and enhancing environmental quality, and the like.

•In the eyes of science, to put it simply, man's creator becomes the vast interwoven fabric of all evolving nature, a tremendously complex concept that includes all the immutable and emergent forces of cosmic causation that control everything from high-energy subnuclear particles to galaxies, not forgetting the causal properties that govern brain function and behavior at individual and social levels. For all of these, science has gradually become our accepted authority, offering a cosmic scheme that renders most others simplistic by comparison and which grows and evolves as science advances.

•Science becomes man's best channel for gaining an intimate understanding of and rapport with those forces that control the universe and created man. This is not to suggest that science take on the functions of religion; but only that there might be mutual and other benefits from a fusion of the two.

•The future of science will be very different depending on whether science is recognized in the public mind to have competence in the realm of values. Reciprocally the future of society also will be very different depending on whether its value perspectives are shaped from science and the worldview of science or by other alternatives that now prevail.

THE KEY TO QUALITY SURVIVAL

Implicit in the foregoing is the conclusion that our top social priority today is to effect a change worldwide in man's sense of value. This translates by hierarchic value theory into a change in what is held most sacred. What is needed, more specifically, is a new ethic, ideology, or theology that will make it sacrilegious to deplete natural resources, to pollute the environment, to overpopulate, to erase or degrade other species, or to otherwise destroy or defile the evolving quality of the biosphere. This is exactly what is found to emerge from our current approach to the theory and prescription of human values. Relying on the kind of truth supported by science we arrive at an ethic that promotes an ultimate respect for nature and its creative principles, including those of its peak thrust into the highest reaches of man's mind, along with corollary value criteria which, if applied worldwide, would promptly set in motion the kinds of corrective legislation and other trends and pressures that are needed to remedy looming global disaster conditions.

On the terms proposed the utility of science takes a different form. Society would look to science not only for new technology and objective knowledge but, more importantly, for the criteria of ultimate value and meaning. Each advance of science brings increased comprehension and appreciation of the nature, meaning, and wonder of the creative forces that move the cosmos and produced man. Even "science as usual" gains, in this context, a heightened social significance and moral support. The special, key role of neuroscience and brain research will be readily apparent.

It remains to further stress a point already implied, namely, that for science to fully qualify and function effectively in this changed role, it will be necessary that we abandon an entire mode of scientific thinking long referred to under the general rubric of "scientific materialism." Moves to abide by the truths of science, as opposed to unproven claims from other sources, have had sporadic support since Francis Bacon. What is new today is the shift in science from reductive physicalism to a holist-mentalist paradigm and the changed interpretations and perspectives that this brings. Among traditional views that consequently require correction is that predicating the impotence of science in regard to value judgments along with much of the doctrine associated with reductive mechanistic determinism that for many decades has characterized science and our scientific outlook. This was the thinking of Karl Marx and is the reason that the more materialistic and animalistic aspects of human nature are put first in Soviet philosophy before man's more idealistic, more spiritual components. The issues at stake are not minor. They involve not only the public image of science, the relation of science to human values, and the kinds of values science upholds, but also some of our more basic concepts in science concerning physical reality, mind and matter, and the nature of causation.

CAUSAL DETERMINISM: THE CENTRAL ISSUE

The issues narrow to opposed views of causation that are basic and central to everything stated thus far. One view holds that the causal forces and laws operating in nature are fully explainable in purely physical terms and are, in principle, ultimately accountable on the basis of quantum theory, the elemental forces of physics, or in some more unifying field theory eventually to be found. Physicalist, i.e. materialist, determinism is assumed to prevail throughout nature in this view and all higher level interactions, including those of the brain, are presumed to be reducible and accountable, in principle, in terms of the ultimate fundamental forces of physics.

Opposed to this long dominant physicalist-behaviorist interpretation is the view which we here uphold and which has recently been gaining increased acceptance particularly in the behavioral sciences. It contends that the higher forces and laws of causation, as seen for example in classical mechanics, in physiology, and in brain function and behavior, cannot be fully explained by the laws of quantum mechanics or by the mechanics or laws of any other ultimate physical force or field. The higher entities and their causal properties and laws of interaction are conceived to be causal realities in their own right and not determined completely (though they are in part) by the causal laws and properties of their components. The larger, higher, more molar properties are perceived to be just as real, just as causal, and in many ways to be of more importance than are the more basic physical properties of their subsidiary components. In this view the fundamental forces of physics are only building blocks used in creating bigger, more competent entities and forces. The patterning of the building components, i.e. their arrangement in space and time, becomes a distinctive key factor in making things what they are, and is not determined solely by the properties of the parts themselves.

To attempt to explain an entity in terms of its parts and then the parts in terms of their parts and so on, results in an infinite regress in which one is left at the end trying to explain everything in terms of next-to-nothing. At each step of the way critical pattern components of causality are lost and the explanation becomes less and less complete at each lower level. To attempt to include, even in principle, the pattern factors, i.e. the space-time components, by invoking the "interactions of the parts," the "organizational relations," etc at each step, sounds good but is an empty lip service. We have no science for the space-time components, no science of the organizational relations and interactions. Particularly we have no science for the collective form in which these are present at each level of the infrastructure. Even the relatively ultra-simple interactions involved in the classic "three-bodies problem" are too much.

Our present view holds further that when a new entity is created the new properties of the entity, or system as a whole, thereafter overpower the causal forces of the component entities at all the successively lower levels in the multi-nested hierarchies of the new infrastructure. In other words whenever an entity joins forces with others to form a new whole, the position that it is forced to take in the universe and its subsequent course through time and space and its eventual fate are thereafter determined more conspicuously by the new properties of the system as a whole than by its own original properties. A degree of self-determinacy is lost to the parts as soon as the higher powers of the new whole become superimposed. Although the causal forces at the lower quantal, atomic, molecular, etc levels in the infrastructure continue to operate in full force as usual they are enveloped, encompassed, overwhelmed, superseded, supervened, and outclassed by the new causal properties that emerge in the whole. Evolution, in the course of compounding new compounds continuously adds new entities and new phenomena that embody new qualities, new causal forces, and principles with new scientific laws and control properties.

The new emergent phenomena, not reducible, in principle, to their parts and deserving to be recognized as causal realities in their own right, are in many respects more powerful and dominant features of reality than are the lower properties of the components. Instead of a universe completely controlled by quantum mechanics and the basic forces of physics, science presents, by this interpretation, a universe controlled by a rich profusion of qualitatively diverse emergent powers that become increasingly complex and competent. Any randomness, chance, caprice, or chaos that may be operating at the quantum level, as modern physics insists, gains little expression because it is effectively superseded and controlled by higher level forces that are anything but random. The higher levels involve much more than mass probabilities. The creative, interlocking web of evolving nature is not blind or chance-like but becomes, as it progresses, rich in irreversible, directional, ever more complex constraints that tend to keep things moving in a trend toward higher and more competent forms.

In the brain, controls at the physico-chemical and physiological levels are superseded by new forms of causal control that emerge at the level of conscious mental processing, where causal properties include the contents of subjective experience. Causal control is thus shifted in brain dynamics from levels of pure physical, physiological, or material determinancy to levels of mental, cognitive, conscious, or subjective determinacy. The flow of nerve impulse traffic and related physiological events in a conscious process is no longer regulated solely by events in kind but becomes caught up in, enveloped, and moved by the higher mental controls, somewhat as the flow of electrons in a television set is moved and differentially patterned by the program content on different channels. Just as the programming variables of a TV monitor have to be included in order to account for the electron flow pattern of the system, so also in the brain the subjective, mental variables of cerebral function have to be included to give a full account of the flow patterns of neural excitation. The mental events of conscious experience and the physico-chemical events of the infrastructure are not conceived to be in a parallelistic relation like that of "two languages," "two logics," or of "two complementary aspects of one and the same situation" in which a "purely physical determinacy of the CNS" is preserved as MacKay (1980) and others would have it. This shift from a causal determinacy that is purely physical to one that includes conscious subjective forces that supersede the physical—in other words the shift from a materialist, reductionist, mechanistic paradigm to a holistic, mentalist paradigm—makes all the difference when it comes to using the "truths" of science as criteria of ethical values.

MARXISM INVERTED

In trying to assess possible social repercussions and the outcome of a societal shift to an ethic founded in science, it would be unfortunate and misleading if one were to rely on Marxism and the Communist World as an example. According to our latest mind-brain theory and its implications, Marxist-Communist doctrine is founded on some basic errors in the interpretation of science and of what science stands for and implies in reference to human nature and to social and worldview perspectives. As a result, the kinds of values upheld in Marxist doctrine are almost the diametric opposite from those which emerge from a scientific approach on our present terms.

If the growing competition between Communist and Free World countries is to continue to be in part a battle for men's minds, it may be worth a few words in closing to point out some of the ideologic value differences that result even though the intent in both cases is to exclude dualist "other-worldly" criteria in favor of the truths of science. Some of the basic starting differences include the following:

1. First and foremost, Marx accepted the long established—but now largely overthrown—view that science, of necessity, leads to and supports a materialist philosophy that rejects subjective mental phenomena as causal and predicates instead a purely materialistic determinancy for the CNS.
2. The doctrine of reductionism was accepted as applied to nature in general and human nature in particular.
3. Also intimately related to the above, Marx failed to recognize the key principle of "downward causation," i.e. the causal control that higher properties in any entity, whether a society or a molecule, invariably impose over the lower properties of their infrastructure.
4. Marxism, further, lacks any theory that serves to resolve the *is-ought* fallacy or the traditional dichotomy that has heretofore kept science and human values separate.
5. Marxism also lacks the "free-will" concepts we have today that free individual and social decision-making from mechanistic determinism.
6. Marx opted for a throughly homocentric value system that makes man the measure of all things and gives precedence to the basic material needs of man over the quality of the biosphere, as well as over man's higher psychological needs. There is no justification in science for this latter choice and it is, in some respects, an inversion of nature that puts the welfare of a part of a system above that of the system as a whole.

Basically, according to Marx, what counts in human affairs and changes the world are man's actions in fulfilling his material needs for subsistence, not man's idealisms, philosophy, or ideology. He emphasized that the materialistic-animalistic needs must come first and that the higher human pursuits are built upon and depend on the more basic components. On the other hand, Marx failed to appreciate that the higher idealistic properties in man and society, once evolved, can then supersede, encompass, control, and take care of the lower material needs; that this is the way of evolving nature and when it comes to rules for progress, works better than the inverse. One of the best refutations of Marxism is Marxism itself in that it was not Marx's actions in satisfying his material needs for subsistence that changed the world, but his philosophy, visionary ideas, and Communist ideology.

A value system that puts its ultimate good in the welfare of the "Party," and at the same time pointedly scorns reverence for nature, does little to help remedy most of our mounting global crisis conditions which today are the overriding concern. There is nothing in Marxist doctrine to help control overpopulation, curb pollution, conserve resources, preserve the environment, protect endangered species, etc. Nature in Marxist materialism is not something to be revered but almost the reverse, i.e. something to be battled and subjugated, transformed, mechanized, and exploited to satisfy man's (mainly material) needs. The forces of nature as interpreted by Marx in the materialist tradition are blind and unprincipled (Bell 1975), not something rich in quality, wonder, and beauty, harmoniously controlled with countless checks and balances, and full of creative strategies, constraints, and principles that have been time-tested for success in creating, preserving, and improving the quality of the biosphere, including man.

In Marxism not nature but technology and production power are idolized. Cathedrals for Marx are factories and skyscrapers, and the beautiful dream is to transform whole continents by industrial progress with "huge new populations springing up as if by magic" (Ryanzanoff 1963). The narrow focus on class conflict in an industrialized society also does nothing for the major ailments of the planet today and again is expressed in terms of the mechanistic determinism of the more material and elemental needs and components in man's makeup at the expense of the higher psychological needs and more idealistic components. The causal power of cognitive ideals that behavioral science recognizes today, was, on principle, dismissed. Where nothing is sacred and there is no higher meaning (beyond that of the "Party"), everything loses meaning. In its homocentric emphasis on the more material needs and aims of man, technology, industry, and production power, combined with its demeaning view of nature, Marxism, in direct contrast to the views we reach above, seems to represent almost the epitome of the worst forces that have caught up with us today to produce most of the adverse crisis conditions that threaten the future. The basic relevance of brain research to all these issues can hardly be overemphasized.

IN SUMMARY

In the context of today's mounting global problems the relative demand for medical, educational, and related social benefits that derive from the neurosciences is diminished. At the same time the human value spin-offs of brain research are thrust into a strategic position of top concern because of their key role as criteria for policy priorities and decision-making guidelines. Recent conceptual developments in the mind-brain sciences rejecting reductionism and mechanistic determinism on the one side, and dualisms on the other, clear the way for a rational approach to the theory and prescription of values and to a natural fusion of science and religion. Science can be upheld as the best route to an increased understanding and rapport with the forces that made and move the universe and created man. The outlines of a value-belief system emerge that include an ultimate respect for nature and the evolving quality of the biosphere, which, if implemented, would set in motion the kind of social change needed to lead us out of the viscious spirals of increasing population, pollution, poverty, energy demands, etc. The strategic importance of neuroscience and the central role of prevailing concepts of the mind-brain relation to all of the foregoing remain evident throughout, as does also the direct relevance of efforts to bring added insight and substantiation of these mind-brain concepts through further advances in brain research.

Literature Cited

Bell, D. 1975. Technology, nature and society. In *The Frontiers of Knowledge*. New York: Doubleday

Brown, L. R. 1979. Consultation column. *InterDependent* 6:1–5

MacKay, D. M. 1980. The interdependence of mind and brain. *Neuroscience* 5: In press

Platt, J. 1969. What we must do. *Science* 166:1115–21

Ryazanoff, D., ed. 1963. *The Communist Manifesto of Karl Marx and Friedrich Engles*. New York: Russel & Russel

Sawhill, J. C. 1979. The role of science in higher education. *Science* 206(4416): 281.

Sperry, R. W. 1965. Mind, brain and humanist values. In *New Views of the Nature of Man*, ed. J. R. Platt. Chicago: Univ. Chicago Press. Condensed in *Bull. At. Sci.* 22:2–6, 1966

Sperry, R. W. 1969. A modified concept of consciousness. *Psych. Rev.* 76:532–36

Sperry, R. W. 1972. Science and the problem of values. *Perspect. Biol. Med.* 16:115–30. Reprinted in *Zygon* 9:7–21, 1974

Sperry, R. W. 1977. Bridging science and values: A unifying view of mind and brain. *Am. Psychol.* 32:237–45. Reprinted in *Zygon* 14:7–21, 1979

Sperry, R. W. 1980. Mind-brain interaction: Mentalism, yes; dualism, no. *Neuroscience* 5:195–206

Ann. Rev. Neurosci. 1981. 4:17–42

MOTOR NERVE SPROUTING ❖11548

M. C. Brown, R. L. Holland, and W. G. Hopkins

University Laboratory of Physiology, Oxford, OX1 3PT, England

INTRODUCTION

At a particular stage in its development each motoneuron in the spinal cord sends out an axon which travels for a distance of many hundreds of times the diameter of its cell body before it branches and contacts hundreds of developing muscle fibers. This remarkable feat of growth can be repeated by mature motoneurons in the adult animal, for if the axons are severed the proximal ends grow and can reinnervate muscle fibers. A different, but in some ways more remarkable potential for additional growth can be demonstrated in the adult when only some of the axons supplying a muscle are interrupted. In this case fine nerve processes—sprouts—appear at the nerve terminals and nodes of Ranvier of the remaining intramuscular nerves, and these also innervate denervated muscle fibers (see Figure 1). The nature of this additional growth of intact neurons is of great interest to neurobiologists studying the development and plasticity of the nervous system. Sprouting of mature axons is a process that has been described throughout the nervous system of vertebrates, for example, in the autonomic nervous system (Murray & Thompson 1957), in the afferent system (Weddell et al 1941), and in the central nervous system (Raisman 1969). However, the accessibility of motor nerves and muscles and the ease with which they can be manipulated has resulted in their being the subject of choice for experimental investigations of nerve growth phenomena.

HISTORY

The early observations and experiments on motor nerve sprouting were reviewed by Edds (1953). It had been shown that partial denervation of a muscle was followed after several weeks by a complete or nearly complete recovery of tension, even in the absence of reinnervation by severed axons

17

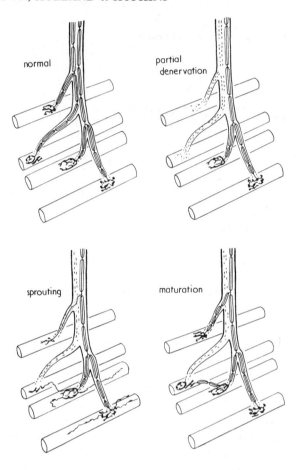

Figure 1   Terminal and nodal sprouting.

Normal: Diagrammatic representation (not to scale) of a branch of an intramuscular nerve with two myelinated axons innervating four muscle fibers. The perineural sheath envelops the axons down to the terminals, which are drawn as they appear after staining with zinc iodide-osmium tetroxide. For simplicity the endoneurium (basement membrane of Schwann cells and collagen within the perineurium) and the perineural sheath and Schwann cells overlying the terminals are not shown.

Partial denervation: The severed axons and their terminals degenerate.

Sprouting: By five days terminal and nodal sprouts have usually appeared and begin to contact previously denervated end-plates.

Maturation: One to two months after partial denervation, sprouts innervating end-plates are myelinated and all other sprouts have usually disappeared. It is not known whether the original shape of each terminal is restored.

(Wehrmacher & Hines 1945, Hines et al 1945, Weiss & Edds 1946). An early observation by Exner (1884) that there were no degenerating muscle fibers within partly denervated muscles was confirmed by Van Harreveld (1945), who also pointed out that a hypertrophy of some of the muscle fibers was not of sufficient magnitude to explain the tension recovery. Among other hypotheses put forward to explain the recovery of function was that of intramuscular ("collateral") nerve growth by the remaining intact nerves (Exner 1884).

A close examination of the intramuscular nerves in the light microscope was made by Edds (1949, 1950) and Hoffman (1950). Both these workers used gold impregnation to stain nerves in partly denervated muscle preparations, and they clearly demonstrated fine nerve outgrowths from the nodes of Ranvier of the remaining intact axons. This aspect of collateral nerve growth is now perhaps better called *nodal sprouting*. Hoffman also used a silver impregnation method and demonstrated fine outgrowths from the intact nerve terminals themselves, and these are now known as *terminal sprouts*. Both terminal and nodal sprouts were observed to innervate denervated muscle fibers and to become myelinated.

The histological observations allowed two important conclusions to be drawn.

1. The sprouting occurred by new growths rather than by the activation of previously nonfunctional nerve branches.
2. The axons sprouted while remaining in contact with muscle fibers.

Weiss & Edds (1946) had suggested that the axons would have to disengage themselves before sprouting.

Edds' review relied heavily on several publications by Hoffman (1950, 1951a,b, 1952, 1953, and Hoffman & Springell 1951), who had attempted to identify the sprouting stimulus (see below). Sprouting in muscle was then little studied for some years, but the last decade has seen a revival of interest in the subject.

SPROUT GROWTH

Sites of Origin

NODAL SPROUTS   Favorable silver-stained preparations of partly denervated muscle show sprouts arising from sharply defined constrictions in the axons, which are almost certainly nodes of Ranvier (Hoffman 1950) (see Figure 1). More than one sprout can arise from a single node (Hoffman 1950). Direct observations by Speidel (1932, 1933, 1935) on live nerves suggested that Schwann cell division on the remaining intact axons could occur and create new nodes from which nodal sprouts arise. Hoffman's

(1952) observation of normal amounts of nodal sprouting in the presence of an inhibitor of mitosis makes it unlikely that a significant number of sprouting nodes arise in this way.

Not all nodes in a nerve of a partly denervated muscle produce visible sprouts. Edds (1949) was first to note that sprouts arose from nodes only within the intramuscular nerves and that no sprouts were observed in the extramuscular nerve. On the other hand, Hoffman's last paper on sprouting (Causey & Hoffman 1955) was an early electron microscopic study of partly "denervated" extramuscular nerves in which large numbers of fine nerve processes were described. Our own unpublished observations show that processes Hoffman categorized as nerves are derived from "denervated" Schwann cells.

However, even within a partly denervated muscle not all nodes produce sprouts. This is clear on inspection of any well-stained, partly denervated muscle and is implicit in Hoffman's 1950 paper and in the more recent work of Slack et al (1979). Unpublished observations (W. G. Hopkins and J. R. Slack) with the electron microscope show that there are few sublightmicroscopic motor nerve nodal sprouts in the intramuscular nerves. It is therefore unlikely that all nodes are sprouting but that only a proportion of the sprouts enlarge enough to become visible by conventional light microscopy.

Hoffman (1950) observed that on the fourth day after partial denervation most of the fine nodal sprouts arose from nodes 100 to 200 μm from the end-plates they were innervating. A closer analysis of nodal sprouting and its time course, made by Slack et al (1979) in a whole mount preparation, has enabled them to make some interesting observations. Sprouts arose only from nodes close to points from which denervated sheaths branched. Sprouts were absent from quite long lengths of intramuscular nerve that lacked such branch points even though those lengths contained degenerating axons and passed close to denervated end-plates. Furthermore, the shortest denervated perineural sheaths were the first to become innervated, in agreement with Hoffman, but even long side branches were found to contain sprouts in preparations examined at later times (16 or more days). In short it appears that the major determinant of which nodes are involved in sprouting seems to be the distance of the node from a denervated end-plate via a vacant perineural sheath. What is not clear is whether every node close to a denervated end-plate that has such a pathway open to it develops a sprout. A possibility (but one that would seem difficult to prove) is that nodal sprouts might represent the actual return of some of the excess branches that neonatal motoneurons possess (Redfern 1970) but that are lost in the first weeks after birth as polyneuronal innervation is eliminated.

TERMINAL SPROUTS   Although first discovered in silver-stained preparations, terminal sprouts are more clearly visualized when the nerve terminals have been selectively stained with zinc iodide-osmium tetroxide (Akert & Sandri 1968). Terminal sprouts first appear as irregular fine extensions of branches of the nerve terminal (see Figure 1). Many such extensions can arise from a single terminal. Terminal sprouts can also arise from the unmyelinated preterminal segment of the axon (Hoffman 1950). The proportion of nerve terminals producing sprouts depends on the method of inducing terminal sprouting (Brown et al 1980), but it is not known whether apparently nonsprouting terminals produce fine outgrowths that are not resolved in the light microscope.

Pathway of Sprout Growth

Hoffman (1950) noted that nodal sprouts were guided to denervated endplates by some elements of the nerve, probably Schwann cells remaining after partial denervation. The minutiae of this guidance remain undescribed. It is not known whether the growth cones of the sprouts advance along bundles of collagen fibrils or along the surface of the "denervated" Schwann cells. The mechanism by which each nodal sprout finds the vacant perineural sheath leading to the nearest denervated end-plate (Slack et al 1979) is also unknown.

Both Hoffman's (1950) and Edds' (1953) diagrams summarizing sprout growth give the impression that sprouts have to traverse barriers surrounding intact and degenerating axons in order to enter the pathway to the denervated end-plates. A fine basement membrane does indeed exist on the surface membranes of Schwann cells and may constitute a physical barrier to nodal sprout growth. A cellular sheath, the perineurium, also surrounds all axons in the intramuscular nerves, but two or more axons in close proximity are enveloped by only one sheath without intervening septa (Burkel 1967). Sprouts remain entirely within this sheath from their nodes of origin to the end-plates they innervate, as shown in Figure 1.

Terminal sprouts appear to grow erratically on the surface of muscle fibers. There is some preference for growth along the long axis of the muscle fibers, which might be due to spatial constraints on neurite growth (Weiss 1955; see also Dunn & Heath 1976). Terminal sprouts resemble "escaped" reinnervating axons (Gutmann & Young 1944); indeed, terminal sprouts, escaped nodal sprouts, and escaped reinnervating axons probably obey the same laws of growth. Schwann cell processes probably accompany terminal sprouts (Duchen 1971) but the relationship between sprout, Schwann cell, muscle fiber surface, and the extracellular matrix has not been investigated.

It is not clear whether terminal sprouts are attracted in some way toward denervated end-plates, but they seem to innervate any they encounter. In the mouse peroneus tertius it appears that most functionally successful terminal sprouts end up at vacated end-plate sites (Brown & Ironton 1978), but this may not be the case in soleus (R. L. Holland, unpublished observations), where new end-plates may be formed more easily.

Maturation of Sprouts

Significant recovery of tension in partly denervated *fast* twitch muscles begins six to seven days after the denervation operation (Brown & Ironton 1978). However, this figure probably overestimates the time taken by the first few nodal sprouts to develop and start transmitting, as nodal sprouting can be seen in such muscles some two to three days earlier (Hoffman 1950).

In the mouse soleus, a *slow* twitch muscle, tension recovery is delayed by a few more days (R. L. Holland, unpublished observations). This perhaps relates to the fact that at six days after partial denervation the soleus shows rather little nodal sprouting, despite having produced terminal sprouts at most intact end-plates (Brown et al 1978b, 1980). As described above, nodal sprouts grow straight to denervated end-plates, but terminal sprouts do not seem to be guided in the same way.

There have been few reported investigations of the functional maturation of sprouts using intracellular recording. Snider & Harris (1979) found a higher quantal content at neuromuscular junctions that had sprouted after eight to nine days of blockade of the nerve with tetrodotoxin (see below). Although they relate this to the increased size of sprouted terminals, they do not exclude the possibility that the release mechanism at inactive terminals is altered. Bennett & Raftos (1977) measured quantal release at end-plates in segmentally denervated and reinnervated axolotl muscles and showed that sprout innervation developed more slowly than reinnervation. This slower development was probably not due to a mismatch between sprouts and myofibers (Bennett et al 1979), but may reflect instead some more subtle difference between sprouting axons and reinnervating axons.

Nodal sprouts are eventually myelinated, as are terminal sprouts that form end-plates (Hoffman 1950). At longer times terminal sprouts that do not form end-plates disappear from partly denervated or blocked preparations (see Figure 1), but the manner of their disappearance has yet to be described.

Limits to Sprout Growth

An upper limit to the rate of total (terminal plus nodal) sprout growth could be determined for each motor unit by the rate of synthesis of various structural components in the nerve cell body. The rate of synthesis may be increased in sprouting motoneurons because biochemical changes associated with chromatolysis occur in the cell bodies following botulinum toxin injection into a muscle (Watson 1969), a situation in which sprouting is known to occur (Duchen & Strich 1968; see below). It is not known whether the cell bodies of intact motoneurons undergo such changes in response to partial denervation, although some of the physiological changes seen in axotomized neurons have been found (Huizar et al 1977).

Hoffman (1952), guided by observations that electrical stimulation of nerves could induce chromatolytic changes in cell bodies (Hyden 1943), investigated the action of stimulation on the rate and extent of sprouting in partly denervated muscles. He found that stimulation of spinal cord and sciatic nerve roots accelerated the rate at which empty perineural sheaths were filled by nodal sprouts in the soleus and gastrocnemius muscles. This was apparently achieved with a stimulation regimen that represented at most a five percent increase in cell firing during the sprouting period. Hoffman (1952) produced an even more marked acceleration of sprouting by feeding rats pyronin in their drinking water (analogous chemicals methylene blue and rhodamine had no effect). Hoffman felt that the acceleration of sprouting in both these cases arose from "modifications of the synthetic processes in the perikarion," but Edds (1953) argued against this. More recently Watson (1976) has suggested that drugs like pyronin may influence the rate of pinocytosis at the terminal, which could change the growth rate "by altering the rate at which membranous vesicles 'cycle.'"

When all sprouts have matured after partial denervation, it is found that motoneurons have on the average increased the number of muscle fibers in their motor units by a maximum of four- to five-fold (Thompson & Jansen 1977). Brown & Ironton (1978b) also showed that this increase is independent of the original motor unit size and type, and it therefore probably represents a growth limit or eventual failure of the neuron to supply more than a five-fold increase in the number of functional terminals. Eventual loss of the sprouting stimulus before all denervated fibers become innervated by sprouts could also limit motor unit expansion. A further constraint may be the division of the muscle into compartments by connective tissue barriers (Kugelberg et al 1970).

A possible determinant of sprout growth that requires systematic analysis is the age of the animal. The motoneurons of newborn rats seem reluctant to sprout (Brown et al 1976, Thompson & Jansen 1977): If a muscle is partly denervated a few days after birth the motor units ultimately formed after the phase of competitive removal of polyneuronal innervation show only a two-fold increase in size instead of the normal five-fold increase that follows partial denervation of adult muscle. Moreover, the two-fold increase could be due to retention of part of the neonatally expanded state of the motor units, rather than to a true sprouting response. Betz et al (1979), however, have evidence that sprouting of neonatal motoneurons occurs to innervate newly formed muscle fibers in a rat lumbrical muscle.

The ability of old neurons to sprout has not been investigated, but Tuffery (1971) has shown that aging end-plates develop a more complex morphology that might be caused by sprouting at the last node of Ranvier before the nerve terminal. This could be the result of an ongoing process of sprouting and possibly also nerve terminal degeneration at end-plates throughout life, as originally proposed by Barker & Ip (1966). Alternatively, cell degeneration in old muscles might cause the observed sprouting (see below).

Competitive Interaction

An interesting question is that of the fate of the sprouts when the severed axons regenerate into a partly denervated muscle. Hoffman (1951a) found histological evidence of double innervation of muscle fibers in such a preparation, in which a sprout and a regenerated axon innervated the fiber at the same place. Confirming this point, Guth (1962) found physiological evidence from tension measurements that sprouts and regenerated axons innervated the same muscle fibers.

Cass et al (1973) investigated the interaction between sprouts and reinnervating axons in axolotl leg muscles. In contrast to the early workers, they found that the regenerating axons inhibited the function of the sprouts. Such a "repression" of the sprouts has been confirmed in other amphibia (Fangboner & Vanable 1974, Bennett & Raftos 1977, Dennis & Yip 1978) and in mammals (Brown & Ironton 1976, 1978, Betz 1977, Thompson 1978).

The functional displacement of the sprouts was shown to depend on the time at which reinnervation by the cut axons occurred. In mammals (Thompson 1978) and in axolotls (Slack 1978) early reinnervation allowed the regenerating axons to displace the sprouts, but after late reinnervation the sprouts could not be displaced. There remains, however, the problem of the persistent multiple innervation described by Hoffman and by Guth. Thompson found that multiple innervation occurred only at early times and resulted in the displacement of the sprout. At later times the reinnervating axons made connections only with denervated muscle fibers, and multiple innervation was not detectable physiologically.

The property of "mature" sprouts that prevents their elimination has not been identified. It might be related to myelination of the sprout, or perhaps to its complete occupation of the end-plate site (Bennett & Raftos 1977). It is also possible that if the regenerating axon has been deprived of its target for some time, it becomes unable to reinnervate effectively (Frank et al 1975).

The mechanism of the competition between the immature sprout and the regenerating axon is obscure. Proposed mechanisms include (*a*) a grappling for a limited amount of synaptic area (Brown & Ironton 1978), (*b*) a competition for a limited amount of trophic substance produced by a muscle fiber (Jansen et al 1978), and (*c*) the proteolytic destruction of one terminal but not the other (O'Brien et al 1978).

It was originally suggested that the nonfunctional sprout remained structurally intact but functionally repressed (Cass et al 1973). However, it is
now known that the sprouts disappear from the muscle (Dennis & Yip
1978). The manner by which the sprouts are removed is unknown, but
presumably it is similar to that which occurs during the elimination of
multiple innervation of mammalian muscle fibers soon after birth (Redfern
1970, Bagust et al 1973, Bennett & Pettigrew 1974, Brown et al 1976). This
has been reported to be by degeneration (Rosenthal & Taraskevich 1977)
and by withdrawal (Korneliussen & Jansen 1976). The ability of neurons
to withdraw processes has been amply demonstrated in living tissue (Speidel
1932) and seems the more likely mechanism for sprout removal. It is
possible that the functionally suppressed but structurally intact sprouts
described by Cass et al (1973) were in an early stage of withdrawal.

INDUCTION OF SPROUTING

General

Motor nerve sprouting was first observed in response to partial denervation,
and the possible consequences that follow partial denervation have all been
postulated at one time or another as the sprouting stimulus (see Figure 2).
The experimental strategy for the identification of the sprouting stimulus

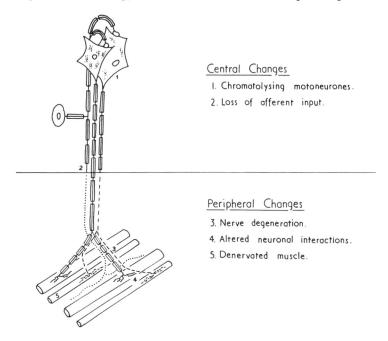

Central Changes

1. Chromatolysing motoneurones.
2. Loss of afferent input.

Peripheral Changes

3. Nerve degeneration.
4. Altered neuronal interactions.
5. Denervated muscle.

Figure 2   Summary of changes following partial denervation.

has been to reproduce as far as possible just one of these changes and then determine whether sprouting is induced. In this section we discuss and evaluate the evidence for and against the various possibilities.

Spread of Signal from Chromatolytic Motoneurons

Axotomized motoneurons undergo characteristic metabolic and structural changes that were originally termed chromatolysis and that are probably associated with the regrowth of the cut axon into the periphery. The possibility of a spread from axotomized, chromatolyzing motoneurons to intact motoneurons was proposed by Rotshenker & McMahan (1976) to explain their observations on the frog cutaneous pectoris muscle. It was found that if the cutaneous pectoris was denervated on one side the proportion of multiply innervated muscle fibers in the contralateral muscle increased. They interpreted this as a sprouting of the contralateral axons in response to a central signal, and the presence of sprouting was confirmed by Reichart & Rotshenker (1979), who stained the nerve terminals in the contralateral muscle with zinc iodide-osmium tetroxide. Rotshenker (1977, 1978, 1979) has shown that contralateral sprouting occurs even if the denervated muscle is excised, and that the onset of sprouting is earlier if the cutaneous pectoris is denervated closer to the spinal cord. He hypothesizes that some signal spreads from the axotomized motoneurons across the spinal cord, the signal being perhaps the same as that which causes the axotomized axons to grow.

Brown et al (1979, 1980) looked for this phenomenon in the mouse by cutting the sciatic nerve in one leg and examining muscles in the contralateral leg. They found no evidence of sprouting. Taking into account the possibility that a central signal might not cross the spinal cord in mice, Brown et al also completely denervated the mouse lower leg except for the soleus muscle. This produced an extensive intermingling of axotomized and intact motoneuron cell bodies (because motoneuron pools overlap extensively in the mouse spinal cord; Biscoe & McHanwell 1979). They found a minimal increase in the sprouting in the remaining, innervated soleus muscle, but in this case the muscles also showed transient rises in acetylcholine sensitivity, and thus the sprouting they observed could have been caused by a peripheral stimulus (see below). Reports of the experiments on the frog cutaneous pectoris have not included measurements of changes in the experimental muscle other than an assessment of sprouting.

If chromatolytic changes in the cell body were the major stimulus to sprouting, one might expect all nerve terminals at an affected motor unit to produce terminal sprouts. This is clearly not the case in rodent muscles, because in favorable, partly denervated preparations with only one remaining motor unit, a proportion of the terminals do not sprout (Betz et al 1980, Brown et al 1980).

Loss of Afferent Input

When a muscle is partly denervated by section of a mixed nerve, afferent input to the cord will be reduced by section of sensory fibers, and conceivably some degeneration of central sensory fibers could also occur. It is unlikely that either of these effects can alone cause sprouting of motor axons, as section of dorsal roots central to the dorsal root ganglion (which causes complete failure of the afferent input and central degeneration of sensory axons) causes no sprouting (Brown et al 1978a).

Peripheral Nerve Degeneration

After partial denervation the remaining intact axons are in close proximity to degenerating nerve throughout their entire course to the muscle fibers, except for the short terminal axon segments. It is therefore plausible that some substance released by degenerating nerve axoplasm or myelin, or cellular changes in the partly denervated nerve, could have a direct effect on the intact nerves to cause sprouting.

Hoffman (1950) and Hoffman & Springell (1951), in a search for a sprout-inducing substance, made extracts of a variety of materials and injected them into rat muscles to determine whether they had any effects on the innervating nerves. Ether extracts of nerves and particularly of the CNS white matter were effective in producing terminal sprouting. After a series of extraction procedures Hoffman concluded that the activity resided in the glyceride of a mono unsaturated fatty acid whose precise identity was unknown. This chemical, "neurocletin," was considered to be a normal constituent of myelin released during Wallerian degeneration, and Hoffman believed that its sprout-inducing capacity lay in its ability to penetrate the nerve cell membrane and "disorganize the axoplasm."

Hoffman could not explain why nodal sprouts are found only close to the denervated tissue and not in the extramuscular nerve where there is abundant nerve degeneration. One possibility is that the difference in structure of the nodes in the extramuscular nerve (Gray & Westrum 1979) makes them refractory to "neurocletin." The nerve terminals also appear to be unresponsive to "neurocletin" because chronic muscle stimulation, which has no effect on nodal sprout growth, inhibits the terminal sprouting produced by partial denervation (Ironton et al 1978, Brown & Holland 1979; see below). A role for neurocletin as a direct stimulus of sprouting therefore remains to be proven. As we discuss below, nerve degeneration probably causes terminal sprouting indirectly by producing a denervation-like change in the muscle, which then stimulates terminal sprouting. Hoffman's extract probably contained the active principle from degenerating nerve that produces these changes.

Another candidate for the signal that causes sprouting is the mitosis of Schwann cells, which does occur during nerve degeneration (Cajal 1928). Hoffman (1952) investigated the possible involvement of Schwann cell mitosis by subjecting rats with partly denervated muscles to a variety of antimitotic drugs. With nitrogen mustard he profoundly inhibited mitosis without affecting sprouting. On this evidence it is unlikely that Schwann cell mitosis is a major sprouting stimulus, or even a prerequisite for sprouting to occur.

Several possible nodal sprout-inducing mechanisms involving cellular changes in the nerves are not excluded by these experiments and require further investigation.

1. Nodal sprouting is stimulated by dedifferentiation of Schwann cells or their basement membranes.
2. Nodal sprouting is stimulated by invasion of the nerves by macrophages.
3. Partial denervation allows access of a sprouting stimulus from muscle.

Altered Neuronal Interactions

It is conceivable that nerves that innervate a particular region of the periphery interact directly in some way to mutually inhibit each other's growth. If some of the nerves were subsequently removed by partial denervation, then the remaining nerves would be released from this inhibition, which would permit sprouting to restore the lost innervation. There are two possible ways in which such inhibition could occur.

1. The nerves normally release from their terminals an agent that directly inhibits nerve growth or that inhibits the action of a sprouting stimulus continually produced by the target tissue (Diamond 1979).
2. The nerves normally deplete by uptake a sprouting stimulus continually produced by the target tissue (Diamond et al 1976).

Both of these possible mechanisms require a functional transport system in the axons to deliver or remove the putative substances. A further hypothesis implicating axoplasmic transport in the control of sprouting would be that nerves maintain normal target tissue properties by the release of trophic substances from their terminals; a target deprived of these substances by transport block would produce a sprouting signal (Diamond 1979).

An investigation of the role of axoplasmic transport in sprouting was undertaken by Aguilar et al (1973), who applied colchicine, an inhibitor of transport, to one nerve root in salamanders. Their finding was that sprouting was stimulated in the nerves of adjacent roots. Their interpretation, that this sprouting was due to transport block, relies upon the assumption that colchicine has no action other than to block transport. However, it is known that this drug can directly cause muscle to undergo biochemical changes normally associated with denervation (Lømo, 1974), and these changes can themselves cause sprouting (see below). Colchicine is also toxic to nerves and may have partly denervated the tissue and thereby induced sprouting. Thus in rabbit cutaneous nerves it is not possible to block axoplasmic transport with colchicine by more than 30% without killing some axons, and that degree of block does not cause sprouting (Jackson & Diamond 1977). However, Diamond and his colleagues after careful investigation claim that at least in salamanders colchicine can cause sprouting without denervation of cutaneous afferents supplying low threshold touch corpuscles (Cooper et al 1977). Unfortunately, the degeneration of other classes of axons was not looked for in this work. Clearly, further experiments with colchicine will only be productive if nerve degeneration can be rigorously excluded.

A Signal from Denervated Muscle

Because one of the most obvious effects of partial denervation is the production of denervated muscle, it is logical to suppose that changes in a muscle following partial denervation might include the production by the muscle of a sprouting stimulus. There is now a convincing amount of evidence to support this hypothesis. Indeed, it has been found that any muscle preparation that for any reason exhibits properties characteristic of denervated muscle also has terminal sprouting present.

Most of the evidence has followed from the observation that the properties of normally innervated mammalian muscle are maintained largely, if not entirely, by activity in the muscle (Lømo & Rosenthal 1972). This means that it has been possible to induce denervation-like changes in a muscle simply by blocking nerve-induced activity, and when this is done sprouting is observed. It has also been possible to prevent or reverse these changes by direct electrical activation of the muscle (Lømo & Westgaard 1976), and in these circumstances sprouting is inhibited. (For a review of the activity-dependent properties of muscle, see Lømo 1976.)

The procedures that reduce or abolish muscle activity and that have been shown to induce terminal sprouting include (a) presynaptic blockade with botulinum toxin (Duchen & Strich 1968), (b) nerve conduction block with tetrodotoxin (Brown & Ironton 1977a), and (c) spinalization (Brown et al 1980). It could be argued that these procedures reduce the transport and/or release of chemicals that might be involved in the control of sprouting (Diamond et al 1976). However, this criticism is unlikely to apply to the sprouting produced by the post-synaptic blocking agent α-bungarotoxin (Holland & Brown 1980), since other post-synaptic blockers are known not to affect axonal transport (Berg & Hall 1975) and α-bungaratoxin does not affect presynaptic release mechanisms (Chang & Lee, 1963). Restoration of muscle activity and inhibition of terminal sprouting by direct electrical stimulation have been shown in botulinum toxin paralyzed preparations (Brown et al 1977, Brown et al 1980) and in partly denervated preparations (Brown & Ironton 1977b, Brown & Holland 1979).

The other evidence that denervated muscle produces a sprouting stimulus comes from a class of muscle preparations in which denervation-like changes have occurred and sprouting has been stimulated even though nerve-induced activity is normal. Brown et al (1978a) showed that terminal sprouting occurred in muscles in which there was degeneration only of sensory nerve fibers. A change to a state characteristic of denervated muscle was indicated because the muscles showed an increased sensitivity to acetylcholine, and it has been shown in other circumstances that "products of nerve degeneration" produce other denervation-like changes in normally innervated muscle fibers (Jones & Vrbova 1974, Cangiano & Lutzemberger 1977). It is not clear whether the products of nerve degeneration have their effect on muscle fibers directly, or whether this effect is mediated by inflammation in response to the degenerating tissue. Indeed, inflammation has been associated with new nerve growth in other parts of the peripheral nervous system (Olsen & Malmfors 1970, Fitzgerald et al 1975). It may be that any disturbance that produces inflammation will also produce denervation-like changes (Lømo 1976) and sprouting.

For each of the preparations discussed in this section in which sprouting has been observed, other changes that could have been the source of the sprouting stimulus have sometimes been present. However, taken together their only common denominator is either inactive muscle or a muscle state showing changes characteristic of denervated muscle. It is therefore concluded that muscle in a denervation-like state in some way stimulates terminal sprout growth.

Nature of the Signal from Muscle

Although a denervation-like change in muscle is clearly indicated as the stimulus for terminal sprouting, its exact nature remains problematic. There are several possibilities: (a) a reduction in the release of a sprout-inhibiting agent that is normally produced by active muscle, (b) a change in the surface of the muscle such that nerve growth is allowed or encouraged, and (c) a chemical released by the muscle after denervation, which triggers or encourages new nerve growth.

A mechanism for the first of these possibilities has been suggested by O'Brien et al (1978) and Vrbova et al (1978), who describe release of proteolytic enzymes at the neuromuscular junction region of muscle fibers in response to acetylcholine. They suggest that the nerve terminal is constantly growing, but that normally growth is balanced by the destruction of the terminal by these enzymes. When the muscle is made inactive by neuromuscular block, or by denervation, the enzymes are no longer released and the terminal consequently starts to grow. These enzymes are also supposed to remove the excess terminal sprouts when the muscle becomes active again. An observation that is hard to reconcile with this theory is that sprouting in a partly denervated muscle is not inhibited by nerve stimulation, even though the innervated fibers are more active than normal (Brown & Holland 1979). Another specific argument against the continual digestion theory is that the release of the enzymes was thought to be mediated by a muscarinic mechanism (O'Brien et al 1978) and so could be inhibited by atropine but not by nicotinic blockers. Therefore, α-bungarotoxin should not have blocked the release of the enzymes and hence should not have produced the sprouting seen by Holland & Brown (1980). Nevertheless the possibility of some other inhibitory mechanism being present in normal active muscle cannot be excluded.

The second possibility is that the signal is a local change on the muscle fiber surface. The motor nerve terminals might be continually probing their local environment, and sprouting would result when the muscle fiber surface around the end-plate became a suitable substrate for neurite growth. There are indeed changes in the surface of denervated and inactive muscle fibers (e.g. an increased number of extrajunctional acetylcholine receptors; see the review by Fambrough 1979), and neurites do have definite preferences for the substrate upon which they grow (Letourneau 1975). In favor of this idea is the evidence that the sprouting signal has very little capacity to spread. Weiss & Edds (1946) reported that denervation of one hemidiaphragm caused absolutely no sign of sprouting in the very closely adjacent motor terminals on the other hemidiaphragm. Our own unpublished observations on three other muscles with adjacent or overlapping fields of innervation have confirmed that the sprouting stimulus has a very short range.

A very specific proposal for the nature of a surface-bound sprouting agent has been made by Pestronk & Drachman (1978), namely that it is the acetylcholine receptor itself that appears in large numbers in denervated and paralyzed muscle fibers. They reported (*a*) that however a muscle was paralyzed, the amount of sprouting was closely correlated with the acetylcholine sensitivity of the muscle; (*b*) that α-bungarotoxin, which blocks transmission by binding with acetylcholine receptors (both junctional and extrajunctional), did not cause sprouting; and (*c*) that they could prevent botulinum-induced sprouting by delivering daily injections of α-bungarotoxin. However, the apparent sprout-inhibiting action of α-bungarotoxin could not be demonstrated by Holland & Brown (1980) using purified toxin: The toxin on its own caused terminal sprouting and had no inhibitory action on sprouting induced by botulinum toxin.

Recent elegant work by McMahan et al (1978) has shown that a surface-contact interaction is important in the formation of nerve terminals following reinnervation. The "receptors" to which the reinnervating axons grow are embedded in the basement membrane, so that it is to molecules in this layer, possibly associated with underlying acetylcholine receptors, that one must look for any surface-bound stimulant to sprout growth.

Although the evidence cited above is suggestive that a sprouting agent might exist on the muscle fiber surface, there are now indications that the stimulus from inactive muscle might be diffusible.

1. Direct electrical stimulation of partly denervated muscle suppresses terminal sprouting, whereas activation of the innervated fibers alone by stimulation of the nerve in the same pattern does not (Brown & Holland 1979); consequently activation of the denervated fibers is required for the inhibitory effect of chronic direct muscle stimulation.
2. In an experiment by Betz et al (1980) one of two nerves supplying certain muscles in the rat foot was blocked with tetrodotoxin; it was subsequently found that both blocked and unblocked terminals had sprouted.

Those who wish to defend the hypothesis of a surface-bound stimulus to sprout growth would have to argue that in both these experiments there might be surface changes present on the active fibers for one reason or another, or that the active nerve terminals came into contact with adjacent inactive muscle fibers.

A direct approach to demonstrate the presence of a diffusible signal would clearly be to extract an agent from denervated muscle, which, if infused into normal muscle, would cause sprouting. Van Harreveld (1947) and Hoffman (1950) both made extracts of denervated muscle, but could not produce significant amounts of sprouting. The most recent experiments in this tradition were performed by Tweedle & Kabara (1977), who injected lipid extracts of denervated and normal gastrocnemius muscles into rat tongues. A significant increase in the percentage of terminals with "collateral sprouting" in the tongues injected with denervated muscle extract seemed to be related to the higher content of lipids in the denervated muscle (Tweedle & Kabara 1978). This sprouting cannot be ascribed to a direct effect of the lipids until the absence of denervation or inflammation in the tongue muscles is demonstrated.

It is possible that a well-designed experiment, using for example either daily injections or continuous infusion of fresh denervated muscle extracts, will produce a substance that does indeed evoke new nerve outgrowths from nerve terminals without causing denervation-like changes in the muscle.

The substance may turn out to be a "specific" stimulant like Nerve Growth Factor, which is important for the development and maintenance of the sympathetic nervous system (Purves & Lichtman 1978). There are of course many other possible candidates including lipids, glycoproteins, or regulatory metabolites.

Nodal Sprouting—Is It Stimulated by Denervated Muscle Fibers?

Nodal sprouts occur only within intramuscular nerves (Edds 1949) "close" to the muscle fibers they innervate (Hoffman 1950). This would seem to give support to the attractive idea that the stimulus for terminal sprouting from denervated muscle is also reaching the appropriate intramuscular nodes and stimulating nodal sprouting. However, further evidence in support of this hypothesis has not been forthcoming. For example, if there were a common signal, the frequencies of terminal and nodal sprouting should in general be correlated, but in all preparations so far examined such a correlation does not exist.

1. Paralyzed muscles produce terminal sprouts but no visible nodal sprouts (Duchen & Strich 1968, Ironton et al 1978).
2. Terminal sprouting in partly denervated muscles is inhibited by chronic stimulation, but nodal sprouting is unaffected (Ironton et al 1978, Brown et al 1980).
3. Different types of muscle have different relative amounts of terminal and nodal sprouting: Mouse soleus, a slow muscle, produces more terminal sprouts and fewer nodal sprouts than faster muscles (Brown et al, 1978b, 1980).
4. For a given type of muscle the amount of terminal sprouting is not correlated with the amount of nodal sprouting, moreover neither correlates with the degree of denervation (Brown et al 1980).

If there is a common stimulus, the apparent lack of a relationship between terminal and nodal sprouting requires explanation. The failure to see nodal sprouts in paralyzed muscles might be either because the signal from muscle cannot reach the nodes when terminal nerve sheaths are all occupied by axons, or because outgrowths from nodes cannot develop without pathways of "denervated" Schwann cells leading to denervated end-plates. A different threshold for terminal and nodal sprouting and/or a competition between terminal and nodal sprouts for a limited supply of growth materials in the axons might produce a complex relationship between the frequencies of terminal and nodal sprouting in the different circumstances examined. Clearly these are all *ad hoc* explanations and only further work will reveal whether terminal and nodal sprouting are activated by the same or different stimuli.

SPROUTING AND NEUROMUSCULAR DISEASE

An extensive study of the changes in the morphological relationship between nerves and muscles in a wide range of neuromuscular diseases was undertaken by Coërs et al (1973). A "terminal innervation ratio," which measures the branching of axons near their terminals, was calculated from assessments of methylene blue-stained biopsies of normal and diseased muscles. Muscle biopsies from patients with neuronal diseases (peripheral neuropathies and lower motor neuron disorders) had higher than normal terminal innervation ratios, presumably a result of nodal sprouting in response to axonal degeneration. Terminal sprouting ("arborization") was assessed qualitatively and less systematically. It was seen "occasionally" in biopsies from individuals with the neuronal diseases and more frequently in muscle diseases (muscular dystrophies, myasthenia gravis, and acquired myopathies).

Terminal sprouting has also been observed in murine muscular dystrophy (Harris & Ribchester 1979), and in this case both denervated muscle fibers (M. C. Brown and R. Ironton, unpublished observations; see also Rowe & Goldspink 1969) and acetylcholine supersensitivity (Howe et al 1977) have been demonstrated. Terminal sprouting is also seen in mice with motor end-plate disease (Duchen 1970), in which there is failure of conduction of impulses near the nerve terminals and raised extrajunctional acetylcholine sensitivity (Duchen & Stefani 1971).

These results are consistent with the hypotheses that nodal sprouting requires nerve degeneration, and that terminal sprouting is produced by denervation changes in muscle.

RELEVANT RESULTS FROM SPROUTING IN OTHER TISSUES

The site of origin of sprouts is often less clear in other tissues. Nodal sprouts have been clearly visualized only by Murray & Thompson (1957) in silver-stained, partly denervated sympathetic preganglionic nerves. Fitzgerald (1963) used silver staining to show that cutaneous sprouts (presumably nodal) arose from the subepidermal nerve plexus rather than from nerve endings. The equivalent of terminal sprouting has not been reported.

Once sprouting has occurred in autonomic ganglia and skin, reinnervation can, as in muscle, lead to a partial or complete elimination of the sprouts (Guth & Bernstein 1961, Roper 1976, Devor et al 1979). Mature sprouts are probably retained following delayed reinnervation (Livingston 1947).

The effect of age on sprouting in skin has recently been investigated. Although axons supplying receptors of most modalities will sprout well in adult rats (Devor et al 1979), Diamond & Jackson (1978) found that the large axons that supply slowly adapting touch receptors were reluctant to sprout in rats older than three weeks.

Nerve stimulation, which appears to accelerate motor nerve sprouting (Hoffman 1952, see above), has similar effects in the sympathetic nervous system. Maehlen & Njä (1979) found an acceleration in the rate at which the number of preganglionic inputs developed in partly denervated guinea pig superior cervical ganglion cells when either the pre- or post-ganglionic nerves were stimulated. The stimulation did not increase the maximum number of sprout inputs and did not advance the initiation of sprouting.

Drugs and toxins that have been used to study sprouting in motor nerves are now being applied to other neuronal systems. Sprouting was not detected using physiological assays in frog cardiac ganglion (Roper & Ko 1978) and rat skin (Betz et al 1980) following conduction block with tetrodotoxin. This is not surprising, for following blockade the detection of terminal sprouts in muscle required histology (Brown & Ironton 1977a) and in any case tissues such as ganglia and skin may not undergo denervation changes to the same extent as muscle when impulse conduction is blocked. A possible role for axoplasmic transport in sprouting in the central nervous system is indicated by the recent results of Goldowitz & Cotman (1980), who produced sprouting with low concentrations of colchicine.

Sprouting can be induced in other parts of the nervous system by producing nerve degeneration (Murray & Thompson 1957, Liu & Chambers 1958, Raisman & Field 1973), but the aspect of degeneration actually giving rise to the sprouting stimulus is unknown. It remains to be seen whether elements of the denervated tracts are important, whether denervated, partly denervated, or inactive postsynaptic neurons produce a sprouting stimulus, and whether the different types of remaining axons respond to the stimulus differently.

RELATION OF SPROUTING TO OTHER ASPECTS OF NERVE GROWTH

The experimental procedures that have an effect on sprouting have been found to influence certain other nerve growth phenomena in a similar way. These phenomena are discussed briefly for their relevance to sprout growth.

One aspect of nerve growth closely analogous to sprouting is that of a foreign nerve growing into a muscle. Elsberg (1917) showed that a mammalian muscle will not accept foreign innervation outside the end-plate region until the muscle is denervated. This observation was first extended by Miledi (1963), who showed that a damaged but still innervated muscle fiber will accept foreign innervation, and that the damage causes the muscle to show denervation-like changes. A muscle paralyzed with botulinum toxin will accept foreign innervation (Fex et al 1966). Duchen & Tonge (1977) described the growth of the foreign nerve, and it appears remarkably similar to the sprouting at botulinized nerve terminals. Frank et al (1975) showed that stimulating a denervated muscle will inhibit it from accepting another nerve, and stimulation inhibits sprouting similarly. It seems therefore, that the signal that stimulates terminal sprouting is the same as that which stimulates the growth of an implanted foreign nerve.

The regrowth of axons following a nerve crush or cut is obviously not in response to a stimulus emanating from the denervated tissue, as the distances involved are too great. This form of axonal growth is possibly analogous to nodal sprouting as both probably require the substrate and guidance provided by some structure remaining in the nerve following denervation.

Paralysis of muscle as we have seen causes terminal sprouting. It also delays the death that occurs normally among the overabundant motoneuron population generated in the developing ventral horn of the spinal cord (Pittman & Oppenheim 1978, Laing & Prestige 1978). It also delays the elimination of the polyneuronal innervation that occurs naturally in neonatal muscle (Benoit & Changeux 1975) and artificially in reinnervated adult muscle (Benoit & Changeux 1978). Conversely, stimulation that prevents terminal sprouting in a partly denervated muscle accelerates the elimination of polyneuronal innervation in the neonate (O'Brien et al 1978).

A possible link between these various findings is the production by uninnervated, inactive, or denervated tissue of an excess of a substance that allows extra motoneurons in the foetus to live, extra terminals in neonatal muscle to survive, and sprouts in adult muscle to grow (see also Purves & Lichtman 1978, Jansen et al 1978). Activity and/or innervation would reduce the supply of this substance to a maintenance level, enough to support only the adult number of motoneurons and only one terminal per muscle fiber, and not enough to support sprouting. It is still too early to say whether such an attractively simple hypothesis will turn out to be correct.

CONCLUSIONS

It is a sobering thought that in the quarter century since Hoffman's pioneering work and Edds' important review only one major new insight into the sprouting process in muscle has occurred: that inactive muscles are responsible for generating a signal for terminal sprouting. The nature of this stimulus is still quite elusive.

The stimulus can be viewed in the context of three possible mechanisms that could control nerve growth at nerve terminals. In the first of these, nerves have an intrinsic tendency to grow but are actively held in check by a diffusible inhibitor: sprouting would then represent disinhibition of growth. In the second mechanism nerves have an intrinsic tendency to grow provided there is a suitable substrate present: sprouting here represents permission to grow. Finally, nerve processes could be quiescent or have an intrinsic tendency to withdraw and require a diffusible substance for growth and stability: sprouting would therefore be a stimulation of growth when an excess of the growth substance was available.

To explain the observed relationship between terminal sprouting and denervation, all three mechanisms could be linked with muscle metabolism: Denervation, or a change to a denervation-like state, could lead to less release of a growth inhibitor, more release of a growth stimulus, or a change in the surface of the muscle. Diffusible substances released by the muscle could also influence nodal sprout growth, but it seems necessary to postulate additional as yet undefined changes in partly denervated intramuscular nerves as a necessary prerequisite for nodal sprouting.

At the moment, no single mechanism seems to be consistent with all the reported observations on sprouting. It may be that growth inhibition, substrate dependence, and growth stimulation mechanisms all exist and play a part in sprouting and in the life of the neuron. Future work will, it is hoped, demonstrate unambiguously the operation of one or more of these mechanisms in specific instances of nerve growth and assess their importance in the development and maintenance of nerve connections.

In Figure 3 we summarize our current ideas on the likely sequence of events involved in stimulation of sprouting in muscle.

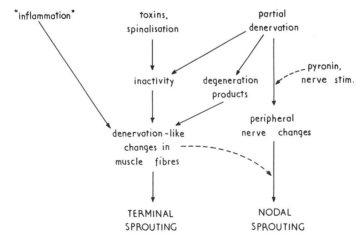

Figure 3 Summary of the most likely causes of sprouting. The *dashed arrows* indicate influences or interactions that require further evidence for confirmation or refutal. For simplicity an arrow showing that degeneration products may produce changes in muscle via an inflammatory effect is omitted. The relevant "changes" in nerve and muscle have yet to be identified.

Literature Cited

Aguilar, C. E., Bisby, M. A., Cooper, E., Diamond, J. 1973. Evidence that axoplasmic transport of trophic factors is involved in the regulation of peripheral nerve fields in salamanders. *J. Physiol. London* 234:449–64

Akert, K., Sandri, C. 1968. An electron-microscopic study of zinc-iodide-osmium impregnation of neurons. I. Staining of synaptic vesicles of cholinergic junctions. *Brain Res.* 7:286–95

Bagust, J., Lewis, D. M., Westerman, R. A. 1973. Polyneuronal innervation of kitten skeletal muscle. *J. Physiol. London* 229:241–55

Barker, D., Ip, M. C. 1966. Sprouting and degeneration of mammalian motor axons in normal and de-afferentated skeletal muscle. *Proc. R. Soc. London B* 163:538–54

Bennett, M. R., McGrath, P. A., Davey, D. F. 1979. The regression of synapses formed by a foreign nerve in a mature axolotl striated muscle. *Brain Res.* 173:451–69

Bennett, M. R., Pettigrew, A. G. 1974. The formation of synapses in striated muscle during development. *J. Physiol. London* 241:515–45

Bennett, M. R., Raftos, J. 1977. The formation and regression of synapses during the reinnervation of axolotl striated muscles. *J. Physiol. London* 265:261–95

Benoit, P., Changeux, J.-P. 1975. Consequences of tenotomy on the evolution of multi-innervation in developing rat soleus muscle. *Brain Res.* 99:345–58

Benoit, P., Changeux, J.-P. 1978. Consequences of blocking the nerve with a local anaesthetic on the evolution of multi-innervation at the regenerating rat neuromuscular junction. *Brain Res.* 149:89–96

Berg, D. K., Hall, Z. W. 1975. Increased extrajunctional acetyl-choline sensitivity produced by chronic post-synaptic neuromuscular blockade. *J. Physiol. London* 244:659–76

Betz, W. J. 1977. Motor nerve terminal sprouting in partially denervated muscle. *Proc. Int. Union Physiol. Sci. Paris, 1977*, 13:73

Betz, W. J., Caldwell, J. H., Ribchester, R. R. 1979. The size of motor units during post-natal development of rat lumbrical muscle. *J. Physiol. London* 297:463–78

Betz, W. J., Caldwell, J. H., Ribchester, R. R. 1980. Sprouting of active nerve terminals in partially inactive muscles of the rat. *J. Physiol. London.* 303:281–97

Biscoe, T. J., McHanwell, S. 1979. Localization of motoneurones supplying hindlimb muscles in the mouse. *J. Physiol. London* 293:37P

Brown, M. C., Goodwin, G. M., Ironton, R. 1977. Prevention of motor nerve sprouting in botulinum toxin poisoned mouse soleus muscles by direct stimulation of the muscle. *J. Physiol. London* 267:42–43P

Brown, M. C., Holland, R. L. 1979. A central role for denervated tissues in causing nerve sprouting. *Nature* 282:724–26

Brown, M. C., Holland, R. L., Ironton, R. 1978a. Degenerating nerve products affect innervated muscle fibers. *Nature* 275:652–54

Brown, M. C., Holland, R. L., Ironton, R. 1978b. Variations in the amount and type of α-motoneurone sprouting following partial denervation of different mouse muscles. *J. Physiol. London* 284:177–78P

Brown, M. C., Holland, R. L., Ironton, R. 1979. Evidence against an intraspinal signal for motoneurone sprouting in mice. *J. Physiol. London* 291:35–36P

Brown, M. C., Holland, R. L., Ironton, R. 1980. Nodal and terminal sprouting from motor nerves in fast and slow muscles of the mouse. *J. Physiol. London.* In press

Brown, M. C., Ironton, R. 1976. The fate of motor axon sprouts in a partially denervated mouse muscle when regenerating nerve fibers return. *J. Physiol. London* 263:181–82P

Brown, M. C., Ironton, R. 1977a. Motor neurone sprouting induced by prolonged tetrodotoxin block of nerve action potentials. *Nature* 265:459–61

Brown, M. C., Ironton, R. 1977b. Suppression of motor nerve terminal sprouting in partially denervated mouse muscles. *J. Physiol. London* 272:70–71P

Brown, M. C., Ironton, R. 1978. Sprouting and regression of neuromuscular synapses in partially denervated mammalian muscles. *J. Physiol. London* 278:325–48

Brown, M. C., Jansen, J. K. S., Van Essen, D. 1976. Polyneuronal innervation of skeletal muscle in new-born rats and its elimination during maturation. *J. Physiol. London* 261:387–422

Burkel, W. E. 1967. The histological fine structure of perineurium. *Anat. Rec.* 158:177–90

Cajal, S. R. 1928. Degeneration and regeneration of the nervous system. Trans. R. M. May. London: Oxford Univ. Press

Cangiano, A., Lutzemberger, L. 1977. Partial denervation affects both denervated and innervated fibers in the mammalian skeletal muscle. *Science* 196:542–45

Cass, D. T., Sutton, T. J., Mark, R. F. 1973. Competition between nerves for functional connections in axolotl muscles. *Nature* 243:201–3

Causey, G., Hoffman, H. 1955. Axon sprouting in partially denervated nerves. *Brain* 78:661–68

Chang, C. C., Lee, C.-Y. 1963. Isolation of neurotoxins from the venom of *Bungarus multicinctus* and their modes of neuromuscular blocking action. *Arch. Intern. Pharmacodyn.* 144:241–57

Coërs, C., Telerman-Toppet, N., Gerard, J.-M. 1973. Terminal innervation ratio in neuromuscular disease. II. Disorders of lower motor neuron, peripheral nerve, and muscle. *Arch. Neurol.* 29:215–22

Cooper, E., Diamond, J., Turner, C. 1977. The effects of nerve section and of colchicine treatment on the density of mechano-sensory nerve endings in salamander skin. *J. Physiol. London* 264:725–49

Dennis, M. J., Yip, J. W. 1978. Formation and elimination of foreign synapses on adult salamander muscle. *J. Physiol. London* 274:299–310

Devor, M., Schonfeld, D., Seltzer, Z., Wall, P. D. 1979. Two modes of cutaneous reinnervation following peripheral nerve injury. *J. Comp. Neurol.* 185:211–20

Diamond, J. 1979. The regulation of nerve sprouting by extrinsic influences. In *The Neurosciences: 4th Study Program*, ed. F. O. Schmitt, F. F. Worden, pp. 937–55. Cambridge, Mass.: MIT Press

Diamond, J., Cooper, E., Turner, C., MacIntyre, L. 1976. Trophic regulation of nerve sprouting. *Science* 193:371–77

Diamond, J., Jackson, P. C. 1978. Do cutaneous nerves sprout in the mammal? *J. Physiol. London* 280:52P

Duchen, L. W. 1970. Hereditary motor end-plate disease in the mouse, light and electron microscope studies. *J. Neurol. Neurosurg. Psychiatry* 33:238–50

Duchen, L. W. 1971. An electron microscopic study of the changes induced by botulinum toxin in the motor end-plates of slow and fast skeletal muscle fibers of the mouse. *J. Neurol. Sci.* 14:47–60

Duchen, L. W., Stefani, E. 1971. Electrophysiological studies of neuromuscular transmission in hereditary motor endplate disease of the mouse. *J. Physiol. London* 212:535–48

Duchen, L. W., Strich, S. J. 1968. The effects of botulinum toxin on the pattern of innervation of skeletal muscle of the mouse. *Q. J. Exp. Physiol.* 53:84–89

Duchen, L. W., Tonge, D. A. 1977. The effects of implantation of a foreign nerve on axonal sprouting usually induced by botulinum toxin in skeletal muscles of the mouse. *J. Anat.* 124:205–15

Dunn, G. A., Heath, J. P. 1976. A new hypothesis of contact guidance in tissue cells. *Exp. Cell Res.* 101:1–14

Edds, M. V. 1949. Experiments on partially deneurotized nerves. *J. Exp. Zool.* 111:211–26

Edds, M. V. 1950. Collateral regeneration of residual motor axons in partially denervated muscles. *J. Exp. Zool.* 113:517–52

Edds, M. V. 1953. Collateral nerve regeneration. *Q. Rev. Biol.* 28:260–76

Elsberg, C. A. 1917. Experiments on motor nerve regeneration and the direct neurotization of paralyzed muscles by their own and by foreign nerves. *Science* 45:318–20

Exner, S. 1884. *Die Innervation des Kehlkopfes. S. B. Akad. Wiss. Wien* 89(3):63–118

Fambrough, D. M. 1979. Control of acetylcholine receptors in skeletal muscle. *Physiol. Rev.* 59:165–227

Fangboner, R. F., Vanable, J. W. 1974. Formation and regression of inappropriate nerve sprouts during trochlear nerve regeneration in *Xenopus laevis. J. Comp. Neurol.* 157:391–406

Fex, S., Sonesson, B., Thesleff, S., Zelená, J. 1966. Nerve implants in botulinum poisoned mammalian muscle. *J. Physiol. London* 184:872–82

Fitzgerald, M. J. T. 1963. Transmedian cutaneous innervation. *J. Anat.* 93:313–22

Fitzgerald, M. J. T., Folon, J. C., O'Brien, T. M. 1975. The innervation of hyperplastic epidermis in the mouse: A light microscopic study. *J. Invest. Dermatol.* 64:169–74

Frank, E., Jansen, J. K. S., Lømo, T., Westgaard, R. H. 1975. The interaction between foreign and original nerves innervating the soleus muscle of rats. *J. Physiol. London* 247:725–43

Goldowitz, D., Cotman, C. W. 1980. Do neurotrophic interactions control synapses formation in the adult rat brain? *Brain Res.* 181:325–44

Gray, E. G., Westrum, L. E. 1979. Marginal bundles of axoplasmic microtubules at nodes of Ranvier within muscle. *Cell Tissue Res.* 199:281–88

Guth, L. 1962. Neuromuscular function after regeneration of interrupted nerve fibers into partially denervated muscle. *Exp. Neurol.* 6:129–41

Guth, L., Bernstein, J. J. 1961. Selectivity in the reestablishment of synapses in the superior cervical sympathetic ganglion of the cat. *Exp. Neurol.* 4:59–69

Gutmann, E., Young, J. Z. 1944. The reinnervation of muscle after various periods of atrophy. *J. Anat.* 78:15–43

Harris, J. B., Ribchester, R. R. 1979. The relationship between end-plate size and transmitter release in normal and dystrophic muscles of the mouse. *J. Physiol. London* 296:245–65

Hines, H. M., Wehrmacher, W. H., Thompson, J. D. 1945. Functional changes in nerve and muscle after partial denervation. *Am. J. Physiol.* 145:48–53

Hoffman, H. 1950. Local reinnervation in partially denervated muscle: A histophysiological study. *Aust. J. Exp. Biol. Med. Sci.* 28:383–97

Hoffman, H. 1951a. Fate of interrupted nerve fibers regenerating into partially denervated muscles. *Aust J. Exp. Biol. Med. Sci.* 29:211–19

Hoffman, H. 1951b. A study of the factors influencing innervation of muscles by implanted nerves. *Aust J. Exp. Biol. Med. Sci.* 29:289–308

Hoffman, H. 1952. Acceleration and retardation of the process of axon-sprouting in partially denervated muscles. *Aust. J. Exp. Biol. Med. Sci.* 30:541–66

Hoffman, H. 1953. The persistence of hyperneurotized end-plates in mammalian muscles. *J. Comp. Neurol.* 99:331–45

Hoffman, H., Springell, P. H. 1951. An attempt at the chemical identification of "neurocletin" (the substance evoking axon sprouts). *Aust. J. Exp. Biol. Med. Sci.* 29:417–24

Holland, R. L., Brown, M. C. 1980. Postsynaptic transmission block can cause motor nerve terminal sprouting. *Science* 207:649–51

Howe, P. R. C., Telfer, J. A., Livett, B. G., Austin, L. 1977. Extra-junctional acetylcholine receptors in dystrophic mouse muscles. *Exp. Neurol.* 56:42–51

Huizar, P., Kuno, M., Kudo, N., Miyata, Y. 1977. Reaction of intact spinal motoneurones to partial denervation of the muscle. *J. Physiol. London* 265:175–91

Hyden, H. 1943. Protein metabolism in the nerve cell during growth and function. *Acta Physiol. Scand.* 6: Suppl. 17, pp. 1–136

Ironton, R., Brown, M. C., Holland, R. L. 1978. Stimuli to intramuscular nerve growth. *Brain Res.* 156:351–54

Jackson, P., Diamond, J. 1977. Colchicine block of cholinesterase transport in rabbit sensory nerves without interference with long-term viability of the axons. *Brain Res.* 130:579–84

Jansen, J. K. S., Thompson, W., Kuffler, D. P. 1978. The formation and maintenance of synaptic connections as illustrated by studies of the neuromuscular junction. *Prog. Brain Res.* 48:3–20

Jones, R., Vrbova, G. 1974. Two factors responsible for the development of denervation hypersensitivity. *J. Physiol. London* 236:517–38

Korneliussen, H., Jansen, J. K. S. 1976. Morphological aspects of the elimination of polyneuronal innervation of skeletal muscle fibers in newborn rats, *J. Neurocytol.* 5:591–604

Kugelberg, L., Edstrom, L., Abbruzzese, M. 1970. Mapping of motor units in experimentally reinnervated rat muscle. *J. Neurol. Neurosurg. Psychiatry* 33:319–29

Laing, N. G., Prestige, M. C. 1978. Prevention of spontaneous motoneurone death in chick embryos. *J. Physiol. London* 282:33–34P

Letourneau, P. C. 1975. Cell to substratum adhesion and guidance of axonal elongation. *Dev. Biol.* 44:92–101

Liu, C. N., Chambers, W. W. 1958. Intraspinal sprouting of dorsal root axons. *Arch. Neurol. Psychiatry* 79:46–61

Livingston, W. K. 1947. Evidence of active invasion of denervated areas in sensory fibers from neighboring nerves in man. *J. Neurosurg.* 4:140–45

Lømo, T. 1974. Neurotrophic control of colchicine effects on muscle? *Nature* 249:473–74

Lømo, T. 1976. The role of activity in the control of membranes and contractile properties of skeletal muscle. In *Motor Innervation of Muscle*, ed. S. Thesleff, pp. 289–321. New York: Academic

Lømo, T., Rosenthal, J. 1972. Control of acetylcholine sensitivity by muscle activity in the rat. *J. Physiol. London* 221:493–513

Lømo, T., Westgaard, R. H. 1976. Control of Ach sensitivity in rat muscle fibers. *Cold Spring Harbor Symp. Quant. Biol.* 40:263–74

Maehlen, J., Njä, A. 1979. Sprouting after partial denervation in the superior cervical ganglion: Effect of preganglionic nerve stimulation. *Acta Physiol. Scand.* 105:18A–19A

McMahan, U. J., Sanes, J. R., Marshall, L. M. 1978. Reinnervation of muscle fiber basal lammia after removal of myofibers. Differentiation of regenerating axons at original synpatic sites. *J. Cell Biol.* 78:176–98

Miledi, R. 1963. Formation of extra nerve-muscle junctions in innervated muscle. *Nature* 199:1191–92

Murray, J. G., Thompson, J. W. 1957. The occurrence and function of collateral sprouting in the sympathetic nervous system of the cat. *J. Physiol. London* 135:133–62

O'Brien, R., Östberg, A., Vrbová, G. 1978. Observations on the elimination of polyneuronal innervation in developing muscle. *J. Physiol. London* 282:571–82

Olson, L., Malmfors, T. 1970. Growth characteristics of adrenergic nerves in the adult rat. Fluorescence, histochemical, and ^{3}H-noradrenaline uptake studies using tissue transplanted to the anterior chamber of the eye. *Acta Physiol. Scand. Suppl.* 348, pp. 1–111

Pestronk, A., Drachman, D. B. 1978. Motor nerve sprouting and acetylcholine receptors. *Science* 199:1223–25

Pittman, R. H., Oppenheim, R. W. 1978. Neuromuscular blockade increases motorneurone survival during normal cell death in the chick embryo. *Nature* 271:364–66

Purves, D., Lichtman, J. W. 1978. Formation and maintenance of synaptic connections in autonomic ganglia. *Physiol. Rev.* 58:821–62

Raisman, G. 1969. Neural plasticity in the septal nuclei of the adult rat. *Brain Res.* 14:25–48

Raisman, G., Field, P. M. 1973. A quantitative investigation of the development of collateral reinnervation after partial denervation of the septal nuclei. *Brain Res.* 50:241–64

Redfern, P. A. 1970. Neuromuscular transmission in newborn rats. *J. Physiol. London* 209:701–9

Reichart, F., Rotshenker, S. 1979. Motor axon terminal sprouting in intact muscles. *Brain Res.* 170:187–89

Roper, S. 1976. Sprouting and degeneration of synaptic terminals in the frog cardiac ganglion. *Nature* 261:148–49

Roper, S., Ko, C.-P. 1978. Impulse blockade in frog cardiac ganglion does not resemble partial denervation in changing synaptic organization. *Science* 202:66–68

Rosenthal, J. L., Taraskevich, P. S. 1977. Reduction of multiaxonal innervation at the neuromuscular junction of the rat during development. *J. Physiol. London* 270:299–310

Rotshenker, S. 1977. Sprouting of motor neurones and synapse formation. *J. Physiol. London* 273:74–75P

Rotshenker, S. 1978. Sprouting of intact motor neurones induced by neuronal lesion in the absence of denervated muscle fibers and degenerating axons. *Brain Res.* 155:354–56

Rotshenker, S. 1979. Synapse formation in intact innervated cutaneous-pectoris muscles of the frog following denervation of the opposite muscle. *J. Physiol. London* 292:535–47

Rotshenker, S., McMahan, U. J. 1976. Altered patterns of innervation in frog muscle after denervation. *J. Neurocytol.* 5:719–30

Rowe, R. W. D., Goldspink, G. 1969. Normal fiber growth in five different muscles in both sexes of mice. II. Dystrophic mice. *J. Anat.* 104:531–38

Slack, J. R. 1978. Interactions between foreign and regenerating axons in axolotl muscles. *Brain Res.* 146:172–76

Slack, J. R., Hopkins, W. G., Williams, M. N. 1979. Nerve sheaths and motoneurone collateral sprouting. *Nature* 282:506–7

Snider, W. D., Harris, G. L. 1979. A physiological correlate of disuse-induced sprouting at the neuromuscular junction. *Nature* 281:69–71

Speidel, C. C. 1932. Studies on living nerves. I. The movement of individual sheath cells and nerve sprouts correlated with the process of myelin sheath formation in amphibian larvae. *J. Exp. Zool.* 61:279–331

Speidel, C. C. 1933. Studies on living nerves. II. Activities of amoeboid growth cones, sheath cells, and myelin segments as revealed by prolonged observation of individual nerve fibers in frog tadpoles. *Am. J. Anat.* 52:1–79

Speidel, C. C. 1935. Studies on living nerves. III. Phenomena of nerve irritation and recovery, degeneration, and repair. *J. Comp. Neurol.* 61:1–82

Thompson, W. 1978. Reinnervation of partially denervated rat soleus muscles. *Acta Physiol. Scand.* 103:81–91

Thompson, W., Jansen, J. K. S. 1977. The extent of sprouting of remaining motor units in partly denervated immature and adult rat soleus muscles. *Neuroscience* 2:523–35

Tuffery, A. R. 1971. Growth and degeneration of motor end-plates in normal cat hindlimb muscles. *J. Anat.* 110:221–47

Tweedle, C. D., Kabara, J. J. 1977. Lipophilic nerve sprouting factor(s) isolated from denervated muscle. *Neuroscience Lett.* 6:41–46

Tweedle, C. D., Kabara, J. J. 1978. *Symp. Pharmacol. Affects Lipids.* AOSC Monogr. No. 5, Chap. 16, pp. 169–78

Van Harreveld, A. 1945. Reinnervation of denervated muscle fibers by adjacent functioning motor units. *Am. J. Physiol.* 144:447–93

Van Harreveld, A. 1947. On the mechanism of the spontaneous reinnervation in paretic muscles. *Am. J. Physiol.* 150:670–76

Vrbova, G., Gordon, T., Jones, R. 1978. *Nerve-Muscle Interaction.* London: Chapman & Hall

Watson, W. E. 1969. The response of motor neurones to intramuscular injection of botulinum toxin. *J. Physiol. London* 202:611–30

Watson, W. E. 1976. *Cell Biology of Brain.* London: Chapman & Hall

Weddell, G., Guttmann, L., Gutmann, E. 1941. The local extension of nerve fibers into denervated areas of skin. *J. Neurol. Psychiatry* 4:206–25

Wehrmacher, W. H., Hines, H. M. 1945. The recovery of skeletal muscle from the effects of partial denervation. *Fed. Proc.* 4:75–76

Weiss, P. A. 1955. Nervous system (Neurogenesis). In *Analysis of Development,* ed. B. M. Willier, P. A. Weiss, V. Hamburger, pp. 346–401. Philadelphia: Saunders

Weiss, P., Edds, M. V. 1946. Spontaneous recovery of muscle following partial denervation. *Am. J. Physiol.* 145:587–607

Ann. Rev. Neurosci. 1981. 4:43–68

DEVELOPMENT OF THE NEUROMUSCULAR JUNCTION: INDUCTIVE INTERACTIONS BETWEEN CELLS

♦11549

Michael J. Dennis

Departments of Physiology and of Biochemistry and Biophysics, University of California, San Francisco, California 94143

INTRODUCTION

A synapse, such as the neuromuscular junction, is the site of intricate and precise interactions between cells. The machinery that mediates these interactions arises through specialization of the presynaptic axon terminal, of the postsynaptic cell membrane, and of the material in the cleft between them. Differentiation of these structures during development must be directed by complex interactions between the component cells, interactions mediated by agents passing in both directions across the synapse and perhaps also laterally between pairs of pre- or postsynaptic cells. In this review I consider what is known of the changes that occur in development and of the intercellular interactions that coordinate this maturation. The information available affords only a fragmentary picture, as is reflected by discontinuities and gaps in the story. It is clear at the outset that we are yet far from understanding the molecular mechanisms that underlie development.

My principal focus is on normal development of the neuromuscular junction in rat, chick, and frog. Where pertinent, however, I draw from studies of synapse formation in cell and tissue culture and from studies of the experimental induction of new synapses on mature muscle fibers.

DIRECT COUPLING BETWEEN EMBRYONIC CELLS

Before considering specific interactions between developing nerve and muscle it is important to consider a general characteristic of embryonic tissues: the presence of low resistance junctions between individual cells (for reviews

43

0147-006X/81/0301-0043$01.00

see Bennett 1973, 1979; Furshpan & Potter 1968). Such junctions may be revealed by the passage of current from the inside of one cell to the inside of another (Sheridan 1968, Blackshaw & Warner 1976), by the passage of tracer molecules, such as fluorescent dyes, from cell to cell (Loewenstein 1976, Goodman & Spitzer 1979), or by electron microscopic visualization of the gap junctions (Kelly & Zacks 1969a, Blackshaw & Warner 1976, Kullberg et al 1977) that mediate coupling. In early embryos cell coupling occurs both within and across tissue boundaries (Potter et al 1966, Sheridan 1968, Warner 1973). During maturation this coupling is lost, apparently with the junctions that cross tissue boundaries disappearing first (Warner 1973). However, at the time when muscles develop the component cells are coupled. Gap junctions, which mediate coupling, have been seen in immature muscles of rat (Kelly & Zacks 1969a), chicken (Sisto-Daneo & Filogamo 1973), and amphibia (Keeter et al 1975, Blackshaw & Warner 1976, Kulberg et al 1977). Also, low resistance coupling between neighboring muscle cells has been shown in embryos of rats (Dennis et al 1980) and amphibia (Blackshaw & Warner 1976, Dennis 1975).

The existence of low resistance junctions between developing muscle cells may have several consequences. First, when a difference in potential exists between the insides of cells that are joined by a low resistance junction, current may flow between them [the basis for the electrical assessment of their existence (Sheridan, 1968)]. Thus a synaptic potential or action potential in one cell will pass, with degradation, to its neighbor (the extent of degradation varies inversely with the efficiency of the coupling). In intercostal muscles of 14 to 16 day rat embryos the electrical coupling is sufficiently strong that an action potential elicited in one or a small cluster of fibers will propagate laterally between fibers across the entire muscle (Dennis et al 1980). Thus at early stages the entire muscle could be exposed to a uniform pattern of electrical activity. In mature rat muscle fibers, electrical activity strongly influences a number of membrane characteristics, such as distribution of acetylcholine receptors (Jones & Vrbová 1970, Lømo & Rosenthal 1972, Lømo & Westgaard 1976), resting potential and transmembrane resistance (Westgaard 1975), and the deposition of acetylcholinesterase at the synapse (Lømo & Slater 1980b). Also, electrical activity has been shown to influence deposition of synaptic cholinesterase in culture (Rubin et al 1980). If activity is also influential in developing muscle, then the spread of excitation between coupled fibers could impose a common regulative influence on all.

The other possibility raised by coupling of embryonic cells is the passage of small organic molecules which themselves may influence gene expression. Gap junctions have been shown to be permeable to amino acids (Rieske et al 1975, Pitts & Simms 1977), sugars (Bennett & Dunham 1970, Rieske et al 1975), and nucleotides (Rieske et al 1975, Pitts & Simms 1977), though they are not permeable to macromolecules (Pitts & Simms 1977, Loewenstein 1976, Flagg-Newton et al 1979). In light of this ability of organic molecules to pass from cell to cell, factors (such as cyclic nucleotides) generated in more mature cells could diffuse into and influence less mature cells.

Electrical coupling between muscle fibers decreases in strength with maturation, and is ultimately lost. In rat intercostal muscles it is relatively weak in the last days of gestation and is lost during the first week after birth (Dennis et al 1980).

Electrical coupling may also occur between nerve terminal and target cell. Fischbach (1972) observed electrical coupling between chick neurons and muscle fibers growing in culture. His observation led to speculation that a transient low resistance junction between axon and muscle fiber might mediate recognition by a growing axon of its appropriate target cell. Indirect support for such an interaction between developing cells comes from a morphological study of central neuronal development in an invertebrate (Lopresti et al 1974). There has, however, been no further evidence for such a phenomenon in neuromuscular development since Fischbach's initial observation. We (Dennis et al 1980) tested for nerve-muscle coupling in rat embryos by blocking all chemical transmission during the period of synapse development, but could not record coupling potentials from muscle fibers upon nerve stimulation. The absence of positive evidence does not exclude this possibility; such junctions might occur earlier or be revealed by some other technique.

NERVE PRECEDES MUSCLE IN THE SYNAPTIC ZONE

In development of skeletal musculature, bundles of axons often arrive in the periphery before muscle cells appear. The motor axons grow into appropriately located masses of undifferentiated mesenchymal cells, which are to give rise to myoblasts and thence to muscle fibers. This precedence of the nerve has been noted in limbs of amphibia (Taylor 1943) and chick (Roncali 1970, Landmesser & Morris 1975, Bennett et al 1980) and in the body wall of rats (Dennis et al 1980). The interval between the arrival of the first axons and the appearance of their target cells is often only a few hours, and may sometimes be less, as in axial muscles of *Xenopus* (Kullberg et al 1977). Even so, the differentiation of myoblasts occurs, from the beginning, in the presence of motor nerve terminals.

This relationship in time of appearance of the synaptic partners raises the possibility that the axon terminals may exert an inductive influence on myogenic cells even before establishing synaptic contact. On the other hand, this is evidence against the possibility that the outgrowing axons are guided to their destinations by attraction to the target muscle fibers (for a review of the generation of neuromuscular specificity, see Landmesser 1980). Finally, this sequence of "nerve before muscle" suggests that different principles may be involved in the reinnervation of mature muscle from those that apply during development.

The ingrowth of the full complement of motor axons to their target zone is quite asynchronous. The first arrive early, but others continue to grow in as the differentiation of muscle fibers takes place (Taylor 1943, Bennett & Pettigrew 1974, Letinsky & Morrison-Graham 1980). In the diaphragm and intercostal muscles of rat embryos, innervation occurs over a five-day span between day 13 and day 18 of gestation (Bennett & Pettigrew 1974, Dennis et al 1980). Similarly, the appearance of muscle fibers is a protracted process. In rats the first muscle fibers appear on day 13 (Dennis et al 1980) and increase in number until at least birth (day 22) (Kelly & Zachs 1969a, Ontell & Dunn 1978) or even after (Betz et al 1979). In chicken limb, contractile muscle fibers first appear at day four in ovo (Landmesser & Morris 1975); the period of their addition has not been documented. Concomitant with this asynchrony of both axonal ingrowth and muscle fiber development is an asynchrony of synapse development.

DIFFERENTIATION OF THE NERVE TERMINAL FOR TRANSMITTER RELEASE

The first physiological sign of interaction between axon and muscle fiber so far described is the appearance of end-plate potentials (EPPs) in the fibers upon activation of the motor nerve. In rat intercostal muscles EPPs have been recorded intracellularly at 14 days of gestation (Dennis et al 1980). This is also the first day that the muscle contracts in response to stimulation. In chick limb muscles a contractile response to nerve stimulation has been reported at six days (stages 26 to 28) (Landmesser & Morris 1975), prior to the time of delineation of individual muscles out of the primitive muscle mass. Synaptic responses were recorded from chick anterior latissimus dorsi muscle (ALD) by Bennett & Pettigrew (1974) at nine days, the earliest time tested. In frog embryos spontaneous miniature end-plate potentials (MEPPs) were recorded from myotomal muscle at stages 21 to 22 (Blackshaw & Warner 1976, Kullberg et al 1977), 24 hr after fertilization, and contractile responses to spinal cord stimulation were observed at stage 24, the earliest examined (Kullberg et al 1977).

In embryonic rat muscles the EPPs fall into fairly distinct categories of fast and slow rates of rise (Bennett & Pettigrew 1974, Dennis et al 1980).

This is due to the low resistance junctions between muscle fibers, discussed above: The faster rising events are true synaptic potentials, whereas those of slow rise are coupling potentials. The coupling potentials arise from primary synaptic responses in nearby fibers, which are communicated with degradation of amplitude and waveform (Dennis 1975, Dennis et al 1980). Stimulation of a nerve bundle often results in a compound response which is a combination of synaptic and coupling potentials.

It is striking that in the rat, and probably also in the chick, the EPPs evoked during the first hours of junctional contact are sufficiently large to produce muscle contraction. That is, before the refinements that occur during maturation, such as the accumulation of postsynaptic receptors and the increase in surface area of nerve-muscle contact, the new junctions serve their purpose: They elicit muscle contraction. One factor that compensates for the lack of maturity of the early contacts is that the muscle cells are small, with high input resistances, and thus require relatively little synaptic current to depolarize them to threshold.

During this initial phase of synapse formation the rate of spontaneous transmitter release, as judged by MEPP frequency, is in most cases very low. Some fibers of 14 to 15 day rat embryos have MEPPs less frequently than 0.1/min, while others have frequencies of up to 1/min (Diamond & Miledi 1962, Bennett & Pettigrew 1974, Dennis et al 1980). Mature junctions, in contrast, have MEPP frequencies of 1/sec. Even the terminals without detectable spontaneous MEPPs usually do release transmitter in response to nerve stimulation (Dennis et al 1980). This initial capacity to release transmitter in response to stimulation, but not spontaneously, is the converse of the "nontransmitting" stage observed early on in *reinnervation* of adult muscle (there MEPPs precede EPPs; Dennis & Miledi 1974) and accentuates the need for caution in generalizing about development from experimental studies of mature tissue. The rate of spontaneous transmitter release remains low in the rat for the duration of gestation and begins to increase only two to three weeks after birth (Diamond & Miledi 1962); it is striking that this increase occurs just at the time that polyneuronal innervation is eliminated (see below). The rate of spontaneous release is also low (2 to 3/min) at developing terminals in chick muscle (Bennett & Pettigrew 1974). These extremely low rates of spontaneous release presumably reflect the fact that the initial contacts between nerve and muscle are small and that the nerve terminals contain few vesicles, as discussed below.

In amphibia, on the other hand, the spontaneous MEPP frequencies in developing myotomal muscle approximate those of mature junctions (1/sec) (Blackshaw & Warner 1976, Kullberg et al 1977). Perhaps the relatively high rates in these axial muscles result because the animals use those muscles early, for swimming, and therefore the synapses must mature rapidly.

In all of the developing systems in which spontaneous release has been studied, the amplitudes of the MEPPs are not normally distributed but instead show skew distributions with a preponderance of small amplitude events. There are several characteristics of developing junctions that contribute to this initial incidence of small amplitude events. First, the developing muscle fibers are electrically coupled to their neighbors; with sufficient coupling a large spontaneous synaptic potential in one fiber will give rise to a small coupling potential in its neighbor (Dennis 1975, Dennis et al 1980). Where this obtains, not all of the spontaneous potentials recorded in a fiber are MEPPs. Second, the developing muscle fibers may not always have a high density of acetylcholine receptors (AChR) on their membrane at the site of the new synapse (see below). Thus, the transmitter released spontaneously from each packet, even if it were of uniform amount, would not always combine with, and activate, the same number of AChR on the postjunctional membrane. The consequence would be irregularity in the amount of postsynaptic current elicited by single packets of transmitter. Finally, in the initial stages of synapse formation, the gap between the nerve terminal and the postsynaptic membrane may vary (see below); this would result in a variable amount of transmitter reaching the postjunctional membrane. In light of these factors it is not unexpected that the spontaneous MEPPs have abnormal amplitude distributions early on. The frequency of spontaneous release is accelerated at these immature junctions by the same changes in the bathing medium [elevated K^+, elevated osmolarity, added La^{3+} (Dennis et al 1980)] that increase MEPP frequency at mature junctions.

The first morphologically identifiable contacts between nerve and muscle have been seen in rat embryos of 16 days of gestation (Kelly & Zachs 1969b), chick embryos of 4 to 6 days (Atsumi 1971, 1977, Sisto-Daneo & Filogamo 1974, 1975), and *Xenopus* larvae of stage 21 (Kullberg et al 1977). These early contacts are quite unspecialized, despite the fact that physiological indications of synaptic function have been detected as much as two days earlier, at least in the rat. Electron micrographs reveal one or a cluster of small diameter axonal processes that contact the myofiber. The group of processes may be loosely enveloped by a supporting cell, presumably a Schwann cell, but they are often incompletely enclosed, and in some instances, in frog for example, the axons may be completely free of accompaniment as they approach the muscle. Some vesicles are seen in the axons, but with no particular concentration at points of neuromuscular contact; occasionally large diameter dense-cored vesicles are present. Sometimes there are localized thickenings of the postsynaptic membrane. There is a striking absence of other morphological specializations characteristic of mature neuromuscular contacts: The terminals lack clustered vesicles, release sites, and mitochondria; the synaptic cleft is of irregular width with only fine wisps of basal lamina or none at all; and there are no postsynaptic folds. It is impressive how little structural specialization seems to be required for function. A problem in interpreting such early ultrastructural findings is that of knowing the physiological state of any particular nerve-muscle contact visualized. It is conceivable that the physiologically active terminals are morphologically specialized but are so small and few that they have been missed.

The morphological features characteristic of mature junctions appear later. In 18 day rat embryos the contacts acquire a 500 to 900 Å gap partially filled by basal lamina, and the muscle membrane lying beneath becomes thickened and more electron-opaque. Very shallow depressions and ridges begin to appear on the postsynaptic membrane; only at the time of birth (day 21 to 22) or later do these become recognizable as developing synaptic folds (Kelly & Zachs 1969b). Likewise, in the chick (Hirano 1967) and the frog (Kullberg et al 1977) the assumption of uniform synaptic structures and the development of folds in the subsynaptic muscle occurs several weeks after establishment of contact.

The fact that synaptic function appears prior to maturation of synaptic structure indicates that transmission between nerve and muscle itself may play a part in induction of the mature structure. This process must involve regulative influences that pass in both directions across the synapse, since the nerve terminals only mature in the presence of appropriate postsynaptic partners.

DIFFERENTIATION OF THE POSTSYNAPTIC MEMBRANE: RECEPTORS

The postsynaptic acetylcholine receptor (AChR) is certainly the most extensively described component of the developing junction, though by no means is it completely understood. The vicissitudes of these receptors are relatively well documented because we have convenient tools with which to study them: neurotoxins that bind to the AChR with high affinity and great specificity; ACh itself, which can be precisely applied; and a physiological action, which can be accurately monitored. Mature neuromuscular junctions have receptors localized in high density on the postsynaptic membrane. Receptor concentrations can be 10^3 to 10^4 times higher in the junctional region than on extrajunctional muscle membrane (Fambrough 1974, Fertuck & Salpeter 1976).

Early studies of the distribution of ACh receptors on chick myofibers in culture (Vogel et al 1972, Fischbach & Cohen 1973, Sytkowski et al 1973) revealed that they were unevenly distributed. Even in the absence of innervation there occur localized regions of high receptor density, "hot spots"

(Fischbach & Cohen 1973). The presence of clusters of receptors on uninnervated muscle cells led to speculation that they might provide specialized sites at which synapses would develop (Fischbach & Cohen 1973, Sytkowski et al 1973). This idea of preferential innervation of receptor clusters, though attractive, has been disposed of by work in two laboratories. In cultures of muscle isolated from *Xenopus* embryos, Anderson & Cohen (1977) bound surface receptors with fluorescent conjugates of α-bungarotoxin (α-BTX) and observed the redistribution of AChRs on individual muscle cells following addition of neurons to the culture. The growing nerve processes showed no predilection for preexisting receptor clusters, and instead induced new patches of fluorescence to appear at the sites of nerve-muscle contact. Frank & Fischbach (1979) physiologically assessed ACh receptor distribution on chick muscle cells in culture. They located spots of high sensitivity on uninnervated myotubes, but found that these were not preferred sites of synapse formation following innervation by spinal neurons. Rather, the ingrowing axons induced new clusters of receptors elsewhere on the muscle membrane.

This question of whether receptor clusters play an instructive role during innervation may have arisen as an artifact of culture, because there is no evidence for clusters on uninnervated muscle cells in vivo. Rather, functional innervation develops at a time when the muscle cells seemingly have a uniform overall sensitivity to ACh (Blackshaw & Warner 1976, Bevan & Steinbach 1977) with no detectable receptor clusters. In axial muscles of rat, clusters of receptors can first be detected by ^{125}I-α-BTX autoradiography at 16 days of gestation (Bevan & Steinbach 1977, Braithwaite & Harris 1979), two days after innervation. These clusters appear in the central region of the muscle fibers, where the synaptic contacts are located (Bennett & Pettigrew 1974, Dennis et al 1980). Likewise in chicken embryos, clusters are first detected at ten days gestation (Burden, 1977a) three to four days after the first signs of synaptic function (Landmesser & Morris 1975). Thus aggregation of receptors normally begins subsequent to innervation.

This suggestion that the nerve influences receptor distribution is strongly reinforced by recent studies of cells in culture that indicate that spinal neurons release a factor(s) that induces insertion and/or clustering of receptors in adjacent postsynaptic membrane. Chick myotubes cocultured with fragments of embryonic spinal cord have a greater average density of receptors on their surface, as assayed by ACh iontophoresis, than do comparable myotubes cultured in isolation (Cohen & Fischbach 1977). Extracts of spinal cord contain a factor, perhaps a small peptide, which when added to the culture medium produces increased insertion of the receptors as judged by overall density, and increased receptor clustering on chick myotube (Jessell et al 1979) and on a cloned rat muscle cell line (Podleski et al

1978). A larger factor has also been found to be secreted by neuroblastoma-glioma hybrid cells, which increases aggregation of receptors on myotubes of mouse, rat, and chick (Christian et al 1978). The latter apparently does not increase the overall receptor density. The isolation of these factors leads toward an understanding of how the motor nerve terminal commands the localization of receptors at its feet.

Some time after the establishment of clusters of receptors in the junctional region, the density of extrajunctional receptors declines. In rats this recession begins in the week preceding birth and extends for one to two weeks after birth (Diamond & Miledi 1962, Bevan & Steinbach 1977). In chick the extrajunctional receptor density drops markedly in the week before hatching, and continues to decline after (Burden 1977a). The decline in extrajunctional receptors is dependent, at least in part, on muscle activity, as the rate of decline can be reduced by paralysis of chick embryos with curare (Burden 1977a). This effect appears to be the same as the well-documented influence of activity on extrajunctional receptors in adult muscle (Jones & Vrbová 1970, Lømo & Rosenthal 1972, Drachman & Witzke 1972, Berg & Hall 1975) and on chick myotubes in culture (Cohen & Fischbach 1973). The AChR thus provides a clearcut case in which the nerve terminal exerts a regulative influence on the postsynaptic cell. In fact, it seems to have a dual influence, causing accumulation of junctional receptors by means of a secreted factor while keeping extrajunctional receptors at bay by means of the electrical activity that it produces in the muscle fiber.

Another kind of evidence against the possibility that ACh receptors play an instructive role during synaptogenesis comes from demonstration that junctions will form even when receptors are blocked. In cultures of neural plate explanted from *Xenopus* embryos, muscle cells differentiate and are innervated (Cohen 1972). Cohen found that inclusion of curare in the culture in sufficient concentration to block neuromuscular transmission did not prevent synapse development: After three to nine days of culture with the drug, synaptic transmission could be detected within 10 min of washing it out. Similarly, in cultures of dissociated muscle and spinal cord from *Xenopus* (Anderson et al 1977), rat (Obata 1977), and chick (Obata 1977, Rubin et al 1980), synapses form in the presence of curare. The reinnervation of adult rat muscle will also occur when ACh receptors have been blocked with α-BTX (Jansen & Van Essen 1975).

ACh receptors at mature neuromuscular junctions have a long half-life in the membrane (11 days) relative to that of extrajunctional receptors (18 to 22 hr) (Berg & Hall 1975, Chang & Huang 1975, Linden & Fambrough 1979, Steinbach et al 1979, Reiness & Weinberg 1980). This raises the possibility that accumulation of receptors at the junction occurs simply because those turn over more slowly and thus remain longer than do receptors in nonjunctional membrane. However, the assumption of a slower turnover rate of junctional receptors occurs considerably later in development than the appearance of junctional clusters. In the extreme case of chick muscle, the reduced rate appears only several weeks after hatching (Burden 1977b). In rat muscle the reduction in junctional turnover rate occurs between 17 and 19 days of gestation (Reiness & Weinberg 1980), whereas the clustering of receptors appears several days earlier.

The ACh receptor and its associated ion channel have physiological properties that also change during development. The most notable of these is the mean time (τ) that channels remain open in response to the transmitter; τ can be measured by noise analysis (Katz & Miledi 1972). In embryonic and newborn rat muscle τ is relatively long (4.6 msec, 21°C); between one and two weeks after birth, it is composed of a mixture of long and short open times; and after two weeks it becomes uniformly short (1.4 msec), as is characteristic of mature junctions (Sakmann & Brenner 1978, Fischbach & Schuetze 1980). The situation appears fundamentally different in chick muscle (Schuetze 1980). There τ is long (4 msec at 23°C) in innervated embryonic as well as uninnervated myotubes in culture, and remains long until at least 18 weeks after hatching, far beyond the time when the junctions have matured as judged by other criteria.

Another developmental change in the ACh receptor is that of its response to d-tubocurarine. In adult muscle curare acts as a nondepolarizing antagonist of acetylcholine receptors. However, it produces depolarization of embryonic rat muscle (Ziskind & Dennis 1978) that is sufficient to elicit muscle contraction. This depolarizing action disappears between the first and second weeks after birth, the time at which receptor gating time also changes.

The acetylcholine receptor is one protein that is localized in the postsynaptic membrane. However, it seems likely that there are others there that are also important for development, but that have not yet been detected for lack of the appropriate probes. Specific antibodies hold promise of being one such tool.

DIFFERENTIATION IN THE SYNAPTIC CLEFT

The synaptic cleft of mature junctions is occupied by basal lamina. This is a filamentous extracellular meshwork of collagen and glycoprotein (Sanes et al 1978) that surrounds the entire myofiber, occupies the gap separating pre- and postsynaptic membranes, and projects into the secondary folds of the postjunctional membrane. Electron microscopy does not reveal any unique variation of its amorphous appearance at the end-plate region. However, at least some of the end-plate specific acetylcholinesterase (AChE), which serves to shorten transmitter action by hydrolysis of ACh, is located within the basal lamina of the cleft (Hall & Kelly 1971, McMahan et al 1978). Furthermore, recent immunological studies by Sanes & Hall (1979) have revealed a distinct class of antigens that are restricted to the synaptic portion of the basal lamina.

In rat intercostal muscle the first rudiments of basal lamina are seen by EM at 16 days gestation between some but not all presumed nerve-muscle contacts (Kelly & Zacks 1969b). From the micrographs of Kelly & Zacks (1969a,b) it appears that the basal lamina may be deposited first in the end-plate region and only later develop in the extrajunctional regions. In chick, basal lamina has been seen at eight days in ovo; here, too, it appears more fully developed at synaptic junctions than elsewhere (Atsumi 1977, Sisto-Daneo & Filogamo 1974).

Junctional acetylcholinesterase first appears at about the same time that the early deposition of basal lamina is detected. In the rat histochemically demonstrable AChE is first seen on day 17 of gestation (Bennett & Petti-grew 1974, Bevan & Steinbach 1977), whereas in the chick, esterase activity is detected at six to seven days (Atsumi 1971). One difficulty with histo-chemical staining for AChE in early muscles is that esterase activity is present in the presynaptic nerve terminal before it appears in the cleft, and the two sources may be confused unless resolution is adequate. In develop-ing cutaneous pectoris muscles of bullfrog tadpoles, AChE staining is in-complete and sometimes absent from the earliest stages (14 to 16) of nerve muscle contact (Letinsky & Morrison-Graham 1980), but is regularly found in association with terminals of later stages. We do not have information as to the time of appearance of the basal lamina in the frog.

Following the establishment of transmitting synapses and the appearance of AChE in the cleft there is a reduction in duration of the synaptic poten-tials. This development of AChE and change in physiology have been carefully correlated for synapses developing in cocultures of chick muscle and spinal cord (Rubin et al 1979). Rubin and colleagues sorted newly formed junctions into categories on the basis of the mean time constant of synaptic current decay (fast < 1.8 msec, intermediate, and slow > 2.6 msec) and then checked each histochemically for the presence of AChE. After four days of coculture, 74% of the synapses with fast decay showed positive AChE staining, whereas none of the slow synapses stained. Furthermore, addition of an inhibitor of AChE caused prolongation of the current flow at fast but not at slow synapses. The intracellular correlate of the decrease in duration of synaptic current flow caused by AChE was a decrease in time to peak of spontaneous and evoked synaptic potentials. During the early stages of innervation of myotomal muscles in *Xenopus* embryos the mean rise time of MEPPs decreased seven-fold in 18 hr; during that same time period AChE activity appeared (Kullberg et al 1980). Such a change in end-plate potential waveform also occurs in rat junctions at the stage of gestation when AChE normally appears, at 17 days (Dennis et al 1980). Not only is the time to peak of synaptic potentials reduced during maturation, but their rate of rise increases (Dennis et al 1980, Kullberg et al 1980). This is due in part to increase in number of the junctional ACh receptors (Bevan & Steinbach 1977, Burden 1977a).

Recent work persuasively indicates that electrical activity of muscle plays an important part in the deposition of AChE in the developing end-plate. In cultures of chick nerve and muscle, Rubin and co-workers (1980) studied the formation of junctions in the presence of curare. Under such conditions synapses did form (see also Cohen 1972) and ACh receptors did cluster at the junctions. However, the junctional form of AChE did not appear, as judged by the findings that (a) the time constant of synaptic current decay remained long, (b) histochemical staining remained negative, and (c) the high molecular weight form of esterase, which is specific for the end-plate (Hall 1973, Vigny et al 1976), did not appear. A crucial factor for the AChE development was muscle activity, as Rubin et al (1980) showed that chronic direct stimulation of the curarized muscles elicited appearance of AChE at the junctions. These experiments clearly distinguish the regulation of ACh receptors from the regulation of AChE at the developing junctions, since receptors still cluster but esterase does not appear in the absence of activity. However, as the authors caution, this does not exclude concomitant involvement of soluble factors, released from the nerve terminal, in the regulation of junctional esterase.

Further indication of the importance of activity in deposition of junctional AChE comes from studies of new end-plates on mature muscle. Lømo & Slater (1980a,b) induced ectopic endplates in rats by implantation of the superficial fibular nerve into soleus muscle. Following lesion of the soleus nerve, functional fibular nerve junctions were established within three to four days, and these had histochemically demonstrable AChE within six to seven days (Lømo & Slater 1980b). These ectopic junctions also induced foci of high ACh sensitivity on the muscle fibers (Lømo & Slater 1980a). When the implanted fibular nerve was cut two to seven days after the initial soleus nerve lesion, the ACh sensitivity remained high at the sites that the fibular sprouts occupied (Lømo & Slater 1980a), but the AChE failed to develop. However, if the muscle was chronically stimulated by implanted electrodes subsequent to this same sequence of nerve lesions, plaques of AChE did appear, coincident with the peaks of ACh sensitivity (Lømo & Slater 1980b). Thus in the first two to seven days of synaptogenesis the nerve induces a persistent local "trace," perhaps in the basal lamina, which primes it for deposition of AChE on the condition of muscle activity. A few days after the initial lesion of the soleus nerve most of the AChE disappears from the original end-plates. However, the "trace" seems to remain, for upon establishment of synaptic transmission, and of muscle activity by the ectopic nerve, AChE also reappears at the old soleus nerve sites even though there are no nerve terminals in the vicinity (Guth et al 1966, Weinberg & Hall 1979).

These experiments (Weinberg & Hall 1979, Lømo & Slater 1980b) also demonstrate that the presence of a nerve terminal is not essential for the appearance of AChE in the end-plate region, which thereby implicates the muscle in its generation. Ultrastructural studies by Wake (1976) of motor end-plate development in chick embryos document a role of muscle fibers in AChE deposition. In day nine muscle cells Wake saw intracellular AChE activity localized in the nuclear envelope, sarcoplasmic reticulum, Golgi complex, and in granules he thought derived from Golgi. As end-plates developed, at day 12, AChE-rich granules accumulated in the sarcoplasm beneath. Then, as the esterase activity appeared in the end-plate, the intracellular levels declined. On this basis he proposed that deposition of AChE in the end-plate resulted from exocytosis of esterase-containing granules (Wake 1976).

The basal lamina does more than simply anchor the cholinesterase in place, as demonstrated by experiments from the laboratory of McMahan on the reinnervation of adult frog muscle. Subsequent to lesion of a mature motor nerve the distal axonal processes degenerate, leaving the muscle fibers without innervation, though the original synaptic site can still be identified by the persistence of folds in the basal lamina and of membrane specializations, such as folds and ACh receptor localization, in the postsynaptic muscle fiber. The proximal stumps of the motor axons sprout, regenerate back to the muscle, and quite accurately reinnervate the original postsynaptic sites on the muscle fibers (Letinsky et al 1976), there forming functional synapses (Miledi 1960). If, at the time of the nerve lesion, the muscle fibers are cut on each side of the end-plate region, they too degenerate and are phagocytosed, leaving behind only the basal lamina sheath. The site of the original synapses can still be identified by the persistence of projections of the basal lamina sheath that formerly occupied junctional folds in the original end-plate and by the continued presence of AChE in the synaptic basal lamina. In time a complete synapse is restored by axonal reinnervation and by regeneration of myofibers within the original basal lamina sheath. Moreover, the new synapses are located precisely at their old sites on the basal lamina (Marshall et al 1977). So the remnants of nerve sheath, Schwann cell, and basal lamina are sufficient to direct restoration of the other synaptic components. Following degeneration of muscle and nerve, the regeneration of myofibers can be prevented by X-irradiation, leaving the axons to return alone. Nevertheless, the nerves grow into the old synaptic site and their terminals differentiate clusters of vesicles in active zones in direct apposition to the folds of the basal lamina, as they occur normally (Sanes et al 1978). This strongly indicates that the basal lamina contains sufficient information to trigger terminal differentiation by the axons; the muscle is not necessary. Finally, if reinnervation is prevented by nerve resection while the myofibers are allowed to regenerate, postsynaptic specialization of the muscle membrane occurs precisely under the old synaptic site on the basal lamina; folds develop and receptors cluster (Burden et al 1979). It seems inescapable that, at least in mature frogs, factors associated with basal lamina can instruct both the regenerating nerve and muscle.

It is impressive that inductive information can be held more or less permanently on an extracellular scaffolding. Clearly we would like to know the molecular form in which this information is stored and when it is deposited. These findings indicate that we would be ill-advised to restrict our attention to cells as we seek the mechanisms by which development proceeds.

CONTROL OF INNERVATION

Skeletal muscles of mature vertebrates usually have only one motor nerve terminal per end-plate and, in most fast muscles, only one end-plate per muscle fiber. However, neonatal muscle fibers are multiply innervated, and may receive functional synaptic contact from as many as six motor axons. This polyneuronal innervation occurs in immature muscles of rat (Redfern 1970), cat (Bagust et al 1973), rabbit (Bixby & Van Essen 1979), chicken (Bennett & Pettigrew 1974), and frog (Bennett & Pettigrew 1975). In every case multiple synaptic contacts occur within a common end-plate, as indicated both morphologically (Atsumi 1971, Bennett & Pettigrew 1974, Brown et al 1976, Korneliussen & Jansen 1976, Riley 1976, 1980, Rosenthal & Taraskevich 1977, Bixby 1980, Letinsky & Morrison-Graham 1980) and physiologically (Brown et al 1976, Dennis et al 1980).

Following the establishment of synaptic contact by one axon, innervation by other axons continues over a period of hours or days. In rat intercostal muscle, after functional synapses first appear on the fourteenth day of gestation, there is a progressive increase in the mean number of inputs per fiber until about the eighteenth day of gestation (Dennis et al 1980). At its peak, the average number of inputs per fiber is slightly more than three, with a range of one to six. Junctions at the earliest stages appear by morphological criteria to receive only a single axon terminal, whereas those at later stages contain multiple axonal profiles (Atsumi 1977, Letinsky & Morris-Graham 1980, Bixby 1980). However, electron micrographs showing clusters of axon terminals in developing junctions must be interpreted with caution for there may be branching of individual axons (cf Bixby 1980), which would be detected only in serial sections.

This polyneuronal innervation disappears two to three weeks after birth in rats (Redfern 1970). As with their establishment, the elimination of synapses is progressive; plots of the mean number of inputs per fiber show a steady decline with age. In rat diaphragm and soleus muscles the mature state of one nerve terminal per muscle fiber is achieved between 15 and 20 days after birth (Redfern 1970, Brown et al 1976, Rosenthal & Taraskevich 1977). Furthermore, the process of elimination is a very orderly one in that ultimately every muscle fiber has one and only one input (Brown et al 1976). The time of onset and the rate of synapse elimination have been shown to vary between different muscles of the rabbit as well as between similar muscles in the rabbit and the rat (Bixby & Van Essen 1979).

Why does polyneuronal innervation occur? What mechanisms underlie its orderly elimination? We cannot yet answer either of these questions, but it is worth considering what can be said. The establishment of multiple synaptic inputs occurs during the period of development when new fibers are rapidly being added to the embryonic muscles (Kelly & Zachs 1969a, Ontell 1977, Ontell & Dunn 1978). During this phase each new myofiber must go through a period, however brief, when it is uninnervated. Perhaps the uninnervated fibers emit a chemical signal that incites the nerve terminals to form synapses (see review by Brown et al 1981, this volume). If one further supposed that at this early stage the nerve terminals are unable to discriminate between uninnervated and innervated fibers, the consequence would be multiple innervation of single muscle fibers. The inability early on to distinguish between fibers with and without innervation could result for a variety of reasons. It must take some time for a muscle fiber to elaborate a signal that indicates it is innervated, or to shut off a signal that indicates lack of innervation. If the delay in signaling were long enough, other axons could make contact in the meantime. Alternatively the electrotonic coupling between embryonic muscle fibers (see above) may prevent the individuation of single fibers necessary for distinction between innervation or a lack thereof.

Another possibility is that the initial excess in innervation arises because motoneurons follow a rigid developmental program. One instruction in such a program could be that between embryonic days 14 and 18 (in rat) the axon terminals are to elaborate as many synapses on muscle fibers as possible. Such a model assigns less of a role in the process to the muscle fibers, aside from serving as targets, yet the consequence would be the same: multiple innervation. I return to the possibility of a developmental program below.

One constraint that obtains during multiple innervation is that all synaptic contracts form at roughly the same site on the muscle fiber. This may result because axons grow in bundles, with those that reach the periphery later following in the track formed by their predecessors. Thus a following axon is likely to arrive at a muscle fiber at the spot where the leader first made synaptic contact, resulting in clustering of synapses. Another influence could also come from the basal lamina. If the basal lamina formed prior to the arrival of following axons, and if there were already distinction between junctional and nonjunctional basal lamina (see above), then those following axons might be induced or constrained to make synaptic contact with the muscle fiber only at the original junction.

What can be said about the retraction of multiple innervation? One potential explanation is that some motoneurons die after establishment of peripheral synapses (Prestige 1976), resulting in loss of excessive contacts. This appears not to be the case, at least in rat embryos where most motoneuronal cell death occurs earlier (embryonic day 14). Brown et al (1976) demonstrated that there is no significant change in the number of motor axons projecting to the soleus muscle during the period of synapse elimination. Involvement of motoneuronal cell death has not been ruled out in the case of chickens, where the period of multiple innervation does in part coincide with the period of motoneuronal cell death. Another possibility is that the establishment of the adult pattern of innervation is due simply to a random retraction by all motoneurons of a proportion of their contacts; yet if retraction were entirely random, one would expect to find some fibers that had no innervation and some fibers with multiple innervation, which has not been observed (Brown et al 1976). Nevertheless, some synaptic terminals are withdrawn, and this retraction continues in at least some muscles, in the presence of uninnervated fibers (Thompson & Jansen 1977). Thompson & Jansen partially denervated soleus muscles of newborn rat pups, and found that those motoneurons that remained intact did retract some of their contacts, though fewer than normal, as judged by reduction of the motor unit size 2 to 20 weeks after denervation. On this basis Jansen and co-workers proposed that the sorting out of the adult innervation pattern involves a combination of spontaneous retraction and competition: All motoneurons must reduce the number of synaptic contacts that they will maintain, but also within each end-plate there is some form of competition between axons that determines the one to remain. In this undefined competition it is presumably the terminals from the most overextended motoneurons which are at a disadvantage and are selectively lost.

A recent report by Betz et al (1980) has raised a question as to whether a motoneuron is obliged to retract any of its synapses in the absence of competition. They took advantage of small lumbrical muscles of the rat that receive input from both the lateral plantar and the sural nerves. When the lateral plantar nerve was cut at birth some muscles were left innervated by only a single motor axon. In these muscles innervated by a single axon they distinguished innervated from denervated fibers by their diameters, and found that the size of the remaining motor unit did not decrease from its initial level (120 fibers per axon) during the period when axons normally retract. Given this apparent absence of retraction they argue that competition is the only factor that dictates synapse withdrawal, and that no withdrawal occurs in its absence. There are two possible explanations for the discrepancy between the conclusion of Betz's group and that of Jansen's as to whether motoneurons must withdraw some contacts. First, retraction and reduction in motor unit size may occur whenever two axons occupy the same end-plate (competition), even though there are other uninnervated fibers nearby. If so, Thompson & Jansen (1977) would expect some reduction of motor unit size in their partially denervated muscles because they always had more than one axon remaining and would thus expect some residual cohabitation of individual end-plates. Alternatively there may be different classes of motoneurons: those that initially make more synaptic contacts than they can ultimately support, and those that are capable of sustaining all of their initial contacts. The former would retract some contacts even in the absence of competition; the latter would not.

To attribute synapse retraction to competition does little to explain it. Do the terminals directly interact with one another, are they indirectly vying for space on the muscle fiber, or do they compete for some substance released by the muscle? We do not know. However, it is clear that there must be some signal to the individual axon terminals that influences their behavior. That signal might be a positive one from the muscle, encouraging the axon to stay, or it might be a negative one from other axons in the same end-plate.

Electrical activity in the muscle fibers may have some influence on the rate of synapse retraction. O'Brien et al (1978) found that chronic stimulation of rat soleus muscles between six and ten days of age produced a significant increase in the rate of synapse elimination. This has also been studied by the converse technique of blocking normal nerve activity. Thompson et al (1979) paralyzed soleus muscles in nine and ten day rat pups by implanting in the sciatic nerve a plastic plug containing tetrodotoxin, which slowly released the toxin and thereby blocked all nerve activity. This had no influence on the rate of elimination of polyneuronal innervation during the first four days of nerve block. At longer times the rate of elimination of multiple inputs was decreased, which was attributed to sprouting of nerves in the paralyzed muscle (Thompson et al 1979). In chick, on the other hand, Srihari & Vrbová (1978) found that paralysis by injection of curare on the seventh and twelfth days of incubation produced an increase in the number of axon profiles per end-plate between the sixteenth and eighteenth days, at a time when the number of profiles normally decreases. Here again, caution is necessary in drawing conclusions about the number of inputs to a fiber from the number of axonal profiles seen in random sections of its end-plate (cf Bixby 1980). An additional complication is that such curare injections have been claimed to prevent motoneuronal cell death in the chick (Pittman & Oppenheim 1979). Thus the contrast in conclusions from these two laboratories on the effects of electrical activity may involve differences between rat and chick, but they may also reflect the fact that activity has some influence in the process of nerve retraction without being a dominant factor.

Several groups have examined end-plate morphology in neonates to determine whether synapse elimination occurs by means of terminal retraction or degeneration and have come to disparate conclusions. In an EM study of rat soleus muscle Korneliussen & Jansen (1976) found an increase in the number of terminal profiles per end-plate section during the first few days after birth, followed by progressive reduction in profiles through the sixteenth day of gestation. The initial increase presumably reflects branching of axons already present, since innervation is complete earlier (Dennis et al 1980). They looked for, but did not find, any sign of axonal degeneration during this period of synapse reduction and so concluded that terminals retract. With silver staining and light microscopy of rat muscle Riley (1976) saw a reduction with postnatal age in number of axons projecting to individual end-plates. Also during this period of synapse elimination he saw some axons that ended in bulbs proximal to end-plates (Riley 1977); these he interpreted as axons retracting from end-plates where they had formerly made synaptic contact. More recently Riley (1980) looked at the fine structure of these end-plates. Again he found no evidence for terminal degeneration, whereas he did find axon terminals that had accumulations of vesicles and membrane sacs proximal to the end-plates. These he interpreted as axons in the process of retracting. In contrast to this conclusion Rosenthal & Taraskevich (1977) found signs of abnormal ultrastructure of some terminals in neonatal end-plates, including condensation of vesicles and increased cytoplasmic density. They also noted Schwann cells impinging into the synaptic cleft, and sometimes enveloping terminal profiles, which they took to be signs of terminal degeneration. A careful ultrastructural study in rabbit (Bixby 1980), which included serial and semiserial sectioning of ten end-plates, revealed no sign of axonal degeneration. Bixby did find that axons could have healthy-looking synaptic contacts in one section and give off irregular protrusions that projected blindly among Schwann cell processes in another. Thus Schwann cell envelopment of axon profiles need not reflect terminal degeneration, as interpreted by Rosenthal & Taraskevich (1977). This underscores the need for caution in interpreting random sections through an end-plate: Multiple terminals may arise from one axon, and they need not all make contact with the postsynaptic membrane. Furthermore, Bixby found regions of post-synaptic membrane, with folds and AChE staining, which were contacted only by Schwann cells. The weight of the evidence thus indicates that reduction of multiple innervation occurs by terminal retraction.

Another factor that must be considered is whether new fibers are being added to the muscles at the time that multiple innervation is declining. Unfortunately this is difficult to reliably assess in perinatal muscles because of the manner in which new fibers develop (Kelly & Zachs 1969a, Ontell 1977, Ontell & Dunn 1978). Rat muscles at birth contain independent myofibers, each enclosed in a basement membrane, but they also contain cell clusters consisting of one large primary myofiber, one or a few smaller secondary myofibers, some satellite myotubes which are still smaller and have irregular myofilaments, and other small cells lacking myofilaments. These clusters are enclosed within one common basement membrane. With time the secondary myofibers and satellite myotubes grow, the clusters break apart, and constituent myofibers each receive their own basal lamina (Kelly & Zachs 1969a, Ontell & Dunn 1978). In the rat extensor digitorum longus muscle these clusters comprise 40% of the basal lamina-bound profiles at two days after birth, 2% of the profiles at five days and 1% of the profiles at eight days. Conversely, the number of independent muscle units increased by 90% during this period, whereas the total number of cells with myofilaments changed little (Ontell & Dunn 1978). However, it is not known whether each of the constituents of the clusters has its own innervation, or whether only the primary myofiber has an end-plate, with the smaller secondary cells coupled to it. If the latter were the case, new synapses would need to be generated on the secondary and tertiary myofibers as they became independent. The only prenatal study of end-plate morphology in the rat (Kelly & Zachs 1969b) does not resolve this question. Our electrophysiological study of prenatal muscles (Dennis et al 1980) indicates that all fibers penetrated had their own innervation before birth, but we may not have recorded from the smallest myotubes in the clusters.

The issue of fiber addition relates to the matter of polyneuronal innervation in that if new fibers were being innervated at the same time that others were losing some of their synaptic inputs, then motoneurons might break off some synaptic contacts and form others with little net change in their motor unit size. In rat lumbrical muscles Betz et al (1979) found a doubling in the number of fibers during the month following birth, and they concluded that synapses were forming de novo on these new fibers. It is unclear whether their muscle fiber counts treated the small satellite myotubes as independent or not.

The suggestion of synaptogenesis after birth raises the intriguing possibility that throughout the period of polyneuronal innervation the synaptic contacts are dynamic. Axon terminals might break and reform contacts regularly before finally establishing permanent contacts. Neither morphological nor physiological studies would reveal such dynamism; perhaps the axonal processes that end in preterminal bulbs (Riley 1977, 1980) are coming rather than going. The fact that the spontaneous MEPP frequency (Diamond & Miledi 1962) and the mean quantal content of EPPs (Dennis et al 1980) remain at very low levels until several weeks after birth suggests that the terminals remain in an immature, and perhaps transient, condition for some time after contacts are first made.

Before embracing the idea of dynamic contacts, however, we must consider some experiments in newborn rats that indicate an inability of terminal axons to sprout. When a rat muscle is partially denervated just after birth, the intact motor units still decline in size (Thompson & Jansen 1977). This retraction may result entirely from competition, as discussed above, but it does occur with uninnervated fibers in the vicinity. Dennis & Harris (1980) were also unable to experimentally induce axonal sprouting in muscles of newborn rats: When the distal halves of intercostal muscles were denervated at birth by intramuscular nerve section, the muscle fibers affected were not reinnervated until at least three weeks later. Similarly, when the nerve was sectioned at one or two weeks of age, reinnervation still only began at three weeks. Following the same denervation in older muscles, on the other hand, reinnervation begins within four to five days. If the postnatal axons can still form junctions, why do they not do so under these experimental conditions?

The inability of motor axons to immediately reform functional contacts following denervation at birth again raises the possibility that motoneurons follow a rigid developmental program. The program might dictate that during the first three weeks after birth the cell must retract excess synaptic contacts, and not form new ones. Thus it may not make new synapses even if presented with stimuli that promote synaptogenesis at a later state in its life cycle.

CONCLUSION

It may be useful to consider the developmental changes described above in terms of the inductive interactions that occur between the component cells of the neuromuscular junction. A number of general types of interaction are evident.

1. A presynaptic neuron can influence its target by means of the post-synaptic electrical activity that it causes. In mature muscle fibers activity

has been shown to play an important role in the regulation of extrajunctional acetylcholine receptors and of membrane electrical properties. Activity has also been demonstrated to be required in the stable deposition of acetylcholinesterase at developing junctions. These studies clearly indicate that electrical activity can influence gene expression (reviewed by Fambrough 1979). Because transmission is established early in the development of the synapse, it is reasonable to expect that transmitted activity may also serve as a signal to developing muscle cells; it may even initiate differentiation of the end-plate, as well as of other regions of the muscle fiber. Although motoneurons may have only limited activity before birth, such signaling need not be extensive; furthermore, the electrical coupling that exists between embryonic muscle fibers could maximize the influence of the excitation that does occur.

2. The presynaptic axon probably exerts an influence on its target cell through release of factors other than the transmitter. For instance, clustering of acetylcholine receptors in the postsynaptic membrane follows establishment of synaptic contact, yet appears to be independent of activity. There is now evidence for the release from spinal neurons of a polypeptide factor (or factors) that induces insertion and clustering of ACh receptors in muscle fiber membrane. The purification of such agents will allow study of the site and mechanisms by which they exert an inductive influence on the recipient cells. The arrival of nerves in the periphery prior to muscle differentiation raises the possibility that motor axons could exert an inductive influence, not only on differentiated muscle cells, but also on the mesenchymal precursors of muscle cells.

3. Some signals must pass in a retrograde direction across the junction, from the muscle cell to the axon. The morphological differentiation of the presynaptic terminal, most of which follows establishment of synaptic contact, depends, at least initially, on the presence of the postsynaptic target cell. This presence might be communicated by a diffusable factor, released by the muscle, or it might be conveyed by molecules anchored on the exterior surface of the muscle membrane. The experiments on reinnervation of basal lamina tubes and redevelopment of muscle fibers within these laminar tubes provide evidence that molecules that convey inductive information to both the pre- and postsynaptic cell may be anchored on this extracellular matrix.

4. Finally, influential signals appear to pass between axon terminals that occupy the same end-plate, for in the course of elimination of polyneuronal innervation, some form of competitive selection occurs that results in one axon terminal alone per end-plate. This could involve either direct lateral signaling between adjacent terminals or mediation of the selection by way of the postsynaptic muscle fiber. Here again, the primary signal could be either chemical or electrical.

Clearly we are yet far from understanding development of the neuromuscular junction, but the information that we have indicates a number of productive avenues of research.

ACKNOWLEDGMENTS

I would like to express my gratitude to Peter Sargent, Lea Ziskind, Ralph Greenspan, and Zach Hall for their many constructive suggestions. Also, I greatly appreciate the patient assistance of Rose Misono and Jeff Bennett in preparation of the manuscript.

Literature Cited

Anderson, M. J., Cohen, M. W. 1977. Nerve-induced and spontaneous redistribution of acetylcholine receptors on cultured muscle cells. *J. Physiol. London* 268:757–73

Anderson, M. J., Cohen, M. W., Zorychia, E. 1977. Effects of innervation on the distribution of acetylcholine receptors on cultured muscle cells. *J. Physiol. London* 268:731–56

Atsumi, S. 1971. The histogenesis of motor neurons with special reference to the correlation of their endplate formation. I. The development of endplates in the intercostal muscle in the chick embryo. *Acta Anat.* 80:161–82

Atsumi, S. 1977. Development of neuromuscular junctions of fast and slow muscles in the chick embryo: A light and electron microscopic study. *J. Neurocytol.* 6:691–709

Bagust, J., Lewis, D. M., Westerman, R. A. 1973. Polyneuronal innervation of kitten skeletal muscle. *J. Physiol.* 229:241–55

Bennett, M. R., Davey, D. F., Ubel, K. E. 1980. The growth of segmental nerves from the brachial myotomes into the proximal muscles of the chick forelimb during development. *J. Comp. Neurol.*

Bennett, M. R., Pettigrew, A. G. 1974. The formation of synapses in striated muscle during development. *J. Physiol. London* 241:515–45

Bennett, M. R., Pettigrew, A. G. 1975. The formation of synapses in amphibian striated muscle during development. *J. Physiol.* 252:203–39

Bennett, M. V. L. 1973. Function of electrotonic junctions in embryonic and adult tissues. *Fed. Proc.* 32:65–75

Bennett, M. V. L. 1979. Electrical transmission: A functional analysis and comparison to chemical transmission. In *Handbook of Physiology: The Nervous System,* pp. 357–416

Bennett, M. V. L., Dunham, P. B. 1970. Sucrose permeability of junctional membrane at an electronic synapse. *Biophys. J.* 10:177

Berg, D. G., Hall, Z. W. 1975. Increased extrajunctional acetylcholine sensitivity produced by chronic post-synaptic neuromuscular blockade. *J. Physiol.* 244:659–76

Betz, W. J., Caldwell, J. H., Ribchester, R. R. 1979. The size of motor units during postnatal development of rat lumbrical muscles. *J. Physiol.* 297:463–78

Betz, W. J., Caldwell, J. H., Ribchester, R. R. 1980. The effects of partial denervation at birth on the development of muscle fibers and motor units in rat lumbrical muscle. *J. Physiol.* In press

Bevan, S., Steinbach, J. H. 1977. The distribution of α-bungarotoxin binding sites on mammalian skeletal muscle developing in vivo. *J. Physiol. London* 267:195–213

Bixby, J. L. 1980. Ultrastructural observations on synapse elimination in neonatal rabbit skeletal muscle. *J. Neurocytol.* In press

Bixby, J. L., Van Essen, D. C. 1979. Regional differences in the timing of synapse elimination in skeletal muscles of the neonatal rabbit. *Brain Res.* 169:275–86

Blackshaw, S. E., Warner, A. E. 1976. Low resistance junctions between mesoderm cells during development of trunk muscles. *J. Physiol.* 255:209–30

Braithwaite, A. W., Harris, A. J. 1979. Neural influence on acetylcholine receptor clusters in embryonic development of skeletal muscles. *Nature* 279:549–51

Brown, M. C., Jansen, J. K. S., Van Essen, D. C. 1976. Polyneuronal innervation of skeletal muscle in new-born rats and its elimination during maturation. *J. Physiol. London* 261:387–422

Brown, M. C., Holland, R. L., Hopkins, W. G. 1981. Motor nerve sprouting. *Ann. Rev. Neurosci.* 4:17–42

Burden, S. 1977a. Development of the neuromuscular junction in the chick embryo: The number, distribution, and stability of acetylcholine receptors. *Dev. Biol.* 57:317–29

Burden, S. 1977b. Acetylcholine receptors at the neuromuscular junction: Developmental change in receptor turnover. *Dev. Biol.* 61:79–85

Burden, S. J., Sargent, P. B., McMahan, U. J. 1979. Acetylcholine receptors in regenerating muscle accumulate at original synaptic sites in the absence of the nerve. *J. Cell Biol.* 82:412–25

Chang, C. C., Huang, M. C. 1975. Turnover of junctional and extrajunctional acetylcholine receptors of the rat diaphragm. *Nature* 253:643–44

Christian, C. N., Daniels, M. P., Sugiyama, H., Vogel, Z., Jaques, L., Nelson, P. G. 1978. A factor from neurons increases the number of acetylcholine receptor aggregates on cultured muscle cells. *Proc. Natl. Acad. Sci. USA* 75:4011–15

Cohen, M. W. 1972. The development of neuromuscular connexions in the presence of *d*-tubocurarine. *Brain Res.* 41:457–63

Cohen S. A., Fischbach, G. D. 1973. Regulation of muscle acetylcholine sensitivity by muscle activity in cell culture. *Science* 181:76–78

Cohen, S. A., Fischbach, G. D. 1977. Relative peaks of ACh sensitivity at identified nerve-muscle synapses in spinal cord-muscle cocultures. *Dev. Biol.* 59:24–38

Dennis, M. J. 1975. Physiological properties of junctions between nerve and muscle developing during salamander limb regeneration. *J. Physiol.* 244:683–702

Dennis, M. J., Harris, A. J. 1980. Transient inability of neonatal rat motoneurons to reinnervate muscle. *Dev. Biol.* 74:173–83

Dennis, M. J., Miledi, R. 1974. Non-transmitting neuromuscular junctions during an early stage of end-plate reinnervation. *J. Physiol.* 239:553–70

Dennis, M. J., Ziskind-Conhaim, L., Harris, A. J. 1980. Development of neuromuscular junctions in rat embryos. *Dev. Biol.* In press

Diamond, J., Miledi, R. 1962. A study of foetal and new-born rat muscle fibers. *J. Physiol.* 162:393–408

Drachman, D. B., Witzke, F. 1972. Trophic regulation of acetylcholine sensitivity of muscle: Effect of electrical stimulation. *Science* 176:514–16

Fambrough, D. M. 1974. Acetylcholine receptors: Revised estimates of extrajunctional receptor density in denervated rat diaphragm. *J. Gen. Physiol.* 64:468–72

Fambrough, D. M. 1979. Control of acetylcholine receptors in skeletal muscle. *Physiol. Rev.* 59:165–227

Fertuck, H. C., Salpeter, M. M. 1976. Quantitation of junctional and extrajunctional acetylcholine receptors by electron microscope autoradiography after ^{125}I-α-bungarotoxin binding at mouse neuromuscular junctions. *J. Cell Biol.* 69:144–58

Fischbach, G. D. 1972. Synapse formation between dissociated nerve and muscle cells in low density cell cultures. *Dev. Biol.* 28:407–29

Fischbach, G. D., Cohen, S. A. 1973. The distribution of acetylcholine sensitivity over uninnervated and innervated muscle fibers grown in cell culture. *Dev. Biol.* 31:147–62

Fischbach, G. D., Schuetze, S. M. 1980. A postnatal decrease in acetylcholine channel open time at rat endplates. *J. Physiol.* In press

Flagg-Newton, J., Simpson, I., Loewenstein, W. R. 1979. Permeability of the cell-to-cell membrane channels in mammalian cell junction. *Science* 205:404–7

Frank, E., Fischbach, G. D. 1979. Early events in neuromuscular junction formation in vitro. Induction of acetylcholine receptor clusters in the postsynaptic membrane and morphology of newly formed synapses. *J. Cell Biol.* 83:143–58

Furshpan, E. J., Potter, D. D. 1968. Low-resistance junctions between cells in embryos and tissue culture. In *Curr. Top. Dev. Biol.* 3:95–127

Goodman, C. S., Spitzer, N. C. 1979. Embryonic development of identified neurons: Differentiation from neuroblast to neurone. *Nature* 280:208–14

Guth, L., Zalewski, A. A., Brown, W. C. 1966. Quantitative changes in cholinesterase activity of denervated sole plates following implantation of nerve into muscle. *Exp. Neurol.* 16:136–47

Hall, Z. W. 1973. Multiple forms of acetylcholinesterase and their distribution in endplate and nonendplate regions of rat diaphragm. *J. Neurobiol.* 4:343–61

Hall, Z. W., Kelly, R. B. 1971. Enzymatic detachment of end-plate acetylcholinesterase from muscle. *Nature New Biol.* 232:62–63

Hirano, H. 1967. Ultrastructural study of the morphogenesis of the neuromuscular junction in the skeletal muscle of the chick. *Z. Zellforsch. Mikrosk. Anat.* 79:198–208

Jansen, J. K. S., Van Essen, D. C. 1975. Reinnervation of rat skeletal muscle in the presence of α-bungarotoxin. *J. Physiol. London* 250:651–67

Jessell, T. M., Siegel, R. E., Fischbach, G. D. 1979. Induction of acetylcholine receptors on cultured skeletal muscle by a factor extracted from brain and spinal cord. *Proc. Natl. Acad. Sci. USA* 76:5397–5401

Jones, R., Vrbová, G. 1970. Effect of muscle activity on denervation hypersensitivity. *J. Physiol. London* 210:144–45P

Katz, B., Miledi, R. 1972. The statistical nature of the acetylcholine potential and its molecular components. *J. Physiol.* 224:665–99

Keeter, J. S., Pappas, G. D., Model, P. G. 1975. Inter- and extramyotomal gap junctions in the axolotl embryo. *Dev. Biol.* 45:21–33

Kelly, A. M., Zacks, S. I. 1969a. The histogenesis of rat intercostal muscle. *J. Cell Biol.* 42:135–53

Kelly, A. M., Zacks, S. I. 1969b. The fine structure of motor endplate myogenesis. *J. Cell Biol.* 42:154–69

Korneliussen, H., Jansen, J. K. S. 1976. Morphological aspects of the elimination of polyneuronal innervation of skeletal muscle fibres in newborn rats. *J. Neurocytol.* 5:591–604

Kullberg, R. W., Lentz, T. L., Cohen, M. W. 1977. Development of the myotomal neuromuscular junction in Xenopus laevis: An electrophysiological and fine-structural study. *Dev. Biol.* 60:101–29

Kullberg, R. W., Mikelberg, F. S., Cohen, M. W. 1980. Contribution of cholinesterase to developmental decreases in the time course of synaptic potentials at an amphibian neuromuscular junction. *Dev. Biol.* In press

Landmesser, L. T. 1980. The generation of neuromuscular specificity. *Ann. Rev. Neurosci.* 3:279–302

Landmesser, L. T., Morris, D. G. 1975. The development of functional innervation in the hind limb of the chick embryo. *J. Physiol.* 249:301–26

Letinsky, M. S., Fischbach, G. D., McMahan, U. J. 1976. Precision of reinnervation of original postsynaptic sites in

muscle after a nerve crush. *J. Neurocytol.* 5:691–718

Letinsky, M. S., Morrison-Graham, K. 1980. Structure of developing frog neuromuscular junctions. *J. Neurocytol.* In press

Linden, D. C., Fambrough, D. M. 1979. Biosynthesis and degradation of acetylcholine receptors in rat skeletal muscles. Effects of electrical stimulation. *Neuroscience* 4:527–38

Loewenstein, W. R. 1976. Permeable junctions. *Gold Spring Harbor Symp. Quant. Biol.* 40:49–63

Lømo, T., Rosenthal, J. 1972. Control of ACh sensitivity by muscle activity in the rat. *J. Physiol.* 221:493–513

Lømo, T., Slater, C. R. 1980a. Acetylcholine sensitivity of developing ectopic nerve-muscle junctions in adult rat soleus muscles. *J. Physiol.* In press

Lømo, T., Slater, C. R. 1980b. Control of junctional acetylcholinesterase by neural and muscular influences in the rat. *J. Physiol.* In press

Lømo, T., Westgaard, R. H. 1976. Control of ACh sensitivity in rat muscle fibers. *Cold Spring Harbor Symp. Quant. Biol.* 40:263–74

Lopresti, V., Macagno, E. R., Levinthal, C. 1974. Structure and development of neuronal connections in isogenic organisms: Transient gap junctions between growing optic axons and lamina neuroblasts. *Proc. Natl. Acad. Sci. USA* 71:1098–1102

Marshall, L. M., Sanes, J. R., McMahan, U. J. 1977. Reinnervation of original synaptic sites on muscle fiber basement membrane after disruption of the muscle cells. *Proc. Natl. Acad. Sci. USA* 74:3073–77

McMahan, U. J., Sanes, J. R., Marshall, L. M. 1978. Cholinesterase is associated with the basal lamina at the neuromuscular junction. *Nature* 271:172–74

Miledi, R. 1960. Properties of regenerating neuromuscular synapses in the frog. *J. Physiol.* 154:190–205

Obata, K. 1977. Development of neuromuscular transmission in culture with a variety of neurons and in the presence of cholinergic substances and TTX. *Brain Res.* 119:141–53

O'Brien, R. A. D., Östberg, A. J. C., Vrbová, G. 1978. Observations on the elimination of polyneuronal innervation in developing mammalian skeletal muscle. *J. Physiol.* 282:571–82

Ontell, M. 1977. Neonatal muscle: An electron microscopic study. *Anat. Rec.* 189:669–90

Ontell, M., Dunn, R. F. 1978. Neonatal muscle growth: A quantitative study. *Am. J. Anat.* 152:539–56

Pittman, R., Oppenheim, R. W. 1979. Cell death of motoneurons in chick embryo spinal cord. IV. Evidence that a functional neuromuscular interaction is involved in the regulation of naturally occurring cell death and the stabilization of synapses. *J. Comp. Neurol.* 187:425

Pitts, J. D., Simms, J. W. 1977. Permeability of junctions between animal cells. Intracellular transfer of nucleotides but not of macromolecules. *Exp. Cell Res.* 104:153–64

Podleski, T. R., Axelrod, D., Ravdin, P., Greenberg, I., Johnson, M. M., Salpeter, M. M. 1978. Nerve extract induces increase and redistribution of acetylcholine receptors on cloned muscle cells. *Proc. Natl. Acad. Sci. USA* 75:2035–39

Potter, D. D., Furshpan, E. J., Lennox, E. S. 1966. Connections between cells of the developing squid as revealed by electrophysiological methods. *Proc. Natl. Acad. Sci. USA* 55:328–36

Prestige, M. C. 1976. Evidence that at least some of the motor nerve cells that die during development have first made peripheral connections. *J. Comp. Neurol.* 170:123–34

Redfern, P. A. 1970. Neuromuscular transmission in new-born rats. *J. Physiol. London* 209:701–9

Reiness, C. G., Weinberg, C. B. 1980. Metabolic stabilization of acetylcholine receptors at newly formed neuromuscular junctions in rat. *Devel. Biol.* In press

Rieske, E., Schubert, P., Kreutzberg, G. W. 1975. Transfer of radioactive material between electrically coupled neurons of the leech CNS. *Brain Res.* 84:365–82

Riley, D. A. 1976. Multiple axon branches innervating single endplates of kitten soleus myofibers. *Brain Res.* 110:158–61

Riley, D. A. 1977. Spontaneous elimination of nerve terminals from the endplates of developing skeletal myofibers. *Brain Res.* 134:279–85

Riley, D. A. 1980. Ultrastructural evidence of axon retraction during the spontaneous elimination of polyneuronal innervation of rat soleus muscle. *J. Neurocytol.* In press

Roncali, L. 1970. The brachial plexus and the wing nerve pattern during early development in chicken embryos. *Monit. Zool. Ital.* 4:81–98

Rosenthal, J. L., Taraskevich, P. S. 1977. Reduction of multiaxonal innervation at the neuromuscular junction of the rat during development. *J. Physiol. London* 270:299–310

Rubin, L. L., Schuetze, S. M. Fischbach, G. D. 1979. Accumulation of acetylcholinesterase at newly formed nerve-muscle synapses. *Dev. Biol.* 69:46–58

Rubin, L. L., Schuetze, S. M., Weill, C. L., Fischbach, G. D. 1980. Regulation of acetylcholinesterase appearance at neuromuscular junctions in vitro. *Nature* 283:264–67

Sakmann, B., Brenner, H. R. 1978. Change in synaptic channel gating during neuromuscular development. *Nature* 276:401–2

Sanes, J. R., Hall, Z. W. 1979. Antibodies that bind specifically to synaptic sites on muscle fiber basal lamina. *J. Cell Biol.* 83:357–70

Sanes, J. R., Marshall, L. M., McMahan, U. J. 1978. Reinnervation of muscle fiber basal lamina after removal of myofibers. Differentiation of regenerating axons at original synaptic sites. *J. Cell Biol.* 78:176–98

Schuetze, S. M. 1980. The acetylcholine channel open time is prolonged at newly formed and mature chick endplates. *J. Physiol.* In press

Sheridan, J. D. 1968. Electrophysiological evidence for low-resistance intercellular junctions in the early chick embryo. *J. Cell. Biol.* 37:650–59

Sisto-Daneo, L., Filogamo, G. 1973. Ultrastructure of early neuro-muscular contacts in the chick embryo. *J. Submicr. Cytol.* 5:219–29

Sisto-Daneo, L., Filogamo, G. 1974. Ultrastructure of developing myo-neural junctions. Evidence for two patterns of synaptic area differentiation. *J. Submicr. Cytol.* 6:219–28

Sisto-Daneo, L., Filogamo, G. 1975. Differentation of synaptic area in slow and fast muscle fibers. *J. Submicro. Cytol.* 7:121–32

Srihari, T., Vrbová, G. 1978. The role of muscle activity in the differentiation of neuromuscular junctions in slow and fast chick muscles. *J. Neurocytol.* 7:529–40

Steinbach, J. H., Merlie, J., Heinemann, S., Bloch, R. 1979. Degradation of junctional and extrajunctional acetylcholine receptors by developing rat skeletal muscle. *Proc. Natl. Acad. Sci. USA* 76:3547–51

Sytkowski, A. J., Vogel, Z., Nirenberg, M. W. 1973. Development of acetylcholine receptor clusters on cultured muscle cells. *Proc. Natl. Acad. Sci. USA* 70:270–74

Taylor, A. C. 1943. Development of the innervation pattern in the limb of the frog. *Anat. Rec.* 87:379–413

Thompson, W., Jansen, J. K. S. 1977. The extent of sprouting of remaining motor units in partly denervated immature and mature rat soleus muscle. *Neuroscience* 2:523–36

Thompson, W., Kuffler, D. P., Jansen, J. K. S. 1979. The effect of prolonged, reversible block of nerve impulses on the elimination of polyneuronal innervation of new-born rat skeletal muscle fibers. *Neuroscience* 4:271–81

Vigny, M., Koenig, J., Rieger, F. 1976. The motor end-plate specific form of acetylcholinesterase: Appearance during embryogenesis and re-innervation of rat muscle. *J. Neurochem.* 27:1347–53

Vogel, Z., Sytkowski, A. J., Nirenberg, M. W. 1972. Acetylcholine receptors of muscle grown in vitro. *Proc. Natl. Acad. Sci. USA* 69:3180–84

Wake, K. 1976. Formation of myoneural and myotendenous junctions in the chick. *Cell Tissue Res.* 173:383–400

Warner, A. E. 1973. The electrical properties of the ectoderm in the amphibian embryo during induction and early development of the nervous system. *J. Physiol.* 235:267–86

Weinberg, C. B., Hall, Z. W. 1979. Junctional form of acetylcholinesterase restored at nerve-free endplates. *Dev. Biol.* 68:631–35

Westgaard, R. H. 1975. Influence of activity on the passive electrical properties of denervated soleus muscle fibres in the rat. *J. Physiol.* 251:683–97

Ziskind, L., Dennis, M. J. 1978. Depolarizing effect of curare on embryonic rat muscle. *Nature* 276:622–23

Ann. Rev. Neurosci. 1981. 4:69–125

CALCIUM CHANNEL ❖11550

Susumu Hagiwara

Department of Physiology, University of California, School of Medicine,
Los Angeles, California 90024

Lou Byerly

Department of Biological Sciences, University of Southern California,
Los Angeles, California 90007

INTRODUCTION

The subject of this review is the voltage-dependent calcium channel. In a wide variety of cells a depolarization of the membrane potential increases the membrane's permeability to Ca^{2+}. We refer to this voltage-dependent permeation mechanism as the Ca channel, even though very little evidence exists to establish its channel nature. We do not discuss Ca currents or fluxes that pass through channels which are not specific for divalent cations, such as the channels opened by acetylcholine. The Ca channel allows the passive movement of Ca^{2+} across the membrane; we do not consider active transport of Ca^{2+}, i.e. when Ca^{2+} is moved in a direction opposite to the electrochemical gradient.

Ca channels have been reviewed several times before (Hagiwara 1973, 1975, Reuter 1973, Baker & Glitsch 1975). In this review we try to give a more general view of Ca channels, comparing data from nerve, muscle, secretory, epithelial, and egg cells, as well as *Paramecium*. We give only a sketchy account of the Ca currents found in vertebrate heart and smooth muscle, since we are not well aware of this literature. We begin by reviewing most of the membranes in which Ca channels have been identified. Next we focus on the problems that have made the study of Ca channels considerably more difficult than the study of the Na channel. Finally, we have chosen to review five properties of Ca channels that are of special interest in explaining either the physiological function of Ca currents or in comparing the Ca currents found in different membranes. These properties are (*a*) saturation, selectivity, and blocking, (*b*) dependence on $[Ca^{2+}]_i$, (*c*) kinetics,

69

(d) molecular nature of Ca conductance, and (e) control by external chemical messengers. We have used data from as wide a variety of Ca currents as possible in discussing each property.

It is probably best to state at the beginning that we are writing with the assumption that the Ca channel is not a unique molecular structure, unlike the Na channel, which shows almost identical selectivity in squid axon, frog nerve, and tunicate egg. Although different environments would be expected to alter the properties of a particular channel (such as voltage dependence and kinetics), different Ca currents show so many differences in fundamental properties that we find it easier to assume that there are more than one type. In certain cases two Ca currents with very different properties are found in the same membrane, e.g. starfish egg. Much of the controversy that surrounds Ca currents is probably due to the fact that conflicting data comes from different membranes. There is a need for the application of a variety of studies on the same membranes so that real differences in Ca channels can be recognized as such, and the properties of particular types of Ca channels can be clearly established.

DISTRIBUTION OF CALCIUM CURRENTS

It is becoming evident that the presence of membrane-potential-gated calcium permeabilities in excitable membranes is more the rule than the exception. Voltage-dependent calcium permeabilities are found in a larger variety of membranes than are voltage-dependent sodium permeabilities, so it would be simpler to list the cases where inward currents are carried by sodium and assert that all the remaining inward currents are carried by calcium. However, in this section we attempt to list most of the cells in which calcium currents have been identified. Because most of these cells were first studied under current-clamp conditions, the calcium currents were recognized by the regenerative, positive-going action potentials (Ca spikes) that they produce.

A positive-going action potential can be produced by a voltage-dependent permeability increase of the membrane to any of various ions, at least theoretically. However, in actuality the possibilities are only Na^+, Ca^{2+}, and Cl^-. Cl action potentials have been found in plant cells and in certain animal cells (Raja electric organ and chick myoballs) under modified extracellular ionic conditions, but the main issue is to distinguish Ca spikes or components of spikes from Na spikes. There are a number of tests used to identify Ca spikes.

1. The overshoot (absolute value of the potential differences between inside and outside the cell reached at the peak of the spike) or the maximum

rate of rise of the spike varies directly with $[Ca^{2+}]_o$. All regenerative response disappears when external Ca^{2+} is replaced by a nonpermeant divalent cation such as Mg^{2+}. Ideally the overshoot increases about 29 mV for a ten-fold increase in $[Ca^{2+}]$, but there are a number of reasons that the increase can be more or less.

2. The overshoot or maximum rate of rise of the spike does not change if external Na^+ is replaced by a large organic cation like $Tris^+$.

3. The influx of calcium measured with radioactive calcium must be sufficient to charge the membrane to the potential reached by the spike. The influx of calcium may also be measured by the fluorescence of the calcium-binding protein aequorin or the increased absorption of calcium-binding dyes like Arsenazo III. However, these methods are difficult to quantify and do not distinguish calcium that has entered through the plasma membrane from calcium released from intracellular stores.

4. The spikes are blocked by polyvalent cations such as Co^{2+}, La^{3+}, Mn^{2+}, Cd^{2+}, and Ni^{2+} at concentrations less than 10 mM. Organic Ca blockers such as verapamil, D600 (methoxy-derivative of verapamil), and nifedipine block some Ca currents at concentrations less than 10 μM; however, they block other Ca currents only at high concentrations where they also block Na currents.

5. The spikes are not blocked by high concentrations of tetrodotoxin (TTX) ($>10^{-6}$ M). This result is necessary, but not sufficient, since TTX-insensitive Na currents have been found in puffer fish nerve and muscle (Kao 1966, Hagiwara & Takahashi 1967a), denervated rat twitch muscle (Redfern et al 1970), developing rat muscle (Kidokoro 1973), tunicate egg cell (Okamoto et al 1976a), and molluscan neurons (Kostyuk et al 1974a, Connor 1979).

6. Spikes persist when Ca^{2+} in the external solution is replaced by Ba^{2+} or Sr^{2+}.

In the following survey of the cells in which Ca currents have been identified, it would have been too lengthy to list which of the above criteria have been demonstrated in every case. Unless stated otherwise, it can be assumed that at least criteria 1, 2, and 4 have been demonstrated.

Invertebrate Muscle

Calcium spikes were first discovered in crustacean muscle. Fatt & Katz (1953) found that crab muscle fibers, which are usually inexcitable in normal medium, produced prolonged action potentials when the Na^+ in the external solution was replaced by quaternary ammonium ions, such as choline, tetraethylammonium (TEA), or tetrabutylammonium (TBA). The TBA-induced action potential is maintained after washing with an isotonic sucrose solution containing only Ca^{2+}, Mg^{2+}, and K^+. They concluded that the inward current was probably carried by an influx of Ca^{2+} or Mg^{2+}, or an efflux of some internal anion. Working with crayfish muscle, Fatt & Ginsborg (1958) showed that the action potential produced by TEA was independent of Na^+ and Mg^{2+}, and concluded that it was produced by an increased membrane permeability for Ca^{2+}. When Ca^{2+} was replaced by Ba^{2+} or Sr^{2+}, all-or-none action potentials were obtained, even without TEA. Werman & Grundfest (1961) also observed that TEA and Ba produced action potentials in lobster muscle; however, they did not consider divalent cations as charge carriers. Hagiwara & Naka (1964) found that Ca spikes could also be observed in crustacean muscle by lowering the intracellular concentration of free Ca^{2+}. Working with giant barnacle muscle fibers (up to 2 mm in diameter), they found that intracellular injection of Ca^{2+}-binding agents [such as sulfate, ethylenediaminetetraacidic acid (EDTA), and citrate] rendered the fibers capable of producing all-or-none spikes, with normal extracellular solutions. The overshoot of such spike potentials increased about 29 mV when $[Ca^{2+}]_o$ was increased ten-fold, as expected for a Ca electrode, and was unaffected by the removal of external Na^+. It was found that Ba^{2+} or Sr^{2+}, but not Mg^{2+}, could substitute for Ca^{2+} in producing an action potential. The influx of ^{45}Ca during one spike was found to be five to ten times the amount needed to charge the membrane capacity by 80 mV (the amplitude of the spike). Crustacean Ca spikes were shown to be insensitive to TTX (10^{-6} to 10^{-5} gm/ml), and blocked by Mn^{2+} (Hagiwara & Nakajima 1966a, Takeda 1967).

Ca spikes have been demonstrated in a number of insect muscles. It was found that muscle fibers of the mealworm larva responded to Ba^{2+} and TEA in a manner very similar to crustacean muscle fibers (Belton & Grundfest 1961). Washio (1972) showed that the local response of locust leg muscle could be converted to an action potential by adding TEA or substituting Ba^{2+} or Sr^{2+} for Ca^{2+}. The overshoot of the action potential increased with increases in the concentration of Ca^{2+}, Sr^{2+}, or Ba^{2+}; but the removal of Na^+ caused no change in the action potential. The action potential was blocked by Mn^{2+}, but not by TTX. Patlak (1976) showed that fly flight muscle has a Ca spike, which is graded in normal saline, but is converted to all-or-none in TEA solution. The action potential of skeletal muscle of beetle larva has been shown to be a Ca spike (Fukuda et al 1977, Fukuda & Kawa 1977). Finally, the ventral longitudinal muscle fibers of the stick insect have a calcium current, which has been studied under voltage-clamp conditions (Ashcroft et al 1979).

Ca spikes have been identified in the muscles of several other invertebrates. Heart muscle of the bivalve *Mytilus* has a Ca spike (Irisawa et al 1967). Twarog (1967) studied the action potential of a catch muscle of

Mytilus and found it was TTX-insensitive. Striated muscle fibers of the larva glochidium of the freshwater mussel *Anodonta* produce a Ca spike (Kidokoro et al 1974), and nematode somatic muscle has an all-or-none Ca spike even in normal saline (Weisblat et al 1976).

The ionic mechanisms underlying the action potential in two protochordate muscles have been examined. Hagiwara & Kidokoro (1971) found that the action potential of amphioxus muscle has both Na and Ca components. After the Na current was eliminated, the muscle produced regenerative potential changes when procaine was added to the saline. This action potential was demonstrated to be due to a Ca current according to all criteria, except for the measurement of Ca influx. In contrast, Miyazaki and co-workers (1972) found the mature muscle cell of tunicate to have a pure Ca spike with no detectable Na component. However, at earlier stages in the process of differentiation, tunicate muscle cells have an action potential with a significant Na component.

Thus, in all invertebrate muscles where the ionic mechanism underlying the regenerative response has been studied, an increased membrane permeability for calcium has been found. Most of these membranes seem to have pure Ca spikes; however, in fly and nematode muscle, secondary regenerative responses are reported that are Na dependent (Patlak 1976, Weisblat et al 1976). The appearance of a Na component in the action potential of protochordate muscle is not surprising, considering the proximity of protochordates to vertebrates; vertebrate skeletal muscle has a nearly pure Na spike (Nastuk & Hodgkin 1950). The primary function of Ca currents in invertebrate muscle appears to be the activation of muscle contraction by directly increasing the concentration of free calcium inside the cell. In some tissues the Ca^{2+} influx during the action potential may be sufficient for contraction, e.g. amphioxus muscle (Hagiwara et al 1971). However, at present it is still unclear how much of the Ca^{2+} necessary for contraction actually comes from outside and how much is released from internal stores. Ca-induced Ca release has not been excluded in invertebrate muscle fibers.

Vertebrate Muscle

Although the action potential in vertebrate twitch muscle is almost a pure Na spike, a weak Ca current has been found in frog skeletal muscle and in embryonic chick and rat muscle. Bianchi & Shanes (1959) demonstrated that while the resting influx of ^{45}Ca in frog sartorius muscle was about equal to that in squid axon, the ^{45}Ca influx during an action potential was 30 times greater for frog muscle than for squid nerve. By removing all Cl^- from the external solution and applying TEA, the Ca current in frog twitch muscle was demonstrated electrically (Beaty & Stefani 1976a,b, Stanfield 1977). This current was identified as a Ca current, since it was not affected by TTX or the removal of external Na^+, was blocked by Co^{2+} and D-600, and disappeared when Ca^{2+} was replaced by Mg^{2+}. Stefani & Uchitel (1976) demonstrated that toad slow muscle, which does not have a Na spike, has an anodal break response in isotonic potassium sulfate that is calcium dependent and is blocked by Co^{2+}.

Embryonic chick skeletal muscle has an action potential with a plateau that appears to be due to a Ca current, since it is TTX insensitive and blocked by Mn^{2+} (Kano et al 1972, Kano & Shimada 1973). In treating chick skeletal muscle cells with colchicine to produce "myosacs," a small action potential is obtained in Na-free solution, the amplitude of which increases with $[Ca^{2+}]_o$ (Fukuda et al 1976). However, the Ca component of the action potential in embryonic chick muscle disappears as the muscle matures (Kano 1975). Kidokoro (1973) demonstrated that the action potential in a clonal rat skeletal muscle cell line had a Ca component. Likewise, this Ca current disappears in the adult rat skeletal muscle, as evidenced by the absence of any regenerative response in a Na-free saline with Ca^{2+} replaced by Ba^{2+} (Kidokoro 1975a). If adult rat and chick muscle fibers were examined under the same conditions used to demonstrate Ca currents in frog skeletal muscle, it is possible that vestigial Ca spikes would also be found in these fibers.

The action potential in vertebrate smooth muscle appears to be a Ca spike in most cases. Studies on guinea pig vas deferens, taenia coli, and ureter have demonstrated the dependence of the spike amplitude on $[Ca^{2+}]_o$. The spike remains when Na^+ is replaced in the external solution; however, it is questionable how completely Na^+ is removed from the diffusion-limited space outside the membrane. These smooth muscle action potentials are insensitive to TTX, but are blocked by Mn^{2+}. Sr^{2+} and Ba^{2+} can substitute for Ca^{2+} in maintaining spike generation. See Reuter (1973) for a review of these studies.

The plateau of the vertebrate heart muscle action potential is related to an increased membrane permeability to Ca^{2+}. This slow inward current has been found in all cardiac tissue investigated and has been shown to satisfy all of the criteria proposed to identify Ca currents. This enormous field of research has been reviewed many times (Reuter 1973, 1979, Trautwein 1973, Weidmann 1974).

The function of the Ca current in vertebrate smooth and heart muscle is presumably the same as that of the Ca current in invertebrate muscle, i.e. to raise $[Ca^{2+}]_i$ and initiate muscle contraction. The large surface to volume ratio of smooth and heart muscle cells suggests that these currents could cause appreciable increases in $[Ca^{2+}]_i$. However, vertebrate skeletal muscle continues contracting for several hours in Ca-free solution, ruling out a role for this Ca current in activating contraction.

Axons

The presence of voltage-dependent Ca currents in the squid giant axon was first detected by the additional influx of ^{45}Ca measured during action potentials (Hodgkin & Keynes 1957). This Ca influx was first demonstrated electrically when Watanabe and co-workers (1967a) obtained long-lasting overshooting action potentials in a Na-free external solution while internally perfusing the axon with sodium salts. Action potentials were produced when Ca^{2+}, Ba^{2+}, or Sr^{2+} was the only external cation; all these action potentials were blocked by TTX (Watanabe et al 1967b). Meves & Vogel (1973) carried out voltage clamp studies on squid axons internally perfused with CsF and sucrose, while the external solution was 100 mM CaCl$_2$ and sucrose. They measured inward currents of 4 to 6 μA/cm^2, which disappeared when Ca^{2+} was replaced by Mg^{2+}. Because these Ca currents were blocked by TTX, they concluded that Ca^{2+} was entering through the Na channels. Surprisingly, the Ca current had a different time course from the Na current; the Ca current was much more prolonged.

The TTX-sensitive Ca current in squid axon is not the type of Ca current on which this review is focused, since it does not flow through a Ca^{2+}-specific channel. However, squid axon also has a Ca influx which appears similar to that of invertebrate muscle. This Ca influx has only been detected by the light produced in axons injected with aequorin (Baker et al 1971). The entry of Ca^{2+} during depolarizing voltage pulse could be divided into an early TTX-sensitive component, plus a late component that was not affected by TTX. This late component of Ca entry was shown not to be mediated by K channels, since it was little changed by injected TEA, which blocked the K current, and was blocked by levels of Mn^{2+}, which had little effect on the K current. Like the Ca current of invertebrate muscle, this late Ca entry was also blocked by Co^{2+}, Ni^{2+}, La^{3+}, and organic Ca antagonists D-600 and iproveratril (Baker et al 1973). Levels of Mn^{+}, Co^{2+}, and Ni^{2+} that block the late Ca uptake do not block the early Ca uptake. Thus, the late Ca current appears to be passing through Ca channels as defined by the criteria at the beginning of this section. Meves & Vogel (1973) did not see this TTX-insensitive Ca current. The aequorin studies indicate that the late Ca current is smaller than the TTX-sensitive Ca current and perhaps too small to be distinguished in voltage clamp experiments. Another possible explanation is that Meves & Vogel perfused with F^{-}; internal F^{-} has been shown to eliminate Ca currents in other preparations (Kostyuk et al 1975d, Takahashi & Yoshii 1978).

In contrast to squid axon, TTX-insensitive Ca spikes have been recorded in *Aplysia* axon, when 4-aminopyridine is added to a Na-free solution (Horn 1978). Na channels of *Aplysia* axon are blocked by TTX, so this Ca current is not passing through Na channels. The overshoot and maximum rate of rise of these spikes increases with $[Ca^{2+}]_o$, and they are blocked by Co^{2+} or Cd^{2+}. Sr^{2+} or Ba^{2+} can substitute for Ca^{2+} in producing action potentials.

It may be that all axons pass through a stage in development when they have a Ca spike. The development of electrical excitability in neuronal processes has been studied for *Xenopus* neurons in culture. At 6 to 11 hours in culture, the neurite action potential appears to be a Ca spike, since it requires Ca^{2+}, is blocked by Co^{2+}, and remains in Na-free solution (Willard 1980). However, at later stages the neurites have Na spikes that are blocked by TTX but are unaffected by Co^{2+}.

Nerve Cell Bodies

The fact that action potentials in nerve cells are not entirely due to inward Na currents was first demonstrated in spinal ganglion cells in frogs (Koketsu et al 1959a,b). These cells produced prolonged action potentials in Na-free solutions when quaternary ammonium ions were present; these action potentials were eliminated when the Ca^{2+} was withdrawn from the Na-free solution. Nishi and co-workers (1965) showed that when Ca^{2+} was replaced by Ba^{2+} in Na-free solutions, action potentials continued and their overshoot increased when $[Ba^{2+}]_o$ was increased. Thus, they concluded that Ca^{2+} (or Ba^+) carries charge across the frog spinal ganglion cell membrane. Also, frog motoneurons have a Ca component to their action potential (Barrett & Barrett 1976). Although the action potential in these neurons is predominantly a Na spike, they often show a slow regenerative depolarizing response when perfused with TTX and TEA. This response is eliminated by adding Mn^{2+} or reducing $[Ca^{2+}]_o$. Rohon-Beard cells (primary sensory neurons) in *Xenopus* larvae have been shown to go through a state in which the somal membrane has a nearly pure Ca spike (Spitzer & Baccaglini 1976, Baccaglini & Spitzer 1977). Later in development a Na component to the spike appears, followed by the loss of the Ca component.

The presence of Ca currents in molluscan ganglion cells was first suggested by studies on the marine pulmonate mollusc *Onchidium* (Oomura et al 1961). These giant nerve cells were found to produce all-or-none action potentials in Na-free solutions; the action potentials were abolished by withdrawal of Ca^{2+}. Likewise, some ganglion cells of the snail *Helix* continued to produce ordinary action potentials for hours after removing Na^+ from the bathing solution (Gerasimov 1964). Ba^{2+} could substitute for Ca^{2+} in producing these action potentials, and the overshoot was linearly related to the logarithm of $[Ba^{2+}]_o$ with a slope close to that expected for a Ba electrode (Gerasimov et al 1965). The possibility that action potentials in Na-free media might be due to a reservoir of Na^+ remaining close to the cell membrane was considered by several researchers (Chamberlain & Kerkut 1967, Moreton 1968, Meves 1968); this possibility was clearly eliminated by later studies in which ganglion cells were completely isolated and the cell membrane directly exposed to the bathing solution (Doroshenko et al 1973). In general, the inward current in molluscan neurons is carried by both Na^+ and Ca^{2+}; however, in some cells the Na current dominates, while in others the Ca current dominates (Kerkut & Gardner 1967). In an identified neuron of *Aplysia* the spike amplitude increased 9.9 mV for a ten-fold increase in $[Na^+]_o$ with $[Ca^{2+}]_o = 11$ mM, and 16.9 mV for a ten-fold increase in $[Ca^{2+}]_o$ with $[Na^+]_o = 15.5$ mM (Junge 1967). However, the spike amplitude increased 59 mV for a ten-fold increase in $[Na^+]_o$ in Ca-free media, and 29.5 mV for a ten-fold increase in $[Ca^{2+}]_o$ in a Na-free media (Geduldig & Junge 1968).

The cell bodies of a number of other invertebrate neurons have been shown to have Ca currents. About half of the neurons examined in the ganglionic X-organ of crayfish were found to have action potentials in Na-free or TTX media (Iwasaki & Satow 1971). The overshoot of the action potential in these media was found to increase with $[Ca^{2+}]_o$ very nearly as expected for a Ca electrode. The regenerative activity recorded from the soma of the I (inverting) neuron involved in processing barnacle visual information appears to be due to voltage-sensitive Ca channels (Stuart & Oertel 1978). TEA prolonged the action potential of leech Retzius cells and made the cells capable of prolonged action potentials in Na-free media (Kleinhaus & Pritchard 1975). The motoneuron cell body of the cockroach normally has only a graded oscillatory response to depolarization, but Pitman (1979) found that extracellular TEA or intracellular EGTA or citrate would convert this response to an all-or-none action potential, which was demonstrated to be a Ca spike. The action potentials of two identified grasshopper neurons have both Na and Ca components (Goodman & Heitler 1979); the Ca current is stronger compared to currents that oppose the regenerative response at earlier developmental stages, when action potentials persist in the absence of Na^+ (Goodman & Spitzer 1979).

Llinas & Hess (1976) recorded dendritic spikes in pigeon cerebellar Purkinje cells, which appeared to be mediated by Ca^{2+} since they were observed with 3-aminopyridine (3-AP) and TTX in the external media and were blocked by Mn^{2+} or Co^{2+}. Chick dorsal root ganglion neurons maintained in cell culture have action potentials with both Ca^{2+} and Na^+ components (Dichter & Fischbach 1977). Overshooting spikes persist in TTX or Na-free media. The size of the spikes varies directly with $[Ca^{2+}]_o$.

Ca currents have been identified in the cell bodies of a number of mammalian neurons. Certain neurons of myenteric plexus have been shown to have Ca spikes (Hirst & Spence 1973, North 1973). Mouse dorsal root ganglion cells cultured in vitro vary in their sensitivities to TTX and Na-free media (Matsuda et al 1976, 1978, Ransom & Holz 1977). Some of these cells have a TTX-insensitive Na current, so only that class of cells that has an action potential in Na-free media is thought to have a Ca spike. This action potential is blocked by Co^{2+} and increases in amplitude with increases in $[Ca^{2+}]_o$. Rat sympathetic neurons either in culture (O'Lague et al 1978) or in situ (McAfee & Yarowsky 1979) have a Ca component to the action potential. Schwartzkroin & Slawsky (1977) identified probable Ca spikes in guinea pigs hippocampal pyramidal neurons. Mouse neuroblastoma cells have an inward current carried by both Na^+ and Ca^{2+} (Spector et al 1973, Moolenaar & Spector 1979).

Thus, it appears that almost all nerve cell bodies have both Na and Ca currents, but that the ratio of the magnitude of the two currents varies considerably, even between neurons of the same ganglion. The function of the Ca current in the somal membrane has not yet been well understood. Perhaps it has no function and is just a "spillover" from the Ca current required in the presynaptic membrane, or, alternatively, it may play a vital role in integrating signals in the soma or controlling metabolic operations of the cell.

Nerve Terminals and Secretory Cells

The presence of extracellular Ca^{2+} is essential for neuromuscular transmission. This led Katz & Miledi (1967a) in their studies on the frog neuromuscular junction to postulate that external Ca^{2+} enters the presynaptic terminal as the first step leading to the release of actylcholine from the nerve terminal. The hypothesis that Ca^{2+} actually enters the presynaptic nerve terminal in a voltage-dependent manner was supported by studies on the giant synapse in the stellate ganglion of squid, where it was shown that when the internal potential of the presynaptic terminal was raised to very high levels, the postsynaptic response was suppressed (Katz & Miledi 1967b). Such high potentials are near or above the reversal potential for Ca^{2+}, so there is no longer a force to drive Ca^{2+} into the cell. This Ca influx was first detected electrically in squid synapse presynaptic axons injected with TEA, where in the presence of TTX prolonged action potentials could be recorded in response to depolarization (Katz & Miledi 1969). Katz & Miledi demonstrated that these responses were due to a Ca current by showing that their amplitude and duration increased when $[Ca^{2+}]_o$ was increased, that Sr^{2+} and Ba^{2+} could substitute for Ca^{2+}, and that Mn^{2+} and, to a lesser extent, Mg^{2+} blocked the response. These TTX-insensitive Ca action potentials showed a marked electronic decrement within 1 mm of the synapse. The presence of this Ca current was further substantiated by the demonstration of an increase in $[Ca^{2+}]_i$ inside the terminal during repetitive synaptic transmission as inferred from the light output of an aequorin-injected axon terminal (Llinas et al 1972).

The presence of voltage-dependent Ca channels in the presynaptic membrane has also been demonstrated for the barnacle photoreceptor. Ross & Stuart (1978) found that external TEA, 3-AP on high-potassium will cause all-or-none spikes to be elicited from the presynaptic terminal of the photoreceptor, either by light or injected current. The peak voltage of these spikes depends on $[Ca^{2+}]_o$, and Ba^{2+} or Sr^{2+} can substitute for Ca^{2+}. Although TTX and Na-free solution have no effect, Co^{2+}, Mn^{2+}, and Mg^{2+} reduce or block the spikes.

Since calcium is required for the secretion of any substance from a cell, whether the substance be a neurotransmitter, a hormone, or mucus, it is not surprising that Ca currents also have been identified in a number of secretory cells. Probably most of the electrically excitable secretory cells have both Na and Ca currents. Matthews & Sakamoto (1975) recorded spikes from mouse pancreatic islet cells and found that the major component of the spike was due to Ca^{2+}. Clonal cells isolated from a rat anterior pituitary tumor have a predominant Ca component to their spike (Kidokoro 1975b), but these cells also have a Na component to the spike (Biales et al 1977). Cultured adrenal chromaffin cells (human or gerbil) have an action potential with a strong Na component, based upon the facts that it is blocked by TTX, but persists in Co^{2+} or Ca-free media (Biales et al 1976). However, working with rat chromaffin cells, Brandt et al (1976) also identified a Ca component to the action potential. Small regenerative action potentials were observed in Na-free saline, and were prolonged when Ca^{2+} was replaced with Ba^{2+}. The presence of voltage-dependent Ca channels has been demonstrated in a rat pheochromocytoma cell line (PC12) by measuring KCl-activated dopamine release (Ritchie 1979). Kater (1977) has studied an invertebrate secretory cell, the mucus-secreting pedal gland cell from the slug *Ariolimax*. These cells have an action potential with both Na and Ca components. External TEA enhances the Ca response so that large regenerative action potentials can be produced in high-calcium, Na-free solutions.

Receptors

Regenerative Ca-dependent responses have been identified in a number of receptor cells. In some cases (barnacle photoreceptor) the voltage-dependent Ca channels are clearly located in the presynaptic terminal; in other cases (afferent nerve terminal to muscle spindle and electroreceptor) the Ca channels are clearly located near the site of origin of the receptor potential;

in still other cases (rod cell and hair cell) it is not clear if the Ca channels are associated with the presynaptic membrane or with the transduction membrane. In this section we include the last two types of cases, where the Ca action potential may function to amplify the receptor potential. Zipser & Bennett (1973) suggested that Ca^{2+} was involved in the electrical response of electroreceptor cells in mormyrid and gymnotid fish, since TTX failed to block these responses. Skate electroreceptor epithelium produces action potentials in response to stimuli that depolarize the lumenal side of the receptor cells. These action potentials are due to an increased calcium permeability of the lumenal membranes, as concluded from the elimination of these responses with Co^{2+} or Ca-free solution (Clusin & Bennett 1977). Photoreceptor membranes, which had been assumed to be passive, have recently been shown to be capable of Ca spikes. Rods of toad retina, when bathed in TEA, generate Ca-dependent oscillations and action potentials (Fain et al 1977). Effects of changing $[Ca^{2+}]_o$ are not conclusive because the resting membrane potential is also changed; however, Sr^{2+} has no effect on the resting potential and the amplitude of the action potential increases as $[Sr^{2+}]_o$ increases. Co^{2+} blocks the action potential, while TTX has no effect. Hudspeth & Corey (1977) occasionally evoked action potentials in hair cells of the bullfrog inner ear by mechanical or electrical stimulation. They suggest these action potentials may be Ca spikes on the basis of their slow time course and their insensitivity to 10 μM TTX. Ito & Komatsu (1979) identified Ca spikes in the afferent nerve terminal of the frog muscle spindle. These action potentials are observed during stretch in Ringer solution with TTX; their amplitude depends on $[Ca^{2+}]_o$ and they are blocked by Co^{2+} and Mn^{2+}.

Egg Cells

Some egg cell membranes, even before fertilization, are electrically excitable; egg cell action potentials have been recently reviewed (Hagiwara & Miyazaki 1977a, Hagiwara & Jaffe 1979). All egg cell action potentials have a Ca component. The first of these action potentials to be studied was in the mature, but unfertilized, egg of the tunicate *Halocynthia*. Miyazaki and co-workers (1972) found that this cell normally had a small resting potential and was inexcitable; however, after the membrane potential had been made more negative (by injecting inward current or by bathing with high calcium saline), prolonged regenerative responses were obtained that had both Na and Ca components. In later voltage-clamp studies Okamoto and co-workers (1976b) divided the Ca current into two components on the basis of voltage dependence and kinetics: (*a*) a larger component that passed through Ca channels similar to those of invertebrate muscle and (*b*) a smaller component that seemed to pass through the Na channels.

Among echinoderms, action potentials have been studied in starfish and sea urchin eggs. The starfish action potential has a definite Ca component. The action potential becomes smaller when the external Na^+ is replaced with either $Tris^+$ or sucrose (*Asterina,* Miyazaki et al 1975; *Mediaster,* Hagiwara et al 1975; *Patiria,* Shen & Steinhardt 1976). However, voltage clamp studies done by Hagiwara and collaborators (1975) indicate that it is unlikely that Na^+ actually carries current. In this study, two inward currents were found; the current activated at more positive potentials was more sensitive to blocking cations like Co^{2+} and Mg^{2+} and appeared to be a Ca current similar to that of invertebrate muscle. The current activated at more negative potential, which was Na-dependent, also decreased when $[Ca^{2+}]_o$ was lowered. It was concluded that Ca^{2+} also carries this current, while Na^+ plays a facilitating role, for the following reasons:

1. The Na-dependent component (determined by the reduction of the inward current when Na^+ was removed) depended on $[Ca^{2+}]_o$.
2. No inward current was detected when Ca^{2+} was replaced by Mg^{2+}.
3. Li^+, Rb^+, and Cs^+ were as effective as Na^+ in maintaining this component of inward current.

Mature, unfertilized sea urchin oocytes have been shown to have Ca currents very similar to those of *Mediaster* (Okamoto et al 1977). Replacing Na^+ with $Tris^+$ reduced the inward current, but replacing Na^+ with $Choline^+$ did not. This can be interpreted to mean that either $Choline^+$ has the same facilitating effect on the Ca current as does Na^+, or that $Tris^+$ or sucrose has a suppressing effect on the Ca current that $Choline^+$ or Na^+ does not.

Voltage-dependent Ca channels have also been identified in eggs of the marine worm *Urechis,* of the polychaete *Chaetopterus,* and of the mouse. The action potential of the polychaete egg is a pure Ca spike, with no Na component (Hagiwara & Miyazaki 1977b). The mouse oocyte shows an anodal break response that is Ca-dependent (Okamoto et al 1977). In voltage-clamp experiments an inward current was measured in Na-free solution that disappeared when Ca^{2+} was replaced by Mn^{2+}; Ba^{2+} and Sr^{2+} could also carry the inward current. Jaffe and co-workers (1979) have identified a Ca spike in the *Urechis* egg. It is unaffected by removing external Na^+, but is reduced by lowering $[Ca^{2+}]_o$ or adding Co^{2+}.

The biological significance of Ca channels in egg cell membranes is still unclear. The regenerative nature of the Ca current boosts the fertilization potential and may help prevent polyspermy (Jaffe 1976). Since an increase in $[Ca^{2+}]_i$ is implicated in activation of egg cells (Steinhardt & Epel 1974), it is intriguing to speculate that these Ca currents might play some direct role in egg activation. However, any function of Ca channels cannot be universal to all egg cells, since some eggs have been found to be electrically unexcitable. The current-voltage relation for the *Drosophila* egg membrane is linear over a wide range of voltages (Miyazaki & Hagiwara 1976), and *Xenopus* egg cells are incapable of producing action potentials (Palmer & Slack 1970, Slack & Warner 1975). In the sea urchin egg the increase in $[Ca^{2+}]_i$ accompanying activation is accomplished by the release of Ca^{2+} from the internal stores rather than by influx (Steinhardt et al 1977).

Paramecium

A regenerative Ca response has been identified in *Paramecium* (Naitoh et al 1972). Normally injection of depolarizing current elicits a graded response. The size of this response is changed very little by removing external Na^+ or by application of TTX, but the overshoot of the response increases 22 to 25 mV for a ten-fold increase in $[Ca^{2+}]_o$. If Ca^{2+} is replaced by Sr^{2+} or Ba^{2+}, the graded response becomes an all-or-none spike. In many ways the *Paramecium* Ca current seems very similar to the Ca currents found in metazoan cells; however, it is less sensitive to blocking cations like Co^{2+}, Mg^{2+}, and La^{3+}. Eckert & Brehm (1979) have recently reviewed the properties of this current. The function of this Ca current is to increase the concentration of Ca^{2+} inside the cilia, causing a reversal of ciliary beating. Therefore, it is not surprising that the Ca channels have been found to be preferentially located in the ciliary surface membrane (Dunlap 1976, 1977, Ogura & Takahashi 1976).

Epithelial Cells

Herrera (1979) studied the bioluminescent epithelial cells (photocytes) of the polychaete *Hesperonoe*. Depolarization elicited two separate regenerative inward currents, a TTX-insensitive, Na-dependent current not associated with light production and a higher-threshold current that produced action potentials even in Na-free solution. This second current was identified as a Ca current, since the overshoot of the action potentials increased 29 mV for a ten-fold increase in $[Ca^{2+}]_o$. Luminescence accompanied each Ca spike, but no light was produced when the cell was depolarized in Ca-free medium. Thus, it was concluded that Ca^{2+} mediates excitation-luminescence coupling in these cells.

An overshooting action potential can be elicited by injected current from the myoepithelial cells of the proventriculus of the polychaete *Syllis spongiphila* (M. Anderson 1979). The action potential is a pure Ca spike since it is reversibly abolished by Ca-free solution, blocked by Co^{2+}, or Ni^{2+} and is unaffected by TTX or low-Na solutions. Mn^{2+}, as well as Ba^{2+} and Sr^{2+}, can substitute for Ca^{2+} in producing this regenerative response. Outer skin pulses of the stolon of the tunicate *Salpa fusiformis* are conducted as overshooting action potentials. P. A. V. Anderson (1979) has shown that these action potentials have a Na component because they are blocked by TTX. Mn^{2+} also blocks the action potentials, suggesting that they may have a Ca component.

DIFFICULTIES WITH BIOPHYSICAL STUDIES

There are a number of reasons that the analysis of Ca currents has proceeded more slowly than did that of the Na current. First, no preparation ideal for biophysical studies of Ca currents has been found. There is no cell to play the role for Ca currents that the squid giant axon has played for the Na current. However, in the last five years, certain preparations have emerged that offer many advantages over the others. Second, once a preparation is found that has a strong Ca current and where satisfactory control of the membrane potential is achieved, there remains the problem of separating the Ca current from the other currents contributing to the net current measured. The fact that Ca^{2+} plays many roles at the membrane besides carrying charge from one side to the other makes this problem of separating currents much more difficult than it was for isolating the Na current. Third, the large assymmetry between the concentrations of Ca^{2+} on the two sides of the membrane puts E_{Ca} at very positive potentials, and makes it nearly impossible to measure. Some of the concepts made familiar through experience with the Na channel, such as the usefulness of a conductance G_{Na} defined by $I_{Na} = G_{Na} (V-E_{Na})$, are probably inappropriate for analyzing the Ca channel. In this section, we want to discuss these difficulties in detail before reviewing certain properties of Ca currents.

Suitability of Preparations

There are four criteria that should be met by any preparation that is to be used for biophysical analysis of the Ca current. We use the word "preparation" since it is not only the cell that is important, but also the state of isolation, the physical configuration, and the electrodes used.

1. The preparation must have a strong Ca current, compared with the leakage and prolonged capacitance-charging currents.
2. All regions of membrane being studied must have the same transmembrane potential.
3. It should be possible to change the membrane potential rapidly compared to the times required for the Ca conductance to change.
4. Finally, it should be possible to control the compositions of the solutions on both sides of the membrane.

The squid giant axon fails criterion 1. Meves & Vogel (1973) were unable to detect electrically a TTX-insensitive Ca current. Their failure may have been partially due to the blocking effect of intracellular F^-, but even without F^- it is doubtful that the TTX-insensitive Ca current would be large enough for accurate measurements. Even the TTX-sensitive Ca currents in squid axon, which are larger than the TTX-insensitive Ca currents as judged by aequorin signals (Baker et al 1971), have a peak amplitude of only 4 to 6 $\mu A/cm^2$ (Meves & Vogel 1973). The time course of the Ca current through Na channels is considerably more prolonged than that of the Na current through Na channels in squid axon (Meves & Vogel 1973). This points out a problem with studies of Ca currents in membranes that have large TTX-insensitive Na currents; it will be difficult to separate Ca current through Ca channels from Ca current through Na channels.

At first it appeared that barnacle giant muscle fibers might become the "squid axon" of Ca currents. Peak inward Ca currents of 0.2 to 2.0 mA/cm^2 were measured, where the area was calculated ignoring any invaginations of the surface membrane (Keynes et al 1973, Hagiwara et al 1974). However, there are deep, long invaginations of the sarcolemma in all crustacean muscle fibers, so that calculations that ignore these invaginations underestimate the sarcolemmal membrane area by a factor of 10 to 20, making the peak Ca current density roughly 20 to 200 $\mu A/cm^2$. These invaginations of the sarcolemma cause barnacle muscle to fail to satisfy criteria 2 and 3. Although the potential across the most superficial regions of the membrane can be controlled, the potential across the invaginated membranes can differ by tens of millivolts from that of the superficial regions. This shows up in the currents recorded during voltage clamp experiments as oscillations following sudden changes of potential. [See examples in Keynes et al (1973) and Hagiwara et al (1974)]. The presence of series resistance also shows up as prolonged capacitance-charging currents, when the measured potential is rapidly stepped to a new potential. This transient current indicates that the potential of the invaginated membranes reaches the new value slowly. These transients have a time constant of a few milliseconds in barnacle muscle, compared to a few microseconds in squid axon. The use of an axial wire in barnacle muscle appears to eliminate any problem with longitudinal voltage control, but in no way helps the control of the voltage of the invaginated membranes. Efforts to reduce the area of membrane from which currents are measured do not solve the problem of invaginated membranes satisfactorily, as demonstrated by the apparent oscillations of current recorded from a 50 μm (diameter) patch of crayfish muscle membrane recorded by Hencek & Zachar (1977). Presumably, voltage clamp studies of insect muscle have the same sort of problems with invaginated membranes as do those of crustacean muscle.

The delayed activation of the Ca current elicited by small depolarizations in frog twitch muscle (Sanchez & Stefani 1978) suggests that this current is also flowing through a membrane that sees a potential difference that is not well controlled by the clamp, perhaps in the T-system membrane.

An "inside-out" situation exists in *Paramecium*. As discussed above, the Ca channels of *Paramecium* are presumably located in the membrane covering the cilia. Due to the small diameter of the cilia and the unknown value of the resistivity of the contents of a cilium, there may be an appreciable amount of resistance in series with the Ca-current carrying membrane. Therefore, the ciliary membrane may escape control during voltage clamp experiments. In *Paramecium,* the series resistance is on the inside of the membrane, while in muscle it is on the outside. If the resistivity of the ciliary contents is assumed to be equal to that of cell cytoplasm, it can be calculated that the measured values of Ca current would cause less than a millivolt potential drop along the length of a cilium. However, this calculation only shows that it is possible that the ciliary membrane is isopotential. If the ciliary resistivity is appreciably higher, neither the magnitude nor the time course of the measured currents will have much meaning, since there could be strong local currents within each cilium.

Multicellular preparations like strands of heart or smooth muscle are extremely difficult subjects for voltage clamp study. Individual cells are small and electrically connected to the surrounding cells. Thus, there are voltage gradients both inside and outside the cells. Only under special conditions can the total membrane that carries current be considered to be at roughly one potential. [See Reuter (1979) for a recent review of the limitations of heart muscle voltage clamps.] Not only do multicellular preparations fail to satisfy criteria 2 and 3, but they also fail on criterion 4. It is nearly impossible to change directly the intracellular composition of the cells and very difficult to control the extracellular solution. The limited access to deeper cells in the muscle strand requires long exchange times and allows accumulation and depletion of ions outside the membrane when large currents flow for extended times. Recently a technique has been developed by Lee and associates (1979) that allows voltage clamp studies of single heart muscle cells; this technique appears to satisfy criteria 2, 3, and 4 much better. Ca currents have recently been recorded using this technique on heart cells (K. S. Lee, personal communication).

It is doubtful if two-microelectrode voltage clamps of squid nerve terminal or molluscan neurons with axons attached satisfactorily satisfy criteria 2 or 3, and they certainly don't satisfy criterion 4. The geometry of both preparations is complicated, allowing the potential of membranes carrying Ca current to differ significantly from the potential of the region where the recording microelectrode is situated. There is considerable series resistance separating distant regions of the axon from the point of potential control, so that both preparations exhibit prolonged capacitance-charging transients (of a millisecond or more). *Aplysia* axon has been shown to have a measurable Ca current (Horn 1978); the Ca currents from this uncontrolled membrane may considerably distort the measured currents, even in Na-free solutions.

It is important to emphasize here that the above problems do not disqualify the work done on these preparations. On the contrary, very meaningful results have been obtained when the limitations of the voltage clamp are carefully taken into consideration for the interpretation of the data.

When molluscan neurons are isolated from surrounding cells and separated from their axon at a point very close to the cell body, as was done by Chen et al (1971) and Geletyuk & Veprinstev (1972), the molluscan neuron becomes almost as satisfactory for voltage clamp studies as is the egg cell. The nearly spherical geometry of both cells assures that the entire membrane is isopotential. The situation may be better for snail neurons than for *Aplysia* neurons, since the latter appear to have considerably more invagination of the surface membrane. The peak Ca current densities of snail neurons ($\sim$100 μA/cm^2, Kostyuk et al 1974a,b) are considerably greater than those of tunicate egg ($\sim$2 μA/cm^2, Okamoto et al 1976b) or starfish egg ($\sim$6 μA/cm^2, Hagiwara et al 1975). However, neither preparation satisfies criteria 3 or 4 completely. Since the membrane capacitance must be charged by passing current through a relatively high resistance microelectrode, a millisecond or more is required to change the membrane potential to a new value. There is no way of making large changes in the composition of the intracellular solutions. Both of these limitations were somewhat overcome by a technique introduced by Krishtal & Pidoplichko (1975) and further developed by Kostyuk et al (1975d, 1977b), Lee et al (1978), and Takahashi & Yoshii (1978), which allows direct access to the interior of the cell so that the intracellular solution can be directly changed and current can be passed through the membrane from macroelectrodes. This lower access resistance makes it possible to charge the membrane capacitance in times of about 100 μsec. Problems caused by an appreciable series resistance between the recording electrodes and the membrane have been met by modifying the technique to include an intracellular microelectrode (Kostyuk et al 1977a) or a second suction pipette (Krishtal 1978) for recording potential. The use of the second suction pipette also improves the rate and efficiency of change of intracellular solutions.

While some preparations are clearly more favorable than others for biophysical studies of the Ca current, certain properties of the Ca current can be reliably determined in the most difficult of preparations. In fact, some of the most interesting properties of Ca currents, e.g. modulation by neurotransmitters, have only been demonstrated in rather difficult preparations. Also, since we want to compare the Ca currents found in different membranes, we try to review the data from as wide a variety of studies as possible.

Separation of Ca Current

Even if the voltage clamp of the membrane is completely successful, the problem of separating the Ca current from the other currents contributing to the measured current is far from trivial. We skip over the complication of separating Ca currents from Na currents, since Na currents have a very different time course and can usually be eliminated by application of TTX or the removal of Na^+ from both sides of the membrane. We focus on the separation of the Ca current from the voltage-dependent and time-dependent outward currents which are normally carried by K^+.

Hodgkin & Huxley (1952a) first outlined the procedure for separating two currents. All possible methods for separating two currents can be seen to follow a slightly generalized form of the Hodgkin-Huxley procedure. The total current is measured at a particular voltage before and after some manipulation that selectively changes one of the two components. Then the two components can be calculated provided the following three conditions are met: (a) the manipulation does not affect one of the current components, so that component is the same in both measured currents; (b) the change in the other component can be expressed as multiplication by a constant k not equal to one, so its time course does not change; (c) at some time, the constant current component is zero, while the manipulated component is not.

Theoretically, there are two approaches to separating the Ca and outward currents by the above method: Either the outward current can be changed or the Ca current can be changed. The first approach has not been successfully applied because the above conditions are not satisfied. There are a number of methods available for reducing outward currents that appear not to change the Ca current, e.g. application of external TEA or 4-AP, replacement of internal K^+ with Cs^+ or $Tris^+$. However, condition (b) is not satisfied because none of the above manipulations reduce the outward currents without changing its time course. In general, the outward current in membranes that have Ca currents appears to be very complicated, consisting of several components: a voltage-dependent K current, a Ca-dependent K current, and a nonspecific outward current (see Meech & Standen 1975, Kostyuk et al 1975a,b,c, Heyer & Lux 1976, Kostyuk et al 1977b, Thompson 1977). Since these components have different time courses, different pharmacological sensitivities, and pass through channels with different selectivities, the various manipulations for reducing outward current usually change the time course of the outward current. Condition (b) would only be satisfied if the outward current could be completely eliminated, i.e. $k = 0$. However, no manipulation has been found that reduces the outward currents to zero. Even when the internal K^+ is replaced by Cs^+ or $Tris^+$, voltage-dependent outward currents remain (Meves & Vogel 1973, Keynes et al 1973, Kostyuk et al 1977b); the outward currents measured by Kostyuk and co-workers (1977b) when internal K^+ is replaced by $Tris^+$ are also clearly time-dependent. Also, condition (c) may not be satisfied by this approach. Ca currents turn on before the outward currents turn on and may never decay to zero, or else may decay with a similar time course to that of some components of the outward current. So there would be no time at which the constant k could be determined.

The second approach to separating the Ca current, namely measuring the total current before and after selectively changing the Ca current, is frequently used but, in general, satisfies the Hodgkin-Huxley conditions no better than did the first approach. The outward currents turn on almost as fast as the Ca current, so there is no time at which pure Ca currents can be determined. However, this approach can still be salvaged if a modification is available for which it is reasonable to assume $k = 0$. There appear to be several such modifications: complete removal of external Ca^{2+}, or a high concentration of a Ca blocker such as Co^{2+}, Cd^{2+}, La^{3+}, or verapamil. Because all inward current disappears for voltage steps to low levels at which outward currents appear not to be activated, k is judged to be zero. Condition (b) is satisfied, of course, if $k = 0$. However, condition (a) is not satisfied, because removing Ca^{2+} or adding polyvalent cations like Co^{2+} and La^{3+} changes the outer surface potential, in general, and thus changes the outward current. This problem can usually be overcome by maintaining high concentrations of divalent cations in the external solution so the surface potential is small. However, condition (a) is still not satisfied in many membranes, because the manipulations that block the Ca current also change the outward current. Treatments that reduce Ca currents—Ca-free solution, Co^{2+}, Mn^{2+}, La^{3+}, verapamil, D600—have been shown to reduce outward currents in a variety of membranes (Meech & Standen 1975, Kostyuk et al 1975c, Kass & Tsien 1975, Hagiwara et al 1975, Heyer & Lux 1976, Thompson 1977). In some of these cases the activation of outward currents seems to depend upon prior Ca^{2+} entry; however, this does not seem to be true in all examples where Ca-current blockers reduce outward currents. Although Kass & Tsien (1975) found that Mn^{2+}, La^{3+}, and D600 decreased the slow outward current of cardiac purkinje fibers, they found that elevating $[Ca^{2+}]_o$, which increased the Ca current, also decreased the slow outward current. Likewise, the reduction of the slow outward current caused by Ca-current blockers in frog muscle fibers appears not to be due to a reduction in Ca^{2+} entry per se (P. T. Palade & W. Almers, personal communication). Although external or internal TEA blocks a large fraction of the outward current, the remaining outward current is still sensitive to external Co^{2+} (Heyer & Lux 1976, Thompson 1977). Even when internal K^+ has been completely replaced by $Tris^+$, 1 mM Cd^{2+}, which blocks the Ca current, strongly reduces the nonspecific, time-dependent outward current in snail neurons (Kostyuk et al 1977b). Are there some membranes where the outward current is not changed by blocking the Ca current? Certainly some membranes have much more Ca-dependent outward current than do others. However, Kass & Tsien's results suggest that the effect of Ca antagonists on the outward current are not entirely mediated by an effect on $[Ca^{2+}]_i$. The outward current may be changed (but the change not detected) even in the cases where the total current after Ca current elimination always remains more positive than the total current with the Ca component.

In spite of the problems pointed out in the paragraph above, this procedure has been used by many researchers to separate Ca currents. We feel these currents are reliable for potentials below +20 to +40 mV, especially when the outward currents have been reduced as much as possible. When the Ca current is blocked, the outward currents are seen not to be activated until the membrane is stepped to considerably more positive potentials than those that activate the Ca current. Since there is no reason to think that Ca-dependent outward currents exist at these less positive potentials, it seems reasonable to apply the above subtraction procedure for these potentials where the Ca current is large compared to the outward current measured after the Ca current is blocked. However, above this voltage range, especially once the total current becomes positive, the validity of the subtraction procedure becomes very doubtful.

The measurement of changes in $[Ca^{2+}]_i$ with aequorin or Arsenazo III can add some information on the influx of Ca^{2+} at large positive membrane potentials (Baker et al 1971, Gorman & Thomas 1978, Ahmed & Connor 1979); but there are many complications in interpreting these results, since the light signal reflects processes other than the passive influx of Ca^{2+}. Perhaps the best hope for isolating the Ca current at large positive potentials is from the tail currents measured when the membrane is suddenly stepped from the positive potential to a voltage where the outward currents will be small, e.g. near the outward current reversal potential (Connor 1977, Adams & Gage 1979b). This procedure will require faster control of membrane potential than has generally been available and also will involve the rather circular problem of determining the reversal potential for the outward current flowing at that potential and time, but it still seems the most promising procedure for determining the properties of Ca currents at large positive potentials.

Problems Surrounding E_{Ca}

The asymmetry of the concentration of Ca^{2+} on the two sides of the membrane creates a unique situation for the Ca current; no other current has been studied that flows down such a large concentration gradient. As a consequence, many studies that have been done for the Na and K channels are at present impossible for the Ca channel. Also, a number of concepts that developed from studies of Na and K currents have been applied without question to the Ca channel. These concepts may not be appropriate to the movement of ions between regions with concentrations that differ by a factor of 10^5.

In squid axon $[Na^+]_o$ is about ten times larger than $[Na^+]_i$, so the reversal potential for Na, E_{Na} (given by the Nernst relation), is about $+60$ mV. When the peak Na current is plotted against membrane potential, the I-V curve is essentially linear for positive potentials, intersecting the voltage axis very near E_{Na}. Hodgkin & Huxley (1952a) interpreted this to mean that the Na conductance, G_{Na}, is maximally activated (and therefore constant) at positive potentials; the peak Na current, I_{Na}, is given by

$$I_{Na} = G_{Na} (V - E_{Na})$$

where V is the membrane potential. The usefulness of this expression for describing the sodium current was further demonstrated when Hodgkin & Huxley (1952b) found that the "instantaneous" I-V relations for the Na current are also linear. If the Na permeation mechanism is not given time to change, G_{Na} is a constant independent of voltage, and I_{Na} is linearly related to V. However, this linearity depends on the fact that the concentrations of Na^+ on the two sides of the membrane are not too different. When external Na^+ is removed, the instantaneous I-V relation for the Na current becomes curved; the slope of the curve, which is G_{Na}, is nearly unchanged for large positive V (outward I_{Na}), but approaches zero for negative V. The reason for the nonlinearity is obvious; there is Na^+ inside the membrane to carry outward current, but almost no external Na^+ to carry inward current. Thus, even though the state of the Na permeation mechanism is not given time to change, G_{Na} becomes a function of voltage and E_{Na} and is no longer very useful for quantitative purposes.

The situation for the Ca current is much like that of the Na channel when Na has been removed from one side, discussed above. When there is one hundred thousand times more Ca^{2+} outside the membrane than inside, it is very difficult to believe the I-V relation for Ca current will be linear, even if the Ca permeation mechanism does become maximally activated at positive potentials. There are many reasons to object to using the constant-field expression (Hodgkin & Katz 1949) to describe the Ca current. However, in the absence of any other simple expression to quantify the dependence of ionic currents on the electrochemical gradient across the membrane, we use it to illustrate how nonlinear the I-V curve probably is for Ca curves, even when permeability is constant. The constant-field expression for Ca current is

$$I_{Ca} = \frac{4F^2 V P_{Ca}}{RT} \left(\frac{[Ca]_i e^{(2VF/RT)} - [Ca]_o}{e^{2VF/RT} - 1} \right)$$ 1.

Where P_{Ca} is the membrane permeability to Ca^{2+} as defined in the constant field theory. Figure 1A shows the I-V relation predicted for $[Ca^{2+}]_o = 10$ mM, $[Ca^{2+}]_i = 10^{-7}$ M, and a constant P_{Ca} such that $I_{Ca} = 10$ nA at +29 mV. It is obvious that the curve is highly nonlinear and that no projection of the I-V curve for voltages more negative than +100 mV would accurately determine E_{Ca}. A second point to be made is that the inward Ca current measured at small positive potentials is almost completely independent of a ten-fold change in $[Ca^{2+}]_i$. Figure 1B shows that increasing $[Ca^{2+}]_i$ to 10^{-6} M reduces the Ca current by only 1% at +58 mV, while the conductance expression $I_{Ca} = G_{Ca}(V - E_{Ca})$ would predict that the Ca current would be reduced by 33% if G_{Ca} is assumed not to change.

As discussed previously, even when the most drastic measures are taken to reduce the outward current activated by positive potentials, the measured current becomes outward at voltages less positive than +80 mV (Meves & Vogel 1973, Keynes et al 1973, Kostyuk et al 1977b, Takahashi & Yoshii 1978, Connor 1979). The I-V curves of Figure 1 suggest that the inward Ca current becomes relatively small at potentials above +60 mV, so that even a small nonspecific outward current would give a net outward current at higher potentials. Since there is no accurate way of separating the Ca current from the outward current, as discussed above, E_{Ca} appears to be essentially unmeasurable by electrical methods. Since E_{Ca} cannot be experimentally determined from current measurements, it is not possible to determine the concentration of Ca^{2+} just inside the membrane. It would be very useful to be able to determine this concentration, since it is thought to control several of the membrane permeabilities. Not being able to measure the reversal potential for the current through the Ca channel also prevents the determination of the relative permeabilities of the various divalent cations that pass through the Ca channel. It might be thought that E_{Ca} could be measured if $[Ca^{2+}]_i$ were increased by a few orders of magnitude; however, as is discussed later, the Ca channel appears to be blocked if

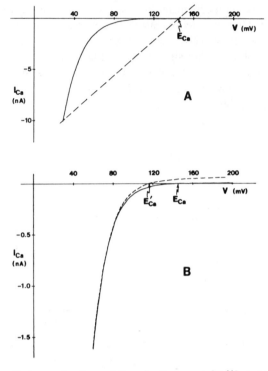

Figure 1 Theoretical current-voltage relations for Ca current. $[Ca^{2+}]_o = 10$ mM. *A*. Predictions of constant-field equation with constant permeability (*solid curve*) and $I_{Ca} = G_{Ca}$ ($V - E_{Ca}$) with constant conductance (*dashed line*). $[Ca^{2+}]_i = 10^{-7}$ M. E_{Ca} is reversal potential for Ca^{2+}. *B*. Predictions of constant-field equation for $[Ca^{2+}]_i = 10^{-7}$ M (*solid curve*) and 10^{-6} M (*dashed curve,*) assuming Ca permeability does not change. E'_{Ca} is Ca reversal potential for higher level of $[Ca^{2+}]_i$.

$[Ca^{2+}]_i$ exceeds 10^{-6} M (Hagiwara & Nakajima 1966b, Kostyuk & Krishtal 1977, Doroshenko & Tsyndrenko 1978, Takahashi & Yoshii 1978). Therefore, the Ca current appears to be inherently a one-way current. It is not known if intracellular Sr^{2+} or Ba^{2+} block the Ca channel; it will be very interesting to see if outward Ba or Sr currents can be measured.

The reliability of I-V relations for the Ca current at large positive potentials is questionable, but still it is interesting to see what data are available for choosing between the linear (*dashed line*) and nonlinear (*solid line*) I-V relations plotted in Figure 1*A*. The presence of outward currents contaminates all the measured Ca currents and causes the I-V relations to tend to curve upward and to cross the voltage axis at values considerably more negative than E_{Ca}. The downward (negative) curvature of the I-V relation predicted for the Ca current by the constant-field equation (*solid line,*

Figure 1A) will be lost, if there is substantial outward current contamination. The tunicate egg membrane has considerably less delayed rectification than most membranes; even without correcting for outward currents, the Ca current I-V relations measured at high positive potentials show a negative curvature (Okamoto et al 1976b). If these I-V curves were corrected by subtracting the currents measured in the presence of 20 mM Co^{2+}, the negative curvature would be much more prominent. Similar negative-curvature I-V relations are obtained for the Ca current of molluscan neurons (Doroshenko et al 1978, Tillotson 1979). In both cases the Ca current has already dropped to a small fraction of its maximum size while still 50 to 100 mV more negative than E_{Ca}. On the other hand, only one study reports a Ca I-V relation that is essentially linear up to E_{Ca}. The results of Hagiwara & Nakajima (1966b) with Ca spikes in barnacle muscle also support the nonlinear I-V relation predicted by the constant-field relation. They found that reducing $[Ca^{2+}]_i$ from 8×10^{-8} to 8×10^{-9} M had no effect on the Ca spike, as would be expected from the nonlinear I-V relation of Figure 1A. In contrast, the linear relation between I_{Ca} and $V-E_{Ca}$ would predict that I_{Ca} at around $V=0$ would increase by about 20% if G_{Ca} does not change. The independence of I_{Ca} on $[Ca^{2+}]_i$ for membrane potentials well below E_{Ca} and $[Ca^{2+}]_i < 10^{-6}$ M has been confirmed in voltage clamp experiments on internally-dialyzed egg cells (Takahashi & Yoshii 1978).

Optical studies of Ca^{2+} influx are difficult to interpret quantitatively due to the uncertain stoichiometry of the Ca binding to the dye or protein and the presence of competing processes; however, they appear to agree better with the predictions of the nonlinear I-V relation of Figure 1A. Baker and co-workers (1971) found that the light response measured in aequorin-injected squid axons dropped rapidly above +20 mV. A plot of the square root of the light response—which should be proportional to $[Ca^{2+}]_i$ assuming 2 to 1 binding—versus membrane potential is very nonlinear, in agreement with the constant-field predictions discussed above. Using arsenazo-loaded molluscan neurons, Ahmed & Connor (1979) found that the plot of absorbance change against membrane potential was again nonlinear in the sense of the constant-field prediction. They point out that intracellular accumulation of Ca^{2+} during the long pulses may be contributing to that nonlinearity.

SATURATION, SELECTIVITY, AND BLOCKING

Saturation

One of the criteria used to identify a Ca current is an increase in inward current as $[Ca^{2+}]_o$ is increased. However, most quantitative studies of the relation between the magnitude of the Ca current and $[Ca^{2+}]_o$ have found that the relation is not linear, but that the Ca current saturates as $[Ca^{2+}]_o$ is increased. This saturation may not be an intrinsic property of the Ca channel, but instead may result from the stabilizing action of Ca^{2+} on the membrane, such as observed by Frankenhaeuser & Hodgkin (1957) for squid axon. If increasing $[Ca^{2+}]_o$ shifts the voltage dependence in the positive direction along the voltage axis, then the membrane permeability for Ca^{2+} at a particular voltage may drop as $[Ca^{2+}]_o$ increases. Also, the concentration of Ca^{2+} at the outer surface of the membrane would not increase as much as does the concentration of Ca^{2+} in the bulk external solution, if increasing $[Ca^{2+}]_o$ reduces the magnitude of a negative surface potential (see Hagiwara 1973). Both of these effects of increasing $[Ca^{2+}]_o$ will tend to cause I_{Ca} to increase less rapidly with $[Ca^{2+}]_o$, and may cause the appearance of saturation. If the intrinsic relation between I_{Ca} and $[Ca^{2+}]_o$ is to be studied without complications from the stabilizing action of Ca^{2+}, precautions must be taken so that the I-V relation for the Ca current does not shift as $[Ca^{2+}]_o$ is changed.

Hagiwara & Takahashi (1967b) studied the dependence of the barnacle muscle Ca current on $[Ca^{2+}]_o$. Using the maximum rate of rise of the action potential as a measure of I_{Ca}, they found that I_{Ca} saturated as $[Ca^{2+}]_o$ was increased and could be described by a Michaelis-Menten or Langmuir type expression:

$$I_{Ca} = I_{Ca}^{max}\frac{C}{C + K'_{Ca}},\qquad\qquad 2.$$

where $C = [Ca^{2+}]_o$, I_{Ca}^{max} is the limiting value that I_{Ca} approaches as $[Ca^{2+}]_o$ becomes very large, and K'_{Ca} is a dissociation constant equal to the value of $[Ca^{2+}]_o$ that gives $I_{Ca} = 1/2\ I_{Ca}^{max}$. They maintained 100 mM Mg^{2+} in the external solution to eliminate the stabilizing action of Ca^{2+}; this appeared to be effective since both spike threshold and the potential of maximum rate of rise remained constant as $[Ca^{2+}]_o$ was varied. They found that K'_{Ca} ranged between 25 and 40 mM. As discussed below, the dissociation constant is affected by the high $[Mg^{2+}]_o$ and would be smaller in the absence of Mg^{2+}. Beirao & Lakshminarayanaiah (1979) studied the same effect on barnacle muscle, but in a voltage-clamp experiment where I_{Ca} was directly measured. They found that I_{Ca} reached a maximum at 0 mV and that this maximum I_{Ca} depended on $[Ca^{2+}]_o$ as predicted by Eq. 2. They obtained an average value of K'_{Ca} of 21 mM, also in the presence of 100 mM Mg^{2+}.

Eggs of the starfish, *Mediaster aequalis,* have two types of Ca channels: Channel I, which is activated at more negative potentials, and Channel II, which is activated at potentials above 0 mV (Hagiwara et al 1975). This membrane has a Ca-dependent outward current that obscures the dependence of I_{Ca} on $[Ca^{2+}]_o$; however, studies of the relation between I_{Ba} and $[Ba^{2+}]_o$ show that both of these channels exhibit saturation of the type described by Eq. 2. Channel II is more saturable, in that K'_{Ba} is 213 mM for Channel I and 73 mM for Channel II. The I-V relations measured at various values of $[Ba^{2+}]_o$ showed almost no shift along the voltage axis, so the addition of Ba^{2+} did not change the stabilizing effect appreciably in starfish egg membrane. The Ca current of tunicate egg also shows saturation (Okamoto et al 1976b). Raising $[Ca^{2+}]_o$ from 10 to 100 mM causes a 10 to 15 mV shift in the Ca current I-V relation. To partially compensate for this stabilization effect, Okamoto and co-workers compared the Ca currents at a higher potential (+25 mV), at which potential they assumed the Ca conductance was fully activated for all $[Ca^{2+}]_o$. These currents saturated with a dissociation constant of about 50 mM.

The Ca current of snail neurons has also been reported to depend on $[Ca^{2+}]_o$ as expected from Eq. 2 (Akaike et al 1978b). In this case, however, the stabilizing effect may make some contributions. Kawa (1979) reported a study of the dependence of the maximum rate of rise of the Ca spike on $[Ca^{2+}]_o$ for a different species of snail. It is unclear what other divalent cations were present or how well the surface potential was maintained constant, but Kawa found that the Ca current, as measured by the maximum rate of rise, depended on $[Ca^{2+}]_o$ as predicted by Eq. 2. The mean dissociation constant was 14 mM.

Saturation of I_{Ca} with $[Ca^{2+}]_o$ has been reported in several additional studies; however, usually no attempt has been made to separate effects due to stabilizing action of Ca^{2+} from a saturation due to a binding site connected with the Ca channel. In *Paramecium tetraurelia*, Satow & Kung (1979) found that the maximum size of I_{Ca} increased with $[Ca^{2+}]_o$ up to about 0.9 mM, but did not change for higher values of $[Ca^{2+}]_o$. However, the I-V curve shifts more than 40 mV to the right as $[Ca^{2+}]_o$ is increased from 0.38 to 5.6 mM; thus, there is some difficulty in interpreting this saturation.

The phenomenon of saturation is not unique to the Ca channel. The Na current also shows a saturation as $[Na]_o$ is increased; if this is attributed to a binding site, the effective dissociation constant for Na^+ is about 400 mM at 0 mV (Hille 1975a,b). Saturation was more quickly discovered for the Ca channel than for the Na channel, because Ca^{2+} normally makes up a small fraction of the total number of cations in the external solution, while Na^+ is the major cation. Thus, the dependence of Ca current on $[Ca^{2+}]_o$ was routinely studied for Ca^{2+} concentrations higher than normal, while the Na current was only studied for Na^+ concentrations less than or equal to the normal concentration.

Selectivity

Usually the selectivity of a channel is specified by the relative permeabilities of that channel for various ions, where the relative permeabilities are calculated from reversal potentials measured for the currents carried by those ions. As discussed above, it has not been possible to measure the reversal potential for currents carried by the Ca channel. Therefore, the selectivity of the Ca channel has only been studied by comparing the magnitudes of the inward currents when various divalent cations are present in the external solution. This sort of a comparison of currents is fraught with many complications. First of all, the currents should be compared with the same voltage across the membrane and the same external surface potential. So, precautions must be taken that the I-V relation does not shift when the external Ca^{2+} is replaced by Ba^{2+}, Sr^{2+}, etc. Second, the ratio of the current carried by some other divalent cation, e.g. Ba^{2+}, to the Ca current will depend, in general, on the concentration chosen at which to compare them and on the presence of other ions that may bind to the Ca channel. I_{Ba}, as well as I_{Ca}, satisfies a relation like Eq. 2, hence

$$\frac{I_{Ba}}{I_{Ca}} = \frac{I_{Ba}^{max}}{I_{Ca}^{max}} \frac{C + K'_{Ca}}{C + K'_{Ba}}. \qquad\qquad 3.$$

The first factor on the righthand side of Eq. 3 is a "mobility" factor, because it gives the ratio of currents when the concentration of the permeant ion, C, is large. The second factor is an "affinity" factor, since it depends only on the concentration C and the dissociation constants. When C is small, the ratio of currents becomes the product of the mobility factor and the ratio of dissociation constants. Thus, I_{Ca} might be greater than I_{Ba} at a low concentration, but I_{Ba} might be the larger at a high concentration. Even for one value of C, the ratio of currents can change due to the presence of other ions that bind to the Ca channel and change the apparent dissociation constants. For example, Hagiwara and co-workers (1974) found that I_{Ba} was normally larger than I_{Ca} in barnacle muscle; however, with 20 mM Co^{2+} in the external solution, I_{Ca} was greater than I_{Ba}. The Co^{2+} increased both apparent dissociation constants by the same factor (see below), making the affinity factor deviate further from one. Thus, the ratio of the currents with and without Co^{2+} allowed a separate ordering of the affinity and mobility factors. They concluded that the affinity sequence was $Ca > Sr > Ba$, whereas the mobility sequence was reversed.

The comparison of the sizes of Ca, Ba, and Sr currents in starfish egg membrane again emphasizes the importance of the concentration of permeant ion used. When C is 10 mM, I_{Ca} is about twice as large as I_{Ba} or I_{Sr} through Channel I; however, when C is increased to 100 mM, Ca, Ba, and Sr currents of Channel I are about the same size (Hagiwara et al 1975). These results are consistent with the conclusion from the barnacle muscle study that Ca^{2+} has a greater affinity for the Ca channel binding site than does Ba^{2+} or Sr^{2+}. At low values of C the current ratios depend on the affinity factor, so I_{Ca} dominates, while at larger values of C the affinity factor approaches one and the mobility factor determines the current ratios.

Okamoto and co-workers (1977) studied the selectivity between Ca^{2+}, Sr^{2+}, and Ba^{2+} of the Ca channels of three different egg cell membranes: tunicate, sea urchin, and mouse. In these studies, no attention was given to the dependence of $I_{Ca} : I_{Sr} : I_{Ba}$ on permeant ion concentrations. Inward current I-V relations were measured for $C = 30$ to 50 mM in sea urchin egg, 30 to 200 mM in tunicate eggs, and $C = 20$ mM in mouse oocyte. In general, the stabilizing effect of the three divalent cations was different, so that the I-V relations were shifted along the voltage axis with respect to one another. Comparing maximum peak inward currents, the average relative sizes of $I_{Ca} : I_{Sr} : I_{Ba}$ were $1 : 0.7 : 0.5$ for sea urchin, $1 : 1.6 : 1.1$ for tunicate, and $1 : 1.4 : 0.7$ in mouse. These authors demonstrated that at least some of the apparent differences in selectivity for these three Ca channels was due to the different stabilizing action of the three cations. Assuming that the Ca current could be described by a constant-field expression like Eq. 1, that the concentration of permeant ion at the surface of the membrane could be calculated from the concentration in bulk solution and the surface potential, and that changes in surface potential were given by shifts in I-V relations, they calculated permeabilities for Ca^{2+}, Sr^{2+}, and Ba^{2+}. $P_{Ca} : P_{Sr} : P_{Ba}$ was $1 : 0.74 : 0.60$ in sea urchin egg, $1 : 0.66 : 0.26$ in tunicate egg, and $1 : 1.12 : 0.73$ in the mouse oocyte. The apparently weaker Ca current in mouse oocyte may be due to the greater amount of contaminating outward current present in this egg; this outward current could be Ca-dependent or partially blocked by Sr^{2+} and Ba^{2+}. Also, it may be that the value of C used for mouse oocyte was larger compared with the corresponding dissociation constants than were the values of C used for sea urchin and tunicate eggs.

Sr^{2+} and Ba^{2+} have been found to carry inward current in all Ca channels studied, so that the ability of Sr^{2+} or Ba^{2+} to substitute for Ca^{2+} in maintaining regenerative responses is one of the criteria used in identifying Ca spikes. However, other than the studies discussed above, there have been few studies that allow the selectivity of different Ca channels to be critically compared. In most cases where Ca^{2+} has been replaced by Sr^{2+} or Ba^{2+}, studies have not been performed to separate the stabilizing effects of the ions from their current-carrying ability, nor to determine both the binding-site affinity and mobility for each ion.

Several other divalent cations have been shown to carry small amounts of current through Ca channels. Mn^{2+} carries inward current in squid giant axon (Yamagishi 1973) and mammalian cardiac muscle (Ochi 1970, 1975, Delahayes 1975). In squid axon it is not clear whether Mn^{2+} passes through Na or Ca channels; however, in heart muscle it appears to pass through the Ca channel, since Mn action potentials are blocked by La^{3+}, but not by TTX (Ochi 1975, 1976). Fukuda & Kawa (1977) demonstrated that larval beetle muscle fibers, which have a pure Ca action potential, can produce an overshooting action potential that lasts for many seconds in solutions that contain only Mn^{2+} and Mg^{2+} as divalent cations. The overshoot and maximum rate of rise of the action potential increase as $[Mn^{2+}]_o$ is increased. In voltage-clamp studies of mouse oocyte (Okamoto et al 1977) and starfish egg (Hagiwara & Miyazaki 1977a), small inward currents were measured when external Ca^{2+} was replaced by Mn^{2+}; however, none were present when Ca^{2+} was replaced by Co^{2+} or Ni^{2+}. Mn^{2+} has recently been shown to pass through Ca channels in myoepithelial cells of a marine polychaete (M. Anderson 1979). The overshoot of these Mn action potentials increased 27 mV per ten-fold increase in $[Mn^{2+}]_o$, and the action potentials were blocked by Co^{2+} and La^{3+}. Fukuda & Kawa (1977) demonstrated that also Cd^{2+}, Zn^{2+}, and Be^{2+} can carry inward current through the Ca channel of larval insect muscle. Mn^{2+} and Cd^{2+} can substitute for Ca^{2+} in carrying inward current through the membrane of frog skeletal muscle fibers (Palade & Almers 1978). Zn^{2+} was also shown to carry inward current in snail neurons (Kawa 1979). Kawa found that the reciprocal of the maximum rate of rise of the Zn action potential was linearly related to the reciprocal of $[Zn^{2+}]_o$, as would be expected if Eq. 2 also holds for the Zn current. From this plot, he estimated the dissociation constant for Zn^{2+} at the Ca channel binding site to be 2.1 mM, about 1/5 to 1/10 of the dissociation constant he found in a similar way for the Ca action potential. At high concentrations of Zn^{2+} and Ca^{2+}, the maximum rate of rise for the Ca action potential was 10 to 20 times that of the Zn action potential. Thus, Kawa concluded that Zn^{2+} has an affinity for the Ca channel binding site that is 5 to 10 times stronger than that of Ca^{2+}, but Ca^{2+} is 10 to 20 times more mobile through the Ca channel.

Blockers

When a second ion competes with Ca^{2+} for the Ca channel binding site, I_{Ca} is given by

$$I_{Ca} = I_{Ca}{}^{max} \frac{C}{C + (1 + \frac{[M]}{K_M})K'_{Ca}},$$

4.

where [M] is the concentration of the second ion and K_M is the corresponding dissociation constant for the binding site (Hagiwara & Takahashi 1967b). If this second ion that competes with Ca^{2+} for the binding site has a low or zero mobility for passing through the Ca channel, the total inward current will be reduced and the ion is considered a blocking ion. Ions, like Mn^{2+}, that have a high binding affinity but a low mobility, can act both as blockers and current carriers for the Ca channel. Using the maximum rate of rise of the barnacle muscle action potential as a measure of I_{Ca}, Hagiwara & Takahashi (1967b) found that the reciprocal of the Ca current increased linearly with the concentrations of various blocking cations, in agreement with Eq. 4. Using the slope of these relationships to estimate the K_Ms, they typically found $K_{Co} : K_{Mn} : K_{Ni} : K_{Mg}$ was $1 : 2.5 : 4 : 46$. From investigation of the blocking effects of a number of polyvalent cations, they concluded that the order of binding affinity was La^{3+}, $UO_2{}^{2+} > Zn^{2+}$, Co^{2+}, $Fe^{2+} > Mn^{2+} > Ni^{2+} > Ca^{2+} > Mg^{2+}$. The value of K_{Ca} calculated in the presence of 100 mM Mg^{2+} includes a factor of $(1 + 100$ mM$/K_{Mg})$. Once K_{Mg} is separately measured, it can be estimated that K_{Ca}, the true dissociation constant for Ca^{2+} that would be measured in the absence of Mg^{2+}, is less than 15 mM.

Many studies of the relative efficacy of the different polyvalent cations in blocking Ca currents have found an order that agrees with the binding affinity sequence given above for barnacle muscle, i.e . $La^{3+} > Co^{2+} > Mn^{2+} > Ni^{2+} > Mg^{2+}$ (Okamoto et al 1976b, Fukuda & Kawa 1977). However, other studies have found somewhat different sequences for the blocking efficacies of polyvalent cations. Van Breemen and co-workers (1973) found the sequence $La^{3+} > Ni^{2+} > Co^{2+} > Mn^{2+} > Mg^{2+}$ for smooth muscle, and Akaike and co-workers (1978b) found $Ni^{2+} > La^{3+} > Co^{2+} > Mg^{2+}$ for snail neurons. Kostyuk and co-workers (1977b) found that Cd^{2+}, an ion not tested on barnacle muscle, was highly effective in blocking the Ca current of snail neurons. They found that the relation between Ca current magnitude and the concentration of Cd^{2+} fit a Langmuir isotherm with $K_{1/2} = 7.2 \times 10^{-5}$ M, where $K_{1/2}$ is the concentration of Cd^{2+} that reduces I_{Ca} to one half. This is in agreement with Eq. 4, but it should be noted that $K_{1/2}$ is not the same as K_M of Eq. 4. $K_{1/2}$ is larger than

K_M and depends on $C = [Ca^{2+}]_o$. In their study of the blocking of snail Ca current Akaike and co-workers (1978b) found $K_{1/2}$ of Cd^{2+} to be about 3 $\times 10^{-3}$ M, indicating a remarkable variation in the sensitivities of snail Ca currents to blocking ions.

Several organic compounds have been found to be potent blockers of the Ca current in heart and smooth muscle; this subject has recently been reviewed (Rosenberger & Triggle 1978). Kohlhardt and co-workers (1972) demonstrated that verapamil (4 μM) and its methyl derivative D–600 (1 μM) caused a drastic reduction in the Ca current measured in voltage-clamp studies on ventricular trabeculae of cat. These levels of D–600 and verapamil did not block the fast Na current. Nifedipine appears to be even more effective than D–600 in blocking the Ca current of heart and smooth muscle (Fleckenstein 1977). Somewhat higher concentrations of organic blockers appear to be required in other smooth muscles: the Ca current of guinea pig taenia colia was greatly reduced, but not eliminated, by 10 μM D-600 (Inomata & Kao 1976). Matthews & Sakamoto (1975) found that 10 μM D-600 blocked the Ca-dependent spikes of rat pancreatic islet cells. The Ca current of frog skeletal muscle fibers is decreased at least four-fold by 1 μM nifedipine or 10 μM D-600 (Palade & Almers 1978). However, these organic Ca current blockers do not appear to be as effective against the Ca currents found in membranes of invertebrate, as well as certain vertebrate, preparations. Baker and co-workers (1973) used 200 μM D-600 to block the late Ca^{2+} entry in squid axon. Kostyuk and co-workers (1977b) found that 500 μM verapamil or 100 μM D-600 considerably reduced, but did not eliminate, the Ca current of snail neuron. Likewise, 100 μM verapamil or D-600 was necessary to suppress by 60% the K^+-induced ^{45}Ca uptake of rat-brain synaptosomes, and 50 μM verapamil had no significant effect on the K^+-induced increase in mepp frequency at the frog neuromuscular junction (Nachshen & Blaustein 1979). Somewhat out of line with these results is the result obtained by Akaike and co-workers (1978b) with snail neurons, that 30 μM verapamil reduced the Ca current by 80%. These results suggest that there are fundamental differences between the Ca channels of different tissues.

As discussed above, none of the Ca current blockers, neither organic compounds nor polyvalent ions, are entirely specific; they block the outward current as well as the Ca current (Kass & Tsien 1975, Kostyuk et al 1975c, 1977b, P. T. Palade & W. Almers, personal communication). At the high concentrations necessary to block Ca currents in certain tissues, the organic blockers appear to inhibit the Na current as well as the Ca current. D-600 (200 μM) almost completely blocked the inward Na current of squid axon (Baker et al 1973). Klee and co-workers (1973) reported that D-600 at concentrations between 100 μM and 2 mM blocked Na spikes as well as Ca spikes and delayed rectification in Aplysia neurons. Using a voltage-sensitive fluorescent dye to measure membrane potential and veratridine to open Na channels, Nachshen & Blaustein (1979) found that the veratridine-induced fluorescence increase in synaptosomes was completely suppressed by 100 μM D-600 or verapamil.

DEPENDENCE OF CALCIUM CURRENT ON $[Ca^{2+}]_i$

The injection of Ca^{2+}-binding agents into cells has been shown to convert graded Ca dependent regenerative responses into all-or-none action potentials (Hagiwara & Naka 1964, Pitman 1979). Hagiwara & Nakajima (1966b) investigated the dependence of barnacle muscle excitability on $[Ca^{2+}]_i$. They controlled $[Ca^{2+}]_i$ by injecting a Ca-buffered EGTA solution along the whole length of the fiber until the fiber diameter had increased by 50%. All-or-none Ca spikes could be elicited when $[Ca^{2+}]_i$ was below 8×10^{-8} M, the Ca response was graded and oscillatory when $[Ca^{2+}]_i$ was between 8×10^{-8} M and 5×10^{-7} M, and no response was obtained at higher concentrations. In this study no attempt was made to determine if this change in excitability was due to intracellular Ca^{2+} inhibiting the Ca current or if it was due to an enhancement of K current; of course, at that time Ca-dependent K currents had not yet been reported. It was not determined if the resting membrane resistance decreased with $[Ca^{2+}]_i$. Meech (1974) found that injecting a Ca-EGTA buffer containing 9×10^{-7} M free calcium reduced the membrane resistance of a snail neuron by 25%.

Experiments done on internally dialyzed snail neurons seemed to promise a clear answer as to whether intracellular Ca^{2+} inhibits currents. Outward currents could be nearly eliminated by replacing intracellular K^+ with nonpermeant cations, while the concentration of free calcium of the dialyzing solution could be changed. According to Kostyuk & Krishtal (1977), when *Limnea* neurons were dialyzed with Tris-aspartate solution containing 10 mM EGTA, the inward Ca current was very sensitive to $[Ca^{2+}]_i$, being reduced by more than half at 2.7×10^{-8} M and completely eliminated at 5.8×10^{-8} M. This suggested the snail neuron Ca current was more sensitive to intracellular Ca^{2+} than was the barnacle muscle Ca current; also, it seemed in conflict with the 4.5×10^{-7} M steady-state Ca activity measured in a *Helix* neuron with a Ca^{2+} sensitive microelectrode (Christoffersen & Simonsen 1977). Using the same technique and solutions on a *Helix* neuron, Doroshenko & Tsyndrenko (1978) found that the Ca current was somewhat less sensitive to $[Ca^{2+}]_i$, being half reduced at 4×10^{-8} M, but still present at 5×10^{-7} M. Akaike and co-workers (1978b) dialyzed *Helix* neurons with Cs salts, but without EGTA. They observed significant Ca currents at 10^{-5} M free calcium in the dialyzing solution. There is, however, a possibility that the concentration of Ca^{2+} at the inner surface of the membrane may not have reached the concentration of the dialyzing solution at the time of measurement. Consequently their result probably provides an upper limit to the level of free Ca^{2+} that blocks the Ca current. Takahashi & Yoshii (1978) dialyzed the tunicate egg with solutions containing 50 mM EGTA and 50 mM DPTA-OH (1,3-diaminopropane-2-ol-N, N'-tetraacetic acid), buffering the free Ca concentration between 10^{-7} and 10^{-3} M, and found that the Ca currents fell to about half their maximum size when the concentration of free calcium of the dialyzing solution was 10^{-5} M. Although the exchange of the solution inside the cell was probably incomplete within their dialyzing time, the Ca ion concentration might have approached that of the dialyzing solution because of the high concentration of buffer. Thus, this result also provides an upper limit to the level of free Ca^{2+} that blocks the Ca current.

In summary, the evidence seems fairly strong that intracellular Ca^{2+} does block Ca currents, but there is considerable uncertainty as to what the critical concentrations are. Studies of the inactivation of Ca currents in *Aplysia* (Tillotson 1979) and *Paramecium* (Brehm & Eckert 1978) have been interpreted in a manner that supports the blocking action of intracellular Ca^{2+} on Ca currents; however, again no evidence is available as to what concentrations of free calcium are required for this blocking action. In *Archidoris* neurons large Ca currents flow for hundreds of milliseconds, showing very little inactivation (Connor 1979). Using a simple model for the internal diffusion and buffering of Ca^{2+} that is consistent with the data obtained from arsenazo-injected neurons, it can be calculated that the concentrations of free calcium at the inner surface of the membrane build up to 10^{-5} to 10^{-4} M (J. Connor, personal communication). This calculation involves many assumptions that may be questioned, but it does make a first estimate as to the extent to which Ca^{2+} accumulates at the inner surface of the membrane. If the results of this calculation are even roughly correct, it would argue that Ca channels are not blocked by transient increases of $[Ca^{2+}]_i$ to levels considerably higher than those that block Ca spikes in barnacle muscle or Ca currents in internally dialyzed snail neurons.

KINETICS

The kinetics of the Ca channel have been seriously explored only in a few cases. Although Ca currents generally activate somewhat slower than Na currents, the process is assumed to be similar to that of the Na current. We review activation briefly. The decay, or inactivation, of the Ca current has actually attracted considerably more interest than has activation of the current. Not only does the Ca channel inactivate much more slowly than the Na channel, but there are a number of reasons to suggest that Ca current inactivation may be fundamentally different.

Activation

Hodgkin & Huxley (1952d) found that the activation of the Na current could be described by a variable, m, raised to the third power, where m is determined by a first-order process. Consequently, the Na current that is activated by a depolarizing voltage step depends on a factor $[1 - \exp(t/\tau)]^3$. Many investigators of Ca currents have followed the Hodgkin-Huxley parameterization and tried to fit the Ca current to a form

$$[1 - \exp(t/\tau)]^X.$$

The values of X so far determined range from 1 to 6 (see Table 1). The wide range of values determined may reflect unsatisfactory space clamp conditions in some of the studies. However, it is worthwhile to emphasize again that it seems quite possible that several different kinds of Ca channels exist and that these channels may have different kinetics for activation. It is generally accepted that the Na channel is a unique structure, common to many different membranes. Yet, even in the Na channel X is found to have different values in different membranes, $X = 2$ for the *Xenopus* node of Ranvier (Frankenhaeuser 1960), while $X = 3$ for squid axon.

Inactivation

Hodgkin & Huxley (1952c) found that the squid Na current was also determined by a second voltage and time-dependent process that they called inactivation and represented by the variable h. They measured the time dependence of h in three ways:

1. Decay of Na current while the membrane was held at one potential.
2. Decrease of Na current magnitude when the test pulse was preceded by a depolarized conditioning pulse of increasing duration.
3. Increase of Na current magnitude when the test pulse was preceded by a hyperpolarizing conditioning pulse of increasing duration.

At those voltages where the rate of inactivation could be measured by more than one of the above procedures, the time constants measured by the different methods agreed. The Ca current also appears to inactivate, although considerably slower than the Na current in most preparations. There is a great deal of disagreement about the rate and extent of Ca current inactivation. Much of this disagreement is due to the problem (discussed earlier) of separating the Ca current from outward currents. When an inward current is seen to decrease in magnitude, the question remains if it

Table 1 Kinetics

Preparation	X	T_D (near V_{Max}) (msec)	Comments and source
Barnacle muscle			
Megabalanus		>500	Int. Cs^+, Keynes et al 1973
Balanus		>200	Ext. TEA, Hagiwara et al 1974
Crayfish muscle	6	20	Current correction, Hencek & Zachar 1977
Squid nerve terminal	5		Llinas et al 1976
Snail neurons			
Limnea	2	40, 220	Int. $Tris^+$, 20°C, Kostyuk et al 1977b
Helix pomatia	2		Int. $Tris^+$, 6.5°C, current correction, Kostyuk et al 1979
Helix pomatia	3	~30, ~1000	Ba current, ext. TEA, Magura 1977
Helix aspersa	1 or 2	5	Int. Cs^+, Akaike et al 1978b
Marine molluscan neurons			
Archidoris		>100	Int. EGTA, ext. TEA, Connor 1979
Aplysia R15		40	Int. Cs^+, Tillotson & Horn 1978
Aplysia R15	1 or 2	70	Ext. TEA, Adams & Gage 1979a
Neuroblastoma (mouse)		35	20–24°C, Moolenaar & Spector 1979
Eggs			
Starfish			
Channel I		25	21–20°C, Ba current, current correction, Hagiwara et al 1975
Channel II		70	
Tunicate	1	140, ∞	15°C, Okamoto et al 1976b
Paramecium			
P. caudatum		1–2	Brehm & Eckert 1978
P. tetraurelia		1	Satow & Kung 1979

is due to inactivation of the Ca current or activation of an outward current.

Most studies of Ca current inactivation have been based on the decay of the inward current recorded when the membrane is held at a constant potential. The T_D values listed in Table 1 give the time required for the current measured near V_{Max}, the potential where Ca current is maximal, to decay to half its maximal value. There is an enormous range of values, from about 1 msec in *Paramecium* to nearly a second in barnacle muscle and some snail neurons. Among the spherical cells (eggs, neuroblastoma, and molluscan neurons) there seems to be fair agreement that the Ca current decays on a time scale of about a hundred milliseconds, with the exception of the Ca current of the snail neurons studied by Akaike and collaborators, which decays about ten times faster. Several studies point out that the Ca current decay is not a single exponential, but that at least two components are present. Kostyuk and co-workers and Magura reported a second component in snail neuron that decayed over five times slower than the first, and Okamoto and co-workers found a second component in tunicate egg that appeared constant on a time scale of hundreds of milliseconds. The data given for starfish egg is taken from Ba currents; in eggs, Ba and Ca currents decay at the same rate (Hagiwara et al 1975, Okamoto et al 1976b, Okamoto et al 1977). This is in contrast to the situation in *Aplysia* and *Paramecium,* where the Ba current appears to decay much more slowly than the Ca current (Brehm & Eckert 1978, Tillotson 1979).

The kinetics of Ca current inactivation have been studied in a few cases by the second or third method of Hodgkin & Huxley. Using a depolarizing conditioning pulse, Magura (1977) found that the peak value of the Ba current inactivated with a time constant of 120 msec at –30 mV, while the long-lasting component of the current inactivated with a time constant of 1.4 sec at –24 mV. Using hyperpolarizing conditioning pulses, the recovery from inactivation in *Aplysia* neurons has been found to also have two components. Tillotson & Horn (1978) found that recovery at –36 mV was a sum of two exponentials, with time constants of 420 msec and 4.4 sec. Adams & Gage (1979a) studied recovery at –42 mV and found one phase with a time constant of about 100 msec and a second with a time constant on the order of seconds. These time constants for recovery from inactivation are measured at considerably more negative potentials than are the time constants for Ca current decay, so it is not possible to compare them. It seems plausible that if recovery were measured at a more positive potential, the time constant of the faster component might agree with the time constant for decay at the same voltage.

Mechanism of Inactivation

Inactivation of the Na current appears to be a process intrinsic to the Na channel itself, depending on voltage and time, but fairly independent of the Na current or the concentrations of ions on either side of the membrane (see Meves 1978). Most researchers have assumed that the nature of Ca current inactivation is similar. However, recently it has been proposed that Ca current inactivation is entirely, or partially, due to accumulation of Ca^{2+} at the inner surface of the membrane (Brehm & Eckert 1978, Tillotson 1979). The hypothesis seems quite plausible in light of the evidence discussed above which indicates that Ca currents are inhibited when $[Ca^{2+}]_i$ rises above 10^{-7} M. Evidence to support this hypothesis has been obtained in *Aplysia* (Tillotson 1979), *Paramecium* (Brehm & Eckert 1978), and insect (F. M. Ashcroft, unpublished result)—animals separated by sufficient evolutionary distance to suggest that the hypothesis might apply universally. This hypothesis immediately provides an explanation (other than voltage-clamp failure) for why the Ca current decays so much faster in *Paramecium* than in *Aplysia* neurons, since Ca^{2+} accumulation would be expected to be greater in the *Paramecium* cilium with its higher surface-to-volume ratio. This problem has been reviewed by Eckert & Brehm (1979).

The fact that the Ca channel experiences almost no inactivation at large positive potentials near E_{Ca}, at which almost no Ca^{2+} enters the cell, has been further demonstrated in both *Aplysia* and *Paramecium*. Clamping the membrane to a large positive potential for several hundred milliseconds immediately before stepping to a test potential where inward Ca currents are observed gave almost the same inward current as was obtained by stepping directly from the holding potential to the test potential. This is consistent with the results of Katz & Miledi (1971) that holding the presynaptic terminal of the squid giant synapse at a large positive potential for one or two seconds did not inactivate the Ca conductance mechanism.

The inward current decays more rapidly with Ca^{2+} outside the membrane than with Ba^{2+}, in *Paramecium* and *Aplysia* neurons. Tillotson, Brehm, and Eckert interpret the slower rate of inactivation of Ba currents to mean that intracellular Ba^{2+} is not as effective as Ca^{2+} in blocking the Ca channel. As mentioned earlier, Sr^{2+} and Ba^{2+} currents decay at the same rate as Ca^{2+} currents in the egg cell membranes of starfish, tunicate, sea urchin, and mouse (Hagiwara et al 1975, Okamoto et al 1977). So the character of the inactivation seen in *Paramecium* and *Aplysia* is not universal to all Ca currents.

It is disturbing that most of the results given to support the dependence of Ca current inactivation on intracellular Ca^{2+} accumulation could equally well be interpreted to indicate that the apparent inactivation of inward current seen in *Aplysia* and *Paramecium* is due to the activation of a Ca-dependent outward current. If a Ca dependent outward current is present, the suppression of the inward current of a second pulse would be expected to depend on the entry of Ca^{2+} during the first pulse, and Ba^{2+} and Sr^{2+} are known to be less effective than Ca^{2+} in evoking the Ca-dependent K current (Hagiwara et al 1975, Gorman & Hermann 1979). However, Tillotson, Brehm, and Eckert have rejected this interpretation for several convincing reasons. Still, the possibility that this is due to a Ca-dependent outward current will not be eliminated until it can be demonstrated that the instantaneous conductance of the membrane is decreasing during the apparent inactivation. Neither of these preparations have a sufficiently fast voltage clamp to allow instantaneous current measurements.

If the inactivation of the Ca channel is due to internal accumulation of Ca^{2+}, instead of a purely voltage-dependent process, there is no reason to expect the rate of onset of inactivation to be the same as the rate of recovery from inactivation, since the accumulation of Ca^{2+} inside the membrane is probably determined by processes different from those that subsequently remove that accumulation. There have been no studies of the onset of inactivation and the recovery from inactivation at the same potential. The slower rates of recovery from inactivation reported for the Ca current of *Aplysia* neurons (Tillotson & Horn 1978, Adams & Gage 1979b) have time constants of several seconds, which are similar to the time constants reported for the decay of the absorption changes measured in arsenazo-injected molluscan neurons (Gorman & Thomas 1978, Ahmed & Connor 1979). The absorption changes would not show the faster process of recovery from inactivation if this corresponds to diffusion of Ca^{2+} away from the membrane toward the interior, rather than a decrease of free calcium inside the cell. A second observation that seems to support the hypothesis that Ca current inactivation is due to Ca^{2+} entry is that most of the Ca currents studied show no appreciable steady-state inactivation at potentials more negative than the lowest potential at which Ca currents are observed.

In conclusion, the evidence for the dependence of Ca current inactivation on Ca^{2+} entry is quite convincing in *Aplysia* and *Paramecium*. However, it remains to be seen how universal this type of inactivation is. The inactivation of Ca current in egg cell membranes appears to be somewhat different. This again suggests there are several different types of Ca channels among different preparations. P. T. Palade and W. Almers (personal communication) recorded Ca currents from frog skeletal muscle fibers, using the Hille-Campbell type voltage clamp. The Ca channels are probably located in the membrane of the transverse-tubular system. They interpret the decay of Ca current as a result of depletion of Ca^{2+} from just outside the tubular system membrane, since no decay was found when the extracellular Ca^{2+} concentration was buffered. Because the normal Ca^{2+} concentration is small in most cases, depletion of extracellular Ca^{2+} should be considered as another mechanism for inactivation.

MOLECULAR NATURE OF CALCIUM CONDUCTANCE

Noise Analysis

In snail neurons, current fluctuations have been observed that appear to be associated with the Ca current. Recording the current from a small patch (30 μ^2) of voltage-clamped membrane Krishtal & Pidoplichko (1977) found fluctuations in the measured current that were larger during a depolarizing voltage pulse that evoked a Ca current than during a hyperpolarizing pulse. When external Ca^{2+} was replaced with Ba^{2+}, the maximal inward current nearly doubled and the current fluctuations became larger. Na and K currents had been eliminated by removing Na^+ from the external solution and internally perfusing the cell with Tris-phosphate. Akaike and co-workers (1978a) also made similar observations, but in the current measured from a larger area of the cell membrane.

If the Ca current is assumed to flow through a number of equivalent, independent unitary conductances, the fluctuations of current may be interpreted as resulting from the opening and closing of individual units. Assuming each unitary conductance fluctuates between two states, one in which no current flows and one where current i flows, the unit current i can be calculated from

$$i = \frac{\sigma_I^2}{\bar{I}(1 - a)} \qquad\qquad 5.$$

where σ_I^2 is the variance of the current about the mean, $\bar{I}$ is the mean current (see Stevens 1972) and a is the mean fraction of unitary conductances that are open and can be estimated from the ratio of the present conductance to the maximum conductance, if it is assumed that the conductance of a single unit is the same at this potential as it is at the potential that gives the maximum conductance. Since this may not be true for the Ca channel (as discussed earlier), it is better to apply Eq. 5 at potentials where $a \ll 1$ and so can be neglected. Krishtal & Pidoplichko (1977) found that the variance for the Ca current was too small compared to background noise for an accurate measurement. However, the variance for the Ba current was four to six times larger than the background and allowed a calculation of the unitary Ba current, i_{Ba}, from Eq. 5. At +15 mV they found i_{Ba} was about 2×10^{-13} A; since maximal Ca currents were about half the size of Ba currents, they concluded that i_{Ca} was about 10^{-13} A. This can be expressed in terms of a unitary conductance by dividing by $(V - E_{Ca})$, which gives a unitary conductance of about 5×10^{-13} S. Akaike and co-workers (1978a) applied Eq. 5 to the difference of variances obtained for the current before and after addition of Ni^{2+} and obtained a unitary Ca conductance of about 10^{-13} S. Maximal Ca currents of about 30 nA are measured in 100 μm snail neurons, giving a current density of one picoampere per square micrometer. The above values for the unitary Ca conductance gives a Ca unitary conductance density of 10 to 50 per μm^2, which is about the same as the densities reported for Na and K channels in squid axon (Conti et al 1975).

Akaike and collaborators (1978a) also analyzed the Ca current fluctuations in terms of power density spectra. They found that the difference in spectra could be fitted to a Lorentzian form $[1 + (f//f_c)^2]^{-1}$, where f_c is the corner frequency that represents the mean open time of the channel. From the f_c and low frequency asymptote of the Lorentzian component of the fitted spectrum, they calculated unitary conductances that ranged between 6×10^{-14} to 6×10^{-13} S.

These measurements of fluctuations of Ca current tend to support the hypothesis that Ca currents flow through channels rather than involving carriers, since most models for carriers predict considerably smaller fluctuations and power density spectra quite different from the Lorentzian form (Szabo 1977). Both Akaike and collaborators (1978a) and Kostyuk (1978) measure power density spectra that can be fitted to a Lorentzian-type spectrum, which is the noise spectrum expected for independent, two-state channels (see Stevens 1972). The fact that the single-channel Ca conductances are calculated to be at least an order of magnitude smaller than the single-channel Na conductance (2 to 8×10^{-12} S, see Conti et al 1976) may not indicate any fundamental difference in the nature of the two channels, but instead, may reflect the inappropriateness of the concept of conductance in describing Ca currents. If $[Ca^{2+}]_i$ were lowered by a factor of 10, the calculated value for the Ca single-channel conductance would become even smaller, although the state of the Ca channel changes in no way. The unitary Ca currents are about 10^{-13} A, which is the size of the Na unitary current that would be measured at $+20$ mV when E_{Na} is $+40$ mV. Due to the very positive value of E_{Ca}, it is probably not meaningful to calculate single-channel conductances for the Ca current.

Displacement Currents

Hodgkin & Huxley (1952d) proposed that the voltage dependence of Na channels might be mediated by the movement of charged particles in the membrane. The displacement current, which probably corresponds to the movement of the gating particles of the Na channel in response to sudden changes in membrane potential, was first observed in squid axon (Armstrong & Bezanilla 1973, Keynes & Rojas 1973). Similar currents associated with Ca channels have been reported in *Aplysia* (Adams & Gage 1976, 1979b) and *Helix* neurons (Kostyuk et al 1977a, 1979). Both groups detected the displacement currents by methods similar to that used for the Na channel. Kostyuk and co-workers (1977a) measured a single outward transient, with no rising phase, at the beginning of the pulse and an inward transient at the end of the pulse. These transient currents do appear to be displacement currents, since the charge carried by the transient at the beginning of the pulse, Q_{on}, is equal in magnitude to that carried by the transient at the end of the pulse, Q_{off}, at least for pulses below $+20$ mV. These transient currents appear to be connected with the Ca current, any transient currents associated with the Na channel that might be present are presumably lost in the initial period of 250 μsec that is blanked out. Adams & Gage (1976, 1979b) analyze only the transient current observed at the beginning of the pulse; they find that this transient has two components, one associated with the Na current and the other with the Ca current. The transient associated with the Ca current has a rising phase of about 1 msec. Different values for the rise time of these assymetry currents could arise from the different structures of the membranes of *Aplysia* and snail neurons; the snail neuron appears to have less invagination of the surface membrane.

Assuming that the gating mechanism for Ca current in *Aplysia* neurons is the same as that in *Helix* neurons, the ratio of Q_{on} to I_{Ca} should be roughly the same in the two preparations. Kostyuk et al (1977a) report a Q_{on} of about 2 X 10^{-11} coul at +20 mV, when the Ca current is about 40 nA with $[Ca]_o$ equal to 10 mM. This gives a ratio of Q_{on}/I_{Ca} of 0.5 msec. Adams & Gage (1979a,b) measure displacement currents in *Aplysia* which have a Q_{on} of about 2 X 10^{-10} coul at +13 mV, when the Ca current is about 200 nA with $[Ca]_o$ equal to 11 mM, giving a Q_{on}/I_{Ca} ratio of 1 msec. This agreement may encourage the conclusion that these displacement currents are related to the gating of the Ca current.

Hodgkin & Huxley (1952d) pointed out the relation that should exist between the magnitude of the gating currents, the voltage dependence of the current and the single-channel conductance. The Ca displacement currents that have been measured are not in agreement with that relation. If a channel is turned on by the movement of Z electronic charges across the complete membrane potential, the voltage dependence of the conductance would be given by

$$G(V) = G_{max}\frac{e^{ZF(V-V_0)/RT}}{1 + e^{ZF(V-V_0)/RT}} \qquad\qquad 6.$$

Fitting this expression to the G vs V curves measured for *Helix* gives values of Z between 2 and 4. Z can also be calculated from the relation

$$\frac{Q_{on}}{I_{Ca}} = \frac{Ze}{i} \qquad\qquad 7.$$

where e is the electronic charge and i is the unitary current. Using the data for *Helix* (Kostyuk et al 1977a, Krishtal & Pidoplichko 1977), Eq. 7 gives a value for Z of about 300. This discrepancy could mean that either the unitary conductance or the gating current is actually two orders of magnitude smaller than reported. At present, it is difficult to say which of the two numbers is more suspect; neither is established beyond doubt. Further research is definitely necessary to reach any firm conclusions about the molecular nature of the Ca channel.

CONTROL OF I_{Ca} BY EXTERNAL CHEMICAL MESSENGERS

Considering the role of Ca^{2+} in mediating transmitter release and muscle contraction, direct control of voltage-dependent Ca currents by chemical messengers would seem to be a very efficient way of modulating synaptic efficacy or muscle contractility. Studies on cardiac muscle and several types of neurons suggest that various transmitters and hormones exert influences directly on the voltage-dependent Ca current. This action of chemical messengers on a voltage-dependent channel is in contrast to the much better studied action of chemical messengers on voltage-independent channels. In cardiac muscle and certain *Aplysia* neurons, where the chemical messenger enhances the current, it appears cyclic nucleotides may act as second messengers; however, there is little evidence to suggest the involvement of cyclic nucleotides in other cases where the chemical messengers depress the Ca current.

Cardiac Muscle

Catecholamines increase the permeability of cardiac muscle membrane to Ca^{2+}. Externally applied noradrenaline increases the magnitude of Ca currents, producing half of the maximal response at 5×10^{-7} M (Reuter 1974). Voltage clamp analysis showed that noradrenaline causes no change in the voltage dependence of the Ca current. Reuter & Scholz (1977) further demonstrated that adrenaline does not change the rate constants associated with activation of the Ca current, nor does it change the apparent reversal potential of the Ca current. The only effect of catecholamines on the Ca current is to increase the size of the current, as though catecholamines increase the number of Ca channels available for activation. This action of adrenaline and noradrenaline is effected via an adrenergic β-receptor; isoprenaline, a β-adrenergic agonist, has the same effect as noradrenaline, and 2×10^{-6} M propranolol, a β-adrenergic blocker, reduces the effect of noradrenaline.

It has been suggested that the adenylate cyclase that is activated by catecholamines is an integral part of the β-receptor (Sutherland et al 1968); therefore, it seems reasonable that the catecholamine-induced increase in cardiac Ca current might result from an increase in intracellular levels of cyclic AMP. Most of the evidence seems to support this hypothesis. Intracellular iontophoresis of cyclic AMP in cardiac Purkinje fibers causes effects similar to the ones observed with extracellularly applied catecholamines, including elevation of the plateau of the action potential (Tsien 1973). Reuter (1974) found that application of dibutyryl cyclic AMP caused an increase in the cardiac Ca current very similar to that caused by noradrenaline, but the effect of the nucleotide was smaller and much more protracted (Reuter 1974). Presumably the dibutyryl cyclic AMP enters the cell and is slowly converted to cyclic AMP. Acetylcholine, which reduces cyclic AMP in heart muscle (Goldberg et al 1973, George et al 1975), decreases the cardiac Ca current (Giles & Tsien 1975, Ikemoto & Goto 1975). Thus, it seems that cardiac Ca channels may require intracellular cyclic AMP to become functional, although this requirement is far from proven.

Aplysia Neurons

Klein & Kandel (1978) have found that external serotonin (5HT) (10^{-5} to 10^{-4} M) causes an increase in the plateau of the action potential recorded from the cell body of certain sensory neurons in *Aplysia*. In the presence of 100 mM TEA, the plateau of the action potential is greatly enhanced, and the action of 5HT to elevate and prolong the plateau is much more obvious. They conclude that 5HT is increasing the voltage-dependent Ca current in these neurons, but they do not establish whether the immediate action of 5HT is to increase the Ca permeability or to decrease the outward current that normally masks the depolarizing effect of the Ca current. A similar enhancement of the plateau was obtained by extracellular application of phosphodiesterase inhibitors or by intracellular injection of cyclic AMP, so Klein & Kandel conclude that the action of 5HT is mediated by an increase in intracellular cyclic AMP. Recent voltage clamp experiments suggest that the immediate action of 5HT on membrane currents is to decrease outward currents, instead of directly enhancing the Ca permeability (M. Klein and E. R. Kandel, personal communication).

Pellmar & Carpenter (1979a,b) have reported that iontophoretic pulses of serotonin on certain *Aplysia* neurons induce a voltage-dependent Ca current. In voltage-clamped neurons, the 5HT induced inward current reaches a peak in 10 to 30 sec after the pulse of serotonin and lasts for 1 to 4 min. The inward current is considered to be a Ca current, since it continues in Na-free solution and is blocked by Co^{2+} and Mn^{2+}. The inward current is maximal when the membrane is clamped to +10 mV; no Ca current is induced by serotonin, when the membrane potential is held at potentials more negative than –40 mV. It is rather difficult to compare this effect of 5HT with that studied by Klein & Kandel (1978), because of the different types of experiments done. Pellmar & Carpenter (1979b) found that phosphodiesterase inhibitors reduced the amplitude of the 5HT-induced, voltage-dependent current, a result that suggests that cyclic AMP may not mediate this action of serotonin. However, cyclic AMP does play some role in controlling the inward current of this cell, since Pellmar (1980) found that intracellular injection of cyclic AMP appeared to mimic the action of 5HT in producing voltage-dependent Ca current.

Dorsal Root Ganglion Neurons

The action potential of embryonic chick dorsal root ganglion cells grown in cell culture has a plateau that is depressed by the external application of various neurotransmitters. The duration of the DRG cell action potential is decreased by γ-aminobutyric acid (GABA), noradrenaline (NA), serotonin (5HT), somatostatin, and [D-Ala^2] enkephalin amide (Dunlap & Fischbach 1978, Mudge et al 1979). Not all transmitters have this action on the duration of the DRG spike. ACh, glycine, substance P, bradykinin, neurotensin, and thyrotropin-releasing hormone are ineffective, and glutamate only produces a small and inconsistent effect. Noradrenaline and 5HT have half their maximal effects at 6×10^{-7} M, while GABA has half its maximal effect at 2×10^{-6} M. When higher concentrations (10^{-5} to 10^{-4} M) of noradrenaline, 5HT, or GABA are applied, return to control spike duration takes several minutes. The action of noradrenaline is mediated by an α-adrenergic receptor; α agonists, phenylephrine and dopamine, are effective, while β agonist, isoprenaline, is not. The effect of noradrenaline is blocked by α-blocker, phentolamine, but not by the β-blocker, propranolol. No experiments have been done to test if cyclic nucleotides are involved in mediating this response.

These transmitters appear to act directly on the Ca current, reducing its magnitude. They have no effect on the resting potential or resting resistance of the cell, and continue to shorten the duration of the action potential in Na-free solutions and solutions containing Ba^{2+}, which blocks K^+ currents. Recent voltage-clamp studies of the action of noradrenaline on these cells supports the interpretation that noradrenaline directly inhibits the Ca current (Dunlap & Fischbach 1979). Noradrenaline decreased the peak inward Ca current without affecting the outward K^+ current measured at the end of the depolarizing voltage pulse. Noradrenaline also decreased the magnitude of the Ca tail currents obtained by rapidly repolarizing the membrane to E_K. The transmitter decreased the magnitude of the Ca currents without affecting their time course, the voltage dependence of activation, or the apparent reversal potential. So, as was the case with catecholamine-induced enhancement of cardiac Ca current, the transmitter seems to control the number of Ca channels available for activation.

Sympathetic Neurons

Catecholamines appear to reduce the magnitude of the voltage-dependent Ca current of post-ganglionic neurons of the rat superior cervical ganglion, in the same way they do in chick DRG neurons. Noradrenaline inhibits the shoulder (or plateau) of the action potential, the magnitude of the hyperpolarizing afterpotential, and the rate of rise of the Ca spike in these cells (Horn & McAfee 1979, 1980). Since the hyperpolarizing afterpotential observed in these cells is thought to be due to a Ca-dependent K conductance (McAfee & Yarowsky 1979), all of the effects of noradrenaline could result from a reduction in the Ca current. Half maximal suppression of the hyperpolarizing afterpotential is produced by 10^{-6} M noradrenaline. The action of noradrenaline is mediated by an α-adrenergic receptor. Dopamine, an α-agonist, is more effective than isoprenaline, a β-agonist. The action of noradrenaline is antagonized by phentolamine, an α-blocker, but not by MJ–1999, a β-blocker. It is not reported if 5HT or GABA have similar effects to those of noradrenaline on these cells. Horn & McAfee suggest that the action of noradrenaline on the Ca current might be mediated by an increase in $[Ca^{2+}]_i$, such as is induced in smooth muscle by noradrenaline (Van Breemen et al 1973). The increase in $[Ca^{2+}]_i$ would then inhibit the Ca current, as has been observed in barnacle muscle, tunicate egg, and molluscan neurons.

SUMMARY

Voltage-dependent Ca channels have been identified in a variety of membranes. Because small Ca currents have been found even in cells that were previously thought to have pure Na action potentials and all excitable cells appear to go through a stage when they can produce Ca action potentials, the Ca channel may be more universal than the Na channel. Ca channels were first identified in invertebrate muscle; all invertebrate muscles that have been studied have a pure Ca spike. Most vertebrate smooth muscle has a pure Ca spike, while cardiac muscle has an action potential with both Na and Ca components. Vertebrate skeletal muscle has a nearly pure Na spike, but a small voltage-dependent Ca current has also been demonstrated in this membrane. The ratio of magnitudes of Ca and Na currents varies considerably in the various regions of the nerve cell membrane. In the axon Na currents predominate; in the nerve terminal Ca currents are much stronger than they are in the axon. Both Na and Ca currents are found in the nerve cell body, but their relative amplitudes vary considerably from cell to cell in one ganglion, ranging from some somas with nearly pure Ca spikes to others with nearly pure Na spikes. Likewise, excitable secretory cells have both Na and Ca currents; however, in some types the Ca component predominates, while in other types the Na component predominates. All excitable egg cell membranes have a voltage-dependent Ca current; some also have a Na current. The receptor potentials in a number of cells appear to be produced by Ca channels, as are the action potentials in certain epithelial cells. Remarkably similar Ca channels have been found in *Paramecium*.

There are a number of problems that have made the biophysical analysis of the Ca channel considerably more difficult than the analysis of the Na channel. First, no preparation has been found that is ideal for the study of the Ca channel. Many of the membranes that have Ca channels have a physical structure such that it is impossible to control the voltage across all regions of the membrane. It has also been difficult to manipulate the solutions on both sides of the membrane. Recently developed techniques have made considerable progress in overcoming these difficulties for certain cells containing potential-dependent Ca currents. Second, it is very difficult to separate Ca currents from the other membrane currents, especially at large positive membrane potentials where the Ca currents are small and the outward currents are large. No method has been found that completely blocks the voltage-dependent (and often time-dependent) outward currents. Since all methods for eliminating the Ca current may also change the outward current, there appears to be no reliable way of measuring the Ca currents at potentials where the outward current is appreciable. Third, the very asymmetrical distribution of Ca^{2+} across the membrane makes inappropriate some of the concepts and experiments typically applied to Na and K currents. Even if the Ca permeability of the membrane becomes voltage-independent at large positive potentials, the Ca conductance is expected to be dependent on voltage as well as the concentrations of Ca^{2+} on both sides of the membrane; so arguments that relate the Ca current to the driving force $(V - E_{Ca})$ become very weak. The highly nonlinear nature of the expected I-V curve suggests that Ca currents will be very small near E_{Ca}— too small to measure given the large voltage-dependent outward currents that flow across the membrane in this range of potentials. Thus, it appears to be practically impossible to measure E_{Ca}.

The selectivity of the Ca channel cannot be measured in terms of relative permeabilities, as it is for other channels, because the reversal potentials for currents through the Ca channel cannot be measured. Instead, the Ca permeation mechanism is characterized in terms of a model in which polyvalent cations bind to a membrane site from which they can then pass through the membrane. The phenomena of saturation, relative current-carrying ability, and competitive blocking of current can all be interpreted with this model. Ions like La^{3+}, Co^{2+}, and Ni^{2+} bind strongly to the site but have almost no ability to cross the membrane, while Ca^{2+}, Ba^{2+}, and Sr^{2+} bind less strongly but cross the membrane easily. Other ions like Mn^{2+}, Zn^{2+}, and Cd^{2+} have a small ability to pass through Ca channels.

The relative ability of polyvalent cations to block Ca current appears to vary somewhat between different tissues. Organic Ca antagonists are over 100 times more effective on some Ca currents than others.

Ca currents appear to be blocked by intracellular Ca^{2+}. In barnacle muscle and snail neurons $[Ca^{2+}]_i = 5 \times 10^{-7}$ M greatly reduces the Ca current; however, it remains to be determined if this concentration of intracellular Ca^{2+} will block all Ca currents. It is an appealing hypothesis that the inactivation of Ca current may be due to the accumulation of Ca^{2+} inside the membrane. There is considerable evidence to support this hypothesis in *Aplysia* neurons and in *Paramecium*. However, the inactivation of divalent currents in egg cells seems to be somewhat at variance with the data for *Aplysia* neurons and *Paramecium*. It may be that the mechanisms for Ca current inactivation are different in different membranes.

Studies of the fluctuations associated with the Ca current seem to agree with the idea that the current is carried via a channel that has two states, closed and conducting. Single-channel conductances of 0.05 to 0.5 pS are calculated. Displacement currents that could be associated with the gating of Ca current have also been measured. The magnitude of these displacement currents relative to the Ca current is not consistent with the measured single-channel conductance and voltage dependence of Ca current activation, using the model for gating currents proposed by Hodgkin & Huxley. More studies are necessary to establish the current-carrying ability of a single Ca channel and to determine the nature of the displacement currents.

One of the most interesting properties of Ca channels is the recently recognized ability of certain external chemical messengers (adrenaline, noradrenalines, serotonin, GABA, somatostatin, and enkephalins) to modulate voltage-dependent Ca currents. Certain of these messengers have been found to enhance the Ca currents in cardiac muscle and certain molluscan neurons and to suppress the Ca currents of dorsal root ganglion and sympathetic neurons. Voltage-clamp studies of cardiac muscle and dorsal root ganglion cells indicate that only the magnitude of the Ca current is changed, while the kinetics and voltage dependence remain the same, as though the number of functional Ca channels changes. In cardiac muscle there is considerable evidence to suggest that the effect is mediated by intracellular cyclic nucleotides. This control of the Ca channel by neurotransmitters is potentially of enormous importance as a mechanism to explain plasticity and modulation at synapses.

The Na channel is generally considered to be a unique structure. Although the time constants, voltage dependence and TTX sensitivity vary from preparation to preparation, the basic kinetics and selectivity remain unchanged. In contrast, there is no compelling evidence to suggest that the Ca channel is also a unique structure. Instead, Ca currents are found to differ in almost every measureable property. Selectivity in terms of relative permeabilities cannot be measured for Ca channels and there is very little agreement as to the kinetics of Ca currents. Some of the most important differences found between Ca currents are the following:

1. The relative efficacy of polyvalent cations to block Ca currents varies between preparations.
2. Organic blockers are much more effective against some Ca currents than others.
3. Some membranes have more than one component of Ca current, i.e. there are Ca currents with different voltage dependencies, different binding affinities and mobilities or different kinetics in the same membrane.
4. The inactivation of Ba currents follows the same time course as that of Ca currents in some membranes, whereas Ba and Ca currents inactivate with different time courses in other membranes.

While future studies may show that some of these differences are no more fundamental than the different TTX-sensitivities of Na channels, at present it seems most objective to give up the prejudice that the Ca channel is like the Na channel and allow that there may be various types of Ca channels. Given the possible existence of various types of Ca channels, it is clearly important to study Ca currents in a variety of tissues, although some are far more favorable for biophysical analysis. Although the analysis of the Na channel has provided a powerful model to follow in studying the Ca channel, it is time to recognize that Ca currents differ from Na currents in fundamental ways that call for unique types of analysis.

Literature Cited

Adams, D. J., Gage, P. W. 1976. Gating currents associated with sodium and calcium current in an *Aplysia* neuron. *Science* 192:783–84

Adams, D. J., Gage, P. W. 1979a. Characteristics of sodium and calcium conductance changes produced by membrane depolarization in an *Aplysia* neurone. *J. Physiol. London* 289:143–61

Adams, D. J., Gage, P. W. 1979b. Sodium and calcium gating currents in an *Aplysia* neuron. *J. Physiol. London* 291: 467–81

Ahmed, Z., Connor, J. A. 1979. Measurement of calcium influx under voltage clamp in molluscan neurones using the metallochromic dye arsenazo III. *J. Physiol. London* 286:61–82

Akaike, N., Fishman, H. M., Lee, K. S., Moore, L. E., Brown, A. M. 1978a. The units of calcium conduction in *Helix* neurones. *Nature* 274:379–81

Akaike, N., Lee, K. S., Brown, A. M. 1978b. The calcium current of *Helix* neuron. *J. Gen. Physiol.* 71:509–31

Anderson, M. 1979. Mn^{2+} ions pass through Ca^{2+} channels in myoepithelial cells. *J. Exp. Biol.* 82:227–38

Anderson, P. A. V. 1979. Epithelial conduction in salps. *J. Exp. Biol.* 80:231–39

Armstrong, C. M., Bezanilla, F. 1973. Currents related to movement of the gating particles of the sodium channels. *Nature* 242:459–61

Ashcroft, F. M., Standen, N. B., Stanfield, P. R. 1979. Calcium currents in insect muscle. *J. Physiol. London* 291:51P–52P

Baccaglini, P. I., Spitzer, N. C. 1977. Development changes in the inward current of the action potential of Rohon-Beard neurones. *J. Physiol.* 271:93–117

Baker, P. F., Glitsch, H. G. 1975. Voltage-dependent changes in the permeability of nerve membranes to calcium and other divalent cations. *Philos. Trans. R. Soc. London B* 270:389–409

Baker, P. F., Hodgkin, A. L., Ridgway, E. B. 1971. Depolarization and calcium entry in squid giant axons. *J. Physiol. London* 218:709–55

Baker, P. F., Meves, H., Ridgway, E. B. 1973. Effects of manganese and other agents on the calcium uptake that follows depolarization of squid axons. *J. Physiol. London* 231:511–26

Barrett, E. F., Barrett, J. N. 1976. Separation of two voltage-sensitive potassium currents, and demonstration of a tetrodotoxin-resistant calcium current in frog motoneurons. *J. Physiol. London* 255:737–74

Beaty, G. N., Stefani, E. 1976a. Calcium dependent electrical activity in twitch muscle fibres of the frog. *Proc. R. Soc. London B* 194:141–50

Beaty, G. N., Stefani, E. 1976b. Inward calcium current in twitch muscle fibres of the frog. *J. Physiol. London* 260:27P

Beirao, P. S., Lakshminarayanaiah, N. 1979. Calcium carrying system in the giant muscle fibre of the barnacle species, *Balanus nubilus. J. Physiol. London* 293:319–27

Belton, P., Grundfest, H. 1961. Comparative effects of drugs on graded responses of insect muscle fibers. *Fed. Proc.* 20:399

Biales, B., Dichter, M., Tischler, A. 1976. Electrical excitability of cultured adrenal chromaffin cells. *J. Physiol. London* 262:743–53

Biales, B., Dichter, M. A., Tischler, A. 1977. Sodium and calcium action potential in pituitary cells. *Nature* 267:172–74

Bianchi, C. P., Shanes, A. M. 1959. Calcium influx in skeletal muscle at rest, during activity, and during potassium contracture. *J. Gen. Physiol.* 42:803–15

Brandt, B. L., Hagiwara, S., Kidokoro, Y., Miyazaki, S. 1976. Action potentials in the rat chromaffin cell and effects of acetylcholine. *J. Physiol. London* 263:417–39

Brehm, P., Eckert, R. 1978. Calcium entry leads to inactivation of calcium channel in *Paramecium. Science* 202:1203–6

Chamberlain, S. G., Kerkut, G. A. 1967.

Voltage clamp studies on snail (*Helix aspersa*) neurones. *Nature* 216:89

Chen, C. F., von Baumgarten, R., Takeda, R. 1971. Pacemaker properties of completely isolated neurons in *Aplysia californica. Nature New Biol.* 233:27–29

Christoffersen, G. R. J., Simonsen, L. 1977. Ca^{2+} sensitive microelectrode: Intracellular steady state measurement in nerve cell. *Acta Physiol. Scand.* 101:492–94

Clusin, W. T., Bennett, M. V. L. 1977. Calcium-activated conductance in skate electroreceptors. *J. Gen. Physiol.* 69:121–43

Connor, J. A. 1977. Time course separation of two inward currents in molluscan neurons. *Brain Res.* 119:487–92

Connor, J. A. 1979. Calcium current in molluscan neurones: Measurement under conditions which maximize its visibility. *J. Physiol. London* 286:41–60

Conti, F., DeFelice, L. J., Wanke, E. 1975. Potassium and sodium ion current noise in the membrane of the squid giant axon. *J. Physiol. London* 248:45–82

Conti, F., Hille, B., Neumcke, B., Nonner, W., Stämpfli, R. 1976. Conductance of the sodium channel in myelinated nerve fibres with modified sodium inactivation. *J. Physiol. London* 262:729–42

Delahayes, J. F. 1975. Depolarization-induced movement of Mn^{2+} across the cell membrane in the guinea pig myocardium. *Circ. Res.* 36:713–18

Dichter, M. A., Fischbach, G. D. 1977. The action potential of chick dorsal root ganglion neurones maintained in cell culture. *J. Physiol. London* 267:281–98

Doroshenko, P. A., Kostyuk, P. G., Krishtal, O. A. 1973. Action of calcium on the somatic membrane of mollusk giant neurons. *Neirofiziologiya* 5:621–27. Transl. 1974, in *Neurophysiology* 5:476–81 (From Russian)

Doroshenko, P. A., Kostyuk, P. G., Tsyndrenko, A. Y. 1978. Reversal potential of the slow component of the inward current in snail neuron membrane. *Neirofiziologiya* 10:206–8. Transl. 1978, in *Neurophysiology* 10:145–47

Doroshenko, P. A., Tsyndrenko, A. Y. 1978. Action of intracellular calcium on the inward calcium current. *Neirofiziologiya* 10:203–5. Transl. 1978, in *Neurophysiology* 10:143–45 (From Russian)

Dunlap, K. 1976. Calcium channels in *Paramecium* confined to ciliary membrane. *Am. Zool.* 16:185

Dunlap, K. 1977. Localization of calcium channels in *Paramecium caudatum. J. Physiol. London* 271:119–33

Dunlap, K., Fischbach, G. D. 1978. Neurotransmitters decrease the calcium component of sensory neurone action potentials. *Nature* 276:837–39

Dunlap, K., Fischbach, G. D. 1979. Neurotransmitter modulation of voltage-sensitive calcium currents in sensory neurons. *Neurosci. Abstr.* 5:291

Eckert, R., Brehm, P. 1979. Ionic mechanisms of excitation in *Paramecium. Ann. Rev. Biophys. Bioeng.* 8:353–83

Fain, G. L., Quandt, F. N., Gerschenfeld, H. M. 1977. Calcium-dependent regenerative responses in rods. *Nature* 269:707–10

Fatt, P., Ginsborg, B. L. 1958. The ionic requirements for the production of action potentials in crustacean muscle fibers. *J. Physiol. London* 142:516–43

Fatt, P., Katz, B. 1953. The electrical properties of crustacean muscle fibers. *J. Physiol. London* 120:171–204

Fleckenstein, A. 1977. Specific pharmacology of calcium in myocardium, cardiac pacemakers, and vascular smooth muscle. *Ann. Rev. Pharmacol. Toxicol.* 17:149–66

Frankenhaeuser, B. 1960. Quantitative description of sodium currents in myelinated nerve fibers of *Xenopus laevis. J. Physiol. London* 151:491–501

Frankenhaeuser, B., Hodgkin, A. L. 1957. The action of calcium on the electrical properties of squid axons. *J. Physiol. London* 137:218–44

Fukuda, J., Furuyama, S., Kawa, K. 1977. Calcium dependent action potentials in skeletal muscle fibres of a beetle larva, *Xylotrupes dichtomus. J. Insect Physiol.* 23:367–74

Fukuda, J., Henkart, M. P., Fischbach, G. D., Smith, T. G. 1976. Physiological and structural properties of colchicine-treated chick skeletal muscle cells grown in tissue culture. *Dev. Biol.* 49:395–411

Fukuda, J., Kawa, K. 1977. Permeation of manganese, cadmium, zinc, beryllium through calcium channels of an insect muscle membrane. *Science* 196:309–11

Geduldig, D., Gruener, R. 1970. Voltage clamp of the *Aplysia* giant neurone: Early sodium and calcium currents. *J. Physiol. London* 211:217–44

Geduldig, D., Junge, D. 1968. Sodium and calcium components of action potentials in the *Aplysia* giant neurone. *J. Physiol. London* 199:347–65

Geletyuk, V. I., Veprintsev, B. N. 1972. Electrical properties of neurons of the mollusc *Lymnea stagnalis* under conditions of tissue culture. *Tsitologiya* 14:1133–39

George, W. J., Ignarro, L. J., Paddock, R. J., White, L., Kadowitz, P. J. 1975. Oppositional effects of acetylcholine and isoproterenol on isometric tension and cyclic nucleotide concentrations in rabbit atria. *J. Cyclic Nucl. Res.* 1:339–47

Gerasimov, V. D. 1964. Effect of ion composition of medium on excitation process in giant neurons of snail. *Fiziol. Zh. SSSR* 50:457. Transl. 1965, in *Fed. Proc.* 24:T371–74 (From Russian)

Gerasimov, V. D., Kostyuk, P. G., Maiskii, V. A. 1965. Effect of bivalent cations on the electrical characteristics of the membrane of giant neurones. *Biofizika* 10:447–53. Transl. A. Crozy, 1966, in *Biophysics* 10:494–502 (From Russian)

Giles, W., Tsien, R. W. 1975. Effects of acetylcholine on membrane currents in frog atrial muscle. *J. Physiol. London* 246:64–66P

Goldberg, N. D., O'Dea, R. F., Haddox, M. K. 1973. Cyclic GMP. *Adv. Cyclic Nucl. Res.* 3:155–223

Goodman, C. S., Heitler, W. J. 1979. Electrical properties of insect neurones with spiking and non-spiking somata: Normal, axotomized, and colchicine-treated neurones. *J. Exp. Biol.* 83:95–121

Goodman, C. S., Spitzer, N. C. 1979. Embryonic development of identified neurones: Differentiation from neuroblast to neurone. *Nature* 280:208–14

Gorman, A. L. F., Hermann, A. 1979. Internal effects of divalent cations on potassium permeability in molluscan neurones. *J. Physiol. London* 296:393–410

Gorman, A. L. F., Thomas, M. V. 1978. Changes in the intracellular concentration of free calcium ions in a pacemaker neurone, measured with the metallochromic indicator dye arsenazo III. *J. Physiol. London* 275:357–76

Hagiwara, S. 1973. Ca spike. *Adv. Biophys.* 4:71–102

Hagiwara, S. 1975. Ca-dependent action potential. In *Membranes—A Series of Advances,* Vol. 3, ed. G. Eisenman, 3:359–81. New York: Dekker

Hagiwara, S., Fukuda, J., Eaton, D. 1974. Membrane currents carried by Ca, Sr, and Ba in barnacle muscle fiber during voltage clamp. *J. Gen. Physiol.* 63:564–78

Hagiwara, S., Henkart, M. P., Kidokoro, Y. 1971. Excitation-contraction coupling in amphioxus muscle cells. *J. Physiol. London* 219:233–51

Hagiwara, S., Jaffe, L. A. 1979. Electrical properties of egg cell membranes. *Ann. Rev. Biophys. Bioeng.* 8:385–416

Hagiwara, S., Kidokoro, Y. 1971. Na and Ca components of action potential in amphioxus muscle cells. *J. Physiol. London* 219:217–32

Hagiwara, S., Miyazaki, S. 1977a. Ca and Na spikes in egg cell membrane. *Prog. Clin. Biol. Res.* 15:147–58

Hagiwara, S., Miyazaki, S. 1977b. Changes in excitability of the cell membrane during "differentiation without cleavage" in the egg of the annelid, *Chaetopterus pergamentaceus*. *J. Physiol. London* 272:197–216

Hagiwara, S., Naka, K. 1964. The initiation of spike potential in barnacle muscle fibers under low intracellular Ca^{2+}. *J. Gen. Physiol.* 48:141–62

Hagiwara, S., Nakajima, S. 1966a. Differences in Na and Ca spikes as examined by application of tetrodoxin, procaine, and manganese ions. *J. Gen. Physiol.* 49:793–806

Hagiwara, S., Nakajima, S. 1966b. Effects of the intracellular Ca ion concentration upon the excitability of the muscle fiber membrane of a barnacle. *J. Gen. Physiol.* 49:807–18

Hagiwara, S., Ozawa, S., Sand, O. 1975. Voltage clamp analysis of two inward current mechanisms in the egg cell membrane of a starfish. *J. Gen. Physiol.* 65:617–44

Hagiwara, S., Takahashi, K. 1967a. Resting and spike potentials of skeletal muscle fibers of salt-water elasmobranch and teleost fish. *J. Physiol. London* 190:499–518

Hagiwara, S., Takahashi, K. 1967b. Surface density of calcium ions and calcium spikes in the barnacle muscle fiber membrane. *J. Gen. Physiol.* 50:583–601

Hencek, M., Zachar, J. 1977. Calcium currents and conductances in the muscle membrane of the crayfish. *J. Physiol. London* 268:51–71

Herrera, A. A. 1979. Electrophysiology of bioluminescent excitable epithelial cells in a polynoid polychaete worm. *J. Comp. Physiol.* 129:67–78

Heyer, C. B., Lux, H. D. 1976. Control of the delayed outward potassium currents in bursting pacemaker neurones of the snail, *Helix pomatia*. *J. Physiol. London* 262:349–82

Hille, B. 1975a. Ionic selectivity of Na and K channels. See Hagiwara 1975, pp. 255–323

Hille, B. 1975b. Ionic selectivity, saturation, and block in sodium channels, a four-barrier model. *J. Gen. Physiol.* 66:535–60

Hirst, G. D. S., Spence, I. 1973. Calcium action potentials in mammalian peripheral neurones. *Nature New Biol.* 243:54–56

Hodgkin, A. L., Huxley, A. F. 1952a. Currents carried by sodium and potassium ions through the membrane of the giant axon of *Loligo*. *J. Physiol. London* 116:449–72

Hodgkin, A. L., Huxley, A. F. 1952b. The components of membrane conductance in the giant axon of *Loligo*. *J. Physiol. London* 116:473–96

Hodgkin, A. L. Huxley, A. F. 1952c. The dual effect of membrane potential on sodium conductance in the giant axon of *Loligo*. *J. Physiol. London* 116:497–506

Hodgkin, A. L., Huxley, A. F. 1952d. A quantitative description of membrane current and its application to conduction and excitation in nerve. *J. Physiol. London* 117:500–44

Hodgkin, A. L., Katz, B. 1949. The effect of sodium ions on the electrical activity of the giant axon of the squid. *J. Physiol. London* 108:37–77

Hodgkin, A. L., Keynes, R. D. 1957. Movements of labelled calcium in squid giant axons. *J. Physiol. London* 138:253–81

Horn, J. P., McAfee, D. A. 1979. Norepinephrine inhibits calcium-dependent potentials in rat sympathetic neurons. *Science* 204:1233–35

Horn, J. P., McAfee, D. A. 1980. Alpha-adrenergic inhibition of calcium-dependent potentials in rat sympathetic neurones. *J. Physiol. London* 301:191–204

Horn, R. 1978. Propagating calcium spikes in an axon of *Aplysia*. *J. Physiol. London* 281:513–34

Hudspeth, A. J., Corey, D. P. 1977. Sensitivity, polarity, and conductance change in the response of vertebrate hair cells to controlled mechanical stimuli. *Proc. Natl. Acad. Sci. USA* 74:2407–11

Ikemoto, Y., Goto, M. 1975. Nature of the negative inotropic effect of acetylcholine on the myocardium. *Proc. Jpn. Acad.* 51:501–5

Inomata, H., Kao, C. Y. 1976 Ionic currents in the guinea-pig taenia coli. *J. Physiol. London* 255:347–78

Irisawa, H., Shigeto, N., Otani, M. 1967. Effect of Na^+ and Ca^{2+} on the excitation of the *Mytilus* (bivalve) heart muscle. *Comp. Biochem. Physiol.* 23:199–212

Ito, F., Komatsu, Y. 1979. Calcium-dependent regenerative responses in the afferent nerve terminal of the frog muscle spindle. *Brain Res.* 175:160–64

Iwasaki, S., Satow, Y. 1971. Sodium- and calcium-dependent spike potentials in the secretory neuron soma of the X-organ of the crayfish. *J. Gen. Physiol.* 57:216–38

Jaffe, L. A. 1976. Fast block to polyspermy in sea urchin egg is electrically mediated. *Nature* 261:68–71

Jaffe, L. A., Gould-Somero, M., Holland, L. 1979. Ionic mechanism of the fertilization potential of the marine worm, *Urechis caupo* (Echiura). *J. Gen. Physiol.* 73:469–92

Junge, D. 1967. Multi-ionic action potentials in molluscan giant neurones. *Nature* 215:546–48

Kano, M. 1975. Development of excitability in embryonic chick skeletal muscle cells. *J. Cell. Physiol.* 86:503–10

Kano, M., Shimada, Y. 1973. Tetrodotoxin-resistant electrical activity in chick skeletal muscle cells differentiated in vitro. *J. Cell. Physiol.* 81:85–90

Kano, M., Shimada, Y., Ishikawa, K. 1972. Electrogenesis of embryonic chick skeletal muscle cells differentiated in vitro. *J. Cell. Physiol.* 79:363–66

Kao, C. Y. 1966. Tetrodotoxin, saxitoxin and their significance in the study of excitation phenomena. *Pharmacol. Rev.* 18:997–1049

Kass, R. S., Tsien, R. W. 1975. Multiple effects of calcium antagonists on plateau currents in cardiac purkinje fibers. *J. Gen. Physiol.* 66:169–92

Kater, S. B. 1977. Calcium electroresponsiveness and its relationship to secretion in molluscan exocrine gland cells. *Neurosci. Symp.* 2:195–214

Katz, B., Miledi, R. 1967a. The release of acetylcholine from nerve endings by graded electric pulses. *Proc. R. Soc. London B* 167:23–38

Katz, B., Miledi, R. 1967b. A study of synaptic transmission in the absence of nerve impulses. *J. Physiol. London* 192:407–36

Katz, B., Miledi, R. 1969. Tetrodotoxin-resistant electrical activity in presynaptic terminals. *J. Physiol. London* 203:459–87

Katz, B., Miledi, R. 1971. The effect of prolonged depolarization on synaptic transfer in the stellate ganglion of the squid. *J. Physiol. London* 216:503–12

Kawa, K. 1979. Zinc-dependent action potentials in giant neurons of the snail, *Euhadra quaestia. J. Membr. Biol.* 49:325–44

Kerkut, G. A., Gardner, D. R. 1967. The role of calcium ions in the action potential of *Helix aspersa* neurones. *Comp. Biochem. Physiol.* 20:147–62

Keynes, R. D., Rojas, E. 1973. Characteristics of the sodium gating current in squid giant axons. *J. Physiol. London* 233:28P

Keynes, R. D., Rojas, E., Taylor, R. E., Vergara, J. 1973. Calcium and potassium systems of a giant barnacle muscle fiber under membrane potential control. *J. Physiol. London* 229:409–55

Kidokoro, Y. 1973. Development of action potentials in a clonal rat skeletal muscle cell line. *Nature New Biol.* 241:158–59

Kidokoro, Y. 1975a. Sodium and calcium components of the action potential in a developing skeletal muscle cell line. *J. Physiol. London* 244:145–59

Kidokoro, Y. 1975b. Spontaneous calcium action potentials in a clonal pituitary cell line and their relationship to prolactin secretion. *Nature* 258:741–42

Kidokoro, Y., Hagiwara, S., Henkart, M. P. 1974. Electrical properties of obliquely striated muscle fiber membrane of *Anodonta* glochidium. *J. Comp. Physiol.* 90:321–38

Klee, M. R., Lee, K. C., Matsuda, Y. 1973. Interaction of D-600 and cobalt with the inward and outward current systems in *Aplysia* neurons. *Pflügers Arch.* 343:R60

Klein, M., Kandel, E. R. 1978. Presynaptic modulation of voltage-dependent Ca^{2+} current: Mechanism for behavioral sensitization in *Aplysia californica. Proc. Natl. Acad. Sci. USA* 75:3512–16

Kleinhaus, A. L., Prichard, J. W. 1975. Calcium dependent action potentials produced in leech Retzius cells by tetraethylammonium chloride. *J. Physiol. London* 246:351–61

Kohlhardt, M., Bauer, B., Krause, H., Fleckenstein, A. 1972. Differentiation of the transmembrane Na and Ca channels in mammalian cardiac fibers by the use of specific inhibitors. *Pflügers Arch.* 335:309–22

Koketsu, K., Cerf, J. A., Nishi, S. 1959a. Effect of quaternary ammonium ions on electrical activity of spinal ganglion cells in frogs. *J. Neurophysiol.* 22:177–94

Koketsu, K., Cerf, J. A., Nishi, S. 1959b. Further observations on electrical activity of frog spinal ganglion cells in sodium-free solutions. *J. Neurophysiol.* 22:693–703

Kostyuk, P. G. 1978. *Calcium channels in the nerve cell membrane.* Presented at Int. Biophys. Congr., Kyoto, Japan

Kostyuk, P. G., Krishtal, O. A. 1977. Effects of calcium and calcium-chelating agents on the inward and outward current in the membrane of mollusc neurones. *J. Physiol. London* 270:569–80

Kostyuk, P. G., Krishtal, O. A., Doroshenko, P. A. 1974a. Calcium currents in snail neurones. I. Identification of calcium current. *Pflügers Arch.* 348:83–93

Kostyuk, P. G., Krishtal, O. A., Doroshenko, P. A. 1974b. Calcium currents in snail neurones. II. The effect of external calcium concentration on the calcium inward current. *Pflügers Arch.* 348:95–104

Kostyuk, P. G., Krishtal, O. A., Doroshenko, P. A. 1975a. Outward currents in isolated snail neurones. I. Inactivation kinetics. *Comp. Biochem. Physiol.* 51C:259–63

Kostyuk, P. G., Krishtal, O. A., Doroshenko, P. A. 1975b. Outward currents in isolated snail neurones. II. Effect of TEA. *Comp. Biochem. Physiol.* 51C:265–68

Kostyuk, P. G., Krishtal, O. A., Doroshenko, P. A. 1975c. Outward currents in isolated snail neurones. III. Effect of verapamil. *Comp. Biochem. Physiol.* 51C:269–74

Kostyuk, P. G., Krishtal, O. A., Pidoplichko, V. I. 1975d. Effect of internal fluoride and phosphate on membrane currents during intracellular dialysis of nerve cells. *Nature* 257:691–93

Kostyuk, P. G., Krishtal, O. A., Pidoplichko, V. I. 1977a. Asymmetrical displacement currents in nerve cell membrane and effect of internal fluoride. *Nature* 267:70–72

Kostyuk, P. G., Krishtal, O. A., Shakhovalov, Y. A. 1977b. Separation of sodium and calcium currents in the somatic membrane of mollusc neurones. *J. Physiol. London* 270:545–68

Kostyuk, P. G., Krishtal, O. A., Pidoplichko, V. I., Shakhovalov, Y. A. 1979. Kinetics of calcium inward current activation. *J. Gen. Physiol.* 73:675–77

Krishtal, O. A. 1978. Modification of Ca channels in nerve cell membrane using EGTA effect. *Dokl. Akad. Nauk SSSR* 238:482–85

Krishtal, O. A., Pidoplichko, V. I. 1975. Intracellular perfusion of Helix giant neurons. *Neirofiziologiya* 7:327–29. Transl. 1976, in *Neurophysiology* 7:258–59 (From Russian)

Krishtal, O. A., Pidoplichko, V. I. 1977. Analysis of fluctuations of the current recorded from small areas of nerve cell soma membrane. *Neirofiziologiya* 9:644–46. Transl. 1978, in *Neurophysiology* 9:488–90 (From Russian)

Lee, K. S., Akaike, N., Brown, A. M. 1978. Properties of internally prefused, voltage-clamped, isolated nerve cell bodies. *J. Gen. Physiol.* 71:489–507

Lee, K. S., Weeks, T. A., Kao, R. L., Akaike, N., Brown, A. M. 1979. Sodium current in single heart muscle cells. *Nature* 278:269–71

Llinas, R., Blinks, J. R., Nicholson, C. 1972. Calcium transient in presynaptic terminal of squid giant synapse: Detection with aequorin. *Science* 176:1127–29

Llinas, R., Hess, R. 1976. Tetrodotoxin-resistant dendritic spikes in avian Purkinje cells. *Proc. Natl. Acad. Sci. USA* 73:2520–23

Llinas, R., Steinberg, I. Z., Walton, K. 1976. Presynaptic calcium currents and their relation to synaptic transmission: Voltage clamp study in squid giant synapse and theoretical model for the calcium gate. *Proc. Natl. Acad. Sci. USA* 73:2918–22

Magura, I. S. 1977. Long-lasting inward current in snail neurons in barium solutions in voltage-clamp conditions. *J. Membr. Biol.* 35:239–56

Matsuda, Y., Yoshida, S., Yonezawa, T. 1976. A Ca-dependent regenerative response in rodent dorsal root ganglion cells cultured in vitro. *Brain Res.* 115:334–38

Matsuda, Y., Yoshida, S., Yonezawa, T. 1978. Tetrodotoxin sensitivity and Ca component of action potentials of mouse dorsal root ganglion cells culture in vitro. *Brain Res.* 154:69–82

Matthews, E. K., Sakamoto, Y. 1975. Electrical characteristics of pancreatic islet cells. *J. Physiol. London* 246:421–37

McAfee, D. A., Yarowsky, P. J. 1979. Calcium-dependent potentials in the mammalian sympathetic neurone. *J. Physiol. London* 290:507–23

Meech, R. W. 1974. The sensitivity of *Helix aspersa* neurones to injected calcium ions. *J. Physiol. London* 237:259–77

Meech, R. W., Standen, N. B. 1975. Potassium activation in *Helix aspersa* neurones under voltage clamp: A component mediated by calcium influx. *J. Physiol. London* 249:211–39

Meves, H. 1968. The ionic requirements for the production of action potentials in *Helix pomatia* neurones. *Pflügers Arch.* 304:215–41

Meves, H. 1978. Inactivation of the sodium permeability in squid giant nerve fibers. *Prog. Biophys. Mol. Biol.* 33:207–30

Meves, H., Vogel, W. 1973. Calcium inward currents in internally perfused giant axons. *J. Physiol. London* 235:225–65

Miyazaki, S., Hagiwara, S. 1976. Electrical properties of the *Drosophila* egg membrane. *Dev. Biol.* 53:91–100

Miyazaki, S., Ohmori, H., Sasaki, S. 1975. Action potential and nonlinear current-voltage relation in starfish oocytes. *J. Physiol. London* 246:37–54

Miyazaki, S., Takahashi, K., Tsuda, K. 1972. Calcium and sodium contributions to regenerative responses in the embryonic excitable cell membrane. *Science* 176:1441–43

Moolenaar, W. H., Spector, I. 1979. The calcium current and the activation of a slow potassium conductance in voltage-clamped mouse neuroblastoma cells. *J. Physiol. London* 292:307–23

Moreton, R. B. 1968. Ionic mechanism of the action potentials of giant neurones of *Helix aspersa*. *Nature* 219:70–71

Mudge, A. W., Leeman, S. E., Fischbach, G. D. 1979. Enkephalin inhibits release of substance P from sensory neurons in culture and decreases action potential duration. *Proc. Natl. Acad. Sci. USA* 76:526–30

Nachshen, D. A., Blaustein, M. P. 1979. The effects of some organic "calcium antagonists" on calcium influx in presynaptic nerve terminals. *Mol. Pharmacol.* 16:579–86

Naitoh, Y., Eckert, R., Friedman, K. 1972. A regenerative calcium response in *Paramecium*. *J. Exp. Biol.* 56:667–81

Nastuk, W. L., Hodgkin, A. L. 1950. The electrical activity of single muscle fibers. *J. Cell. Comp. Physiol.* 35:39–73

Nishi, S., Soeda, H., Koketsu, K. 1965. Effect of alkali-earth cations on frog spinal ganglion cell. *J. Neurophysiol.* 28:457–72

North, R. A. 1973. The calcium-dependent slow after-hyperpolarization in myenteric plexus neurones with tetrodotoxin-resistant action potentials. *Br. J. Pharmacol.* 49:709–11

Ochi, R. 1970. The slow inward current and the action of manganese ions in guinea-pig's myocardium. *Pflügers Arch.* 316:81–94

Ochi, R. 1975. Manganese action potentials in mammalian cardiac muscle. *Experientia* 31:1048–49

Ochi, R. 1976. Manganese-dependent propagated action potentials and their depression by electrical stimulation in guinea-pig myocardium perfused by sodium-free media. *J. Physiol. London* 263:139–56

Ogura, A., Takahashi, K. 1976. Artificial deciliation causes loss of calcium-dependent responses in *Paramecium*. *Nature* 264:170–72

Okamoto, H., Takahashi, K., Yamashita, N. 1977. Ionic currents through the membrane of the mammalian oocyte and their comparison with those in the tunicate and sea urchin. *J. Physiol. London* 267:465–95

Okamoto, H., Takahashi, K., Yoshii, M. 1976a. Membrane currents of the tunicate egg under the voltage-clamp condition. *J. Physiol. London* 254:607–38

Okamoto, H., Takahashi, K., Yoshii, M. 1976b. Two components of the calcium current in the egg cell membrane of the tunicate. *J. Physiol. London* 255:527–61

O'Lague, P. H., Potter, D. D., Furshpan, E. J. 1978. Studies on rat sympathetic neurons developing in cell culture. Dev. Biol. 67:384–403

Oomura, Y., Ozaki, S., Maeno, T. 1961. Electrical activity of a giant nerve cell under abnormal conditions. *Nature* 191:1265–67

Palade, P. T., Almers, W. 1978. Slow Na and Ca currents across the membrane of frog skeletal muscle fibers. *Biophys. J.* 21:168a

Palmer, J. F., Slack, C. 1970. Some bio-electric parameters of early *Xenopus* embryos. *J. Embryol. Exp. Morphol.* 24:535–53

Patlak, J. B. 1976. The ionic basis for the action potential in the flight muscle of the fly, *Sarcophaga bullata*. *J. Comp. Physiol.* 107:1–11

Pellmar, T. C. 1980. A transmitter induced calcium current. *Fed. Proc.* In press

Pellmar, T. C., Carpenter, D. O. 1979a. Voltage-dependent calcium current induced by serotonin. *Nature* 277:483–84

Pellmar, T. C., Carpenter, D. O. 1979b. Involvement of cyclic AMP in voltage-dependent calcium current elicited by serotonin. *Neurosci. Abstr.* 5:596

Pitman, R. M. 1979. Intracellular citrate or externally applied tetraethylammonium ions produce calcium-dependent action potentials in an insect motoneurone cell body. *J. Physiol. London* 291:327–37

Ransom, B. R., Holz, R. W. 1977. Ionic determinants of excitability in cultured mouse dorsal root ganglion and spinal cord cells. *Brain Res.* 136:445–53

Redfern, P., Lundh, H., Thesleff, S. 1970. Tetrodotoxin resistant action potentials in denervated rat skeletal muscle. *Eur. J. Pharmacol.* 11:263–65

Reuter, H. 1973. Divalent cations as charge carriers in excitable membranes. *Prog. Biophys. Mol. Biol.* 26:1–43

Reuter, H. 1974. Localization of beta adrenergic receptors, and effects of noradrenaline and cyclic nucleotides on action potentials, ionic currents and tension in mammalian cardiac muscle. *J. Physiol. London* 242:429–51

Reuter, H. 1979. Properties of two inward membrane currents in the heart. *Ann. Rev. Physiol.* 41:413–24

Reuter, H., Scholz, H. 1977. The regulation of the calcium conductance of cardiac muscle by adrenaline. *J. Physiol. London* 264:49–62

Ritchie, A. K. 1979. Catecholamine secretion in a rat pheochromocytoma cell line: Two pathways for calcium entry. *J. Physiol. London* 286:541–61

Rosenberger, L., Triggle, D. J. 1978. Calcium, calcium translocation and specific calcium antagonists. In *Calcium in Drug Action,* ed. G. B. Weiss, pp. 3–31. New York: Plenum

Ross, W. N., Stuart, A. E. 1978. Voltage sensitive calcium channels in the presynaptic terminals of a decrementally conducting photoreceptor. *J. Physiol. London* 274:173–91

Sanchez, J. A., Stefani, E. 1978. Inward calcium current in twitch muscle fibers of the frog. *J. Physiol. London* 283:197–209

Satow, Y., Kung, C. 1979. Voltage sensitive Ca-channels and the transient inward current in *Paramecium tetraurelia. J. Exp. Biol.* 78:149–61

Schwartzkroin, P. A., Slawsky, M. 1977. Probable calcium spikes in hippocampal neurons. *Brain Res.* 135:157–61

Shen, S., Steinhardt, R. A. 1976. An electrophysiological study of the membrane properties of the immature and mature oocyte of the batstar, *Patiria miniata. Dev. Biol.* 48:148–62

Slack, C., Warner, A. E. 1975. Properties of surface and junctional membranes of embryonic cells isolated from blastula stages of *Xenopus laevis. J. Physiol. London* 248:97–120

Spector, I., Kimhi, Y., Nelson, P. G. 1973. Tetrodotoxin and colbalt blockade of neuroblastoma action potentials. *Nature New Biol.* 246:124–26

Spitzer, N. C., Baccaglini, P. I. 1976. Development of the action potential in embryo amphibian neurons in vivo. *Brain Res.* 107:610–16

Stanfield, P. R. 1977. A calcium dependent inward current in frog skeletal muscle fibers. *Pflügers Arch.* 368:267–70

Stefani, E., Uchitel, O. D. 1976. Potassium and calcium conductance in slow muscle fibers of the toad. *J. Physiol. London* 255:435–48

Steinhardt, R. A., Epel, D. 1974. Activation of sea urchin eggs by a calcium ionophore. *Proc. Natl. Acad. Sci. USA* 71:1915–19

Steinhardt, R., Zucker, R., Schatten, G. 1977. Intracellular calcium release at fertilization in the sea urchin egg. *Dev. Biol.* 58:185–96

Stevens, C. F. 1972. Inferences about membrane properties from electrical noise measurements. *Biophys. J.* 12:1028–47

Stuart, A. E., Oertel, D. 1978. Neuronal properties underlying processing of visual information in the barnacle. *Nature* 275:287–90

Sutherland, E. W., Robinson, G. A., Butcher, R. W. 1968. Some aspects of the biological role of adenosine 3',5'-mono-phosphate (cyclic AMP). *Circulation* 37:279–306

Szabo, G. 1977. Electrical characteristics of ion transport in lipid bilayer membranes. *Ann. NY Acad. Sci.* 303:266–80

Takahashi, K., Yoshii, M. 1978. Effects of internal free calcium upon the sodium and calcium channels in the tunicate egg analyzed by the internal perfusion technique. *J. Physiol. London* 279:519–49

Takeda, K. 1967. Permeability changes associated with the action potential in procaine-treated crayfish abdominal muscle fibers. *J. Gen. Physiol.* 50:1049–74

Thompson, S. H. 1977. Three pharmacologically distinct potassium channels in molluscan neurones. *J. Physiol. London* 265:465–88

Tillotson, D. 1979. Inactivation of Ca conductance dependent on entry of Ca ions in molluscan neurons. *Proc. Natl. Acad. Sci. USA* 76:1497–1500

Tillotson, D., Horn, R. 1978. Inactivation without facilitation of calcium conductance in caesium-loaded neurones of *Aplysia. Nature* 273:312–14

Trautwein, W. 1973. Membrane currents in cardiac muscle fibers. *Physiol. Rev.* 53:793–835

Tsien, R. W. 1973. Adrenaline-like effects of intracellular iontophoresis of cyclic AMP in cardiac Purkinje fibers. *Nature New Biol.* 245:120–22

Twarog, B. M. 1967. Excitation of *Mytilus* smooth muscle. *J. Physiol. London* 192:857–68

Van Breemen, C., Farinas, B. R., Casteels, R., Gerba, P., Wuytack, F., Deth, R. 1973. Factors controlling cytoplasmic Ca^{2+} concentration. *Philos. Trans. R. Soc. London B* 265:57–71

Washio, H. 1972. The ionic requirements for the initiation of action potentials in insect muscle fibers. *J. Gen. Physiol.* 59:121–34

Watanabe, A., Tasaki, I., Lerman, L. 1967a. Bi-ionic action potentials in squid giant axons internally perfused with sodium salts. *Proc. Natl. Acad. Sci.* 58:2246–52

Watanabe, A., Tasaki, I., Singer, I., Lerman, L. 1967b. Effects of tetrodotoxin in excitability of squid giant axons in sodium-free media. *Science* 155:95–97

Weidmann, S. 1974. Heart: Electrophysiology. *Ann. Rev. Physiol.* 36:155–69

Weisblat, D. A., Byerly, L., Russell, R. L. 1976. Ionic mechanisms of electrical activity in somatic muscle of the nematode *Ascaris lumbricoides. J. Comp. Physiol.* 111:93–113

Werman, R., Grundfest, H. 1961. Graded and all-or-none electrogenesis in arthropod muscle. II. The effects of alkaliearth and onium ions on lobster muscle fibers. *J. Gen. Physiol.* 44:997–1027

Willard, A. L. 1980. Electrical excitability of outgrowing neurites of embryonic neurones in cultures of dissociated neural plate of *Xenopus laevis. J. Physiol. London* 301:115–28

Yamagishi, S. 1973. Manganese-dependent action potentials in intracellularly perfused squid giant axons. *Proc. Jpn. Acad.* 49:218–22

Zipser, B., Bennett, M. V. L. 1973. Tetrodotoxin resistant electrically excitable responses of receptor cells. *Brain Res.* 62:253–59

Ann. Rev. Neurosci. 1981. 4:127–62
Copyright © 1981 by Annual Reviews Inc. All rights reserved

INTERACTIONS BETWEEN ❖11551
AXONS AND THEIR SHEATH CELLS

Garth M. Bray, Michael Rasminsky, and Albert J. Aguayo

Neurosciences Unit, The Montreal General Hospital, and Department
of Neurology, McGill University, Montreal, Quebec, H3G 1A4, Canada

INTRODUCTION

The demonstration in the early 1950s that myelin is formed by the surface membranes of sheath cells was a turning point in the understanding of axon-sheath cell relationships. The subsequent development of ultrastructural, electrophysiologic, biochemical, and immunologic techniques has made possible the identification of many facets of the complex interdependencies in fibers of both the central and peripheral nervous systems. Furthermore, it is now recognized that nerve fibers both influence and are influenced by cellular and extracellular components of their immediate environment.

In this selective review, we discuss structural and functional relationships of axons and sheath cells in normal and abnormal nerves. Our concern is primarily with axons and Schwann cells in the peripheral nervous system but we also touch on the more complicated relationships involving axons, oligodendrocytes, and other neuroglia in the central nervous system.

STRUCTURE-FUNCTION CORRELATES OF
NORMAL AXON-SHEATH CELL RELATIONSHIPS

In the following section we review axon-sheath cell relationships in mature and developing nerve fibers, emphasizing the structural specializations that are presumed to be functionally significant.

Myelinated Fibers

When a segment of axon membrane is depolarized to threshold for generation of an action potential, there is a transient explosive rise in permeability to sodium ions, which flow inward through the membrane down concentra-

127

tion and voltage gradients. As this current advances through the axoplasm, it depolarizes adjacent segments of axolemma. In unmyelinated fibers, an impulse is propagated continuously along an axon as each segment of membrane generates inward current as a consequence of its depolarization. Conduction velocity in unmyelinated fibers is limited by the large capacitance of the axon membrane that attenuates forward axial flow of current within the axoplasm. The development of myelination is an evolutionary solution to the need for an increase in the velocity of conduction without an excessive increase in fiber diameter.

In myelinated fibers, membrane depolarization leads to generation of inward membrane current only at nodes of Ranvier (Huxley & Stämpfli 1949, Tasaki 1959). Conduction is saltatory from node to node with internodal losses of axial current limited by the relatively low capacitance of myelin. The membrane currents generated at nodes are of much larger density per unit area of membrane than those generated in unmyelinated axons and, as pharmacologic techniques have shown (Ritchie and Rogart 1977), there is a correspondingly high density of sodium channels at nodes of Ranvier. Following passage of an impulse, energy must be expended by the fiber to extrude sodium ions; the restriction of sodium entry to nodes of Ranvier in myelinated fibers thus reduces the metabolic demands on the fiber and is a further evolutionary consideration favoring myelination.

The essential morphologic substrates for saltatory conduction are obvious by light microscopic examination of myelinated fibers. Visualized longitudinally, these fibers show two repeating zones along their lengths: the *nodes of Ranvier* where axonal ensheathment is interrupted and the myelin-ensheathed *internodes*. An internode consists of two small *paranodal regions* and a longer *internodal segment*. Each of these three regions shows characteristic ultrastructural specializations that reflect different axon-sheath cell relationships.

NODES OF RANVIER   The nodal axolemma, delimited by the outer terminal loops of the two adjacent sheath cells (Figure 1), is characterized by distinctive ultrastructural specializations that presumably relate to the function of this region as the site of action potential generation. The nodes of Ranvier, in contrast to the internodal regions, have a population of large particles that cleave with the external surface (E-face) of the freeze fractured axolemma (Rosenbluth 1976, Kristol et al 1978). It is likely that these *E-face particles* are intramembranous components of the sodium channels known to be present in high density at nodal membranes. Further indirect evidence relating the E-face particles and sodium channels comes from correlative studies of the neurogenic electric organ of the knifefish *Sternar-chus albifrons* in which high densities of E-face particles are found at the

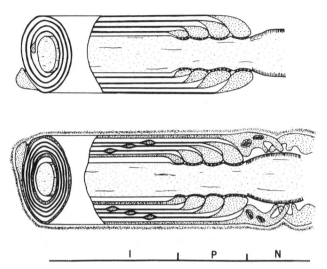

I   P   N

Figure 1 Schematic representation of axon-sheath cell relationships at internodal (*I*), paranodal (*P*), and nodal (*N*) regions of myelinated fibers in the central (*upper*) and peripheral (*lower*) nervous systems. The diagram illustrates (*a*) the apposition of sheath cell plasma membranes to form compact myelin, (*b*) the persistence of sheath cell cytoplasm at Schmidt-Lanterman incisures as well as the inner and outer ensheathing layers in the peripheral nervous system, at the inner and outer tongue processes (seen in diagramatic cross-section) in the central nervous system, and at the paranodal loops in both, (*c*) the narrow paranodal space containing transverse septa, (*d*) the nodal subaxolemmal thickening, and (*e*) the continuous basal lamina and the mitochondria-rich processes of the outer layer of Schwann cell cytoplasm that extend over the nodal region in the peripheral nervous system.

excitable, but not at the inexcitable, nodes of Ranvier (Kristol et al 1977). Smaller collections of E-face particles are also present in the paranodal axolemma between the axo-glial contacts (Rosenbluth 1976); their presence suggests that the actual area of specialized membrane may be greater than the anatomically defined nodal axolemma.

Electrophysiological measurements of membrane current density, pharmacological assays of sodium channel density, and morphologic estimates of E-face particle densities have been made on various peripheral and central nerve fibers. However, because it has not been possible to perform all these measurements on the same population of fibers, the density of E-face particles cannot be precisely related to sodium channel densities.

A distinct population of particles has also been described in the protoplasmic surface (P-face) of freeze fractured nodal membranes (Ellisman 1979). Ellisman suggests that these *P-face particles* are components of the Na^+-K^+ATPase moiety. This enzyme, which maintains the ionic gradients across axon membranes, is present in high concentrations at nodal regions (Wood et al 1977).

Beneath the nodal axolemma there is a dense layer of granular material approximately 20 nm thick (Peters 1966), the *subaxolemmal undercoating,* which has distinct cytochemical properties (Quick & Waxman 1977) and consists of 5 nm filaments (Ellisman 1977). This filamentous material may be part of the cytoskeleton (Ellisman 1977) and is perhaps involved in anchoring the special nodal particles; indeed, in less complex cell systems, it has been demonstrated that the linkage of membrane components and cytoplasmic filaments immobilizes clustered surface particles (Singer et al 1978).

The extent to which the development and maintenance of the nodal differentiation of axons is dependent upon sheath cell contacts remains unresolved although two possibilities have been proposed. Rosenbluth (1976) suggested that glial cell contact is necessary for the specialization of axonal membranes at nodes. Observations in myelin-deficient mutant mice provide support for this hypothesis. In jimpy mice, where central myelin is lacking, regions resembling nodal specializations were not observed (Rosenbluth 1977). In dystrophic mice, where axon segments in the spinal roots lack Schwann cell ensheathment (Bradley & Jenkison 1973) or are only partially enveloped by undifferentiated Schwann cells (Perkins et al 1980a), Rosenbluth (1979) observed aberrant paranode-like contacts between axons and undifferentiated Schwann cells or the Schwann cells of adjacent myelinated fibers. At the sites of such contacts, patches of the subaxolemmal granular material resembling that seen in normal nodes of Ranvier were observed. Ellisman (1976), on the other hand, suggested that nodal specializations can develop independently of sheath cell contact. Evidence for this hypothesis comes from the demonstration of small patches of node-like accumulations of P-face (Ellisman 1976, 1979) and E-face particles (Bray et al 1979) on segments of the unensheathed axons in dystrophic mouse spinal roots. Physiologic studies of experimental demyelination are consistent with both these hypotheses. When the axon-Schwann cell relationship in myelinated fibers is disrupted by segmental demyelination with diphtheria toxin, nodal concentrations of sodium channels appear to be lost (Bostock & Sears 1976, 1978, Bostock 1980). This observation might suggest that sodium channels are locally confined to the nodal region of axon by an influence of Schwann cells. On the other hand, physiologic observations on fibers recovering from lysolecithin demyelination (Bostock 1980) indicate that local accumulation of sodium channels may precede remyelination. This implies that axon membranes may determine the location of nodes and paranodal attachments. Freeze fracture studies of developing, regenerating, and remyelinating nerves may help to clarify this issue.

PARANODAL REGIONS   At paranodes, the axon is contacted by the sequential termination of the myelin lamellae, which separate at their major dense lines to accomodate terminal loops of sheath cell cytoplasm (Figure 1). At these points of contact between axons and the spiral paranodal loops there are specializations within the axon and glial cell membranes as well as in the space between them. The periaxonal space at the paranode region is reduced to approximately 3 nm and is crossed by bands or septa (Peters 1966, Hirano & Dembitzer 1967; Schnapp et al 1976). Within the paranodal sheath cell and axon membranes, examined by freeze-fracture, there is a regular pattern of obliquely oriented ridges and grooves (Livingstone et al 1973, Schnapp et al 1976) that are in register with the transverse septa.

The appearance of the specialized paranodal contacts between axons and sheath cells suggests several possible functions. In many respects, although not all, the axoglial junctions resemble septate junctions (Schnapp et al 1976) that generally function as sites of intercellular adhesion in invertebrates (Gilula 1974). It has also been postulated that the axoglial junctions might restrict E-face particles to the nodal regions of the axolemma (Rosenbluth 1976). Finally, it has been suggested that these junctions might be pathways of low resistance ionic coupling between axons and sheath cells (Livingstone et al 1973), although there is still no evidence for such a gap-junction-like function for paranodes. Nevertheless, the paranodal junctions clearly represent unique specializations that reflect an interaction between axons and sheath cells.

INTERNODAL REGIONS   In the internodal regions, a space approximately 20 nm wide separates the axon from the sheath cell. In freeze-fracture replicas of the internodal axolemma, most particles cleave with the P-face, and the E-face is relatively particle-poor. There are no specialized ultrastructural features along these membranes, although there may be a few "junctional spots" that resemble paranodal attachments (Schnapp et al 1976).

The compaction of glial membranes into myelin provides the internodal axon membrane with insulation of low capacitance; in the peripheral nervous system, the thickness of the normal myelin sheath usually falls within the theoretical optimum for maximum conduction velocity (Waxman 1978). The effectiveness of myelin as an insulator also demands electrical isolation of the nodal and internodal extracellular spaces. The small extracellular space within the internode is divided into two compartments. The *intramyelinic compartment,* represented by the potential space between the intraperiod lines of the myelin sheath, is closed by a continuous system of tight junctions that extend both longitudinally along the inner and outer mesaxons and spirally around the margins of the Schmidt-Lanterman incisures and the paranodal loops (Schnapp & Mugnaini 1975). The internodal *periaxonal compartment* between the axon and myelin-forming sheath cell may communicate with the extracellular environment at the nodes of Ranvier; microperoxidase (Feder et al 1969) and lanthanum (Hirano & Dembitzer 1969) but not horseradish peroxidase (Hirano et al 1969) can pass through the paranodal junctions. However, this narrow channel at the paranodes presumably presents an extremely high resistance to electrical current flow. Thus, during the transmission of nerve impulses in normal myelinated fibers the major route for internodal current leakage from the interior of the axon to the exterior of the fiber is through the membrane capacitance of the axolemma and myelin and not through this resistive pathway at the paranodes.

In the internodal regions of peripheral myelinated fibers, the inner and outer myelin lamellae are not compacted but enclose thin layers (5 to 15 nm) of organelle-free cytoplasm with intervening pockets of organelle-rich cytoplasm (Mugnaini et al 1977). In the central nervous system these inner and outer layers form compact myelin except at the narrow "tongue" processes (Figure 1). In the peripheral nervous system the inner and outer layers of sheath cell cytoplasm are connected by spiral cytoplasmic channels, the Schmidt-Lanterman incisures (Figure 1). This complex continuity between the various cytoplasmic compartments of the myelin-forming sheath cells may serve a role in the transfer of axonally transported phospholipids to myelin. Radioautographic studies suggest that phospholipids, but not proteins or glycoproteins, are transported along axons to the subsurface cisternae; the phospholipids are subsequently observed in the adaxonal cytoplasm of Schwann cells, in the Schmidt-Lanterman incisures, and eventually in myelin (Droz et al 1979). Conversely, labeled proteins move from Schwann cells to the axoplasm, although this has only been demonstrated convincingly in squid giant axons (Lasek et al 1977, Gainer et al 1977).

MYELINATED FIBERS IN THE CENTRAL AND PERIPHERAL NERVOUS SYSTEMS   Although there are certain differences between central myelin, formed by oligodendrocytes, and peripheral myelin, formed by Schwann cells (Peters et al 1976), the basic features of central and peripheral myelinated fibers are similar: Both show repeating zones of nodes, internodes, and paranodes. However, whereas the nodal axolemma in the central nervous system is completely bare, in the peripheral nervous system it is often covered by interdigitating processes from the outer layer of Schwann cell cytoplasm (Berthold 1968). In addition, the basal lamina of peripheral nerve fibers is continuous across the nodal region and the space surrounding the nodal axolemma contains a mucopolysaccharide that binds cations

(Landon & Langley 1971). The functional significance of this nodal gap substance is unknown; it is possible that, together with the mitochondria-rich external layer of Schwann cell cytoplasm in the paranodal region and the paranodal Schwann cell processes (Berthold & Skoglund 1967), it contributes to the maintenance of the ionic environment of the node (Landon & Hall 1976).

Unmyelinated Nerve Fibers

A different but no less complex relationship between axons and Schwann cells exists in peripheral unmyelinated (Remak) fibers (Ochoa 1976, Aguayo et al 1976a). These fibers consist of several small caliber axons embedded partially or completely within a Schwann cell or its processes. The longitudinal configuration of unmyelinated fibers varies continuously due to the regrouping of Schwann cell processes and axons (Aguayo et al 1976a). No distinct junctions or other membrane specializations have been recognized in unmyelinated fibers; the processes of adjacent Schwann cells overlap and interdigitate without forming special contacts, although increased density in the Schwann cell cytoplasm of such regions may be apparent in electron micrographs (Eames & Gamble 1970). By freeze-fracture, the distribution of intramembranous particles along the axons of unmyelinated fibers resembles that of the internodal segments of myelinated fibers with most particles fracturing with the P-face and few with the E-face.

In the central nervous system unmyelinated fibers are totally unensheathed and often grouped in bundles that are partitioned by astrocytic processes (Peters et al 1976). Although the membranes of central unmyelinated axons are more closely apposed than those in peripheral nerves, their large capacitance probably ensures electrical isolation of the adjacent bare axons adequate to preclude cross-excitation (Clark & Plonsey 1970, 1971).

Development

During the development of peripheral and central nerve fibers, there are several axon-sheath cell interactions, including the migration and growth of neurons and glial cells, the proliferation of sheath cells to yield a number appropriate for the ensheathment of all axons, and finally the differentiation and maturation of nerve fibers.

Factors controlling the *migration* of neurons and glial cells are complex and only partially understood. In the central nervous system, radially oriented glial processes guide the migration of postmitotic neurons (Caviness & Rakic 1978) and may also direct growing axons. In certain mammalian and submammalian species, glial cells have surface grooves that may provide directional guidance for developing axons. (Singer et al 1979, Silver &

Sidman 1980). In the peripheral nervous system, such guiding channels have not been identified in vivo although the outgrowth of neurites can be patterned by adhesive substrates in vitro (Letourneau 1975). In addition, factors such as the chemotactic effect of the peripheral target structures and various trophic substances are likely to be important (Varon & Bunge 1978). The best-characterized trophic substance is nerve growth factor (NGF), a specific protein necessary for the survival and growth of certain neurons. Of particular significance in the context of axon-sheath cell interactions is the observation that nonneuronal cells (presumably Schwann cells) can sustain dorsal root ganglion neurons in vitro, even in the absence of NGF (Burnham et al 1972).

Several mechanisms are probably involved in the matching of axon and sheath cell populations. Important among these are the stimulation and curtailment of Schwann cell multiplication as well as the loss of redundant cells. In vitro studies suggest that there are several Schwann cell *mitogens.* The first of these to be demonstrated was contact with axons (Wood & Bunge 1975, McCarthy & Partlow 1976a,b). Homogenates of neurites will also stimulate Schwann cells to divide (Hanson & Partlow 1977, Salzer et al 1977). Extracts of pituitary gland or brain (Raff et al 1978a) or cyclic AMP induction (Raff et al 1978b) have similar effects in vitro. Thus, circulating factors may also stimulate sheath cell multiplication in vivo. The *curtailment* of Schwann cell proliferation is probably influenced by axons as well as by contacts between Schwann cells themselves (Abercrombie & Johnson 1946, Aguayo et al 1976d). Finally, the adjustment of axons and sheath cell populations involves the *death* of redundant axons (Reier & Hughes 1972, Aguayo et al 1973, Fraher 1974) and Schwann cells (Berthold 1973). In the developing central nervous system, it is presumed that axons also influence the proliferation of oligodendrocytes (Skoff 1976), although this relationship is more difficult to document.

During development of the peripheral nervous system, Schwann cell processes segregate axons into increasingly smaller groups (Peters & Muir 1959, Webster 1971, Webster et al 1973). However, the ensuing *differentiation* of Schwann cells into the sheaths of myelinated or unmyelinated fibers is controlled by axons; the undifferentiated Schwann cells are bipotential (Aguayo et al 1976a,b, Weinberg & Spencer 1976). Although sheath cells in both the central and peripheral nervous systems respond to a common axonal signal to form myelin (Aguayo et al 1978a, Weinberg & Spencer 1979), tissue culture studies indicate that axonal signals may be more important for the differentiation of Schwann cells than oligodendrocytes. When isolated Schwann cells are cultured, they stop producing immunocytochemically detectable amounts of the major myelin proteins. In contrast, oligodendrocytes in dissociated cultures continue to make these proteins in the absence of neurons (Mirsky et al 1980). Intrinsic properties of oligodendrocytes and Schwann cells must also be responsible for differences in central and peripheral myelin. Although the precise molecular mechanisms responsible for the differentiation of sheath cells are unknown, glycoproteins are likely candidates as intercellular links in the initiation of myelinogenesis (Quarles et al 1979).

In mature nerves, the size of the myelin sheath, as reflected by internodal length and myelin thickness, is proportional to the diameter of the axon with which it is associated (Vizoso & Young 1948, Bernstein 1966, Friede & Samorajski 1967, Murray & Blakemore 1980). In other words, the axons of myelinated fibers influence both the radial and longitudinal growth of the sheath cell plasma membrane. Although more difficult to document, comparable influences presumably exist in peripheral unmyelinated fibers where Schwann cell territories, indicated by internuclear distances, gradually increase during maturation (Peyronnard et al 1973). In contrast to these expressions of axonal influences on ensheathing cells, there is also evidence, based on experimental studies of normal and abnormal remyelination, that sheath cells influence axonal diameter (Raine et al 1969, Aguayo et al 1979a). Thus, during primary development one can envisage axons and sheath cells increasing in size, each one influencing the other.

The physiological properties of developing nerve fibers have not been studied with contemporary single fiber recording techniques. Early evidence (Landon & Hall 1976, Berthold 1978) indicates that conduction velocity in developing cat myelinated fibers increases with the formation of myelin, but the appearance of a normal refractory period and of the ability of nerves to sustain activity at high frequencies is better correlated with maturation of the morphological and histochemical properties of the nodal region.

EXPERIMENTAL STRATEGIES: METHODS AND MODELS

Because the structures that form the peripheral and central nervous systems are so complexly interwoven, it is difficult to define the exact role of individual cellular and extracellular components. In the following section, some of the research strategies used to overcome these difficulties are described.

Tissue Culture

In vitro studies of axon-sheath cell relationships have been possible since the development of techniques that permit the separation of neurites, glial cells, and fibroblasts (Wood 1976, McCarthy & Partlow 1976a, Mirsky et al 1980). With these technical achievements, neurons and sheath cells have been recombined to study Schwann cell multiplication (Wood & Bunge

1975, McCarthy & Partlow 1976a,b), axonal ensheathment (Bunge & Bunge 1978), and the role of axons in the production of Schwann cell basal lamina (Williams et al 1976). In vitro techniques have also been used to study the pathogenetic mechanisms in nerves of myelin-deficient mouse mutants (Cornbrooks et al 1980, Okada et al 1980).

Nerve Grafts

Experiments using radioactive, immunologic, and cytopathologic markers established that when a segment of one nerve is grafted between the cut ends of another, Schwann cells in the donor segments survive, multiply, and eventually ensheath axons that regenerate from the proximal stump of the recipient nerve (Aguayo et al 1979b). The boundaries between Schwann cells in the host and grafted segments usually remain well delineated in these regenerated nerves, an observation that indicates that there is no substantial migration of Schwann cells (Aguayo et al 1979a). Thus, regenerated grafted nerves represent in vivo combinations of host axons and transplanted Schwann cells, each originating from different experimental animals or from animal and human nerves. It is possible to study interactions between axons and Schwann cells using immune-suppressed animals as recipients and selecting host and donor nerves in a variety of combinations. Such combinations have been used to study Schwann cell differentiation (Aguayo et al 1976b), multiplication (Aguayo et al 1979a), and the pathogenesis of some hereditary disorders of myelination in man (Aguayo et al 1977b, Aguayo et al 1978b, Dyck et al 1979) and experimental animals (Aguayo et al 1977a, 1979a, Scaravilli et al 1980). In addition, regeneration and myelination have been studied in experimental combinations of peripheral nerve axons and transplanted optic nerves (Aguayo et al 1978a, Weinberg & Spencer 1979), as well as in peripheral nerve segments grafted into rat spinal cords (Richardson et al 1980).

Schwann cells grown in tissue culture can also be transplanted into peripheral nerves (Aguayo et al 1979c) and into the demyelinated spinal cords of quaking and normal mice (Duncan et al 1981). In such grafts, the cultured Schwann cells will ensheath and remyelinate central and peripheral axons, which indicates once again that there is an axonal signal for ensheathment and myelination common to both the central and peripheral nervous systems.

Myelin Deficient Mutants

Myelin deficient mutant animals provide natural models of disturbed cell interactions (Guénet 1980). Several mutant mice have been well characterized.

1. In dystrophic mice (*dystrophia muscularis*), axons are bare and lack Schwann cells and myelin ensheathment particularly in segments of the spinal roots (Bradley & Jenkison 1973).
2. Trembler mice have Schwann cells that form little or no myelin (Ayers & Anderson 1973, Low 1977).
3. In quaking mice, there is a generalized deficit of myelin in both the central and peripheral nervous systems (Sidman et al 1964, Friedrich 1975, Suzuki & Zagoren 1977).
4. Jimpy mice are characterized by an extreme lack of central myelin (Skoff 1976).
5. In shiverer mice, central myelin is structurally abnormal and contains no detectable levels of myelin basic protein (Privat et al 1979, Dupouey et al 1979).
6. Twitcher mice show extensive, patchy demyelination in the central and peripheral nervous systems as a result of galactosylceramidase deficiency (Kobayashi & Suzuki 1980).

Mouse Chimaeras

Primary chimaeras—composite organisms in which two or more genotypically distinct cell lines coexist throughout development and in the mature animal (Mullen 1977, Peterson et al 1979)—provide a unique opportunity to assess interactions during primary development. With this technique (which avoids many of the experimental complications caused by surgical manipulation and transplantation as well as the artificial conditions of tissue culture) eight-cell pre-implantation embryos from the myelin deficient mutant mouse, Trembler, and those of normal mice have been aggregated to produce chimaeras whose peripheral nerves contain some Schwann cells that express the Trembler genotype and others that are normal (Rayburn et al 1980). It is anticipated that further studies of such chimaeras will provide new insights into the mechanisms involved in normal and abnormal myelination.

Cell Markers

Cell markers and new immunologic techniques for the identification of different populations of cells, including neurons, Schwann cells, fibroblasts, astrocytes, and oligodendroglial cells, have been used for the study of cell interactions in tissue culture, animal models of disease, and human tissues (Le Douarin 1973, Brockes et al 1977, Livett 1978, Itoyama et al 1980, Mirsky et al 1980).

Demyelination and Remyelination

Study of the structural and functional alterations caused by experimental demyelination has provided insights into the role of some of the components of normally myelinated fibers as well as an appreciation of the pathogenetic mechanisms in demyelinating disorders. Demyelination can be produced by many techniques but a few have been particularly useful in correlative studies of structure and function. The intraneural injection of diphtheria toxin (Webster et al 1961, Cavanagh & Jacobs 1964, Allt & Cavanagh 1969), lysolecithin (Hall & Gregson 1971), or antigalactocerebroside serum (Saida et al 1979) produces, with varying time course, paranodal demyelination that may progress to disruption of complete internodal segments of myelin. Depending on the extent of demyelination, there is either reconstitution of damaged paranodal regions or the development of intercalated internodal segments (Allt 1969).

Electrophysiological Techniques

The relationship between myelination and the distribution of electrical excitability in various types of pathologically ensheathed axons has been explored using a technique for recording membrane current from multiple sites along single nerve fibers in vivo (Rasminsky & Sears 1972, Bostock & Sears 1976, 1978, Rasminsky et al 1978). Foci of excitable membrane are identified as sites of inward membrane current generation. This technique has permitted the recognition of both saltatory (Rasminsky & Sears 1972) and continuous (Bostock & Sears 1976, 1978) conduction in rat spinal root fibers demyelinated with diphtheria toxin.

Voltage clamp techniques, previously only feasible for the study of nonmammalian nerve fibers, have now been used to study normal (Chiu et al 1979, Brismar 1980) as well as pathological mammalian myelinated fibers (Brismar 1979, Chiu & Ritchie 1980).

RESULTS OF EXPERIMENTAL MANIPULATIONS OF AXON-SHEATH CELL RELATIONSHIPS

In the following sections, we review some of the structural and electrophysiological changes that occur in response to pathologic disturbances of axonsheath cell relationships in experimental animals.

Axon Interruption and Regeneration

In the mammalian peripheral nervous system, the responses distal to axonal interruption (Wallerian degeneration) have long been recognized as examples of axon-sheath cell interdependencies (for review see Thomas 1975). When myelinated fibers are interrupted, degeneration of the distal axon segment leads to fragmentation of myelin sheaths. Concomitantly, Schwann cells divide (Bradley & Asbury 1970), probably triggered by products of the degenerating myelin; axon degeneration in unmyelinated (Romine et al

1976) or premyelinated (Salzer & Bunge 1980) nerve fibers produces little Schwann cell proliferation. During Wallerian degeneration in the central nervous system, most of the dividing cells in adult animals are astrocytes or microglia rather than oligodendrocytes (Skoff 1975).

Effective regeneration of interrupted axons depends upon several events: survival of the neuronal perikaryon, sprouting of axons in the proximal stump and their subsequent elongation, axon ensheathment and increase in axon calibre, and, finally, the reestablishment of terminal contacts and the loss of redundant axon branches. In the peripheral nervous system, this sequence of events leads to the restoration of nerve fiber structure (Cragg & Thomas 1964, Diamond & Jackson 1980) and function (Burgess & Horch 1973, Burgess et al 1974). When nerve fibers are transected in the mammalian central nervous system, there is axonal sprouting, but significant elongation does not occur and regeneration is abortive (Ramón y Cajal 1928). Although many mechanisms probably contribute to the failure of effective regeneration in the central nervous system (Cotman 1978), experimental evidence suggests that interactions between axons and the neuroglial environment may play an important role. When peripheral nerves are grafted into the transected spinal cords of rats, axons from neurons within the cord grow along these Schwann cell grafts (Richardson et al 1980). Furthermore, axons that regenerate normally when ensheathed by Schwann cells fail to grow into central glia. Two examples may be cited. First, when segments of optic nerve are grafted between the cut ends of sciatic or sural nerves in rats or mice (Aguayo et al 1978a, Weinberg & Spencer 1979), few regenerating PNS axons enter the CNS graft and those that do rarely penetrate it for more than a few millimeters; in contrast, PNS axons will regenerate through peripheral nerve grafts with little difficulty (Aguayo et al 1979b). Second, injured dorsal root axons regenerate as far as the PNS-CNS boundary but most do not enter the spinal cord (Stensaas et al 1979; T. Carlstedt, C. S. Perkins, K. Mizuno, A. J. Aguayo, unpublished observations). Thus, it is possible that differences in the elongation of axons after injury in the central and peripheral nervous systems are less dependent upon intrinsic properties of the axons themselves than upon the different populations of neuroglia with which they are associated. In contrast to the situation in mammals, CNS regeneration in amphibians and fish is more effective. It has been suggested that in these submammalian species glial cells may provide channels that permit appropriate axonal regrowth (Singer et al 1979, Michel & Reier 1979).

The differentiation of myelinated and unmyelinated nerve fibers is another aspect of regeneration that has been investigated in peripheral nerves. When axons regenerate from the proximal stump of crushed or transected nerves, they are eventually sorted and ensheathed in a manner that recapitulates development. Using nerve transplantation techniques (Aguayo et al 1976b,c), as well as cross-union between myelinated and unmyelinated nerves (Weinberg & Spencer 1975, 1976, Aguayo et al 1976c), it has been demonstrated that Schwann cells in regenerating nerves are signaled by axons to differentiate into the sheaths of unmyelinated (Remak) or myelinated fibers. Axons also appear to be necessary for the survival of Schwann cells; if regenerating axons are prevented from entering the distal stumps of transected nerves, columns of denervated Schwann cells persist for only a limited period of time (Weinberg & Spencer 1978).

Genetic Alterations of Myelination

Mutant mice with disorders of myelination in the central and peripheral nervous systems present a spectrum of major alterations in axon-sheath cell interactions (Aguayo et al 1979a, Guénet 1980). In the following section, some of the manifestations of these altered relationships are discussed.

TREMBLER MICE   In Trembler, a dominantly inherited disorder that results from a mutation on chromosome 11, myelin is either abnormally thin, poorly compacted, or totally absent in the peripheral nervous system (Ayers & Anderson 1973, 1976, Low 1977). The general appearance of peripheral nerves in this mutant is similar to that observed in certain human hypomyelinating neuropathies (Lyon 1969, Kennedy et al 1977). Because of persistent multiplication, populations of Schwann cells in Trembler nerves increase rapidly during the first month after birth to levels nearly ten times those of normal controls, in which Schwann cell multiplication is negligible after two weeks of age. In addition to the developmental inability to form appropriately thick myelin sheaths, there is also evidence for myelin breakdown in Trembler nerves. In contrast to these abnormalities of myelinated fibers, the morphology and density of Schwann cells is normal in unmyelinated (Remak) fibers of Trembler peripheral nerves (Perkins et al 1980a).

Experimental nerve grafts in mature animals (Aguayo et al 1979b), in vitro cultures of Schwann cells (Cornbrooks et al 1980), and Trembler-normal chimaeras (Rayburn et al 1980) have demonstrated that the abnormalities in Trembler nerves are due to a primary disorder of Schwann cells rather than axons. In the nerve graft experiments, the myelin deficit found in Trembler nerves was reproduced by Trembler Schwann cells transplanted into nerves of normal mice, whereas in reciprocal experiments, axons from Trembler animals were myelinated fully by grafted normal Schwann cells (Figure 2). The heightened proliferation of Schwann cells was also reproduced in regenerated Trembler grafts but not in the proximal or distal stumps of recipient normal nerves (Perkins et al 1981a). Because

PROXIMAL   GRAFT   DISTAL

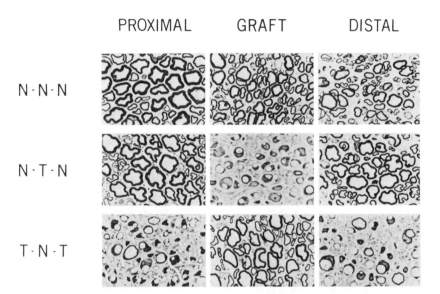

Figure 2 Cross-sections of proximal stumps, grafts, and distal stumps of regenerated nerves 4 months after grafting. In normal combinations (N-N-N), the graft and distal stump contains many regenerated fibers that resemble the intact fibers in the proximal stump. When Trembler mouse nerves are grafted into normal nerves (N-T-N), the proximal and distal stumps are myelinated as in the control N-N-N nerve, but fibers in the graft lack myelin or are hypomyelinated. Normal nerve segments grafted into Trembler nerves (T-N-T) show a marked myelin deficit in the host stumps but the grafted segment is normally myelinated. (Phase micrographs, X 500. Reproduced from Aguayo et al 1979a, with permission of the Annals of the New York Academy of Sciences.)

collagen and other elements in the neural environment influence myelination (Bunge & Bunge 1978), the possibility that these factors also contribute to the pathogenesis of the Trembler abnormality could not be excluded from the nerve graft experiments in which collagen, fibroblasts, and perineural cells were transplanted together with the Schwann cells. The primacy of the Schwann cell abnormality has now been confirmed by tissue culture experiments utilizing pure preparations of Trembler Schwann cells which, when combined with normal axons, reproduced the Trembler abnormality (Cornbrooks et al 1980). Finally, the Trembler ↔ normal chimaeras demonstrate that the Trembler phenotype is expressed even by Schwann cells that have undergone their entire primary development in association with normal Schwann cells and axons (Rayburn et al 1980; Figure 3).

Although Schwann cells in unmyelinated fibers of Trembler mice are phenotypically normal, the characteristic Trembler abnormality of hypomyelination and increased Schwann cell proliferation was replicated after transplantation of the Trembler cervical sympathetic trunk, an un-

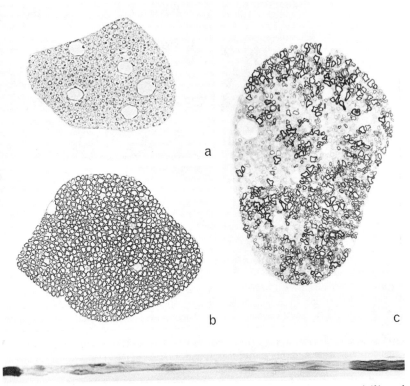

Figure 3 Transverse sections of lumbar ventral roots from Trembler (*a*), normal (*b*), and Trembler ↔ normal chimaera (*c*) mice. The characteristic myelin deficit seen in Trembler nerves is reproduced by fibers grouped in several irregular patches in the chimaera root. A single teased nerve fiber (*d*) from the chimaera is ensheathed by both Trembler and normal myelin-producing Schwann cells. (Magnifications: *a*, *b* X 150, *c* X 240, *d* X 640. Photographs courtesy of H. Rayburn.)

myelinated nerve, into the richly myelinated sural nerves of normal mice (Perkins et al 1981b). When cervical sympathetic trunks from normal animals are grafted into normal sural nerves, the regenerated nerve grafts contain fibers that are normally myelinated (Aguayo et al 1976b). Thus, although the abnormal gene is present in all Trembler Schwann cells, the Trembler phenotype is only expressed when these cells are challenged by axons to form myelin. This observation indicates that Trembler Schwann cells are able to ensheath axons in unmyelinated (Remak) fibers and surround single axons as premyelinated fibers, but are unable to differentiate beyond the stage of primary ensheathment to myelin formation.

Rates of Schwann cell multiplication, determined in nerves that are myelinated in normal animals, are increased in Trembler nerves (Perkins et al 1981a). However, the number of Schwann cells in these nerves remains stable, although at levels greater than those of controls (Perkins et al 1981a). The absence of further increases in the numbers of Trembler Schwann cells, despite their continuing multiplication, suggests an increased rate of Schwann cell death. The trigger for Schwann cell proliferation in these Trembler nerves may be the breakdown products of myelin (Hall & Gregson 1975). Alternatively, if the putative mitogen present on the surface of axons (Wood & Bunge 1975) is exposed by recurrent demyelination and the death of ensheathing cells, the enhanced Schwann cell mitosis could also be triggered by axons. The presence of this effect does not make it necessary to postulate an abnormality of the Trembler axons themselves because Trembler Schwann cells continue to divide even when experimentally transplanted into normal nerves (Perkins et al 1981a).

Transplantation experiments using Trembler and normal nerves have also provided insights into Schwann cell influences on axon diameter (Aguayo et al 1979a). The caliber of Trembler axons in unoperated nerves is nearly one half that of normal axons but returns to normal when the axons of this mutant are ensheathed by normal Schwann cells in nerve grafts. Conversely, when Trembler Schwann cells ensheath normal axons, axon diameters are decreased within the grafted segment. These findings suggest that axon diameters depend not only on influences that arise from the perikaryon and peripheral fields of innervation but also on local interactions with their ensheathing cells.

Functional consequences of the introduction of normal myelin-forming cells into Trembler nerves have also been documented (Pollard & McLeod 1980). Intact Trembler sciatic nerves conduct impulses very slowly but the conduction velocity within the graft is greatly increased three months after transplantation of normal Schwann cells into Trembler nerves.

Thus, the Trembler mutant mouse is an important example of a genetic disorder in which the cellular interdependencies in peripheral nerve fibers are disrupted by the presence of a primary genetic disorder involving one component of these fibers, the Schwann cell. Although the Schwann cell's failure to differentiate beyond primary ensheathment of axons is expressed as an inability to form and sustain myelin, this defect only becomes apparent when Trembler Schwann cells associate with axons that require myelination. Once the Trembler Schwann cell has been challenged to form myelin, a sequence of secondary changes is set in motion: demyelination, Schwann cell multiplication, Schwann cell death, and reduction of axon caliber.

QUAKING MICE   Animals affected with the quaking autosomal recessive mutation on chromosome 17 show generalized hypomyelination which, unlike that in Trembler mice, involves both the central and peripheral nervous systems (Sidman et al 1964, Samorajski et al 1970, Friedrich 1975, Suzuki & Zagoren 1977).

In the spinal roots of quaking mice, axon populations are normal but Schwann cell numbers are increased as a result of continuing proliferation (Aguayo et al 1979b). Schwann cells in unmyelinated (Remak) fibers of quaking mice are morphologically normal and are not increased in number. Experiments involving nerve grafting have shown that transplanted quaking Schwann cells fail to produce sufficient myelin when they ensheath normal axons; conversely, normal Schwann cells myelinate quaking mouse axons normally (Aguayo et al 1979b). These observations suggest that in quaking, as in Trembler, a primary Schwann cell abnormality is responsible for the peripheral neuropathy. Indirect evidence for a similar alteration in oligodendroglial cells is provided by the demonstration that CNS axons become normally myelinated when rat Schwann cells, cultured in vitro, are transplanted into the spinal cords of quaking mice (Duncan et al 1981). The formation of normal myelin by these exogenous Schwann cells suggests that CNS axons or more general conditions in the animal are not responsible for the myelin deficit in the quaking central nervous system.

The defect that leads to the hypomyelination in quaking peripheral nerves probably involves a later step in the formation of myelin than that which is affected in Trembler mice. In quaking nerves, the deficiency of myelin is less marked and becomes apparent only when, with maturation, a thicker myelin sheath is required. The generalized character of the myelin deficiency in quaking mice also suggests that the affected gene is normally involved in myelin formation by both Schwann cells and oligodendrocytes.

DYSTROPHIC MICE   Dystrophic mice have a mutation on chromosome 10 that affects several tissues including muscles and peripheral nerves. In the peripheral nerves, there is a spectrum of alterations in axonal ensheathment. The naked axons in the spinal roots of these animals represent the ultimate failure in the development of axon-Schwann cell relationships (Bradley & Jenkison 1973). The naked axonal segments tend to occur in groups so that the membranes of adjacent axons are closely apposed. In addition to the focal absence of Schwann cells, some axons are surrounded by inappropriately thin myelin (Bradley & Jenkison 1973), some have long nodal gaps (Jaros & Bradley 1979), and most show discontinuities in their basal laminae (Madrid et al 1975). Some fibers in the dorsal spinal roots are myelinated by oligodendroglial cells (Weinberg et al 1975), while in both the ventral and dorsal roots there are many undifferentiated cells that continue to divide but may die prematurely (Perkins et al 1980a).

Although the abnormality in the dystrophic spinal roots can be regarded as a localized failure of Schwann cells to ensheath axons, the specific mechanisms responsible for this disturbance have not been determined. When dystrophic spinal roots are transplanted into peripheral nerves of normal or dystrophic mice, the degree of myelination after regeneration is indistinguishable from that produced by transplantation of normal spinal roots (Aguayo et al 1979a). Furthermore, the undifferentiated cells that continue to multiply but fail to ensheath axons normally in the dystrophic spinal roots (Perkins et al 1980a) will myelinate axons in nerve grafts (Perkins et al 1980b). This suggests that the pathogenesis of the dystrophic spinal root abnormality may be more complex than that of Trembler or quaking mutants. Perhaps local axonal factors or extracellular elements such as the collagen matrix are involved (Bunge & Bunge 1978).

JIMPY MICE   In jimpy, an X-linked mutation, there is a severe lack of CNS myelin but peripheral nerves appear normal (Sidman et al 1964). Affected animals rarely survive beyond four weeks of age. The precise mechanisms leading to this abnormality are unknown. Although most studies suggest that oligodendroglial cells are responsible for the deficiency of myelin, astrocyte responses and axonal influences may play a role. Oligodendrocytes are reduced in number and some show signs of active degeneration (Meier & Bischoff 1975, Privat et al 1972). Astrocytic hyperplasia may also prevent the development of normal axon-oligodendrocyte interactions in jimpy mice (Skoff 1976). Finally, the demonstration of a smaller diameter of axons in jimpy optic nerves (Webster & Sternberger 1980) suggests that jimpy neurons may fail to stimulate oligodendroglial cells to myelinate normally. Alternatively, it is possible that the changes in axonal diameter may be secondary to a glial cell deficit, such as that seen in Trembler nerves.

SHIVERER MICE   In shiverer, a recessively inherited disorder, myelin in the central nervous system is poorly compacted, lacks major dense lines, and contains no detectable levels of myelin basic protein (Bird et al 1977, Privat et al 1979, Dupouey et al 1979). Although myelin basic protein is also absent in the peripheral nervous system of these animals, the morphology of myelin in peripheral nerve fibers is essentially normal (Kirschner & Ganser 1980). Thus, although myelin basic protein appears to be a prerequisite for the compaction and stability of CNS myelin, it may fulfill a different function in the peripheral nervous system.

In the central nervous system of shiverer mice, there is also an abnormal proliferation of paranode-like junctions along axons (Rosenbluth 1979). The elucidation of the cytopathogenesis of this striking abnormality of axon-sheath cell contacts should contribute to the understanding of the mechanism involved in the development of normal paranodes.

TWITCHER MICE   Homozygous twitcher mice appear normal at birth but lose weight, develop a tremor, become progressively weak, and die by three months of age. The central nervous system shows demyelination, gliosis, and axonal degeneration. In peripheral nerves, there is also demyelination and remyelination. Macrophage-like cells containing a variety of inclusions are seen in both the central and peripheral nervous systems (Duchen et al 1980). These histopathologic findings as well as the demonstration that galactosyl ceramidase is deficient in these animals (Kobayashi & Suzuki 1980) indicate that twitcher represents a murine counterpart of the human and canine globoid leukodystrophies.

Transplantation of twitcher nerve segments into normal litter mates has resulted in a replication of the abnormalities that are characteristic of the peripheral nerve in this mutant (Scaravilli et al 1980).

AXON-SHEATH CELL INTERACTIONS IN MUTANT MICE   The abnormalities in the mutant mice discussed in the preceding paragraphs illustrate alterations at different stages of ensheathment and myelination. In dystrophic mice, some axonal segments remain totally unensheathed because their initial association with Schwann cells is defective. In Trembler mice, primary ensheathment is possible but myelin formation is limited or absent. In quaking mice, the initial stages of myelin formation are normal but maturation of myelin sheaths is defective. Finally, in shiverer mice there is a defective production of a specific myelin protein, and in twitcher mice a lysosomal enzyme is deficient.

In addition to these defects at specific stages of ensheathment and myelination during primary development, a common characteristic of these mutants is the inability to sustain the myelin that is formed. Thus, to varying extents, most of these mutant animals show evidence of myelin breakdown.

Physiological Consequences of Disordered Axon-Sheath Cell Relationships

The development of electrophysiologic techniques that permit the recording of external longitudinal current in single nerve fibers, together with the introduction of methods that allow determinations of ion channel distributions in axon membranes, has given new insight into the effects of sheath cells on axonal function. In this section we shall review the effects of demyelination and remyelination on conduction of nerve impulses, the relationships between myelination and ion channel distribution in axon membranes, and the electrical instability of axon membranes in demyelinated and abnormally myelinated axons.

CONDUCTION OF NERVE IMPULSES IN DEMYELINATED AND REMYELINATED NERVE FIBERS   In nerve fibers demyelinated by diphtheria toxin, loss of the insulation provided by myelin results in decreased conduction velocity and, as an extreme consequence, conduction block (McDonald 1963, Rasminsky & Sears 1972). In most diphtheria-toxin demyelinated fibers, conduction remains saltatory (Rasminsky & Sears 1972, Bostock & Sears 1978). Although it has not been feasible to determine the structural correlates of the specific conduction changes in individual fibers, it is likely that slow saltatory conduction occurs when there is partial paranodal demyelination or remyelination, whereas continuous conduction may reflect the disruption of whole internodal segments of myelin.

In contrast to the conduction changes that develop several days after the injection of diphtheria toxin, conduction slowing and block occur within a few hours after the topical application of antigalactocerebroside serum (Rasminsky & Sumner 1980). In this situation, where the paranodal regions are damaged during the first few hours after exposure to the antiserum (Brown et al 1980), conduction block is probably due to an increase in the nodal surface area associated with the paranodal demyelination. In the more chronic demyelination produced by diphtheria toxin, the precise relationship between structure and function is much less clear, perhaps because of a reorganization of axon membrane properties (see below).

In addition to changes in the velocity and mode of impulse conduction, demyelinated nerve fibers also show increases in refractory period, inability to transmit trains of impulses, posttetanic depression, and susceptibility to conduction block with small changes in metabolic environment (Rasminsky 1978a).

After remyelination, internodal distances are shorter than normal (Fullerton et al 1965, Gledhill et al 1973) but conduction velocities return to approximately normal (Morgan-Hughes 1968, Kraft 1975). These observations are consistent with computer simulations that suggest that within broad limits about the normal, internodal length has relatively little influence on conduction velocity (Brill et al 1977). In the CNS, remyelination is also accompanied by the restoration of secure conduction (Smith et al 1979).

DISTRIBUTION OF ION CHANNELS IN ABNORMALLY ENSHEATHED AXON MEMBRANES   The density of *sodium channels* at nodes of Ranvier associated with the high density inward membrane current carried by sodium ions during the rising phase of the action potential has now been determined by ^{3}H-saxitoxin binding studies. With this technique, sodium channel density in normal rabbit peripheral nerves has been estimated to be 10,000 to 12,000/μm^2 for nodes of Ranvier (Ritchie & Rogart 1977, Ritchie 1979), less than 25/μm^2 for internodal axon membranes (Ritchie & Rogart 1977), and 110/μm^2 for unmyelinated (Remak) fibers (Ritchie, Rogart & Strichartz 1976). Although unmyelinated fibers are able to conduct impulses, it is not known if the low density of sodium channels along the internodal axon membranes would permit impulse conduction if the myelin sheath were removed.

The following lines of evidence suggest that the density of sodium channels in axon membrane does indeed undergo local changes as a consequence of alterations in axon ensheathment:

1. Continuous conduction is observed in some fibers segmentally demyelinated by diphtheria toxin (Bostock & Sears 1976, 1978). This may require increased numbers of internodal sodium channels, either by the redistribution of preexistent channels or by the insertion of new ones. In the continuously conducting portions of fibers demyelinated with diphtheria toxin, peaks of inward membrane current density corresponding to the sites of former nodes are not found (Bostock & Sears 1976, 1978, Bostock 1980), which indicates that the high density of sodium channels previously present in the nodal axolemma disappears at this stage of demyelination.

2. In fibers that retain slow but saltatory conduction after exposure to diphtheria toxin and are presumably paranodally demyelinated, nodal capacitance is increased, reflecting the greater area of axolemma in the nodal region. However, because the rise time of action potentials at such nodes is not invariably increased, Bostock & Sears (1978) suggest that increased numbers of sodium channels at nodes may compensate for paranodal demyelination. Such remodeling of the paranodal region may depend on the rate, extent, or specific type of demyelination; increases in nodal sodium currents are not observed when the nodal axon surface area is increased acutely (Chiu & Ritchie 1980).

3. Continuous conduction is observed in the axon segments that lack Schwann cell ensheathment, and therefore myelin, in the spinal roots of dystrophic mice (Rasminsky et al 1978). In these experiments, the spatial resolution for the successive measurements of membrane current was approximately 250 μm. More closely spaced sites of inward membrane current generation, perhaps corresponding to the node-like patches of E-face particles seen in freeze-fracture replicas of the unensheathed axons in dystrophic roots (Bray et al 1979), would not have been identified.

4. During the early stages of regeneration after crush injury when most axons are not myelinated, conduction is continuous rather than saltatory

(Bostock et al 1977, Sears 1979). As with studies of the dystrophic mouse axons, the spatial resolution of these experiments was about 250 μm, leaving some question about the use of the term "continuous conduction" with respect to regenerating axons. Nonetheless, if sodium channels in myelinated fibers are indeed virtually absent from mature internodal axon membranes (Ritchie & Rogart 1977), the presence of electrical excitability prior to myelination indicates that a redistribution of these channels may occur with myelination.

5. In remyelinated or regenerated nerves, internodal distances are shorter than normal, requiring the formation of new segments of nodal membrane. The development of new foci of excitable membrane has now been demonstrated in peripheral nerve fibers following lysolecithin demyelination (Bostock 1980; H. Bostock, S. Hall and K. J. Smith, personal communication, 1980). New foci of inward membrane current generation are observed four days after the injection of lysolecithin at a time when demyelinated axons are ensheathed by a thin layer of Schwann cell cytoplasm but no myelin (Hall & Gregson 1971). Initially these foci are separated by about 150 μm. Over the next three to four weeks, the average distance between these sites gradually increases, further indicating that remodeling of the axon membrane occurs as remyelination progresses. The demonstration of the very close spacing of foci of inward membrane current in the lysolecithin-demyelinated axons was dependent upon technical improvement of the recording system to a spatial resolution of about 100 μm, but even under these conditions, conduction in some diphtheria-demyelinated axons still appeared to be continuous (Bostock 1980), which suggests that sodium channels in such axons are either evenly distributed or aggregated in patches less than 100 μm apart.

The distribution of *potassium channels* in myelinated nerve fibers is also influenced by axon-Schwann cell relationships. In amphibian nerves, the nodal axolemma contains potassium channels. During the action potential at nodes of Ranvier, a late phase of outward potassium current follows the early inward sodium current (Hille 1977). In mammalian nerves, on the other hand, potassium channels are present along the paranodal and/or internodal axolemma but not at nodes. In studies of voltage-clamped normal mammalian nodes, outward potassium current is small or absent (Horáckova et al 1968, Chiu et al 1979, Brismar 1980). In addition, 4-aminopyridine, a specific blocker of potassium channels, does not alter the shape of the propagated action potential in normal mammalian myelinated fibers (Sherratt et al 1980). However, in mammalian nerves, potassium channels are present in unmyelinated (Remak) fibers (Sherratt et al 1980) and are observed in myelinated fibers after experimental lesions that presumably disrupt the paranodal junctions. During acute osmotic changes, mammalian nodes of Ranvier suddenly develop large, late potassium currents concommitant with the increase in nodal capacitance, which signifies the exposure of a large area of axon membrane in the nodal region (Chiu & Ritchie 1980). Furthermore, axons demyelinated with diphtheria toxin show a late outward current that is blocked by 4-aminopyridine (Sherratt et al 1980). The potassium currents that have been observed in voltage-clamp studies of nodes of Ranvier in alloxan diabetic rats (Brismar 1979) suggest that axon-Schwann cell contacts may also be disturbed in these nerves.

These physiologic and pharmacologic studies of sodium and potassium channels suggest that in mammalian myelinated fibers the axon membrane is divided into two distinct domains: (a) the nodal region, specialized to generate action potentials, characterized by a high density of sodium channels and a low density of potassium channels and (b) the internode, characterized by low densities of sodium channels and, at least in its paranodal regions, higher densities of potassium channels. The significance of the internodal potassium channels is at present unclear.

Plasticity of ion channel distribution may ultimately prove to have important clinical implications. Redistribution of sodium channels in demyelinated axons could represent a recovery mechanism following acute demyelination in multiple sclerosis (Ritchie & Rogart 1977, Bostock & Sears 1978). Use of potassium channel blocking drugs such as 4-aminopyridine to prolong nodal action currents could be expected to enhance conduction in demyelinated axons while having no deleterious effect on normally myelinated axons (Sherratt et al 1980). Clinical trials of such agents have not been reported.

ABNORMAL EXCITABILITY   An important physiologic property of normally myelinated nerve fibers is the resting stability of their nodal axon membranes. Impulses are normally generated only at specialized portions of nerve cells—the initial segments of axons and sensory terminals. However, disorders of axon ensheathment, either congenital or acquired, are associated with an instability of axon membranes manifested by ectopic impulse generation. In the spinal roots of dystrophic mice, the following three types of such abnormal impulse generation have been recognized (Huizar et al 1975, Rasminsky 1978b, 1980a,b):

1. Impulses may arise *spontaneously* in mid-root and be propagated toward both the spinal cord and the periphery (Figure 4). Such spontaneous activity may be generated as single impulses, short bursts with frequencies up to 100 Hz, or continuous activity persisting for several minutes (Rasminsky 1978b).
2. Some ectopic sites are stimulated to generate bursts of impulses by the passage of a single impulse along the same fiber, i.e. *autoexcitation.* The

interval between passage of the stimulating impulse and initiation of a subsequent burst of impulses from the ectopic site may be more than 100 msec (Rasminsky 1980a).

3. Other impulses are generated ectopically by *ephaptic transmission* (cross talk) between adjacent single fibers (Figure 4). In some instances, the direction of transmission is clearly from a naked axon to one which is myelinated. At least some ephapses are capable of transmitting impulses at frequencies up to 70 Hz (Rasminsky 1980a).

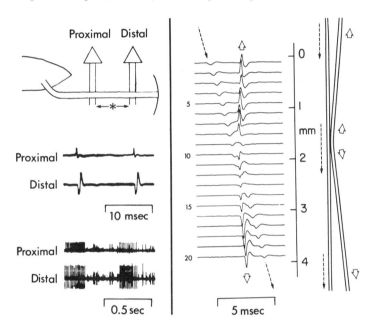

Figure 4   Ectopic excitation and ephaptic transmission in dystrophic mouse ventral roots. *Left:* Recordings of spontaneous activity are made biphasically from pairs of wire electrodes applied to the root; upward and downward deflections reflect impulses in single fibers propagating toward and away from the spinal cord, respectively. Impulses originating in mid-root (*) travel in opposite directions and propagate past the two pairs of recording electrodes (*upper traces*). Bursting of the single unit shown in the *upper traces* is shown at slower sweep speed in the *lower traces*. *Right:* Biphasic recordings are obtained at intervals of 200 μm along the root. From top to bottom, the recordings are progressively further away from the spinal cord. An impulse in one fiber propagates away from the spinal cord (downward deflection in record 1 with progressively greater latency at each successive recording site) (*dashed arrows*). A second fiber is ephaptically excited in mid-root near recording site 9, and an impulse in this fiber (*open arrows*) is propagated back toward the spinal cord (upward deflections with progressively greater latencies in records 8 to 1) and towards the periphery (downward deflection with faster rise times and progressively greater latencies in records 10 to 20). [Adapted from Rasminsky (1978b, 1980a) with permission of the Annals of Neurology and the Journal of Physiology.]

Demyelinated axons do not demonstrate the same level of spontaneous ectopic activity as dystrophic mouse spinal root axons, but both peripheral and central nervous system demyelination causes an increased sensitivity of axons to mechanical stimulation (Howe et al 1977; W. I. McDonald and K. J. Smith, personal communication, 1980). Spontaneous activity and ephaptic transmission are also seen in peripheral nerve fibers of Trembler mice (Rasminsky 1980b) and in experimental traumatic neuromas in rats and mice (Wall & Gutnick 1974, Govrin-Lippmann & Devor 1978, Seltzer & Devor 1979, David & Aguayo 1980).

At present there is no convincing explanation for the increased excitability of abnormally myelinated fibers, but the phenomenon presumably reflects changes in axon membrane properties secondary to abnormal axon-sheath cell interactions. For example, it is possible that the node-like patches of E-face particles seen in the unensheathed axons of dystrophic mouse spinal roots (Bray et al 1979) could be implicated in the ectopic generation of nerve impulses (Rasminsky 1980b). Whatever its explanation, the increased excitability of pathological nerve fibers is also of interest in a clinical context. Demyelination of peripheral or central axons is frequently associated with positive sensory and motor symptoms (Rasminsky 1978b, 1980a,b, McDonald 1980). Small groups of abnormally myelinated peripheral or central nerve fibers could act as pathologic amplifiers of normally and ectopically generated impulses both spatially by ephaptic excitation and temporally by autoexcitation and reexcitation via reciprocal ephapses.

NERVE FIBERS AND THEIR METABOLIC ENVIRONMENT

In previous sections we have reviewed some of the relationships between axons and their sheath cells in both the central and peripheral nervous systems. In addition to these reciprocal interactions within nerve fibers, there is an accumulating body of evidence, much of it admittedly inconclusive, that indicates that nerve fibers influence and, in turn, are influenced by their environment.

Extracellular Potassium

In the central nervous system, increases in extracellular potassium concentration due to neuronal activity cause the membrane potentials of glial cells to fall (Kuffler & Nichols 1966, Somjen 1975). Although the Schwann cells investing invertebrate peripheral axons act as passive diffusion barriers to limit the flow of potassium ions away from axons following their activity (Frankenhaeuser & Hodgkin 1956, Adelman et al 1973), potassium-mediated interactions between axons and glial cells have not been investigated in the mammalian peripheral nervous system. Some of the activity-dependent conduction properties of axons such as intermittent transmission at branch points and other regions of low safety factor and postexcitation supranormality and subnormality (Swadlow et al 1980) may reflect changes in extracellular potassium concentration. However, the extent to which glial cells may play an active role in the control of the extracellular ionic milieu of neurons and axons remains controversial (Somjen 1975, 1979, Varon & Somjen 1979, Stewart & Rosenberg 1979).

Neurotransmitters

Glial cells may also modulate neuronal activity by influencing the levels of transmitter substances in the vicinity of synapses. High affinity uptake and release of several putative neurotransmitters have been described for glial cells in various CNS and PNS tissues as well as glial cells in tissue culture (Varon & Somjen 1979, Stewart & Rosenberg 1979). Of particular interest in the context of axon-glia interactions are the demonstrations of high affinity uptake of several neurotransmitters by Schwann cells in dorsal root ganglia and dorsal roots (Schon & Kelly 1974, Roberts & Keen 1974a,b), although the function of receptors in regions where neurotransmission does not occur remains obscure.

A specific neurotransmitter, acetylcholine, has been implicated in a curious interaction between squid giant axons and their surrounding Schwann cells (Villegas 1978). Axonal impulse activity is associated with an initial phase of glial depolarization, presumably due to increased extracellular potassium concentration, followed by a phase of hyperpolarization (Villegas 1972) caused by acetylcholine (Villegas 1974) and prevented by blocking nicotinic receptors (Villegas 1973, 1975). Acetylcholine is present in much higher concentrations in Schwann cells investing the squid giant axons than in the axons themselves (Villegas & Jenden 1979); acetylcholine receptors with high affinity for I^{125}-bungarotoxin have been localized autoradiographically to the Schwann cell surface membrane (Rawlins & Villegas 1978); acetylcholinesterase is present in the axon membrane (Villegas & Villegas 1974). Thus, Villegas (1978) suggests that axonal activity stimulates Schwann cells to release acetylcholine, which in turn causes hyperpolarization of the Schwann cell. The nature of the stimulus that triggers these Schwann cells to release acetylcholine has not been determined, but Villegas (1978) speculates that it may be mediated by glutamate released by the axon during activity. Properties of the glial cells associated with the giant axons of the crayfish are similar to those of the squid Schwann cell (Lieberman et al 1979).

The functional significance of this complex interaction between axons and Schwann cells in invertebrate giant axons remains to be elucidated. Axon-Schwann cell relationships in these nerves differ strikingly from those of vertebrate axons. Notably, Schwann cells form a pallisade around the invertebrate giant axons rather than individually investing the total axon circumference, as occurs in vertebrate myelinated and unmyelinated fibers. Nevertheless, it would be of great interest to determine if there is any physiological interaction between axons and Schwann cells in vertebrates analagous to that characterized for invertebrate giant axons by Villegas and his collaborators.

CONCLUDING REMARKS

The study of nerve fibers can be viewed historically as moving through several broad stages. Myelinated and unmyelinated fibers were initially considered to be simple conductors composed of axons surrounded by passive sheaths. Subsequently, it became possible to delineate the various surface membrane specializations that form the basis of axonal ensheathment. Further investigations of structure and function, many of which are discussed in this review, established the extent of diverse interactions and mutual dependencies that involve the component cells of nerve fibers as well as their immediate environment. It is likely that future investigations will focus on the molecular basis of these relationships.

ACKNOWLEDGMENTS

The authors acknowledge with appreciation the assistance of M. Attiwell, J. Laganière, J. Trecarten, and W. Wilcox. Research carried out in the authors' laboratories in the Montreal General Hospital Research Institute was supported by the Medical Research Council, the Muscular Dystrophy Association of Canada, the Multiple Sclerosis Society of Canada, and Dysautonomia Association of America.

Literature Cited

Abercrombie, M., Johnson, M. L. 1946. Quantitative histology of Wallerian degeneration. I. Nuclear population in rabbit sciatic nerve. *J. Anat. London* 80:37–50

Adelman, W. J., Palti, Y., Senft, J. P. 1973. Potassium ion accumulation in a periaxonal space and its effect on the measurement of membrane potassium ion conductance. *J. Membr. Biol.* 13:387–410

Aguayo, A. J., Terry, L. C., Bray, G. M. 1973. Spontaneous loss of axons in sympathetic unmyelinated nerve fibres of the rat during development. *Brain Res.* 54:360–64

Aguayo, A. J., Bray, G. M., Terry, L. C., Sweezey, E. 1976a. Three-dimensional analysis of unmyelinated fibres in normal and pathologic autonomic nerves. *J. Neuropathol. Exp. Neurol.* 35:136–51

Aguayo, A. J., Charron, L., Bray, G. M. 1976b. Potential of Schwann cells from unmyelinated nerves to produce myelin: A quantitative ultrastructural and autoradiographic study. *J. Neurocytol.* 5:565–73

Aguayo, A. J., Epps, J., Charron, L., Bray, G. M., 1976c. Multipotentiality of Schwann cells in cross-anastomosed and grafted myelinated and unmyelinated nerves: Quantitative microscopy and radioautography. *Brain. Res.* 104:1–20

Aguayo, A. J., Peyronnard, J. M., Terry, L. C., Romine, J. S., Bray, G. M. 1976d. Neonatal neuronal loss in rat superior cervical ganglia: Retrograde effects on developing preganglionic axons and Schwann cells. *J. Neurocytol.* 5:137–55

Aguayo, A. J., Attiwell, M., Trecarten, J., Perkins, S., Bray, G. M. 1977a. Abnormal myelination in transplanted Trembler mouse Schwann cells. *Nature* 265:73–75

Aguayo, A. J., Kasarjian, J., Skamene, E., Kongshavn, P., Bray, G. M. 1977b. Myelination of mouse axons by Schwann cells transplanted from normal and abnormal human nerves. *Nature* 268:753–55

Aguayo, A. J., Dickson, R., Trecarten, J., Attiwell, M., Bray, G. M., Richardson, P. R. 1978a. Ensheathment and myelination of regenerating PNS fibres by transplanted nerve glia. *Neurosci. Lett.* 9:97–104

Aguayo, A. J., Perkins, C. S., Duncan, I. D., Bray, G. 1978b. Human and animal neuropathies studied in experimental nerve transplants. In *Peripheral Neuropathies,* ed. N. Canal, N. Pozza, pp. 37–48. Amsterdam: Elsevier/-North-Holland Biomed. Press

Aguayo, A. J., Bray, G. M., Perkins, C. S. 1979a. Axon-Schwann cell relationships in neuropathies of mutant mice. *Ann. NY Acad. Sci.* 317:512–31

Aguayo, A. J., Bray, G. M., Perkins, C. S., Duncan, I. D. 1979b. Axon-sheath cell interactions in peripheral and central nervous system transplants. *Soc. Neurosci. Symp.* 4:361–83

Aguayo, A. J., Bunge, R. P., Duncan, I. D., Wood, P. M., Bray, G. M. 1979c. Rat Schwann cells, cultured in vitro, can ensheath axons regenerating in mouse nerves. *Neurology* 29:589

Allt, G. 1969. Repair of segmental demyelination in peripheral nerves: An electron microscope study. *Brain* 92:639–46

Allt, G., Cavanagh, J. B. 1969. Ultrastructural changes in the region of the node of Ranvier in the rat caused by diphtheria toxin. *Brain* 92:459–68

Ayers, M. M., Anderson, R. M. 1973. Onion bulb neuropathy in the Trembler mouse: A model of hypertrophic interstitial neuropathy (Déjerine-Sottas) in man. *Acta Neuropathol.* 25:54–70

Ayers, M. M., Anderson, R. M. 1976. Development of onion bulb neuropathy in the Trembler mouse. *Acta Neuropathol.* 36:137–52

Bernstein, J. J. 1966. Relationship of corticospinal tract growth to age and body weight in the rat. *J. Comp. Neurol.* 127:207–18

Berthold, C.-H. 1968. Ultrastructure of the node-paranode region of mature feline ventral lumbar spinal-root fibres. *Acta Soc. Med. Ups.* 73: Suppl. 9, pp. 37–70

Berthold, C.-H. 1973. Local demyelination in developing feline nerve fibres. *Neurobiology* 3:339–52

Berthold, C.-H. 1978. Morphology of normal peripheral axons. In *Physiology and Pathobiology of Axons,* ed. S. G. Waxman, pp. 3–63. New York: Raven

Berthold, C.-H., Skoglund, S. 1967. Histochemical and ultrastructural demonstration of mitochondria in the paranodal region of developing feline spinal roots and nerves. *Acta Soc. Med. Ups.* 72:37–70

Bird, T., Farrel, D. F., Sumi, S. H. 1977. Genetic developmental myelin defect in "shiverer" mouse. *Trans. Am. Soc. Neurochem.* 8:153

Bostock, H. 1980. Conduction changes in demyelination. In *Abnormal Nerves and Muscles as Impulse Generators,* ed. W. Culp, J. Ochoa. New York: Oxford Univ. Press. In press

Bostock, H., Feasby, T. E., Sears, T. A. 1977. Continuous conduction in regenerating myelinated nerve fibres. *J. Physiol. London* 269:P88–89

Bostock, H., Sears, T. A. 1976. Continuous conduction in demyelinated mammalian nerve fibres. *Nature* 263:786–87

Bostock, H., Sears, T. A. 1978. The internodal axon membrane: Electrical excitability and continuous conduction in segmental demyelination. *J. Physiol. London* 280:273–301

Bradley, W. G., Asbury, A. K. 1970. Duration of synthesis phase in neurilemma cells in mouse sciatic nerve during degeneration. *Exp. Neurol.* 26:275–82

Bradley, W. G., Jenkison, M. 1973. Abnormalities of peripheral nerves in murine muscular dystrophy. *J. Neurol. Sci.* 18:227–47

Bray, G. M., Cullen, M. J., Aguayo, A. J., Rasminsky, M. 1979. Node-like areas of intramembranous particles in the unensheathed axons of dystrophic mice. *Neurosci. Lett.* 13:203–8

Brill, M. H., Waxman, S. G., Moore J. W., Joyner, R. W. 1977. Conduction velocity and spike configuration in myelinated fibres: Computed dependence on internode distance. *J. Neurol. Neurosurg. Psychiatry* 40:769–74

Brismar, T. 1979. Potential clamp experiments on myelinated nerve fibres from alloxan diabetic rats. *Acta Physiol. Scand.* 105:384–86

Brismar, T. 1980. Potential clamp analysis of membrane currents in rat myelinated fibres. *J. Physiol. London* 298:171–84

Brockes, J. P., Fields, K. L., Raff, M. C. 1977. A surface antigenic marker for rat Schwann cells. *Nature* 266:364–66

Brown, M. J., Sumner, A. J., Saida, T., Asbury, A. K. 1980. The evolution of early demyelination following topical application of anti-galactocerebroside serum in vivo. *Neurology* 30:371

Bunge, R. P., Bunge, M. B. 1978. Evidence that contact with connective tissue matrix is required for normal interaction between Schwann cells and nerve fibres. *J. Cell Biol.* 78:943–50

Burgess, P. R., English, K. B., Horch, K. W., Stensaas, L. J. 1974. Patterning in the regeneration of type I cutaneous receptors. *J. Physiol. London* 236:57–82

Burgess, P. R., Horch, K. W. 1973. Specific regeneration of cutaneous fibres in the cat. *J. Neurophysiol.* 36:101–14

Burnham, P. A., Raiborn, C., Varon, S. 1972. Replacement of nerve growth factor by ganglionic nonneuronal cells for the survival in vitro of dissociated ganglionic neurons. *Proc. Natl. Acad. Sci. USA* 69:3556–60

Cavanagh, J. B, Jacobs, J. M. 1964. Some quantitative aspects of diphtheritic neuropathy. *Br. J. Exp. Pathol.* 45: 309–22

Caviness, V. S. Jr., Rakic, P. 1978. Mechanisms of cortical development: A view from mutations in mice. *Ann. Rev. Neurosci.* 1:297–326

Chiu, S. Y., Ritchie, J. M. 1980. Potassium channels in nodal and internodal axonal membrane of mammalian myelinated fibres. *Nature* 284:170–71

Chiu, S. Y. Ritchie, J. M., Rogart, R. B., Stagg, D. 1979. A quantitative description of membrane currents in rabbit myelinated nerve. *J. Physiol. London* 292:149–66

Clark, J. W., Plonsey, R. 1970. A mathematical study of nerve fiber interaction. *Biophys. J.* 10:937–57

Clark, J. W., Plonsey, R. 1971. Fiber interaction in a nerve trunk. *Biophys. J.* 11:281–94

Cornbrooks, C., Cochran, M., Mithen, F., Bunge, M. B., Bunge, R. P. 1980. Myelination in explants of Trembler dorsal root ganglion grown in culture. *Trans. Am. Soc. Neurochem.* 11:110

Cotman, C. W., ed. 1978. *Neuronal Plasticity.* New York: Raven. 335 pp.

Cragg, B. G., Thomas, P. K. 1964. The conduction velocity of regenerated peripheral nerve fibres. *J. Physiol. London* 171:164–75

David, S., Aguayo, A. J. 1980. Spontaneous activity generated in sciatic neuromas—a [^{14}C]2-Deoxyglucose uptake study in the rat. *Soc. Neurosci. Abstr.* 6:439

Diamond, J., Jackson, P. C. 1980. Regeneration and collateral sprouting of peripheral nerves. In *Nerve Repair and Regeneration,* ed. D. L. Jewett, H. R. McCarroll Jr., pp. 115–27. St. Louis, Mo.: Mosby

Droz, B., Brunetti, M., DiGiamberardino, L., Koenig, H. L., Porcellati, G. 1979. Transfer of phospholipid constituents to glia during axonal transport. *Soc. Neurosci. Symp.* 4:344–60

Duchen, L. W., Eicher, E. M., Jacobs, J. M., Scaravilli, F., Teixeira, F. 1980. A globoid cell type of leucodystrophy in the mouse: The mutant twitcher. In *Neurological Mutations Affecting Mylenation, INSERM Symp. Ser.,* ed. N. Baumann, 14:107–15. Amsterdam: Elsevier/North Holland

Duncan, I. D., Aguayo, A. J., Bunge, R. P., Wood, P. 1980. Transplantation of rat Schwann cells grown in tissue culture into the mouse spinal cord. *J. Neurol. Sci.* In press

Dupouey, P., Jacque, C., Bourre, J. M., Cesselin, F., Privat, A., Baumann, N. 1979. Immunochemical studies of myelin basic protein in shiverer mouse devoid of major dense line of myelin. *Neurosci. Lett.* 12:113–18

Dyck, P. J., Lais, A. C., Sparks, M. F., Oviatt, K. F., Hexum, L. A., Steinmuller, D. 1979. Nerve xenografts to apportion the role of axon and Schwann cell in myelinated fiber absence in hereditary sensory neuropathy, type II. *Neurology* 29:1215–21

Eames, R. A., Gamble, H. J. 1970. Schwann cell relationships in normal human cutaneous nerves. *J. Anat.* 106:417–35

Ellisman, M. H. 1976. The distribution of membrane molecular specializations characteristic of the node of Ranvier is not dependent upon myelination. *Neurosci. Abstr.* 2:410

Ellisman, M. H. 1977. High voltage electron microscopy of cortical specializations associated with membranes at nodes of Ranvier. *J. Cell Biol.* 75:A108

Ellisman, M. H. 1979. Molecular specializations of the axon membrane at nodes of Ranvier are not dependent upon myelination. *J. Neurocytol.* 8:719–35

Feder, N., Reese, T. S., Brightman, M. W. 1969. Microperoxidase, a new tracer of low molecular weight. A study of the interstitial compartments of the mouse brain. *J. Cell Biol.* 43:A35–36

Fraher, J. P. 1974. A numerical study of cervical and thoracic ventral nerve roots. *J. Anat.* 118:127–42

Frankenhaeuser, R., Hodgkin, A. L. 1956. The after-effects of impulses in the giant nerve fibres of *Loligo. J. Physiol. London* 131:341–76

Friede, R. L., Samorajski, T. 1967. Relation between the number of myelin lamellae and axon circumference in fibres of vagus and sciatic nerves of mice. *J. Comp. Neurol.* 130:223–31

Friedrich, V. L. Jr. 1975. Hyperplasia of oligodendrocytes in quaking mice. *Anat. Embryol.* 147:259–71

Fullerton, P. M., Gilliatt, R. W., Lascelles, R. G., Morgan-Hughes, J. A. 1965. The relation between fiber diameter and internodal length in chronic neuropathy. *J. Physiol. London* 178:P26–28

Gainer, H., Tasaki, I., Lasek, R. J. 1977. Evidence for the glia-neuron protein transfer hypothesis from intracellular perfusion studies of giant squid axons. *J. Cell Biol.* 74:524–30

Gilula, N. B. 1974. Junctions between cells. In *Cell Communication,* ed. R. P. Cox, pp. 1–29. New York: Wiley

Gledhill, R. F., Harrison, B. M., McDonald, W. I. 1973. Pattern of remyelination in the CNS. *Nature* 244:443–44

Govrin-Lippmann, R., Devor, M. 1978. Ongoing activity in severed nerves: Source and variation with time. *Brain Res.* 159:406–10

Guénet, J.-L. 1980. Mutants of the mouse with an abnormal myelination: A review for geneticists. See Duchen et al 1980, pp. 11–23

Hall, S. M., Gregson, N. A. 1971. The in vivo and ultrastructural effects of injection of lysophosphatidyl choline into myelinated peripheral nerve fibres of the adult mouse. *Cell Sci.* 9:769–89

Hall, S. M., Gregson, N. A. 1975. The effects of mitomycin C on the process of remyelination in the mammalian peripheral nervous system. *Neuropathol. Appl. Neurobiol.* 1:149–70

Hanson, G. R., Partlow, L. M. 1977. Stimulation of thymidine incorporation by cell sonicates. *Trans. Am. Soc. Neurochem.* 8:142

Hille, B. 1977. Ionic basis of resting and action potentials. In *Handbook of Physiology,* Sec. 1, ed. E. R. Kandel, 1:99–136. Bethesda, Md.: Am. Physiol. Soc.

Hirano, A., Becker, N. H., Zimmerman, H. M. 1969. Isolation of the periaxonal space of the central myelinated fiber with regard to diffusion of peroxidase. *J. Histochem. Cytochem.* 17:512–16

Hirano, A., Dembitzer, H. M. 1967. A structural analysis of the myelin sheath in the central nervous system. *J. Cell Biol.* 34:555–67

Hirano, A., Dembitzer, H. M. 1969. The transverse bands as a means of access to the periaxonal space of the central myelinated nerve fibre. *J. Ultrastruct. Res.* 28:141–49

Horáckova, M., Nonner, W., Stämpfli, R. 1968. Action potentials and voltage clamp currents of single rat Ranvier nodes. *Int. Physiol. Congr. 24th.* 7:198

Howe, J. F., Loeser, J. D., Calvin, W. H. 1977. Mechanosensitivity of dorsal root ganglia and chronically injured axons: A physiological basis for the radicular pain of nerve root compression. *Pain* 3:25–41

Huizar, P., Kuno, M., Miyata, Y. 1975. Electrophysiological properties of spinal motoneurones of normal and dystrophic mice. *J. Physiol. London* 248:231–46

Huxley, A. F., Stämpfli, R. 1949. Evidence for saltatory conduction in peripheral myelinated nerve fibres. *J. Physiol. London* 108:315–39

Itoyama, Y., Sternberger, N. H., Kies, M. W., Cohen, S. R., Richardson, E. P. Jr., Webster, H. D. 1980. Immunocytochemical method to identify myelin basic protein in oligodendroglia and myelin sheaths of the human nervous system. *Ann. Neurol.* 7:157–66

Jaros, E., Bradley, W. G. 1979. Atypical axon-Schwann cell relationships in the common peroneal nerve of the dystrophic mouse: An ultrastructural study. *Neuropathol. Appl. Neurobiol.* 5:133–47

Kennedy, W. R., Sung, J. H., Berry, J. F. 1977. A case of congenital hypomyelination neuropathy. *Arch. Neurol. Chicago* 34:337–45

Kirschner, D. A., Ganser, A. L. 1980. Compact myelin exists in the absence of basic protein in the shiverer mutant mouse. *Nature* 283:207–10

Kobayashi, T., Suzuki, K. 1980. Biochemistry of twitcher mouse: An authentic murine model of human globoid cell leukodystrophy. See Duchen et al 1980, pp. 253–57

Kraft, G. H. 1975. Serial nerve conduction and electromyographic studies in experimental allergic neuritis. *Arch. Phys. Med. Rehabil.* 56:333–40

Kristol, C., Akert, K., Sandri, C., Wyss, U. R., Bennett, M. V. L., Moor, H. 1977. The Ranvier nodes in the neurogenic electric organ of the knifefish *Sternarchus:* A freeze-etching study on the distribution of membrane associated particles. *Brain Res.* 125:197–212

Kristol, C., Sandri, C., Akert, K. 1978. Intermembranous particles at the nodes of Ranvier of the cat spinal cord: A morphometric study. *Brain Res.* 142:391–400

Kuffler, S. W., Nichols, J. G. 1966. The physiology of neuroglial cells. *Ergeb. Physiol.* 57:1–90

Landon, D. H., Hall, S. 1976. The myelinated nerve fibre. In *The Peripheral Nerve,* ed. D. N. Landon, pp. 1–105. London: Chapman & Hall

Landon, D. H., Langley, O. K. 1971. The local chemical environment of nodes of Ranvier: A study of cation binding. *J. Anat.* 102:419–32

Lasek, R. J., Gainer, H., Barker, J. L. 1977. Cell-to-cell transfer of glial proteins to the squid giant axons. The glia-neuron protein transfer hypothesis. *J. Cell Biol.* 74:501–23

Le Douarin, N. 1973. A biological cell labelling technique and its use in experimental embryology. *Dev. Biol.* 30:217–22

Letourneau, P. C. 1975. Cell to substratum adhesion and guidance of axonal elongation. *Dev. Biol.* 44:77–91

Lieberman, E. M., Villegas, J., Villegas, G. M. 1979. The nature of the membrane potential of Schwann cells of crayfish ventral nerve cord. *Biophys. J.* 25:A305

Livett, B. G. 1978. Immunohistochemical localization of nervous system specific proteins and peptides. *Int. Rev. Cytol. Suppl.* 7:53–237

Livingstone, R. B., Pfenninger, K., Moor, H., Akert, K. 1973. Specialized paranodal and interparanodal glial axonal junctions in the peripheral and central nervous system: A freeze-etching study. *Brain Res.* 58:1–24

Low, P. A. 1977. The evolution of "onion bulbs" in hereditary hypertrophic neuropathy of the Trembler mouse. *Neuropathol. Appl. Neurobiol.* 3:81–92

Lyon, G. 1969. Ultrastructure study of a nerve biopsy from a case of early infantile chronic neuropathy. *Acta Neuropathol.* 13:131–42

Madrid, R. E., Jaros, E., Cullen, M. J., Bradley, W. G. 1975. Genetically determined defect of Schwann cell basement membrane in dystrophic mouse. *Nature* 257:319–21

McCarthy, K. D., Partlow, L. M. 1976a. Preparation of pure neuronal and nonneuronal cultures from embryonic chick sympathetic ganglia: A new method based on both differential cell adhesiveness and the formation of homotypic neuronal aggregates. *Brain Res.* 114:391–414

McCarthy, K. D., Partlow, L. M. 1976b. Neuronal stimulation of [^{3}H]thymidine incorporation by primary cultures of highly purified non-neuronal cells. *Brain Res.* 114:415–26

McDonald, W. I. 1963. The effects of experimental demyelination on conduction in peripheral nerve: A histological and electrophysiological study. II: Electrophysiological observations. *Brain* 86:501–24

McDonald, W. I. 1980. Clinical consequences of conduction defects produced by demyelination. See Bostock 1980. In press

Meier, C., Bischoff, A. 1975. Oligodendroglial cell development in jimpy mice and controls. An electron-microscopic study in the optic nerve. *J. Neurol. Sci.* 26:517–28

Michel, M. E., Reier, P. J. 1979. Axonalependymal associations during early regeneration of the transected spinal cord in *Xenopus laevis* tadpoles. *J. Neurocytol.* 8:529–48

Mirsky, R., Winter, J., Abney, E. R., Pruss, R. M., Gavrilovic, J., Raff, M. C. 1980. Myelin specific proteins and glycolipids in rat Schwann cells and oligodendrocytes in culture. *J. Cell Biol.* 84:483–94

Morgan-Hughes, J. A. 1968. Experimental diphtheritic neuropathy, a pathological and electrophysiological study. *J. Neurol. Sci.* 7:157–75

Mugnaini, E., Osen, K. K., Schnapp, B., Friedrich, V. L. Jr. 1977. Distribution of Schwann cell cytoplasm and plasmalemmal vesicles (caveolae) in peripheral myelin sheaths. An electron microscopic study with thin sections and freeze-fracturing. *J. Neurocytol.* 6:647–68

Mullen, R. J. 1977. Genetic dissection of the CNS with mutant-normal mouse and rat chimaeras. *Soc. Neurosci. Symp.* 2:47–65

Murray, J. A., Blakemore, W. F. 1980. The relationship between internodal length and fibre diameter in the spinal cord of the cat. *J. Neurol. Sci.* 45:29–41

Ochoa, J. 1976. The unmyelinated nerve fibre. See Landon & Hall 1976, pp. 106–58

Okada, E., Bunge, R. P., Bunge, M. B. 1980. Abnormalities expressed in long term cultures of dorsal root ganglia from the dystrophic mouse. *Brain Res.* 194:455–70

Perkins, C. S., Aguayo, A. J., Bray, G. M. 1981a. Schwann cell multiplication in Trembler mice. *Neuropathol. Appl. Neurobiol.* In press

Perkins, C. S., Aguayo, A. J., Bray, G. M. 1981b. Behavior of Schwann cells from Trembler mouse unmyelinated fibers transplanted into myelinated nerves. *Exp. Neurol.* In press

Perkins, C. S., Bray, G. M., Aguayo, A. J. 1980a. Persistent multiplication of axon associated cells in spinal roots of dystrophic mice. *Neuropathol. Appl. Neurobiol.* 6:83–91

Perkins, C. S., Bray, G. M., Aguayo, A. J. 1980b. Evidence for undifferentiated Schwann cells in the spinal roots of dystrophic mice. *Can. J. Neurol. Sci.* 7:123

Peters, A. 1966. The node of Ranvier in the central nervous system. *Q. J. Exp. Physiol.* 51:229–36

Peters, A., Muir, A. R. 1959. The relationship between axons and Schwann cells during development of peripheral nerves in the rat. *Q. J. Exp. Physiol.* 44:117–30

Peters, A., Palay, S. L., Webster, H. D. 1976. *The Fine Structure of the Nervous System.* Philadelphia: Saunders. 406 pp. 2nd ed.

Peterson, A. C., Frair, P. M., Rayburn, H. R., Cross, D. P. 1979. Development and disease in the neuromuscular system of muscular dystrophic ↔ normal mouse chimaeras. *Soc. Neurosci. Symp.* 4:258–73

Peyronnard, J.-M., Aguayo, A. J., Bray, G. M. 1973. Schwann cell internuclear distances in normal and regenerating unmyelinated nerve fibers. *Arch. Neurol. Chicago* 24:56–59

Pollard, J. D., McLeod, J. G. 1980. Nerve grafts in the Trembler mouse—An electrophysiological and histological study. *J. Neurol. Sci.* 46:373–83

Privat, A., Jacque, C., Bourre, J. M., Dupouey, P., Baumann, N. 1979. Absence of the major dense line in myelin of the mutant mouse 'shiverer'. *Neurosci. Lett.* 12:107–12

Privat, A., Robain, O., Mandel, P. 1972. Aspects ultrastructuraux du corps calleux chez la souris Jimpy. *Acta Neuropathol.* 21:282–95

Quarles, R. H., McIntyre, L. J., Sternberger, N. H. 1979. Glycoproteins and cell surface interactions during myelinogenesis. *Soc. Neurosci. Symp.* 4:322–43

Quick, D. C., Waxman, S. G. 1977. Specific staining of the axon membrane at nodes of Ranvier with ferric ion and ferrocyanide. *J. Neurol. Sci.* 31:1–11

Raff, M. C., Abney, E., Brockes, J. P., Hornby-Smith, A. 1978a. Schwann cell growth factors. *Cell* 15:813–22

Raff, M. C., Hornby-Smith, A., Brockes, J. P. 1978b. Cyclic AMP as a mitogenic signal for cultured rat Schwann cells. *Nature* 273:672–73

Raine, C. S., Wisniewski, H., Prineas, J. 1969. An ultrastructural study of experimental demyelination and remyelination. II. Chronic experimental allergic encephalomyelitis in the peripheral nervous system. *Lab. Invest.* 21:316–27

Ramón y Cajal, S. 1928. *Degeneration and Regeneration of the Nervous System,* Vol. 2. London: Oxford Univ. Press

Rasminsky, M. 1978a. Physiology of conduction in demyelinated axons. See Berthold 1978, pp. 361–76

Rasminsky, M. 1978b. Ectopic generation of impulses and cross-talk in spinal nerve roots of "dystrophic" mice. *Ann. Neurol.* 3:351–57

Rasminsky, M. 1980a. Ephaptic transmission between single nerve fibres in the spinal nerve roots of dystrophic mice. *J. Physiol. London.* 305:151–69

Rasminsky, M. 1980b. Ectopic excitation, ephaptic excitation and autoexcitation in peripheral nerve fibers of mutant mice. See Bostock 1980. In press

Rasminsky, M., Kearney, R. E., Aguayo, A. J., Bray, G. M. 1978. Conduction of nervous impulses in spinal roots and peripheral nerves of dystrophic mice. *Brain Res.* 143:71–85

Rasminsky, M., Sears, T. A. 1972. Internodal conduction in undissected demyelinated nerve fibres. *J. Physiol. London* 227:323–50

Rasminsky, M., Sumner, A. 1980. Development of conduction block in single rat spinal root fibers locally exposed to antigalactocerebroside serum. *Neurology* 30:371

Rawlins, F. A., Villegas, J. 1978. Autoradiographic localization of acetylcholine receptors in the Schwann cell membrane of the squid nerve fiber. *J. Cell Biol.* 77:371–76

Rayburn, H. R., Peterson, A. C., Aguayo, A. J. 1980. Development and deployment of Schwann cells in peripheral nerves of trembler ↔ normal mouse chimaeras. *Soc. Neurosci. Abstr.* 6:660

Reier, P. J., Hughes, A. 1972. Evidence for spontaneous axon degeneration during peripheral nerve maturation. *Am. J. Anat.* 135:147–52

Richardson, P. M., McGuinness, U. M., Aguayo, A. J. 1980. Axons from CNS neurons regenerate into PNS grafts. *Nature* 284:264–65

Ritchie, J. M. 1979. Sodium channels in muscle and nerve. In *Current Topics in Nerve and Muscle Research,* ed. A. J. Aguayo, G. Karpati, pp. 210–19. Amsterdam: Excerpta Medica

Ritchie, J. M., Rogart, R. B. 1977. The density of sodium channels in mammalian myelinated nerve fibres and nature of the axonal membrane under the myelin sheath. *Proc. Natl. Acad. Sci. USA* 74:211–15

Ritchie, J. M., Rogart, R. B., Strichartz, G. P. 1976. A new method for labelling saxitoxin and its binding to non-myelinated fibres of the rabbit vagus, lobster walking leg and garfish olfactory nerves. *J. Physiol. London* 261:477–94

Roberts, P. J., Keen, P. 1974a. High affinity uptake system for glutamine in rat dorsal roots but not in nerve endings. *Brain Res.* 67:352–57

Roberts, P. J., Keen, P. 1974b. (^{14}C) Glutamate uptake and compartmentation in glia of rat dorsal sensory ganglion. *J. Neurochem.* 23:201–9

Romine, J. S., Bray, G. M., Aguayo, A. J. 1976. Schwann cell multiplication after crush injury of unmyelinated fibres. *Arch. Neurol. Chicago* 33:49–54

Rosenbluth, J. 1976. Intramembranous particle distribution at the node of Ranvier and adjacent axolemma in myelinated axons of the frog brain. *J. Neurocytol.* 5:731–45

Rosenbluth, J. 1977. Absence of nodal and paranodal axolemmal membrane specializations in freeze-fracture replicas of jimpy mouse brain and spinal cord. *Soc. Neurosci. Abstr.* 3:335

Rosenbluth, J. 1979. Abnormalities of axolemmal membrane specializations in myelin deficient "shiverer" mice. *Soc. Neurosci. Abstr.* 5:432

Saida, K., Saida, T., Brown, M. J., Silberberg, D. H. 1979. In vivo demyelination induced by intraneural injection of anti-galactocerebroside serum. *Am. J. Pathol.* 95:99–110

Salzer, J. L., Bunge, R. P. 1980. Studies of Schwann cell proliferation. I. An analysis in tissue culture of proliferation during development, Wallerian degeneration and direct injury. *J. Cell Biol.* 84:739–52

Salzer, J. L., Glaser, L., Bunge, R. P. 1977. Stimulation of Schwann cell proliferation by a neurite membrane fraction. *J. Cell Biol.* 75:A118

Samorajski, T., Friede, R. L., Reimer, P. R. 1970. Hypomyelination in the quaking mouse. A model for the analysis of disturbed myelin formation. *Neuropathol. Exp. Neurol.* 29:507–23

Scaravilli, F., Jacobs, J. M., Teixeira, F. 1980. Quantitative and experimental studies on the twitcher mouse. See Duchen et al 1980, pp. 115–23

Schnapp, B., Mugnaini, E. 1975. The myelin sheath: Electron microscopic studies with thin sections and freeze-fracture. In *Golgi Centennial Symposium: Perspectives in Neurobiology,* ed. M. Santini, pp. 209–33. New York: Raven

Schnapp, B., Peracchia, C., Mugnaini, E. 1976. The paranodal axo-glial junction in the central nervous system studied with thin sections and freeze-fracture. *Neuroscience* 1:181–90

Schon, F., Kelly, J. S. 1974. Autoradiographic localization of [^{3}H]GABA and [^{3}H]glutamate over satellite glial cells. *Brain Res.* 66:275–88

Sears, T. A. 1979. Nerve conduction in demyelination, amyelination and early regeneration. See Ritchie 1979, pp. 181–88

Seltzer, Z., Devor, M. 1979. Ephaptic transmission in chronically damaged peripheral nerves. *Neurology* 29:1061–64

Sherratt, R. M., Bostock, H., Sears, T. A. 1980. Effects of 4-aminopyridine on normal and demyelinated mammalian nerve fibers. *Nature* 283:570–72

Sidman, R. L., Dickie, M. M., Appel, S. H. 1964. Mutant mice (quaking and jimpy) with deficient myelination in the central nervous system. *Science* 144:309–11

Silver, J., Sidman, R. L. 1980. A mechanism for the guidance and topographic patterning of retinal ganglion cell axons. *J. Comp. Neurol.* 189:101–11

Singer, M., Nordlander, R. H., Egar, M. 1979. Axonal guidance during embryogenesis and regeneration in the spinal cord of the newt: The blueprint hypothesis of neuronal pathway patterning. *J. Comp. Neurol.* 185:1–21

Singer, S. J., Ash, J. F., Bourguignon, L. Y. W., Heggeness, M. H., Louvard, D. 1978. Transmembrane interactions and the mechanisms of transport of proteins across membranes. *J. Supramol. Struct.* 9:373–89

Skoff, R. P. 1975. The fine structure of pulse labeled (^{3}H-thymidine cells) in degenerating rat optic nerve. *J. Comp. Neurol.* 161:595–611

Skoff, R. P. 1976. Myelin deficit in the jimpy mouse may be due to cellular abnormalities in astroglia. *Nature* 264: 560–62

Smith, K. J., Blakemore, W. F., McDonald, W. I. 1979. Central remyelination restores secure conduction. *Nature* 280:395–96

Somjen, G. G. 1975. Electrophysiology of neuroglia. *Ann. Rev. Physiol.* 37:163–90

Somjen, G. G. 1979. Extracellular potassium in the mammalian central nervous system. *Ann. Rev. Physiol.* 41:159–77

Stensaas, L. J., Burgess, P. R., Horch, K. W. 1979. Regenerating dorsal root axons are blocked by spinal cord astrocytes. *Soc. Neurosci. Abstr.* 5:684

Stewart, R. M., Rosenberg, R. N. 1979. Physiology of glia: Glial-neuronal interactions. *Int. Rev. Neurobiol.* 21:275–309

Suzuki, K., Zagoren, J. C. 1977. Quaking mouse: An ultrastructural study of the peripheral nerves. *J. Neurocytol.* 6: 71–84

Swadlow, H. A., Kocsis, J. D., Waxman, S. G. 1980. Modulation of impulse conduction along the axonal tree. *Ann. Rev. Biophys. Bioeng.* 9:143–79

Tasaki, I. 1959. Conduction of the nerve impulse. In *Handbook of Physiology,* Sec. 1, ed. J. Field, H. W. Magoun, N. E. Hall, 1:75–121. Washington DC.: Am. Physiol. Soc.

Thomas, P. K. 1975. Nerve injury. In *Essays on the Nervous System,* ed. R. Bellairs, E. G. Gray, pp. 44–70. London: Oxford Univ. Press

Varon, S. S., Bunge, R. P. 1978. Trophic mechanisms in the peripheral nervous system. *Ann. Rev. Neurosci.* 1:327–61

Varon, S. S., Somjen, G. G. 1979. Neuronglial interactions. *Neurosci. Res. Prog. Bull.* 17:1–239

Villegas, G. M., Villegas, J. 1974. Acetylcholinesterase localization in the giant nerve fiber of the squid. *J. Ultrastruct. Res.* 46:149–63

Villegas, J. 1972. Axon-Schwann cell interaction in the squid nerve fibre. *J. Physiol. London* 225:275–96

Villegas, J. 1973. Effects of tubocurarine and eserine on the axon-Schwann cell relationship in the squid nerve fibre. *J. Physiol. London* 232:193–208

Villegas, J. 1974. Effects of acetylcholine and carbamylcholine on the axon and Schwann cell electrical potentials in the squid nerve fibre. *J. Physiol. London* 242:647–59

Villegas, J. 1975. Characterization of acetylcholine receptors in the Schwann cell membrane of the squid nerve fibre. *J. Physiol. London* 249:679–89

Villegas, J. 1978. Cholinergic properties of satellite cells in the peripheral nervous system. In *Dynamic Properties of Glial Cells,* ed. E. Schoffeniels, G. Franck, D. B. Towers, L. Hertz, pp. 207–215. Oxford: Pergamon

Villegas, J., Jenden, D. J. 1979. Acetylcholine content of the Schwann cell and axon in the giant nerve fiber of the squid. *J. Neurochem.* 32:761–66

Vizoso, A. D., Young, J. Z. 1948. Internode length and fibre diameter in developing and regenerating nerves. *J. Anat.* 82: 110–34

Wall, P. D., Gutnick, M. 1974. Ongoing activity in peripheral nerves: The physiology and pharmacology of impulses originating from a neuroma. *Exp. Neurol.* 43:580–93

Waxman, S. G. 1978. Variations in axonal morphology and their functional significance. See Berthold 1978, pp. 169–90

Webster, H. D., 1971. The geometry of peripheral myelin sheaths during their formation and growth in rat sciatic nerves. *J. Cell Biol.* 48:348–67

Webster, H. D., Martin, J. R., O'Connell, M. F. 1973. The relationships between interphase Schwann cells and axons before myelination: A quantitative electron microscopic study. *Dev. Biol.* 32:401–16

Webster, H. D., Spiro, D., Waksman, B., Adams, R. D. 1961. Phase and electron microscopic studies of experimental demyelination. II. Schwann cell changes in guinea pig sciatic nerves during experimental diphtheritic neuritis. *J. Neuropathol. Exp. Neurol.* 20:5–34

Webster, H. D., Sternberger, N. H. 1980. Morphological features of myelin formation. See Duchen et al 1980, pp. 73–87

Weinberg, H. J., Spencer, P. S. 1975. Studies on the control of myelinogenesis. I. Myelination of regenerating axons after entry into a foreign unmyelinated nerve. *J. Neurocytol.* 4:395–418

Weinberg, H. J., Spencer, P. S. 1976. Studies on the control of myelinogenesis. II. Evidence for neuronal regulation of myelin production. *Brain Res.* 113: 363–78

Weinberg, H. J., Spencer, P. S. 1978. The fate of Schwann cells isolated from axonal contact. *J. Neurocytol.* 7:555–69

Weinberg, H. J., Spencer, P. S. 1979. Studies on the control of myelinogenesis. III. Signalling of oligodendrocyte myelination by regenerating peripheral axons. *Brain Res.* 162:273–79

Weinberg, H. J., Spencer, P. S., Raine, C. S. 1975. Aberrant PNS development in dystrophic mice. *Brain Res.* 88:532–37

Williams, A. K., Wood, P., Bunge, M. B. 1976. Evidence that the presence of Schwann cell basal lamina depends upon interactions with neurons. *J. Cell Biol.* 70(2):A138

Wood, J. G., Jean, D. H., Whitaker, J. N., McLaughlin, B. J., Albers, R. W. 1977. Immunocytochemical localization of the sodium, potassium activated ATPase in knifefish brain. *J. Neurocytol.* 6:571–81

Wood, P. M. 1976. Separation of functional Schwann cells and neurons from normal peripheral nerve tissue. *Brain Res.* 115:361–75

Wood, P. M., Bunge, R. P. 1975. Evidence that sensory axons are mitogenic for Schwann cells. *Nature* 256:662–64

Ann. Rev. Neurosci. 1981. 4:163–94

STRENGTH AND WEAKNESS ❖11552
OF THE GENETIC APPROACH
TO THE DEVELOPMENT OF THE
NERVOUS SYSTEM

Gunther S. Stent

Department of Molecular Biology, University of California, Berkeley,
California 94720

THE GENETIC APPROACH

In the mid-1960s, at the time of the triumphant culmination of molecular
biological research in the cracking of the genetic code and the elucidation
of the mechanism of protein synthesis, research projects began to be formu-
lated that attempted to wed the disciplines of genetics and developmental
neurobiology. In particular, the idea gained currency that the deep biological problem posed by the metazoan nervous system, namely how its cellular
components and their precise interconnections arise during ontogeny,
could, and even should, be approached by focusing on genes. For instance,
Seymour Benzer, one of the main founders of molecular biology and the
person more responsible than any other for the transformation of the classical abstract concept of the gene into its modern molecular-genetic version,
outlined such a project in his Lasker Award Lecture entitled "From Gene
to Behavior" (Benzer 1971). Similarly, Sydney Brenner, another leading
pioneer of molecular biology, set forth the merits of genetic methodology
for the solution of neurological problems in an essay entitled "The Genetics
of Behavior" (Brenner 1973). Since then, the genetic approach to neurobiology has gained many adherents and has given rise to a substantial literature, including numerous reviews (Bentley 1976, Pak & Pinto 1976, Dräger
1977, Ward 1977, Hall & Greenspan 1979, Kankel & Ferrus 1979, Mullen
& Herrup 1979, Palka 1979, Quinn & Gould 1979, Macagno 1980, Anderson et al 1980). The purpose of this article is not to add one more account

163

0147-006X/81/0301-0163$01.00

of recent advances in neurogenetics to that already extensive corpus of reviews, but rather to attempt a critical appraisal of the genetic approach.

To make that appraisal it is convenient to consider two related, yet distinct conceptual aspects of the genetic approach to developmental neurobiology: the *ideological* and the *instrumental* aspects. The ideological aspect of the genetic approach confronts us with the basic belief that the structure and function of the nervous system, and hence the behavior of an animal, is specified by its genes, which are held to "contain the information for the circuit diagram" (Benzer 1971). Admittedly, it cannot be the case that the genes really contain enough information that, if we could only decipher it, would allow us to draw a neuron-by-neuron schematic of the nervous system (Brindley 1969). But, all the same, the circuit might somehow be implicit in the genetic information, since the kind of purely quantitative information-theoretical arguments advanced by Horridge (1968) against the notion of genetic determination of the nervous system are clearly invalid (Stent 1978). And so the ideological aspect presents the discovery of how genes play their determinative role in the genesis of the nervous system and its behavioral output as the ultimate goal of developmental neurobiology. As Hall & Greenspan (1979) express this ideology: "The single gene approach to behavioral genetics and neurogenetics starts out with the knowledge that genes blatantly specify the assembly of the nervous system and the components that underlie the function of the cells in that system. Our 'only' questions, then revolve around trying to find out *how* specific genes control neurobiological phenomena." Indeed, this discovery is not merely a distant goal but one which has already been brought within grasp, since "work with single gene mutations in a variety of organisms is yielding insights on how genes build nerve cells and specify the neural circuits which underlie behaviour" (Quinn & Gould 1979).

By contrast, as seen from its instrumental aspect, the genetic approach does not necessarily entail a belief in genetic specification of the nervous system or present as its goal the discovery of how specific genes control neurobiological phenomena. Instead, from the instrumental aspect the genetic approach appears as the study of differences in neurologic phenotype between animals of various genotypes, without any particular interest (other than methodologic) in the concept of genetic specification. For instance, Mullen & Herrup (1979) find that "the rationale for using genetic mutations involving the nervous system is two-sided. On one side, they are exciting and unique means for lesioning the system at the level of cell interactions and to help expose the ground rules of the developmental process. On the other side, they are models for inherited diseases that unfortunately are as common in human beings as they are in mice." Similarly, Pak & Pinto (1976) see the genetic approach merely as "one of many techniques that could be applied to the nervous system and that . . . alone is unlikely to solve any of the major problems in neurobiology. One of the major contributions that the mutant approach could make may well be to provide convenient experimental preparations to which other techniques can be applied. Thus the ultimate success of the genetic approach is likely to depend on the further development and application of suitable other techniques."

In order to assess the strength and weakness of the genetic approach from both of its aspects, we consider a by no means exhaustive but, it is hoped, representative sample collection of studies in which genetic techniques have been brought to bear on problems connected with the development of the nervous system. These studies make use of two distinct, indeed diametrically opposite, procedures. Under the first, less common procedure, differences in neurologic phenotype are noted among the individuals of a genetically homogenous population. Here the aim is to ascertain the degree of the precision with which developmental process gives rise to the structures of the nervous system. Under the second, more common procedure, differences in neurologic phenotype are noted among the individuals of a genetically heterogeneous population. Here the aim is to establish a causal link between an observed difference in neurologic phenotype and known differences in genotype, in the hope that this link can be accounted for in terms of developmental mechanisms. We first consider some results obtained by the study of genetically homogeneous populations.

PHENOTYPIC VARIANCE

The variance of phenotypes among individuals of a single species has long been appreciated. One part of such phenotypic variance is attributable to intraspecific variations in genotype, or genetic polymorphisms, that are of importance for evolutionary biology. But another part of phenotypic variance simply reflects what Waddington (1957) called "developmental noise." The study of this noise—its sources and the physiological processes that reduce its consequences—seems to be an underdeveloped area of contemporary developmental biology, even though an understanding of developmental noise bids fair to lead to significant insights into developmental mechanisms. The nervous system is particularly suitable for the study of developmental noise, in that here the existence of subtle phenotypic variance can be more readily assessed, and can have more pronounced functional consequences, than in many other organs. To ascertain the degree to which developmental noise actually occurs in the formation of some part of the nervous system, it is necessary to examine a set of isogenic animals (i.e. all sharing exactly the same genotype). Any variance found under these conditions can then be attributed to noise rather than to genetic polymorphisms.[1] The aspects of specifically identifiable neuronal structures whose phenotypic variance may be examined from individual to individual include (a) the number of neurons of which the structure is composed, (b) the relative position of these neurons within the structure, (c) the morphology of their axonal and dendritic processes, and (d) the pattern of their synaptic connections.

The nervous system of presumptively isogenic individuals of the nematode *Caenorhabditis* has been examined from these aspects of phenotypic variance (Ward et al 1975, Ware et al 1975, Albertson & Thomson 1976, White et al 1976). As far as cell number is concerned, the *Caenorhabditis* nervous system was found to be highly invariant, consisting of exactly 258 identifiable neurons. This constant neuron number is attributable to an invariant pattern of cell cleavage and cell death during development. The location of the cell bodies of these neurons, however, is subject to some significant variation, due to a random element in the migration of neuroblast precursor cells (Sulston 1976, Sulston & Horvitz 1977). Furthermore, the morphology and synaptic connection pattern of the nematode neurons manifest a considerable degree of phenotypic variance. Although the major geometrical features of the processes of each indentifiable neuron are quite constant, the fine details of their branching patterns are quite variable. Similarly, although the overall topology of the neuronal network is quite constant, the synaptic sites possessed by individual identified neurons are quite variable with regard to their number and location.

Another detailed study of phenotypic variance was carried out on the visual system of isogenic individuals in parthogenetic clones of the Crustacean *Daphnia* (Macagno et al 1973). The *Daphnia* compound eye consists of 22 facets, or ommatidia. As is generally true for arthropod compound eyes, every ommatidium contains eight primary photoreceptor cells, each of which occupies a characteristic position within the ommatidium. The axons of the eight receptor cells project as a bundle to an underlying layer of 110 secondary visual neurons, the optic lamina. There the eight photoreceptor axons of each bundle make synaptic contact with five contiguous laminar interneurons that form an optic cartridge. The axons of the second-order laminar neurons in turn project to and make synaptic contacts with an underlying layer of about 320 third-order visual neurons, the medulla.

[1]It should be noted that in the context of phenotypic variance the concept of "genotype" has to refer to the genetic constitution of the zygote, or the germ line, rather than of the somatic tissues. This distinction is important because during development, a variety of processes, such as chromosome loss or somatic recombination, may give rise to differences in genetic constitution between somatic cells. But in so far as random somatic generation of genetic diversity does give rise to phenotypic variance, it would merely represent one source of developmental noise.

Examination of more than 100 individuals of an isogenic *Daphnia* clone revealed only one that did not have exactly 22 bundles of eight photoreceptor axons; that specimen had one nine-axon bundle (Macagno 1980). No variance was observed in the position of the photoreceptors within the ommatidial array, and, in 100 optic cartridges examined, the cell body of only one second-order neuron was seen to be in an atypical location (Levinthal et al 1975). Analysis of the fine structural details of the synaptic contacts between the eight photoreceptor axons and the five second-order laminar neurons showed that there exist "strong" and "weak" connections. Strong connections consist of 30 or more synaptic contacts between a photoreceptor axon terminal and a laminar neuron, whereas weak connections consist of no more than five such synaptic contacts. The pairs of first- and second-order neurons linked by strong connections were found to be the same from specimen to specimen, although the number of actual synaptic contacts between any given strongly linked pair varied from about 30 to 100. By contrast, the identity of the neuronal pairs found to be linked via weak connections showed a high degree of variance.

A low variance in the projection pattern of compound eye photoreceptor axons to the lamina was established also in nonisogenic specimens of the fly *Calliphora* (Horridge & Meinertzhagen 1970). (In order to establish some upper limit to the developmental noise in the formation of some neuronal structure the population of animals studied need not be isogenic, since true isogenicity would only serve to reduce phenotypic variance.) Here two of the eight photoreceptors of each ommatidium project their axons directly to the optic medulla without forming synaptic contacts in the lamina. Of the remaining six photoreceptors, each projects its axon to a different optic cartridge in the lamina. At that cartridge, the axon meets the axon terminals of five other photoreceptors, each from a different neighboring ommatidium. This optical system is constructed so that all six photoreceptors whose axons project to the same cartridge receive light from the same point in the visual field. Examination of the projection pattern of about 650 photoreceptors, originating in 120 different *Calliphora* ommatidia, failed to turn up a single case of an axon entering the wrong optic cartridge, i.e. a cartridge in which an axon would have met axons reporting incident light from a different point in the visual field.

The number and morphology of identifiable second-order neurons contacted by axons of ocellar (rather than ocular) photoreceptors have been examined for their phenotypic variance in isogenic individuals of parthenogenetic clones of the locust *Schistocerca* (Goodman 1974, 1976, 1977, 1978). Of these ocellar interneurons, more than 70 can be individually identified from animal to animal. Study of 11 parthogenetic locust clones showed that one particular interneuron was present in duplicate in about one-half to one-third of the individuals belonging to three such clones. Among the individuals of the remaining eight clones, duplications of that interneuron were rarely seen. No individuals lacking that particular inter- neuron occurred in any of these parthogenetic clones, although in some nonisogenic populations of *Schistocerca* that interneuron is not infrequently absent. A survey of the location of the cell bodies of the identified interneu- rons showed that among the individuals of one clone the relative position of a particular interneuron was more variable than was the case in the other clones, and the major branching pattern of the axon of a third identified interneuron showed significant abnormalities in the majority of the individ- uals of two of the clones. The major branching pattern of the other interneu- rons was, however, quite invariant in these two clones. As for the fine branching pattern of the axons, it was found to be highly variable for all the identified ocellar interneurons. That variance was equally great for inter- and intra-clone comparisons. These differences in phenotypic vari- ance between individual locust clones indicate that the genotype can influ- ence the degree of developmental noise to which a particular neuron or neuronal structure is subject.

A large phenotypic variance was found to obtain in the properties of higher order visual interneurons of the locusts *Schistocerca* and *Locusta* (Pearson & Goodman 1979). Although the locusts examined were merely highly inbred, rather than authentically isogenic, it seems most likely that the variance observed is attributable largely to developmental noise rather than to genetic polymorphism. Even the major branches of the interneu- ronal axons were found to be so variable from individual to individual that no "normal" pattern could be described. This morphological variability of the interneurons was matched by a corresponding variability in its func- tional synaptic connections to a set of motorneurons innervating leg and wing muscles; that is to say, functional connections were present in some individuals that were absent in many others.

Thus, these few studies of phenotypic variance show that if the nervous system and its components are examined at progressively finer levels, some point is eventually reached at which the effects of developmental noise become manifest. In some cases that point is reached already at the rela- tively gross level of cell number, whereas in other cases it is reached only at the level of fine branching of neuronal processes and individual synaptic sites. One can account for this fact by resorting to the facile, all-purpose evolutionary explanation that in the development of any part of the nervous system just that amount of developmental noise is tolerated which is still compatible with adaptive function. The nematode, whose nervous system of only 258 cells generates a behavior of complexity not far behind that shown by animals with nervous systems consisting of hundreds or thousands of times more neurons, can probably ill-afford any departure from the normal cell number. Similarly, it can be argued that for the compound eye to provide its high-resolution image of the visual surround for effective guidance of visual behavior, the connections between primary ommatidial photoreceptors and optic cartridges must be of high precision. By contrast, because the ocellus appears to respond mainly to changes in overall light intensity (Chappell & Dowling 1972) and provides little spatial resolution of the visual surround (Wilson 1978), it can be imagined that here less precision and more developmental noise can be tolerated in the establishment of its central neuronal connections. Among the mechanisms that appear to be available for reducing the effect of developmental noise to acceptable levels is an initial overproduction of neurons and synaptic connections followed by functional testing and elimination of the unwanted excess (Stent 1978). But resort to any mechanism for the reduction of developmental noise represents, in the parlance of sociobiology, an investment on the part of the animal, which it would be wise to avoid unless it were absolutely necessary.

NEUROLOGIC MUTANTS

We now consider results obtained from the study of genetically heterogeneous populations and, in particular, from the comparison of normal animals and conspecific genetic mutants exhibiting abnormal neurologic phenotypes. Let us consider one of many possible examples illustrating this approach: the role of sensory afference in the development of the central nervous system of crickets, such as *Acheta* or *Teleogryllus*. From the rear of the abdomen of the cricket there projects a pair of sensory appendages, the cerci, that carry several thousand chemo- and mechanosensory organs. One type of organ consists of a filiform hair that is set into directionally selective vibration by sound or wind stimulation. The hairs are innervated by the endings of sensory neurons whose axons project via the cercal nerve to the abdominal ganglion, where they form synaptic contacts, particularly with the ipsilateral Medial Giant Interneuron (MGI) of the ventral nerve cord. Two sound- and wind-insensitive cricket mutants were isolated by Bentley (1973, 1975, 1977). In one mutant some of the filiform hairs are absent and in the other they are missing altogether, which is the reason that exposure of the cerci of these mutants to the normally adequate mechanical stimulus evokes no response. Despite their lack of functional hairs, the mutant cerci carry normal-looking sensory cells. Moreover, intracellular recordings taken from the MGI of the mutants showed a normal postsynaptic response following direct electrical stimulation of the hairless sensory cells. The hairless sensory cells are therefore electrophysiologically functional, albeit inactive because they are unresponsive to natural stimuli. Upon neuroanatomical examination of these mutants Bentley found, however, that the dendrites of their MGIs are stunted and significantly thinner than normal. What developmental inferences can be made from these findings? First, it follows that the development of functional mechanosensory cells on the cerci does not require the postembryonic presence of functional hair cells. Second, development of functional central synaptic contacts between the sensory axon terminals and the MGI does not require a normal electrical activity pattern on the part of the presynaptic sensory cells. And third, normal structural development of the MGI dendrites does require a normal afferent synaptic input pattern from their presynaptic sensory cells.

Some good will is actually necessary for the acceptance of the third, and most important, conclusion regarding the morphogenetic role of sensory afference; since all tissues, and not just the cerci, of the mutant crickets are of mutant genotype, it is formally possible that the abnormal development of the MGI represents a direct morphogenetic effect of the gene mutation on interneuron development. In that case the inferred indirect effect of the absence of functional cercal hairs mediated via the lack of normal activity of the cercal sensory cells and their consequent lack of normal afferent synaptic input to the developing MGI might not exist; however, idependent, nongenetic experiments make that possibility unlikely. A very similar stunted anatomy of MGI dendrites was found in genetically normal crickets that were raised throughout their postembryonic development with their cerci having been surgically removed (Murphey et al 1975). Here sensory deprivation achieved by surgical means caused a similar morphogenetic effect on the central nervous system as that observed under analogous deprivation achieved by gentic means. It is fortunate that Bentley troubled to ascertain the state of the cercal hairs of his sound- and wind-insensitive cricket mutants, instead of proceeding directly to an examination of their neuroanatomy. Otherwise some believers in the genetic approach might have concluded from the abnormal MGI morphology that the mutation is located in a gene that specifies the structure of the cricket central nervous system. That the abnormal neurologic phenotype engendered by a particular gene mutation is the result of a cascade of pleiotropic effects initiated by a (usually unknown) primary effect of the mutation is a feature inhering in nearly all the neurogenetic examples to be considered below.

GENETIC MOSAICS

In view of the difficulty of making straightforward developmental interpretations of the phenotypes of mutant animals whose tissues are all of mutant genotype, it is fortunate that there exists a genetic method that can be called on to ascertain whether the neurologic abnormality of structure or function A is really a direct result of the mutation or merely a secondary, cascading result of the abnormal phenotype of B. This method consists of the production of *genetic mosaics,* i.e. of animals, some of whose cells are of mutant genotype and others are of normal genotype. The mosaic method is especially useful in situations where the kind of nongenetic, surgical intervention cannot be carried out that, in the case of the cricket mutants, gave support to the interpretation of an influence of the activity of the presynaptic sensory cell on the morphology of the postsynaptic interneuron. It is not accidental that the use of genetic mosaics for developmental neurobiology has thus far been limited mainly to *Drosophila* (see Hall et al 1976, Palka 1979 for review) and to the mouse (see Mullen & Herrup 1979 for review), because they are some of the few animals for which the diversity of mutants with altered neurologic phenotype and the very high level genetic technology required for the experimental production of mosaics is presently available.

Drosophila

In *Drosophila,* two different methods for the production of genetic mosaics can be used. One of these consists of the experimental generation of gynanders, or individuals whose tissues form a mosaic of male and female cells. For this purpose a female zygote of XX genotype is produced in which one of the pair of X sex chromosomes is ring-shaped rather than, as is normally the case, rectilinear. There is a high probability that the ring-X chromosome is lost during one of the first few mitotic nuclear divisions in the developing early embryo, thus giving rise to a daughter nucleus that carries only a single X chromosome, or is of XO genotype. (Although in *Drosophila* the normal male genotype is XY, zygotes of XO genotype also develop the male phenotype.) In this way there arises in the female embryo a clone of male cells that eventually gives rise to a patch of contiguous male tissue in the post-embryonic animal. The position of the boundary between male and female tissues in the gynandromorph mosaic is highly variable, because in *Drosophila* embryogenesis the first nuclear division in the fertilized egg is oriented at random with respect to the future body axes. In order to identify the patch of male tissue, the rectilinear X chromosome usually carries a mutation whose expression is recessive to the dominant allele carried by the ring-X chromosome subject to elimination. Thus, only the male cells of XO genotype show the recessive mutant phenotype.

The second method for generating mosaic flies relies on the induction of somatic recombination of homologous chromosomes during nuclear divisions at stages of early embryogenesis. For this purpose, a heterozygote is produced that carries a recessive mutation on one chromosome and the dominant allele on the homolog of that chromosome. Exposure to low doses of X rays of the heterozygote nuclei at a stage of the mitotic cycle when their chromosomes have already replicated (i.e. are present as sister chromatids) induces crossing-over between a pair of homologous non-sister chromatids. Such somatic crossing over can result in a diploid homozygote nucleus in which both homologous chromosomes carry the recessive mutant locus. In this way there arises a clone of homozygotic cells that eventually produces a patch of contiguous tissue in which the recessive mutant phenotype is expressed, with the remainder of the body showing the phenotype of the dominant allele. As in the case of the gynanders, the position of the boundary between the two types of tissue in somatic recombinant mosaic flies is highly variable; however, in contrast to the gynander method, the somatic recombinant method permits experimental control over the size of the mosaic patch, in that the patch will be generally smaller the later the developmental stage at which the embryo is X-irradiated. Moreover, whereas the gynander method is limited to studying the developmental effects of mutant loci on the X-chromosome, the somatic recombinant method is applicable also to gene mutations carried on the three other (autosomal) *Drosophila* chromosomes.

FATE MAPPING   At an early developmental stage, the *Drosophila* embryo consists of a single, superficial layer of 5000 to 10,000 cells, the blastoderm. As was first realized by Sturtevant (1929), the distribution of the boundary between male and female tissues in post-embryonic *Drosophila* gynanders should provide an indication of the spatial distribution of the precursor cells of various body parts over the blastoderm; for, so Sturtevant reasoned, the more frequently the male-female tissue boundary happens to fall between any two given body parts in a population of gynanders, the more distant should the precursor cells of these body parts, or landmarks, be on the blastoderm. By observing the pairwise frequency of the male-female difference between an ensemble of such landmarks, a two-dimensional map can be constructed that represents the relative disposition of the landmark precursor cells over the blastoderm. In order to apply such blastoderm fate-mapping to landmarks of the *Drosophila* nervous system, a recessive mutation leading to the loss of the histochemically identifiable enzyme acid phosphatase in neurons of male genotype was introduced into the X-chromosome (Kankel & Hall 1976). By these means it is possible to show that the precursor cells of the central nervous system lie in the ventral area of the blastoderm, with the precursors of the supraesophageal, subesophageal, and optic parts of the head ganglion and the thoracic and abdominal parts of the thoracicoabdominal ganglion being disposed in an antero-posterior sequence. It was also possible to estimate the number of blastoderm cells that constitute the precursor ensemble of individual ganglia, by searching for the smallest patch of male or female tissue that can be observed within a single mosaic ganglion. Since no mosaic patches smaller than one-third to one-tenth of the total ganglionic mass were seen, it can be concluded that only a few blastoderm cells (from three to ten) are the precursors of each major ganglion on the right or left side of the body.

Fate-mapping studies also showed that the precursor cells of the *Drosophila* compound eye are situated in the anterior part of the blastoderm, but at a level more dorsal than that of the precursors of the ganglia of the central nervous system. How do these ocular precursor cells eventually give rise to the highly regular array of ommatidia? Early studies on the development of the arthropod compound eye had led to the conclusion that the eight primary photoreceptor cells of each ommatidium form a clone, i.e. are the descendants of a single founder cell (Bernard 1937). Both radiological and genetic experiments later seemed to provide support for this view of the development of the 700 to 800 ommatidia of the *Drosophila* compound eye (Becker 1957); however, more recent findings with genetic mosaics of *Drosophila* produced by the somatic recombination method argue against this view (Hofbauer & Campos-Ortega 1976, Ready et al 1976). In these studies heterozygous fly embryos carrying a recessive white (i.e. pigmentless) eye color mutation and its dominant red (i.e. normally pigmented) allele were X-irradiated at an early developmental stage to induce somatic recombination. Among the resulting adult flies there were many with a mosaic eye, having one patch of ommatidia with normally pigmented photoreceptors and another patch of ommatidia with pigmentless photoreceptors. Contrary to the expectations of the view of the ommatidium as a single cell clone, the border between pigmented and pigmentless patches was found to be formed by ommatidia that were themselves of mosaic character. That is to say, of the eight photoreceptors within a single ommatidium, some were pigmented and others not. Hence they could not all be the descendants of a single founder cell. Thus the morphogenetic selection of eight photoreceptor cells for formation of a given ommatidium is not strictly determined by their developmental line of descent. It is nevertheless still possible that the *Drosophila* photoreceptors do arise as eight-cell clones but that the members of a single clone are not constrained to participate in the formation of the same ommatidium (Campos-Ortega & Hofbauer 1977).

ORGANIZATION OF THE OPTIC LAMINA   How, in the development of the arthropod visual system, is the low phenotypic variance achieved in the connection of photoreceptor axons with the second-order visual neurons of the optic cartridges? A partial answer to this question is that the ommatidia of the compound eye play an organizational role in the development of the underlying optic lamina. One of the most convincing demonstrations of this developmental interaction was provided by the use of *Drosophila* mosaics (Meyerowitz & Kankel 1978). For this purpose, two recessive mutations, *rough* eye and *glass* eye, and one dominant mutation, *glued* eye, were studied. The phenotype attributable to any one of these three mutations consists of severe abnormalities in the structure of both the ommatidia and the optic cartridges. In order to ascertain whether the abnormal phenotype of one of these components of the arthropod visual system is merely a secondary consequence of the abnormal phenotype of the other, mosaic flies were produced by both the gynander and the somatic recombination methods. In these mosaics either the ommatidia or the optic lamina were of normal genotype, while the other component was of mutant genotype. (Here the genotype of the photoreceptor cells was identified by means of an eye color mutation linked genetically to the eye structure mutation; the genotype of the optic lamina neurons was identified by means of a linked mutation leading to loss of acid phosphatase.) The result of this experiment was that in all cases in which a patch of the eye was of mutant genotype, both the ommatidia of that patch and the underlying optic cartridges were of abnormal, i.e. structurally disrupted, phenotype; but in the genotypically normal remainder of the mosaic eye, both the ommatidia and their correspondent optic cartridges were of normal phenotype, even when the underlying optic lamina neurons were of mutant genotype. Thus, it can be inferred that the effect of the three eye structure mutations studied here is more proximal to the development of the ommatidia than it is to that of the optic lamina. More importantly, it can be concluded that the normal development of the optic cartridges depends on an interaction of the neurons of the optic lamina with a structurally normal array of ommatidia and, presumably, on a normal projection pattern of photoreceptor axons.

BEHAVIORAL MOSAICS   As was shown by Hotta & Benzer (1972, 1976), the genetic mosaic method can be used also to pinpoint the neural structures, or "foci," that are in control of behavioral routines. One such routine that was examined in *Drosophila* gynanders is the wing vibration associated with male courting behavior. These studies showed that mosaic flies with male heads and female bodies extend their wings toward females, as if in courtship. But the actual wing vibration is produced only if, in addition to the male head, there is some male tissue present also in the thoracic body segments. The neurologic interpretation of these findings was in some doubt, however, because the sexual genotype of head and thorax had been ascertained by examining the male-female boundary of the external cuticle of mosaic flies rather than of the underlying neural tissues. Subsequently, courting behavior was studied also in gynanders whose neurons were scored for their male or female genotype by the acid phosphatase method. In this way it was shown that for there to be any vestige of male courting behavior, such as extension of the wing toward the female, it is necessary that the dorso-posterior part of the gynander head ganglion is of male genotype (Hall 1979), but for wing vibration to occur, some male tissue must also be present in the thoracic ganglion (von Schilcher & Hall 1979). Thus, it can be concluded that one part of the male-specific neural circuits required for this behavioral routine (presumably the "command" component) is located in the head ganglion and another part (presumably the "motor pattern generator" is located in the thoracic ganglion of the ventral nerve cord. Within the context of developmental neurobiology these findings imply that the formation of a particular neural network can depend on the local genotype of the neurons that compose it, rather than on body-wide, e.g. hormonal, influences.

HOMEOTIC MUTANTS   There exists a class of mutants of *Drosophila*, termed *homeotic*, in which one part of the body of the fly is transformed into another (Lewis 1963, Ouweneel 1976). One example of such a mutant is *Antennapedia*, in which the distal part of the antenna is transformed into a mesothoracic (or middle) leg. Another is *bithorax-postbithorax* (*bx pbx*), a double mutant that shows a transformation of the meta- (or hind) thorax into a second meso- (or middle) thorax and is thus endowed with a second pair of wings in place of the pair of halteres, balancing organs that aid the fly in navigation. Since their abnormally positioned, or ectopic, structures are innervated, homeotic mutant flies are highly promising objects for the study of the mechanisms that govern the formation of neural connections. For instance, application of a sugar solution to the homeotic leg of an *Antennapedia* mutant evokes the proboscis-extension reflex also evoked by stimulation of normally positioned legs, but not by stimulation of normal antennae (Deak 1976, Stocker 1977). This indicates that despite its ectopic attachment to the head, the homeotic leg carries chemoreceptors that make functional connections in the central nervous system capable of eliciting a normal motor output. Anatomical study of the sensory fibers projecting centrally from the ectopic leg show that they enter the head ganglion via the antennal nerve and ramify in the head ganglion in a manner similar to the normal antennal sensory fibers (Stocker et al 1976). The sensory fibers from the ectopic leg do not appear to project rearward to the thoracicoab-dominal ganglion that is the ordinary destination of the sensory fibers from the normally positioned mesothoracic leg. The sensory fibers from the normally positioned leg do not in turn project frontward to the head ganglion. In view of this neuroanatomical hiatus, it seems a considerable challenge to try to explain how such a functional pathway from sensory receptors to motor neurons responsible for the proboscis extension reflex manages to be formed in the case of the ectopic *Antennapedia* leg.

The anatomy of wing sensory fibers of *bx pbx* mutants has also been studied in considerable detail (Ghysen 1978, Palka et al 1979). As for the normal (mesothoracic) wings, they carry three anatomically different types of mechanosensory structures: bristles and large and small campaniform sensilla (or sc), which project their axons into the nerve cord via a mesothoracic nerve. The bristles send fine sensory axons into the ventral region of mesothoracic segment of the thoracicoabdominal ganglion. The large sc send thick axons that ramify in the ventral region of all three thoracic segments of that ganglion. And the small sc send axons into a dorsal tract of the nerve cord, where they bifurcate and project both frontward to the head ganglion and rearward to the metathoracic segment of the thoracicoabdominal ganglion. By contrast, the normal (metathoracic) halteres carry only small sc, which send axons via the (posteriorly situated) haltere nerve into the dorsal nerve tract. There they project frontward to the head ganglion and ramify along the way in both metathoracic and mesothoracic segments. No axonal projections are manifest from normal halteres to the ventral region of the nerve cord, the destination of the sensory axons of wing bristles and large sc.

The homeotic (metathoracic) wing of the *bx pbx* mutant resembles the normal wing in carrying bristles as well as large and small sc. The axons of these homeotic sensory structures all enter the central nervous system at the site normally occupied by the haltere nerve. Once in the central nervous system, the central projections of the axons of the sensory structures on the homeotic wings form a pattern that would be expected of posterior serial homologs of the corresponding sensory structures on the normal wings. Thus the bristles on the homeotic wing send their fine axons to the ventral region of their own metathoracic segment, where they form a pattern similar to that formed in the mesothoracic segment by the bristle axons from the normal wing. The thick axons of the large sc on the homeotic wing ramify in the ventral region of all three thoracic segments of the thoracicoabdominal ganglion, where they overlap with the axon endings of the large sc on the normal wing. And the axons from the small sc on the homeotic wings enter the dorsal tract of the nerve cord, ramify within the metathoracic segment, and project frontward to the head ganglion, just as do the axons of the small sc on the normal wings. In this last case, of course, the axons of the small sc on the homeotic wings can also be said to project centrally as if they still belonged to halteres. Indeed, the pattern of ramification of the small sc axons on the homeotic wing in the metathoracic segment is indistinguishable from that of small sc on halteres in normal flies.

How is one to account for these findings with *bx pbx* mutants? As far as the large sc on the homeotic wing are concerned, one could imagine that their sensory axons are destined to seek the same central target neurons as the homologous axons coming from the large sc on the normal wing, even though the former enter the nerve cord on an abnormal (posterior) level. The same explanation can be given for the small sc on the homeotic wing, although in this case an even simpler explanation is available, namely that the small sc on the homeotic wing are destined for the same central targets as the small sc of the absent haltere. But a more complicated explanation is required for the bristles on the homeotic wing, whose axons end in the metathoracic segment of the thoracicoabdominal ganglion not normally reached by axons from any bristle cells. One likely explanation is that in the metathoracic segment there normally arise neurons that are appropriate targets for wing bristle axons but are not ordinarily reached by them because bristle cell axons of the normal wing are somehow destined to remain in the segment at which they entered the nerve cord. But there is also an alternative explanation that could be considered for the observed projection pattern of the bristle cells on the homeotic wing: Contrary to the general finding that the proximal phenotypic effect of homeotic mutations is confined to tissues derived from imaginal disks [of which the nerve cord is not one (Ghysen 1978)], it would still be formally possible that, as a result of their own mutant genotype, the neurons of the metathoracic segment of the nerve cord take on the quality of mesothoracic neurons. In that case, the central projection pattern of the homeotic wing sensory cells would reveal little about the mechanisms underlying the normal case. In order to test for this possibility, Palka et al (1979) produced mosaic flies with patches of homeotic bithorax wings embedded within an otherwise normal haltere, and with a central nervous system of non-homeotic genotype. Neuroanatomical analysis of the sensory projections from such homeotic wing patches to the genotypically normal metathoracic segment of the thoracicoabdominal ganglion showed the presence of fine axons in the ventral region, characteristic of the sensory fibers from homeotic wing bristle cells. It follows, therefore, that in the case of *bx-pbx* mutant flies it is not the mutant genotype of the central nervous system that is responsible for directing bristle cell fibers to a metathoracic region that is ordinarily never reached by such axons.

Mouse

In the early 1960s a technique became available by means of which mice (and by now, some other mammals) can be produced whose tissues are mosaics of cells of two different genotypes (Tarkowski 1961, Mintz 1962, 1965). With this technique, an early embryo, or morula, consisting of only a few cells is removed from the oviduct of a pregnant female mouse and dissociated into its constituent cells by enzyme treatment. These cells are mixed with a second set of dissociated cells similarly obtained from another pregnant female mouse carrying a morula of a different genotype. Upon incubation for some hours, the mixed cells reassociate, continue to cleave, and form a blastocyst. This blastocyst is then reimplanted into the uterus of a female host animal. There the artificial hybrid blastocyst continues normal embryonic development and is eventually brought to term as a morphologically normal mouse, some of whose cells are derived from one pair of parents and others from another pair of parents. The mosaic character of such tetraparental mice is readily manifest if the two parental pairs differed from each other genetically in some external characteristic, such as coat color. In mosaic mice, the actual proportion of the total cell population descended from either of the two mixed morulas varies within wide limits; some mosaic specimens are 50/50 mixtures of the two cell genotypes, in others one genotype contributes as little as 1% of the total, and yet others show no evidence of mosaicism at all. This variability reflects a developmental indeterminacy, regarding which few morula cells contribute any descendants at all to the post-natal mouse and the developmental fate of the progeny of any morula cell that does make a contribution. This method can thus generate a large variety of mosaic patterns. For use of such mosaic mice in the context of developmental neurobiology, cells of two morulas are mixed: one is of normal genotype and the other is of a mutant genotype that gives rise to some particular neurologic abnormality. In addition, the two morulas are made to differ at some other genetic locus, so as to produce a histologically detectable but neurologically neutral difference between the nerve cells derived from either source. Two neutral cell-marker loci favorable for this purpose are the structural genes for β-glucoronidase (Condamine et al 1971, Feder 1976, Mullen 1977a,b) and β-galactosidase (Dewey et al 1976), at which mutations affect the intensity of staining produced by an enzyme-specific histologic reaction.

RETINAL DEGENERATION MUTANTS   The first neurological use of mosaic mice was made in the study of a retinal degeneration, or *rd,* mutation that leads to the postnatal degeneration of the retinal photoreceptor cells (Mintz & Sanyal 1970, Wegmann et al 1971). Here mixing of morulas of normal and *rd* genotype resulted in mice whose retinae consisted of a mosaic of structurally normal patches and patches where the photoreceptors had degenerated. The neutral cell marker showed that the pigment epithelium underlying the photoreceptors was also of genotypically mosaic character. But the phenotype of the photoreceptors in any retinal patch was found to be unrelated to the genotype of the underlying pigment epithelium.

Thus it can be inferred that the degenerative effect of the *rd* mutation is expressed directly within individual photoreceptor cells, rather than being the result either of a retina-wide influence of the presence of the mutant gene in some other tissue or of a local influence of its presence in the contiguous pigment epithelium. This finding contrasts with a seemingly similar photoreceptor degeneration phenotype in the rat. In that case retinal disorganization was found to be a secondary effect of the presence of the mutant genotype in the underlying pigment epithelium (La Vail & Mullen 1976, Mullen & La Vail 1976). Another instance of an indirect, or secondary effect was uncovered upon study of a muscular dystrophy, or *dy,* mutation. Here a mosaic mouse was found with a muscle of predominantly dystrophic phenotype, whose muscle fibers, according to the neutral cell marker, were of predominantly normal genotype. Conversely, in another mosaic mouse, a muscle of predominantly normal phenotype was found whose muscle fibers were, according to the neutral cell marker, of predominantly *dy* mutant genotype. Thus, dystrophy appears to be attributable to some disturbance of a normal muscle function by action of the *dy* mutation at a distant, extra-muscular site (Peterson 1974).

CEREBELLAR MUTANTS   Thus far mosaic mice have found their most extensive neurological application in the study of the dozen or so mutations that are known to perturb the development of the cerebellum and give rise to easily recognizable deficits in locomotory behavior (Sidman et al 1965). One such case examined by this method is the *pcd* mutation that causes postnatal degeneration of all the Purkinje cells in the cerebellar cortex. Here it was found that in mosaic mice of mixed *pcd* and normal genotype some Purkinje cells degenerate and some survive. The neutral cell marker indicated that all the surviving Purkinje cells were of normal genotype and that all the mutant Purkinje cells degenerated (Mullen 1977a). Hence it appears that the cerebellar effect of the *pcd* mutation resembles the retinal effect of the *rd* mutation, in that its degenerative action is expressed within individual cells of mutant genotype. It is to be noted, however, that the effect of the *pcd* mutation is not restricted to the cerebellar Purkinje cells, since it engenders also the degeneration of photoreceptor, mitral, and sperm cells (Mullen & La Vail 1975, Mullen et al 1976).

Reeler   The effects of the two other well-known, cerebellar mutations, *reeler* (*rl*) and *staggerer* (*sg*), have also been studied within the mosaic mouse context. The *reeler* mutation, rather than leading to degeneration of the Purkinje cells, prevents their migration during development of the embryonic cerebellum. This migration normally proceeds outward from the deep ventricular zone of origin of the Purkinje cells to the cerebellar cortex, where their cell bodies eventually form a single layer below the molecular layer formed by the parallel fibers and above the granule cell layer. But in *reeler* mutants, the vast majority of Purkinje cells fails to migrate outward and remains below the granule cell layer, which is itself composed of far fewer than the normal number of granule cells (Hamburgh 1963, Caviness 1977, Mariani et al 1977).

Upon production of *reeler*/normal mosaic mice, specimens were found with a cerebellum whose gross morphology was intermediate between the fully normal and the fully *reeler* phenotype: Some Purkinje cells had migrated to their appropriate positions in the cerebellar cortex whereas other Purkinje cells had remained in the underlying deep zone. As shown by the neutral cell marker, many of the normally positioned cells were of the mutant *reeler* genotype and many of the abnormally positioned cells were of the normal genotype (Mullen & Herrup 1979). This finding shows that its failure to migrate outward is the result of an indirect, secondary effect of the *rl* mutation on the Purkinje cell, with the primary effect being exerted at another, extrinsic site. In thus exerting its effect indirectly, or transcellularly, the *reeler* mutation seems to resemble the *dy* muscular dystrophy mutation. However, the interspersion of phenotypically normal and abnormal Purkinje cells makes it appear that each of these cells is positioned individually during histogenesis of the cerebeller cortex and that the signals mediating the transcellular effect of the *rl* mutation aborting normal migration reach the target cells, not in the form of a homogenous field, such as would be produced by diffusion of a chemical substance, but as a fine-grained mosaic. Two possible explanations of such a fine-grained mosaic of transcellular signals are that the migration of each Purkinje cell is individually stimulated by early synaptic contact with the axon terminals of another type of neuron (Caviness & Sidman 1972) or is individually guided by contact with a particular glial cell (Caviness & Rakic 1978). Here the eventual position of a Purkinje cell would depend, not on its own *reeler* genotype, but on that of the presumptive presynaptic stimulatory neuron or guiding glial cell with which it happens to come into contact (Mullen 1977b).

As is the case for the *pcd* (Purkinje cell degeneration) mutation, the effect of the *rl* mutation is not confined to the cerebellar Purkinje cells. In *reeler* mice the structure of the cerebral cortex is also highly abnormal in that here, too, neurons fail to migrate during development. Just as do the Purkinje cells, the neurons of the cerebral cortex arise in the deep ventricular zone and later migrate outward. The nature of the migration in the cerebral cortex is such that the first neurons to arise remain as the deepest layer, with neurons arising at progressively later developmental stages and migrating outward past the neurons that have arisen before them. Thus the normal cerebral cortex is composed of a set of cell layers of which the most superficial layer contains the last, and the deepest layer the first, neurons to have arisen. Because in the *reeler* mutant this migration is aborted, the structure of its cerebral cortex is inside-out, with the most superficial layer containing the first, and the deepest layer the last neurons to have arisen (Caviness & Sidman 1973). Despite this spatio-temporal inversion of the normal structure of the cortical cell layers, the afferent nerve fibers of the optic radiation manage to project to cell layers of appropriate age in the visual cortex of *reeler* mutants (Caviness 1977). Moreover, neurophysiological study of the visual neurons of the *reeler* cortex showed that both their receptive field structure and their retinotopic organization is quite normal (Dräger 1977). No studies have yet been reported of the visual cortex of *reeler*/normal mosaics, but the present findings with *reeler* mutants nevertheless allow at least one significant conclusion regarding cortical development: The eventual identity of cortical neurons with respect to both morphology and function is more closely related to their "birthdates" than to their positions along the radial (inside-outside) axis. And, as a corollary, it can be concluded that the formation of neuronal synaptic contacts within the cortex is not so much governed by cell position along the radial axis as it is by cell identity.

Staggerer   The *staggerer* mutation leads to a poorly developed cerebellum nearly devoid of the normal granule cell layer. Although during embryonic development of *staggerer* mice the granule cells arise more or less normally on the outer surface of the cerebellum, the granule cells subsequently degenerate during their inward postnatal migration across the molecular and Purkinje cell layers and fail to reach the deeper zone at which the granule cell layer is ordinarily present. The morphology of the Purkinje cells of the *staggerer* cerebellum is also abnormal in that (*a*) they do not form their characteristic single-cell layer and (*b*) they have dendritic trees that are very sparsely developed, whose much fewer dendrites lack the spines at which the granule cell axons, or parallel fibers, normally make synaptic contacts (Landis & Sidman 1978, Landis & Reese 1977, Sotelo & Changeux 1974, Yoon 1977). Upon production of *staggerer*/normal mosaic mice, individuals were found that, as in the case of *reeler*/normal mosaics, had a cerebellum whose gross morphology was intermediate between fully normal and fully mutant phenotypes (Mullen & Herrup 1979). In these mosaics, there was present a nearly normal and well-developed granule cell layer, but the Purkinje cells formed a patchwork of cells, of which some had abnormal positions and sparse and spineless dendrites and others had normal positions and dendritic trees. The status of the neutral cell marker indicated that the Purkinje cells with abnormal positions and sparse dendrites were of *staggerer* mutant genotype and that cells with normal positions and normal dendrites were of normal, non-mutant genotype. Thus, in the directness of its effect on Purkinje cell development, the *sg,* or *staggerer* mutation resembles the *pcd,* or Purkinje cell degeneration mutation, and differs from the indirectly acting *rl,* or *reeler,* mutation. Because of the small size of the granule cells, it has not been possible thus far to establish their genotype by means of a neutral cell marker. Thus it has not been ascertained whether the normally appearing granule cell layer of the *staggerer* mosaic mice comprises cells of both normal and *staggerer* mutant genotype; however, it seems probable on other grounds (Landis & Sidman 1978, Messer & Smith 1977) that the death of granule cells in *staggerer* mutant mice is a secondary consequence of the primary effect of the *sg* mutation on Purkinje cell development. In other words, the postnatal survival and normal inward migration of granule cells appears to require their interaction with Purkinje cells having normal positions and dendrite structure, a requirement that would explain also the granule cell deficiency of the *reeler* cerebellum. Hence one would expect that the granule cell layer of *staggerer* mosaic mice actually contains cells of normal phenotype that nevertheless carry the *staggerer* genotype.

THE SIAMESE CAT

Finally, as our last example, we consider a familiar mammalian mutant phenotype, the Siamese cat (see Guillery et al 1974 for review). The visual pathway of mammals is generally arranged such that the visual cortex receives visual input from the nasal part of the contralateral retina and from the temporal part of the ipsilateral retina. To produce this projection pattern, the optic nerve axons of ganglion cells originating in the nasal part of the retina cross over, or decussate, at the optic chiasm to the contralateral lateral geniculate nucleus (LGN), whereas the axons originating in the temporal part of the retina do not cross over at the optic chiasm and proceed to the ipsilateral LGN. In the ordinary domestic cat, as in other carnivores and in primates, the retinal line of demarcation for decussation of ganglion cell axons is just midway between the nasal and temporal edges. This is a necessary correlate of the fact that these species have binocular vision over nearly the whole of their visual field, since here left nasal and right temporal half retinas both receive visual input from the left half of the field, whereas right nasal and left temporal half retinas both receive visual input from the right half of the field.

The feline LGN consists of three layers, of which the outer two layers receive the axons projecting from the contralateral (nasal) retina and the middle layer receives the axons from the ipsilateral (temporal) retina. The entire projection occurs in a retinotopic order: The retinal ganglion cells located near the retinal midline project to the medial edge of each LGN layer; ganglion cells located at progressively more nasal sites in the (contralateral) retina project to progressively more lateral sites in the outer two LGN layers, and ganglion cells located at progressively more temporal sites in the (ipsilateral) retina project to progressively more lateral sites in the middle LGN layer. The axons of the higher order LGN neurons of these layers in turn travel to the ipsilateral visual cortex via the optic radiation, where they form a *binocular* retinotopic projection: Axons reporting visual input from corresponding parts of the contralateral visual field received through either eye project to common sites on the visual cortex, in a characteristic nasotemporal order.

As was noted by Guillery (1969), this normal projection pattern is significantly perturbed in the visual pathway of Siamese cats. As far as the projection from retina to LGN is concerned, the line of demarcation for decussation of retinal ganglion cell axons is displaced from the dorsoventral retinal midline toward the temporal edge of the retina: Some retinal ganglion cell axons from the temporal half of the retina project to the contralateral instead of ipsilateral LGN. There their endings occupy those positions of the middle layer that, in normal cats, are occupied by the ganglion cell axons from corresponding locations in the ipsilateral retina (Guillery & Kaas 1971, Shatz 1977a, Kaas & Guillery 1973). As for the projection from LGN to visual cortex, that projection is reorganized in Siamese cats in one of two quite distinct modes. Under one of these modes the axons from the contralaterally (i.e. abnormally) innervated part of the LGN middle layer pass to the cortical area normally innervated by them, but in an order that is the reverse of normal. The rest of the projection from the LGN is compressed in a normal retinotopic order onto the remainder of the visual cortex. The result of this reorganization is the reconstitution of a normal nasotemporal sequence of visual field representation, but with two domains of monocular rather than binocular visual input (Hubel & Wiesel 1971, Shatz 1977b). Under the other reorganization mode, the anatomical pattern of the projection from LGN to visual cortex is apparently normal, but the whole of the input from the LGN middle layer is functionally suppressed. The result of this other reorganization mode is that these Siamese cats can see only the nasal part of their visual field (Kaas & Guillery 1973).

These findings suggest that in the development of the mammalian visual system innervation of the LGN layers is guided mainly by the position of origin of retinal ganglion cell axons. That is to say, neither the ipsi- or contralateral origin of a retinal axon nor the visual coherence of the eventual projection appears to play a determinative role in the establishment of the normal retinotopic order on the LGN. By contrast, the innervation of the visual cortex cannot be guided mainly by the position of origin of the LGN cell axons but must be governed by a process that demands visual coherence of axons making functional connections at the same cortical site.

Although the genome of the Siamese cat certainly differs from that of ordinary cats in more than one genetic locus, it has been possible to identify —by induction rather than by conventional genetic crosses—the mutant gene whose presence is responsible for the abnormal decussation mode of retinal ganglion cell axons at the optic chiasm. It is the gene which encodes the structure of the enzyme tyrosinase that catalyzes a reaction step in the biosynthesis of the dark pigment melanin. In the Siamese cat this gene carries a mutation that renders the mutant protein unable to carry out its catalytic function at 37°C and thus prevents melanin synthesis at body temperature (Searle 1968). It is this mutation which is responsible for the characteristic Siamese coat color, namely the lightly pigmented body fur framed by black hair patches on the tips of the ears, the paws, and the snout. Despite the lack of any demonstration that in crosses of Siamese to normal cats neurologic phenotype invariably segregates with coat color, there can be little doubt that it really is the mutant tyrosinase gene (and the lack of melanin formation at 37°C) which is responsible for the abnormal retinogenicular projection pattern. For it turns out that the impaired projection of axons from the temporal retina to the ipsilateral LGN is a general property of albino (i.e. melanin-deficient) mutants of mammalian species, such as tigers, mice, rats, ferrets, and mink. Some of these mutants owe their lack of dark pigment to a mutation in a gene that encodes an enzyme other than tyrosinase in the pathway of melanin synthesis (Guillery 1974). But what is the possible connection between the formation of melanin and the ingrowth of optic nerve axons to the right or left LGN? And why does the absence of pigment produce their aberrant decussation, particularly in view of the fact that the retinal ganglion cells themselves do not normally contain noticeable amounts of melanin? The answers to these questions are not yet available, but the most likely developmental link between optic nerve axon projection pattern and melanin is provided by the layer of melanin-containing cells that form the pigment epithelium that underlies the retina and shields the photoreceptors from stray light. Thus, the absence of pigment from that epithelium would engender some retinal disturbance during the outgrowth of the optic nerve axons that causes some of them to end up on the wrong side of the brain. And as we already saw in the case of the photoreceptor degeneration mutant of the rat, an abnormal genotype of the pigment epithelium can have a disruptive influence on normal development of the retina.

CONCLUSION

The Instrumental Aspect

The sample collection of experimental findings presented in the foregoing indicates that the genetic approach to developmental neurobiology offers considerable strength, even though, as Pak & Pinto (1976) have remarked, "no major breakthrough has yet been achieved as a result of this approach. Nevertheless the studies that have been carried out to date show sufficient promise to be encouraging." Instead of major breakthroughs, the results of neurogenetic studies can be seen to have provided further support to (or placed on a more secure basis) notions generally held by developmental neurobiologists. Thus the detailed neuroanatomical comparisons of members of isogenic clones have brought into clearer focus the quantitative aspects of developmental noise, of whose existence few embryologists could have had any doubt. Similarly, the general proposition that presynaptic sensory neurons can play a morphogenetic or organizational role in the development of postsynaptic sensory interneurons, as demonstrated by the cases of the sound- and wind-insensitive cricket mutants, the *Drosophila* eye mutants, and the visual cortex of the Siamese cat, lies squarely in the conceptual mainstream of developmental neurobiology. And this is also the case for the converse proposition, applicable to the development of motor systems that, as demonstrated by the case of the granule cells in the *reeler* and *staggerer* mouse cerebellum, survival of the presynaptic neuron is contingent on contact with an appropriate target cell. Finally, the conclusions devolving from *Drosophila* homeotic mutants (that axons of peripheral sensory neurons can make appropriate central terminations despite an abnormal point of entry into the central nervous system) and from the innervation pattern of the LGN of the Siamese cat (that at their cerebral terminus sensory cell axons are recognized according to their relative positions of origin) had also been previously reached on other grounds (Frank et al 1977). Study of the Purkinje cells in the *reeler* mutants revealed a previously unknown feature of histogenetic cell migration, however, namely that migrating cells can be positioned individually by contact with a fine-grained, possibly cellular, mosaic rather than a homogeneous morphogenetic field. It is difficult to see how this insight could have been reached by any method other than genetic mosaics. In any case, the neurologic mutants have been of great value in unravelling the underlying mechanisms in the particular cases to which they pertain and in providing working material for the developmental neurobiologist. Some of the most fascinating instances among this working material, but possibly also the most difficult for future analysis, are the male/female behavioral mosaics of *Drosophila*. As for fate mapping, i.e. tracing the origin of the nervous system to specific sites of the blastula, this idea was first conceived and realized experimentally by direct observation of leech embryos more than a century ago by Whitman (1878). Nevertheless, the extension of genetic methodology to fate mapping in embryos of species such as *Drosophila*, whose complexity precludes direct observational methods, constitutes an important technical advance for developmental neurobiology. All these remarks pertain, of course, to the instrumental aspect of the genetic approach, to ignore which, according to Pak & Pinto (1976), "may well be to neglect one of the potentially most powerful tools in the study of the nervous system."

One limitation of the instrumental aspect of the genetic approach to neural development should be noted, namely that, thus far at least, its contributions have furthered the understanding mainly of systems such as the arthropod compound eye or the mammalian visual system and cerebellum, for which prior neuroanatomical, neurophysiological, and behavioral investigations had already provided a fairly high level of understanding. The reason for this is that whereas for any approach to the nervous system, be it anatomical, physiological, or behavioral, the investigator must bring some preunderstanding to his material, this requirement is much more demanding in the case of the genetic approach. As Kankel & Ferrus (1979) have pointed out, "the devising of screening procedures and the subsequent analyses of isolated mutations imply that one already knows at least some of the relevant questions to ask. As always, the latter is the most difficult of problems and is one for which genetics alone provides no unique answer."

The Ideological Aspect

Whereas the main strength of the genetic approach lies in the instrumental aspect, its main weakness derives from the ideological aspect. For the viewpoint that the structure and function of the nervous system of an animal is specified by its genes provides too narrow a context for actually understanding developmental processes and thus sets a goal for the genetic approach that is unlikely to be reached. Here "too narrow" is not to mean that a belief in genetic specification of the nervous system necessarily implies a lack of awareness that in development there occurs an interaction between genes and environment, a fact of which all practitioners of the genetic approach are certainly aware. Rather, "too narrow" means that the role of the genes, which, thanks to the achievements of molecular biology, we now know to be the specification of the primary structure of protein molecules, is at too many removes from the processes that actually "build nerve cells and specify neural circuits which underlie behavior" to provide an appropriate conceptual framework for posing the developmental questions that need to be answered. In this regard the ideological aspect of the genetic approach resembles the quantum mechanical approach to genetics (Jordan 1938) that had some vogue in the 1930s and 1940s. Since, so it was then thought, the structure and function of genes is obviously determined by the atoms of which they are composed, the goal of genetic research should be to give an account of heredity in terms of interatomic forces that govern the formation of chemical bonds. Eventually the problem of the mechanism of heredity was solved at the macromolecular rather than atomic level, although atomic interactions such as the hydrogen bond did turn out to have a crucial part in the story. Similarly, the insights into developmental mechanisms thus far available suggest that the solution to the problem of development lies at a cellular and intercellular rather than a genetic level, although genes will undoubtedly figure in some crucial part, but only a part, of that solution.

WHAT IS A PROGRAM? Those who speak of a genetic specification of the nervous system, and hence of behavior, rarely spell out what it actually is that they have in mind. On information-theoretical grounds, the genes obviously cannot embody a neuron-by-neuron circuit diagram; and even if they did, the existence of an agency that reads the diagram in carrying out the assembly of the component parts of a neuronal Heathkit would still transcend our comprehension. So a seemingly more reasonable view of the nature of the specification of behavior would be that the genes embody, not a circuit diagram, but a *program* for the development of the nervous system (Brenner 1973, 1974). But this view is rooted in a semantic confusion about the concept of "program." Once that confusion is cleared up, it becomes evident that development from egg to adult is unlikely to be a programmatic phenomenon. Development belongs to that large class of regular phenomena that share the property that a particular set of antecedents generally leads, via a more or less invariant sequence of intermediate steps, to a particular set of consequents; however, of the large class of regular phenomena, programmatic phenomena form only a small subset, almost all the members of which are associated with human activity. For membership of a phenomenon in the subset of programmatic phenomena it is a necessary condition that, in addition to the phenomenon itself, there exists a second thing, the "program," whose structure is isomorphic with, i.e. can be brought into one-to-one correspondence with, the phenomenon. For instance, the on-stage events associated with a performance of "Hamlet," a regular phenomenon, are programmatic since there exists Shakespeare's text with which the actions of the performers are isomorphic. But the no less regular off-stage events, such as the actions of house staff and audience, are mainly nonprogrammatic, since their regularity is merely the automatic consequence of the contextual situation of the performance. One of the very few regular phenomena independent of human activity that can be said to have a programmatic component is the formation of proteins. Here the assembly of amino acids into a polypeptide chain of a particular primary structure is programmatic because there exists a stretch of DNA polynucleotide chain—the gene—whose nucleotide base sequence is isomorphic with the sequence of events that unfolds at the ribosomal assembly site. However, the subsequent folding of the completed polypeptide chain into its specific tertiary structure lacks programmatic character, since the three-dimensional conformation of the molecule is the automatic consequence of its contextual situation and has no isomorphic correspondent in the DNA.

When we extend these considerations to the regular phenomenon of development we see that its programmatic aspect is confined mainly to the assembly of polypeptide chains (and of various species of RNA). But as for the overall phenomenon, it is most unlikely—and no credible hypothesis has as yet been advanced how this could indeed be the case—that the sequence of its events is isomorphic with the structure of any second thing, especially not with the structure of the genome. The fact that mutation of a gene leads to an altered neurologic phenotype shows that genes are part of the causal antecedents of the adult organism, but does not in any way indicate that the mutant gene is part of a program for development of the nervous system.

But are not polemics about the meaning of words such as "program" just a waste of time for those who want to get on with the job of finding out how the nervous system develops? As J. H. Woodger (1952) showed in his Tarner Lectures, "Biology and Language," published shortly before Watson and Crick's discovery of the DNA double helix and Benzer's reform of the gene concept, semantic confusion about fundamental terms, such as "gene," "genotype," "phenotype," and "determination," had become the bane of classical genetics. It would be well to avoid reconstituting that confusion in the context of developmental biology and to remember Woodger's advice (Woodger 1952, p. 6) that "an understanding of the pitfalls to which a too naive use of language exposes us is as necessary as some understanding of the artifacts which accompany the use of microscopical techniques."

DEVELOPMENT AS HISTORY   The notion of genetic specification of the nervous system is not only defective at the conceptual level but also represents a misinterpretation of the knowledge already available from developmental studies, including those that have resorted to the genetic approach.

As Székely (1979) has pointed out, we know enough about its mode of establishment already to make it most unlikely that neuronal circuitry is, in fact, pre-specified; rather, all indications (including those provided by the study of phenotypic variance reviewed here) point to stochastic processes as underlying the apparent regularity of neural development. That is to say, development of the nervous system, from fertilized egg to mature brain, is not a programmatic but a historical phenomenon under which one thing simply leads to another. To illustrate the difference between programmatic specification and stochastic history as alternative accounts of regular phenomena, we may consider the establishment of ecological communities upon colonization of islands (Simberloff 1974), or growth of secondary forests (Whittaker 1970). Both of these examples are regular phenomena, in the sense that a more or less predictable ecological structure arises via a stereotypic pattern of intermediate steps, in which the relative abundances of various types of flora and fauna follow a well-defined sequence. The regularity of these phenomena is obviously not the consequence of an ecological program encoded in the genome of the participating taxa. Rather it arises via a historical cascade of complex stochastic interactions between various biota (in which genes play an important role, of course) and the world as it is.

Although, compared to all the other cases of neurologic mutants considered here, our present knowledge of the genotypic differences between Siamese and normal cats is very crude, and the production of Siamese/normal mosaic cats has yet to be achieved, this case illuminates better than any other the strength of the instrumental and the weakness of the ideological aspect of the genetic approach to the development of the nervous system. Here mutation to loss of function of a single gene participating in a known metabolic pathway leads to a series of obviously interrelated changes in the structure and function of the central nervous system (in addition to the change in coat color). From the instrumental aspect, the existence of the Siamese cat (and of other mammalian albino mutants) is likely to be of enormous help in the elucidation of the mechanisms underlying the development of the mammalian visual pathway. First, study of the role of the presence of melanin in the retinal pigment epithelium in the pattern of optic fiber decussation at the chiasm should throw light on the mechanism that normally determines whether a retinal ganglion cell axon projects to the contra- or ipsilateral LGN. Second, study of the mode of (inappropriate) innervation of the middle layer of the LGN from the contralateral retina should provide valuable information on how the relative retinal position of origin of ganglion cell axons is normally recognized upon arrival at the LGN. And third, study of the rearranged cortical projection pattern of

LGN axons to the visual cortex should reveal insights into the hitherto quite mysterious processes by which coherence of binocular visual input plays its role in the establishment of cortical innervation by afferent neurons. As for the ideological aspect, however, the case of the Siamese cat reveals the conceptual poverty of the notion that genes specify the neural circuits that underlie behavior. All that the gene mutated in the Siamese cat can usefully be said to "specify" is the amino acid sequence of an enzyme that takes part in melanin synthesis, presumably in retinal epithelial cells. That gene is evidently one of the causal antecedents in the developmental history of the mammalian visual system, in that the absence of the polypeptide chain it specifies sets off a cascade of dysfunctional, albeit specific aberrations, which eventually lead to a specific reorganization of the feline brain. But despite these specific cerebral aberrations there is, in this paradigmatic case, no trace of genetic specification of neural circuitry.

Summary Appraisal

By way of a summary appraisal it can be said that the genetic approach is certainly of great practical and technical significance. First, within the context of neurophysiology, the availability of neurologic mutants can be of great assistance for the analysis of nerve cell functions and networks. For instance, an abnormal behavior and a concomitant abnormal neural structure of a mutant animal can obviously provide insights into how the normal circuitry generates normal behavior. Second, within the context of developmental neurobiology, the perturbation of developmental processes by mutant genes can help us to recognize the functional relations that create the normal pathways that lead to the genesis of the adult nervous system. Finally, within the context of psychology and medicine it is of the utmost importance to understand the hereditary contribution to differences in behavior. For instance, if it could be shown that schizophrenia is attributable to a particular mutant gene, then the value of this knowledge would be in no way diminished by the realization that this gene does not "specify" brain development. But from the conceptual point of view the examples considered show that the focus on genetic specification is not likely to be a very fruitful approach to explaining the development of the nervous system. Rather, they indicate that the goal of developmental neurobiology ought not to be phrased as the understanding of how genes build nerve cells and specify the neural circuits that underlie behavior, but as the discovery of epigenetic functional relations, or algorithms. The horizon of that epigenetic approach to developmental neurobiology extends far beyond the genes and must encompass the universe of nonprogrammatic, contextually governed intra- and intercellular interactions that underlie the historical phenomenon of metazoan ontogeny.

ACKNOWLEDGMENTS

I thank John Palka and David Weisblat for helpful criticisms of the manuscript. My research has been supported by NIH grant NS12818 and NSF grant BN577-19181.

Literature Cited

Albertson, D. G., Thomson, J. N. 1976. The pharynx of Caenorhabditis elegans. Philos. Trans. R. Soc. London Ser. B 275:299-325
Anderson, H., Edwards, J. S., Palka, J. 1980. Developmental neurobiology of invertebrates. Ann. Rev. Neurosci. 3:97-139
Becker, H. J. 1957. Über Röntgenmosaikflecken und Defektmutationen am Auge von Drosophila melanogaster und die Entwicklungsphysiologie des Auges. Z. Indukt. Abstamm. Vererbungsl. 88:333-73
Bentley, D. 1973. Postembryonic development of insect motor systems. In Developmental Neurobiology of Arthropods, ed. D. Young, pp. 147-77. Cambridge: Cambridge Univ. Press
Bentley, D. 1975. Single gene cricket mutations: Effects on behavior, sensilla, sensory neurons and identified interneurons. Science 187:760-64
Bentley, D. 1976. Genetic analysis of the nervous system. In Simpler Networks and Behavior, ed. J. C. Fentress, pp. 126-39. Sunderland, Mass: Sinauer Assoc.
Bentley, D. 1977. Development of insect nervous systems. In Identified Neurons and Behaviour of Arthropods, ed. G. Hoyle, pp. 461-81. New York: Plenum
Benzer, S. 1971. From gene to behavior. J. Am. Med. Assoc. 218:1015-22
Bernard, F. 1937. Recherches sur la morphogénèse des yeux composés d'arthropodes. Bull. Biol. Fr. Belg. 23: Suppl. XXIII, pp. 1-162
Brenner, S. 1973. The genetics of behaviour. Brit. Med. Bull. 29:269-71
Brenner, S. 1974. The genetics of Caenorhabditis elegans. Genetics 77:71-94
Brindley, G. S. 1969. Nerve net models of plausible size that perform many simple learning tasks. Proc. R. Soc. London Ser. B 174:173-91
Campos-Ortega, J. A., Hofbauer, A. 1977. Cell clones and pattern formation in the lineage of photoreceptor cells in the compound eye of Drosophila. Wilhelm Roux's Arch. Dev. Biol. 181:227-44
Caviness, V. S. 1977. Reeler mouse mutant: A genetic experiment in developing mammalian cortex. In Society for

Neuroscience Symposia, ed. W. M. Cowan, J. A. Ferendelli, 2:27-46. Bethesda, Md: Soc. Neurosci.
Caviness, V. S. Jr., Rakic, P. 1978. Mechanisms of cortical development: A view from mutations in mice. Ann. Rev. Neurosci. 1:297-326
Caviness, V. S. Jr., Sidman, R. L. 1972. Olfactory studies of the forebrain in the reeler mutant mouse. J. Comp. Neurol. 145:85-104
Caviness, V. S. Jr., Sidman, R. L. 1973. Time of origin of corresponding cell classes in the cerebral cortex of normal and reeler mutant mice: An autoradiographical analysis. J. Comp. Neurol. 148:141-51
Chappell, R. L., Dowling, J. E. 1972. Neural organization of the median ocellus of the dragonfly. I. Intracellular electrical activity. J. Gen. Physiol. 60:121-47
Condamine, H., Custer, R. P., Mintz, B. 1971. Pure-strain and genetically mosaic liver tumors histochemically identified with the β-glucoroindase marker in allophenic mice. Proc. Natl. Acad. Sci. USA 68:2032-36
Deak, I. I. 1976. Demonstration of sensory neurons in the ectopic cuticle of spineless-aristapedia, a homeotic mutant of Drosophila. Nature 260:252-54
Dewey, M. J., Gervais, A. G., Mintz, B. 1976. Brain and ganglion development from two genotypic classes of cells in allophenic mice. Dev. Biol. 50:68-81
Dräger, V. C. 1977. Abnormal neural development in mammals. In Function and Formation of Neural Systems, ed. G. S. Stent, pp. 111-38. Berlin: Dahlem Konferenzen
Feder, N. 1976. Solitary cells and enzyme exchange in tetraparental mice. Nature 263:67-69
Frank, E., Fillenz, M., Jansen, J. K. G., Kandel, E. R., Kuffler, S. W., Landmesser, L. T., Lømo, T., McMahan, U. J., Nicholls, J. G., Parnas, I., Patterson, P. H., Purves, D. 1977. Formation and maintenance of neural connections. See Dräger 1977, pp. 225-52
Ghysen, A. 1978. Sensory neurones recognize defined pathways in Drosophila

central nervous system. *Nature* 274: 869–72

Goodman, C. S. 1974. Anatomy of locust ocellar interneurons: Constancy and variability. *J. Comp. Physiol.* 95:185–201

Goodman, C. S. 1976. Constancy and uniqueness in a large population of small interneurons. *Science* 193:502–4

Goodman, C. S. 1977. Neuron duplications and deletions in locust clones and clutches. *Science* 197:1384–86

Goodman, C. S. 1978. Isogenic grasshoppers: Genetic variability in the morphology of identified neurons. *J. Comp. Neurol.* 182:681–705

Guillery, R. W. 1969. An abnormal retinogeniculate projection in Siamese cats. *Brain Res.* 14:739–41

Guillery, R. W. 1974. Visual pathways in albinos. *Sci. Am.* 230:44–54

Guillery, R. W., Casagrande, V. A., Uberdorfer, M. D. 1974. Congenitally abnormal vision in Siamese cats. *Nature* 252:195–99

Guillery, R. W., Kaas, J. H. 1971. A study of normal and congenitally abnormal retinogeniculate projections in cats. *J. Comp. Neurol.* 143:73–100

Hall, J. C. 1979. Control of male reproductive behavior by the central nervous system of Drosophila: Dissection of a courtship pathway by genetic mosaics. *Genetics* 92:437–57

Hall, J. C., Gelbart, W. M., Kankel, D. R. 1976. Mosaic systems. In *The Genetics and Biology of Drosophila,* ed. M. Ashburner, E. Novitski, pp. 265–314. New York: Academic

Hall, J. C., Greenspan, R. J. 1979. Genetic analysis of Drosophila neurobiology. *Ann. Rev. Genetics* 13:127–95

Hamburgh, M. 1963. Analysis of the postnatal developmental effects of "reeler," a neurological mutation in mice. A study in developmental genetics. *Dev. Biol.* 8:165–85

Hofbauer, A., Campos-Ortega, J. A. 1976. Cell clones and pattern formation: Genetic eye mosaics in *Drosophila melanogaster. Wilhelm Roux's Arch. Dev. Biol.* 179:275–89

Horridge, G. A. 1968. *Interneurons,* p. 321. San Francisco: Freeman

Horridge, G. A., Meinertzhagen, I. A. 1970. The accuracy of the patterns of connexions of the first- and second-order neurones of the visual system of Calliphora. *Proc. R. Soc. London Ser. B* 175:69–82

Hotta, Y., Benzer, S. 1972. Mapping of behavior in Drosophila mosacis. *Nature* 240:527–35

Hotta, Y., Benzer, S. 1976. Courtship in *Drosophila* mosaics: Sex-specific foci of sequential action patterns. *Proc. Natl. Acad. Sci. USA* 73:4154–58

Hubel, D. H., Wiesel, T. N. 1971. Aberrant visual projections in the Siamese cat. *J. Physiol. London* 218:33–62

Jordan, P. 1938. Zur Frage einer spezifischen Anziehung zwischen Genmolekülen. *Phys. Z.* 39:711–14

Kaas, J. H., Guillery, R. W. 1973. The transfer of abnormal visual field representations from dorsal lateral geniculate nucleus to visual cortex in Siamese cats. *Brain Res.* 59:61–95

Kankel, D. R., Ferrus, A. 1979. Genetic analyses of problems in the neurobiology of Drosophila. In *Neurogenetics: Genetic Approaches to the Nervous System,* ed. X. Breakefield, pp. 27–66. New York: Elsevier/North Holland

Kankel, D. R., Hall, J. C. 1976. Fate mapping of nervous system and other internal tissues in genetic mosaics of *Drosophila melanogaster. Dev. Biol.* 48: 1–24

Landis, D. M. D., Reese, T. S. 1977. Structure of the Purkinje cell membrane in staggerer and weaver mutant mice. *J. Comp. Neurol.* 171:247–60

Landis, D. M. D., Sidman, R. L. 1978. Electron microscopic analysis of postnatal histogenesis in the cerebellar cortex of staggerer mutant mice. *J. Comp. Neurol.* 179:831–64

LaVail, M. M., Mullen, R. J. 1976. Role of pigment epithelium in inherited retinal degeneration analyzed with experimental mouse chimeras. *Exp. Eye Res.* 23:227–45

Levinthal, F., Macagno, E. R., Levinthal, C. 1975. Anatomy and development of identified cells in isogenic organisms. *Cold Spring Harbor Symp. Quant. Biol.* 40:321–31

Lewis, E. B. 1963. Genes and developmental pathways. *Am. Zool.* 3:33–56

Macagno, E. R. 1980. Genetic approaches to invertebrate neurogenesis. *Curr. Top. Dev. Biol.* 16:In press

Macagno, E. R., Lopresti, V., Levinthal, C. 1973. Structure and development of neuronal connections in isogenic organisms: Variations and similarities in the optic system of Daphnia magna. *Proc. Natl. Acad. Sci. USA* 70:57–61

Mariani, J., Crepel, F., Mikoshiba, K., Changeux, J.-P., Sotelo, C. 1977. Anatomical, physiological and biochemical studies of the cerebellum from reeler mutant mouse. *Philos. Trans. R. Soc. London Ser. B* 281:1–28

Messer, A., Smith, D. M. 1977. In vitro behavior of granule cells from staggerer and weaver mutants of mice. *Brain Res.* 130:13–23

Meyerowitz, E. M., Kankel, D. R. 1978. A genetic analysis of visual system development in *Drosophila melanogaster. Dev. Biol.* 62:63–93

Mintz, B. 1962. Formation of genotypically mosaic mouse embryos. *Ann. Zool.* 2:432

Mintz, B. 1965. Genetic mosaicism in adult mice of quadriparental lineage. *Science* 148:1232–33

Mintz, B., Sanyal, S. 1970. Clonal origin of the mouse visual retina mapped from genetically mosaic eyes. *Genetics* 64: 43–44 (Suppl.)

Mullen, R. J. 1977a. Site of *pcd* gene action and Purkinje cell mosaicism in cerebella of chimeric mice. *Nature* 270:245–47

Mullen, R. J. 1977b. Genetic dissection of the CNS with mutant-normal mouse and rat chimeras. In *Society for Neuroscience Symposia,* ed. W. M. Cowan, J. A. Ferendelli, 2:47–65. Bethesda, Md: Soc. Neurosci.

Mullen, R. J., Eicher, E. M., Sidman, R. L. 1976. Purkinje cell degeneration, a new neurological mutation in the mouse. *Proc. Natl. Acad. Sci. USA* 73:208–12

Mullen, R. J., Herrup, K. 1979. Chimeric analysis of mouse cerebellar mutants. See Kankel & Ferrus, 1979, pp. 173–96

Mullen, R. J., LaVail, M. M. 1975. Two new types of retinal degeneration in cerebellar mutant mice. *Nature* 258:528–30

Mullen, R. J., LaVail, M. M. 1976. Inherited retinal distrophy: Primary defect in pigment epithelium determined with experimental rat chimeras. *Science* 192:799–801

Murphey, R. K., Mendenhall, B., Palka, J., Edwards, J. S. 1975. Deafferentation slows the growth of specific dendrites of identified giant interneurons. *J. Comp. Neurol.* 159:407–18

Ouweneel, W. J. 1976. Developmental genetics of homeosis. *Adv. Genet.* 18:179–248

Pak, W. L., Pinto, L. H. 1976. Genetic approach to the study of the nervous system. *Ann. Rev. Biophys. Bioeng.* 5:397–448

Palka, J. 1979. Mutants and mosaics: Tools in insect developmental neurobiology. *Soc. Neurosci. Symp.* 4:209–27

Palka, J., Lawrence, P. A., Hart, H. S. 1979. Neural projection patterns from homeotic tissue of *Drosophila* studied in *bithorax* mutants and mosaics. *Dev. Biol.* 69:549–75

Pearson, K. G., Goodman, C. S. 1979. Correlation of variability in structure with variability in synaptic connection of an identified interneuron in locusts. *J. Comp. Neurol.* 184:141–65

Peterson, A. C. 1974. Chimera mouse study shows absence of disease in genetically dystrophic muscle. *Nature* 248:561–64

Quinn, W. G., Gould, J. L. 1979. Nerves and genes. *Nature* 278:19–23

Ready, D. F., Hanson, T. E., Benzer, S. 1976. Development of the *Drosophila* retina, a neurocrystalline lattice. *Dev. Biol.* 53:217–40

Searle, A. G. 1968. *Comparative Genetics of Coat Color in Mammals,* pp. 146–47. New York: Academic

Shatz, C. J. 1977a. A comparison of visual pathways in Boston and Midwestern Siamese cats. *J. Comp. Neurol.* 171: 205–28

Shatz, C. J. 1977b. Anatomy of interhemispheric connections in the visual system of Boston Siamese and ordinary cats. *J. Comp. Neurol.* 173:497–518

Sidman, R. L., Green, M. C., Appel, S. H. 1965. *Catalog of the Neurological Mutants of the Mouse.* Cambridge, Mass: Harvard Univ. Press

Simberloff, D. S. 1974. Equilibrium theory of island biogeography and ecology. *Ann. Rev. Ecology Syst.* 5:161–82

Sotelo, C., Changeux, J.-P. 1974. Transsynaptic degeneration "en cascade" in the cerebellar cortex of staggerer mutant mice. *Brain Res.* 67:519–26

Stent, G. S. 1978. *Paradoxes of Progress,* pp. 169–89. San Francisco: Freeman

Stocker, R. F. 1977. Gustatory stimulation of a homeotic mutant appendage, *Antennapedia,* in *Drosophila melanogaster. J. Comp. Physiol.* 115:351–61

Stocker, R. F., Edwards, J. S., Palka, J., Schubiger, G. 1976. Projections of sensory neurons from a homeotic mutant appendage, *Antennapedia,* in *Drosophila melanogaster. Dev. Biol.* 52:210–20

Sturtevant, A. H. 1929. The claret mutant type of *Drosophila simulans:* A study of chromosome elimination and cell lineage. *Z. Wiss. Zool.* 135:325–56

Sulston, J. E. 1976. Post-embryonic development in the ventral cord of *Caenorhabditis elegans. Philos. Trans. R. Soc. London Ser. B* 275:287–97

Sulston, J. E., Horvitz, H. R. 1977. Post-embryonic cell lineages of the nematode, *Caenorhabditis elegans. Dev. Biol.* 56:110–56

Székely, G. 1979. Order and plasticity in the nervous system. *Trends Neurosci.* 2:(10)245–48

Tarkowski, A. K. 1961. Mouse chimera developed from fused eggs. *Nature* 190:857–60

von Schilcher, F., Hall, J. C. 1979. Neural topography of courtship song in sex mosaics of *Drosophila melanogaster. J. Comp. Physiol.* 129:85–95

Waddington, C. H. 1957. *The Strategy of the Genes.* London: Allen & Unwin

Ward, S. 1977. Invertebrate neurogenetics. *Ann. Rev. Genet.* 11:415–50

Ward, S., Thomson, H., White, J. G., Brenner, S. 1975. Electron microscopical reconstruction of the anterior sensory anatomy of the nematode *Caenorhabditis elegans. J. Comp. Neurol.* 160:313–38

Ware, R. W., Clark, D., Crossland, K., Russell, R. L. 1975. The nerve ring of the nematode *Caenorhabditis elegans.* Sensory input and motor output. *J. Comp. Neurol.* 162:71–110

Wegmann, T. G., LaVail, M. M., Sidman, R.

L. 1971. Patchy retinal degeneration in tetraparental mice. *Nature* 230:333–34

White, J. G., Southgate, E., Thomson, J. N., Brenner, S. 1976. The structure of the ventral nerve cord of *Caenorhabditis elegans. Philos. Trans. R. Soc. London Ser. B* 275:327–48

Whitman, C. O. 1878. The embryology of *Clepsine. Q. J. Microsc. Sci.* 18:215–315

Whittaker, R. H. 1970. *Communities and Ecosystems.* New York: Macmillan. 158 pp.

Wilson, M. 1978. The functional organisation of locus ocelli. *J. Comp. Physiol.* 24:297–316

Woodger, J. H. 1952. *Biology and Language.* Cambridge: Univ. Press. 364 pp.

Yoon, C. H. 1977. Fine structure of the cerebellum of "staggerer-reeler," a double mutant of mice affected by staggerer and reeler condition. II. Purkinje cell anomalies. *J. Neuropathol. Exp. Neurol.* 36:427–39

Ann. Rev. Neurosci. 1981. 4:195–225

THE BIOLOGY OF MYASTHENIA GRAVIS

♦11553

Daniel B. Drachman

Johns Hopkins University, School of Medicine, Department of Neurology, Baltimore, Maryland 21205

INTRODUCTION

Myasthenia gravis (MG) is a neuromuscular disorder of man characterized by weakness and fatigability of skeletal muscles. The clinical features, first noted in the seventeenth century by Thomas Willis (1672), were thoroughly described by 1900 (Erb 1879, Campbell & Bramwell 1900). The neuromuscular junction was implicated in the disease process more than 40 years ago, because of the similarity between MG and curare poisoning (Oppenheim 1901) and the remarkable response of many patients to anticholinesterase drugs (Remen 1932, Walker 1934). The exact site and nature of the defect remained elusive, however, until the development and application of a new set of tools, neurotoxins from elapid snake venoms (Lee 1972), which permitted the specific identification of the acetylcholine receptor (AChR) abnormality in 1973 (Fambrough et al). Since that time, it has become clear that the basic defect in MG is a reduction of available AChRs at neuromuscular junctions, brought about by an antibody-mediated autoimmune attack. Our understanding of the pathogenesis of MG has advanced with remarkable rapidity and has begun to serve as a rational basis for the design of treatment. This review focuses on the close relationship between the basic pathogenetic mechanisms and the clinical manifestations and treatment of MG.

CLINICAL FEATURES

MG is not a rare disease: it affects approximately 4 to 6 individuals per 100,000 population (Kurtzke & Kurland 1977). The cardinal features of MG consist of weakness and fatigue of skeletal muscles. Only the motor

195

0147-006X/81/0301-0195$01.00

system is impaired; sensation, reflexes, coordination, and other neural functions remain normal. Although any muscle can be affected, certain typical patterns of weakness are commonly seen. Often the elevators of the lids and the external ocular muscles are selectively involved, which results in ptosis (drooping of the lids) and diplopia (double vision). More severe involvement of the cranial musculature is manifested by a characteristic "snarling" facial expression and impairment of the voice, speech, and ability to chew and swallow. Generalized weakness may occur, and often affects the proximal muscles of the limbs and the neck extensors. In severe cases, the patient's life may be endangered by weakness of the muscles of respiration and swallowing. The muscle strength rapidly fatigues on repeated or sustained contraction, and may improve after rest. Similarly, repetitive electrical stimulation of motor nerves results in progressive weakness of muscle contractions ("Jolly test"), and a decline in the amplitude of evoked muscle action potentials (decremental responses) (Ozdemir & Young 1976). Cholinesterase inhibitors produce improvement in the patient's motor power and in the response to nerve stimulation. Conversely, minute doses of neuromuscular blocking agents, such as d-tubocurarine, greatly exaggerate the myasthenic weakness and decremental responses (Rowland et al 1961). These physical findings and electrophysiological and pharmacological tests have long been used in the clinical diagnosis of MG.

The severity and course of MG are highly variable. In some patients, the weakness remains localized to a few muscle groups, while in others it progresses to become generalized. Fluctuations in the severity of the weakness may occur spontaneously, or in response to apparently unrelated infections, or to endocrine, metabolic, or emotional changes. With modern treatment, most myasthenic patients are able to lead full, active lives, although there is still a great need for more effective and specific modes of treatment.

THE NEUROMUSCULAR JUNCTION IN MYASTHENIA GRAVIS

Walker's discovery of the beneficial effects of anticholinesterase drugs first established the neuromuscular junction as the general site of the defect in MG (Walker 1934). Further confirmation of the abnormality of neuromuscular transmission came from physiological studies, such as those of Harvey & Masland (1941), who described the decremental response to repetitive nerve stimulation in MG and noted its similarity to the changes seen in curarized subjects.

In 1964 Elmqvist et al applied microelectrode techniques to the investigation of muscles from myasthenic patients. They made the important observation that the amplitude of miniature endplate potentials (mepps) was reduced to about 20% of normal at neuromuscular junctions of muscles from MG patients. This reduction of mepp amplitudes was originally thought to be due to a decrease in the number of ACh molecules per quantum, resulting from some abnormality in the motor nerve terminals. The concept of a presynaptic defect was widely accepted for many years; evidence from physiological, pharmacological, and morphological studies did not further define the precise site of the defect (Fields 1971).

AChR Deficit

On theoretical grounds, it seemed possible that a decrease in the number of available AChRs could also result in reduced amplitudes of mepps. When the availability of α-bungarotoxin (α-BuTx) as a specific probe for receptors made direct measurement possible, we undertook an investigation of this question (Fambrough et al 1973). The method used depends on the specific, quantitative, and virtually irreversible binding of α-BuTx to AChRs of skeletal muscles. The toxin can be labeled with radioactive iodine, and the number of AChR sites can be determined from the amount of bound radioactivity (Hartzell & Fambrough 1972). The neuromuscular junctions of myasthenic patients bound only 11% to 30% as much radioactivity as those of normal individuals, which indicates that they had a markedly reduced number of AChR sites. The reduction of available AChRs is not attributable to treatment of the myasthenic patients with anticholinesterase drugs, although very high doses of these agents have been shown to reduce junctional AChRs in experimental animals (Fambrough et al 1973, Chang et al 1973).

The original observation of reduced numbers of AChRs at neuromuscular junctions has now been confirmed in several laboratories, by a variety of techniques using α-BuTx-binding (Drachman et al 1976, Engel et al 1977, Ito et al 1978, Lindstrom & Lambert 1978) or electrophysiological measurements (Albuquerque et al 1976b) (Table 1). Indeed, recent data suggest that the radiometric measurement of junctional AChRs in muscle biopsies may be clinically valuable as one of the most sensitive diagnostic tests for MG (A. Pestronk and D. B. Drachman, in preparation).

These findings raised the question of whether the reduction of AChRs per se could account for the clinical and physiological abnormalities in MG, or whether it represented a secondary response to some other defect. To resolve this question, Satyamurti et al (1975) developed an experimental animal model in which the number of available AChRs in rats was reduced by specific pharmacological blockade with α-cobra toxin. This model reproduced all the characteristic features of human MG: the animals were weak, and showed decremental responses on repetitive nerve stimulation, typical

Table 1   Reduced AChR sites at neuromuscular junctions of myasthenia patients

Method		Results		Ref.
		Normal	MG	
^{125}I-α-BuTx bind- ing to N-M jcts	Scintillation counting Autoradiog- raphy	3.7 × 10^7 AChR sites/N-M jct 3.75/4 arbitrary units	0.54 × 10^7/ N-M jct 1.54/4 arbi- trary units	Fambrough et al 1973
^{125}I-α-BuTx bind- ing to N-M jcts	γ-counting	1.54 × 10^7 AChR sites/N-M jct	0.49 × 10^7/ N-M jct	Ito et al 1978
Peroxidase α-BuTx	Electron mi- croscopy	AChR surface/pre- synaptic mem- brane surface 3.06	0.98	Engel et al 1977
^{125}I-α-BuTx bind- ing	γ-counting	28.6 × 10^{-14} mol AChR/g muscle	10.3 mol/g	Lindstrom & Lambert 1978
ACh iontophoresis	ACh sensi- tivity	2302 mV/nC	675 mV/nC	Albuquerque et al 1976b

of the pattern in MG. The decremental responses were exaggerated by small doses of d-tubocurarine, and were markedly improved by administration of anticholinesterase agents. Post-tetanic responses, thought to be particularly characteristic of MG (Desmedt 1973), were also reproduced by the AChR blockade model. Furthermore, α-cobra toxin has been shown to reduce the amplitude of mepps (Chang & Lee 1966). The α-cobra toxin model strongly supported the hypothesis that a decrease of available AChRs is sufficient to account for the clinical and physiological defects of human MG.

Morphological studies of neuromuscular junctions have provided an- other line of evidence consistent with a postsynaptic defect (Zacks et al 1962, Woolf 1966, Bergman et al 1971, Engel & Santa 1971). The post- synaptic membranes showed sparse, shallow folds, with markedly simplified geometric patterns. Although the motor nerve terminals were often elon- gated and reduced in diameter, they contained normal numbers of structur- ally intact ACh vesicles (Woolf 1966, Engel & Santa 1971). These studies did not distinguish whether the pre- or postsynaptic abnormality was pri- mary, but it now seems clear that the changes in the postsynaptic membrane are the important ones, and reflect the underlying AChR deficit.

The physiological abnormalities of neuromuscular transmission appear to be accounted for by the AChR deficit. No additional disorders of pre- or post-synaptic function have been found that might further impair neuro- muscular transmission in MG. The resting release of ACh from myasthenic muscles did not differ significantly from control values (Molenaar et al 1979). The potassium-evoked ACh release was initially even greater than that of controls; this finding is apparently consistent with the reportedly raised ACh content of myasthenic intercostal muscles (Molenaar et al 1979). The number of ACh quanta liberated by a nerve impulse was estimated to be approximately normal (Elmqvist et al 1964, Lindstrom & Lambert 1978), or increased (Cull-Candy et al 1980) at myasthenic nerve terminals. Finally, the AChRs remaining at MG junctions were reported to function normally in terms of current flow and open times of ACh-induced ion channels (Cull-Candy et al 1979).

Correlation of AChR Deficit With Clinical Features

Many of the clinical features of MG that were formerly poorly understood can now be explained satisfactorily in terms of the reduction of available AChRs at neuromuscular junctions. The basic principle is that the amplitude of endplate potentials (epps) depends on the number of interactions between ACh molecules and AChR molecules, which is a matter of probability (Katz & Miledi 1972). Only a proportion of the ACh molecules released in response to a nerve impulse eventually interact with the AChRs; only a small fraction of the 20 to 40 million AChRs at a neuromuscular junction are activated at one time. A reduction of either the amount of ACh released or the number of AChRs will result in a reduction of the epp amplitude. At the normal junction, the number of interactions is more than necessary to produce an epp sufficiently large to trigger a muscle action potential; the excess above threshold is termed the "safety margin" of neuromuscular transmission (Waud 1971). In MG, the decreased number of AChRs reduces the number of interactions, which results in epps of diminished amplitude. At some muscle fibers, the epps are below threshold and fail to trigger muscle action potentials. This has been directly demonstrated by intracellular recording in vitro (Elmqvist et al 1964, Albuquerque et al 1976b). Failure of transmission at individual muscle fibers has also been recorded in the intact patient by the technique of single fiber electromyography (Stålberg et al 1976). When transmission fails at many junctions the power of the whole muscle is reduced, which is manifested clinically as weakness.

Neuromuscular fatigue is the single most characteristic feature of MG. The patient is unable to sustain or repeat muscular contractions. The electrical counterpart of this phenomenon is the decrement of muscle action potentials evoked by repetitive stimulation of the motor nerve (Ozdemir & Young 1976). These phenomena result from the reduced safety margin at myasthenic junctions and the normal decline of evoked ACh release, which is termed "presynaptic rundown." During repeated nerve stimulation the amount of ACh released per impulse normally declines after the first few impulses, since the nerve terminal is not able to sustain its original release rate (Thies 1965, Barrett & Magleby 1976). In the myasthenic patient, junctions with reduced numbers of AChRs ("low safety margins") therefore fail progressively, resulting in the activation of fewer and fewer muscle fibers by successive nerve impulses. In normal individuals, the safety margin is sufficiently great so that decremental responses and fatigue do not occur at rates of stimulation below 40 to 50 Hz. At very high stimulation rates, however, the decline of ACh release may be so pronounced that transmission failure occurs even at normal neuromuscular junctions, which results in decremental responses and neuromuscular fatigue similar to those found in MG.

The beneficial effect of anticholinesterase agents is also consistent with the underlying AChR deficit. The primary action of these drugs is to reduce the hydrolysis of ACh by the enzyme acetylcholinesterase (Koelle 1975). Although this does not repair the primary deficiency of AChRs, it permits the ACh released by the nerve terminals to act repeatedly over a longer time period (Katz & Miledi 1973). Thus, the total number of interactions between ACh and receptors is increased, which gives rise to larger epps. This increase may be sufficient to raise the amplitude of the epps above threshold, thereby improving both the patient's strength and the electrical response to repetitive nerve stimulation.

Curare (d-tubocurarine), a postsynaptic blocking agent, has been used in a diagnostic test for MG (Rowland et al 1961). A small dose of curare (usually one-tenth the average curarizing dose) is administered intravenously. In the normal individual, the small amount of curare blocks only a fraction of the available AChRs, leaving more than enough to maintain neuromuscular transmission; however, in the myasthenic patient, blockade of even a small number of AChRs is sufficient to reduce the precarious safety margin even further, thus bringing out both the clinical and the electrophysiological manifestations of weakness and fatigue.

AUTOIMMUNE PATHOGENESIS OF MG

The possibility that MG might be an autoimmune disease was originally suspected on the basis of indirect evidence, consisting of a high rate of abnormalities of the thymus gland (Castleman 1966), the association of MG with other presumed autoimmune diseases (Simpson 1960), and reduced complement levels in some myasthenic patients (Nastuk et al 1960). The discovery that a proportion of myasthenic patients had serum antibody directed against skeletal muscle presented a particularly intriguing clue (Strauss et al 1960). The antimuscle antibody binds to skeletal muscle in a striated pattern, possibly at the level of the sarcoplasmic reticulm (Mendell et al 1973), and has been shown to cross-react with the "myoid" or muscle-like cells of the thymus (van der Geld & Strauss 1966). This relation may have certain implications for the origin of the autoimmune reaction in MG, but there is no evidence that the antimuscle antibody itself contributes to the pathogenesis of MG.

This circumstantial evidence strongly suggested an abnormality of the immune system in MG; however, further understanding of the immunologic disorder awaited the advent of a direct technique for studying AChRs. An important clue that the autoimmune attack might be directed against AChRs came from the experiments of Patrick & Lindstrom (1973), who were attempting to raise antibodies to pure AChR. They found that rabbits immunized with AChR isolated from the electric organs of eels developed marked muscular weakness and respiratory insufficiency. The rabbits showed typical decremental responses on repetitive nerve stimulation and improved after treatment with anticholinesterase agents. The effects of AChR immunization of a variety of animal species from frogs (Nastuk et al 1979) to monkeys (Tarrab-Hazdai et al 1975) have now been extensively studied, and similarities between the animal model of "experimental allergic myasthenia gravis" (EAMG) and the human disease have been described in detail (Sugiyama et al 1973, Granato et al 1976, Heilbronn et al 1976, Penn et al 1976, Sanders et al 1976). The animals manifest the typical reduction of mepp and epp amplitudes and show a decrease of available AChR sites (Green et al 1975, Lambert et al 1976). Ultrastructural studies (Engel et al 1976) show destructive changes and active cellular infiltration at the postsynaptic membrane during the acute phase of EAMG. During the late chronic phase, a simplification of the postsynaptic membrane more closely resembling the picture of MG has been observed. Differences between EAMG and the human disease include the nature of the immunizing event itself and the early development of an acute episode of weakness in the animals. Although the immunizing event in human MG remains unknown at present, it surely does not involve exposure to a massive dose of exogenous AChR in adjuvant. The acute phase of the animal disease does not have a parallel in human MG, and some of the immune mechanisms involved in the animal and human conditions are undoubtedly different. A detailed discussion of EAMG is beyond the scope of the present review; however, the animal model provided evidence that an autoimmune reaction directed against AChRs could produce many of the features of MG.

Anti-acetylcholine Receptor Antibody

Based on the knowledge of a receptor deficit in MG, and the analogy to the experimental animal model, the search for antibodies directed against AChRs in the human disease was soon begun. Antireceptor antibody was identified by several different methods (Almon et al 1974, Appel et al 1975,

Bender et al 1975, Lindstrom et al 1976, Mittag et al 1976), all of which depend on α-BuTx for their specificity. Almon et al (1974) first demonstrated that serum immunoglobulin from approximately one-third of a group of myasthenic patients was capable of blocking α-BuTx binding to rat AChR in vitro. Bender et al (1975), using frozen sections of human muscle as a source of AChR, found that 68% of myasthenic sera inhibited α-BuTx binding. More sensitive radioimmunoassays, based on the binding of antibody to AChR labeled with ^{125}I-α-BuTx, have now been developed (Appel et al 1975, Lindstrom et al 1976). Receptor-binding antibodies are detected in up to 87% of sera from myasthenic patients (Lindstrom et al 1976). Human AChR (from surgically amputated limbs) or primate AChR have proven to be the most appropriate sources of antigen (McAdams & Roses 1980). AChR from other species gives unpredictably lower titers because of problems of antibody cross-reactivity (Oda et al 1980); however, it should be emphasized that the antibody titers correspond only loosely with the patient's clinical status, even when human AChR is used (Almon et al 1974, Appel et al 1975, Bender et al 1975, Lindstrom et al 1976, Mittag et al 1976, McAdams & Roses 1980).

Pathogenicity of Myasthenic Immunoglobulin

The question of whether the circulating antibodies are pathogenic or merely represent a secondary response to AChR damage caused by some other agent is of paramount importance in understanding the pathogenesis of MG. A well-known "experiment in nature" suggested that a circulating factor—possibly immunoglobulin—might be pathogenic. Approximately one out of every six infants born to myasthenic mothers manifests transient signs of MG during the first few postnatal weeks (Namba et al 1970). Numerous attempts to transfer MG from humans to experimental animals or nerve-muscle preparations had failed (Nastuk et al 1959, Nastuk & Strauss 1961, Namba et al 1976); however, the possibility of a circulating factor assumed new importance in the light of the discovery of antibodies to AChR. In most studies, the exposure to myasthenic serum had been brief, lasting only minutes or hours. Based on the idea that more prolonged exposure to antibody might be necessary to produce the effect, Toyka et al (1975) undertook a passive transfer experiment in which physiologic levels of human myasthenic IgG were maintained in recipient mice for a period of days to weeks. An immunoglobulin fraction prepared from the serum of myasthenic patients was injected daily into mice, and produced typical myasthenic changes in the recipient animals. The mean amplitude of mepps and the mean number of AChR sites per neuromuscular junction were reduced by more than 50% in the diaphragms of MG immunoglobulin-treated animals. Marked clinical weakness and decremental responses on repetitive nerve stimulation were seen in some animals. The earliest time at which mepp amplitudes were decreased was 16 to 21 hr after injection of MG immunoglobulin, and the decrease was usually pronounced by 3 or 4 d. Serum immunoglobulin from 94% of the patients tested produced typical features of MG in recipient mice. Further purification of serum immunoglobulin fractions from the patients showed that IgG alone produced the same myasthenic features as whole immunoglobulin, whereas IgM had no effect (Toyka et al 1977). These passive transfer experiments, which have now been repeatedly confirmed (Pagala et al 1977, Howard & Sanders 1978), have established the important role of antibodies in the pathogenesis of MG.

Recently, ultrastructural studies have added further confirmation of the role of antibodies in MG. By means of high resolution electron microscopy, Rash and his colleagues (1976) have found material resembling IgG in configuration and dimensions ("fuzzy coats") in the region of junctional AChRs. Engel et al (1979) have visualized IgG at postsynaptic membranes of human myasthenic neuromuscular junctions using an immunoperoxidase technique.

EFFECTS OF MYASTHENIC ANTIBODIES ON ACETYLCHOLINE RECEPTORS

Theoretically, myasthenic patients' IgG may reduce the number of available AChRs by several possible mechanisms:

1. It may alter the turnover of AChRs, either by increasing the rate of degradation, or by decreasing the rate of synthesis.
2. It may block the active site of the receptor.
3. It may damage the AChRs, possibly in conjunction with complement and/or cellular elements.

The evidence now available suggests that accelerated degradation of AChRs, blockade, and damage may all be involved, but the relative role of each mechanism has yet to be determined.

Accelerated Degradation of AChRs

Studies of the effects of myasthenic IgG on receptor degradation have been carried out in a rat skeletal muscle tissue culture system, using a modification of the method of Devreotes & Fambrough (1975). The AChRs are first labeled with ^{125}I-α-BuTx. As the labeled receptors undergo degradation, the attached ^{125}I-α-BuTx is broken down to iodotyrosine, which appears in the culture medium. The rate of degradation can be calculated from the rate of release of ^{125}I into the medium and is normally approximately 4% per hour for rat skeletal muscle.

When immunoglobulin from myasthenic patients was added to the cultures, the AChR degradation rate increased up to two- to three-fold, as compared with cultures treated with control immunoglobulin (Appel et al 1977, Bevan et al 1977, Kao & Drachman 1977b). The acceleration of degradation was triggered by IgG alone, without requiring other humoral or cellular components of the immune system. The addition of complement did not alter the rate of degradation (Drachman et al 1980). Both the normal AChR degradation process (Devreotes & Fambrough 1975) and the acceleration produced by myasthenic immunoglobulin are temperature dependent and partially inhibitable by dinitrophenol (Appel et al 1977, Kao & Drachman 1977b), which suggests that they involve energy-dependent processes of the muscle cells. Similar results have also been reported using immunoglobulin preparations from animals with EAMG.

SELECTIVITY OF DEGRADATION OF AChRs WITH BOUND IgG
Theoretically, the myasthenic IgG may increase AChR degradation by (a) accelerating the muscle cell's overall receptor-degrading mechanism or (b) altering only AChRs to which IgG is bound, so that these receptors are *selectively degraded* at a more rapid rate.

To distinguish between these possibilities, cultures were prepared in which one set of AChRs was directly exposed to myasthenic patients' IgG for 2 hr; a second set of AChRs was allowed to develop immediately after the IgG exposure (Drachman et al 1978a). Each set of receptors was separately labeled with ^{125}I-α-BuTx, and its degradation rate independently followed. The results showed that only the set of AChRs directly exposed to myasthenic IgG (and presumably having bound IgG) was degraded at an accelerated rate two to three times normal. In contrast, the second set of receptors (in identically treated cultures, but without bound IgG) were degraded at the control rate. This suggested that the binding of IgG altered the receptors in some way that caused them to be preferentially selected for degradation.

CROSS-LINKING OF AChRs BY MYASTHENIC IgG   IgG molecules are known to be Y-shaped, with two arms capable of binding to identical antigenic sites, thereby linking them together. Certain actions of IgG, such as the induction of "capping" of lymphocytes are known to require cross-linking (Taylor et al 1971). Drachman et al (1978b) have recently shown that the ability of myasthenic patients' IgG to induce accelerated degradation of AChRs depends on its capacity to cross-link the receptors. To test the cross-linking hypothesis, they prepared pure IgG, divalent F (ab')$_2$

fragments, and monovalent Fab fragments from myasthenic and control sera by standard purification and enzymatic cleavage methods. When added to muscle cultures, the IgG and divalent F (ab')$_2$ fragments produced equally accelerated rates of AChR degradation (Fig. 1A, B). In contrast, the monovalent Fab fragments failed to accelerate the degradation rate (Fig. 1C), although they bound to AChRs of cultured skeletal muscle. When a second "piggyback" antibody directed against the Fab fragments was added, AChR degradation was accelerated (Fig. 1D). The effect of the second antibody was to cross-link the Fab fragments and the AChRs to which they were attached. To determine whether direct contact of an antibody with the AChR was necessary for accelerated degradation, another method of cross-linking was also tested. In this experiment, only α-BuTx was directly attached to the AChRs of cultured muscle. Antibodies prepared against α-BuTx were used to cross-link the α-BuTX-AChR complexes, which again resulted in a two- to three-fold acceleration of the rate of degradation of AChRs (Fig. 1E). These findings clearly demonstrated that cross-linking of AChRs by antibodies from myasthenic patients is the factor that triggers their rapid degradation; this has now been confirmed by studies in several laboratories, using antibody fragments derived from sera of EAMG animals (Lindstrom 1979, Tarrab-Hazdai et al 1979).

ENDOCYTOSIS OF AChRs   The normal process of AChR degradation begins with endocytosis, or "internalization" of the receptors, which are then broken down within the muscle cell by lysosomal enzymes (Devreotes & Fambrough 1976, Libby et al 1980). The antibody-triggered mechanism of accelerated degradation is also thought to involve endocytosis and enzymatic lysis.

There is increasing evidence that endocytosis rather than lysosomal degradation may be the step that determines the rate of degradation of AChRs:

1. As described above (Drachman et al 1978a), two different rates of degradation can occur simultaneously in a single muscle culture; the antibody-bound receptors are degraded more rapidly than those without bound antibody. This fits the concept of a selective increase in endocytosis brought about by the antibody rather than an overall increase in the lysosomal degradation rate.

2. Morphological studies suggest that the antibody-induced acceleration of AChR degradation involves a preliminary step of redistribution of the receptors within the muscle membrane. Autoradiography (Tarrab-Hazdai et al 1979) and fluorescence microscopy (Lennon 1978) have shown that the addition of anti-AChR serum causes AChRs to aggregate in patches. Freeze fracture electron microscopy gives a picture of the molecular events that take place after the addition of myasthenic IgG (D.W. Pumplin & D.B.

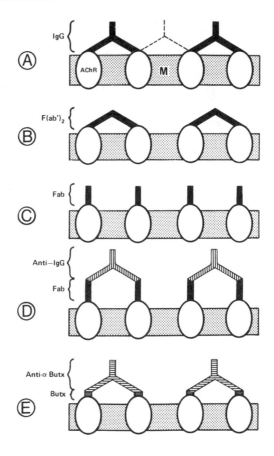

Figure 1   Diagrammatic representation of cross-linking of receptors by the various antibody fragments. M denotes muscle membrane; AChR, acetylcholine receptor; and α-BuTx, α-bungarotoxin. Acceleration of degradation was observed in the experimental conditions shown in panels *A, B, D,* and *E;* Fab alone (*C*) failed to accelerate acetylcholine-receptor degradation. [Reprinted by permission of the *New England Journal of Medicine,* 298:1120 (1978).]

Drachman, in press), because it permits visualization of the individual AChRs as intramembranous particles (Cohen & Pumplin 1979). By about 1 hr, many AChR particles cluster in tightly-packed groups of up to 60. Within a cluster, the interparticle distances are consistent with the spread of an IgG molecule, which suggests that the packing of AChRs may result from cross-linking by the anti-AChR antibodies. The individual clusters tend to aggregate fairly close to one another, but are not actually contiguous. In regions where many clusters are aggregated, shallow depressions 100 nm in diameter appear in the surface membrane. These depressions are probably the "coated pits" that are believed to serve as vehicles for the transport of a variety of surface components to the interior of the cell (Orci et al 1978). The aggregated clusters disappear from the membrane over the course of hours.

Taken together, the morphological findings suggest that the antibody may induce clustering of AChRs, aggregation of clusters, and endocytosis by way of pits within the muscle membrane.

3. The role of lysosomal enzymes in the degradation process has been explored by the use of the enzyme inhibitors, leupeptin and antipain. When added to normal or antibody-treated skeletal muscle cultures, these agents retard the release of degradation products into the medium (Drachman et al 1980, Libby et al 1980); however, the enzyme inhibitors do not slow the rate of loss of AChRs from the surface membrane of the muscle.

The above evidence supports the concept that the endocytotic step is the critical one for the degradation of both normal and antibody-treated AChRs; the lysosomal enzyme system is capable of disposing of all the AChRs presented to it, unless it is specifically inhibited.

EFFECT OF MYASTHENIC IMMUNOGLOBULIN ON AChRs AT INTACT NEUROMUSCULAR JUNCTIONS   There are important differences between AChRs at neuromuscular junctions and those at extrajunctional sites of cultured or denervated muscle. Junctional receptors have far slower turnover rates (half-life of approximately 10 d, as compared to 18 hr for extrajunctional receptors), different immunological reactivity, and different physical and pharmacological properties (Brockes et al 1976). These differences raise questions about the applicability of the results of tissue culture experiments to the situation at intact neuromuscular junctions. Since junctional AChRs can be labeled with ^{125}I-α-BuTx in vivo and their degradation rates can be determined from the loss of bound radioactivity (Berg & Hall 1975, Chang & Huang 1975), the effect of myasthenic immunoglobulin on AChR turnover could be tested directly in the intact animal.

Stanley & Drachman (1978) showed that immunoglobulin from myasthenic patients accelerated the rate of loss of labeled AChRs in the intact mouse diaphragm approximately three-fold, as in the tissue culture system. The loss of bound radioactivity was due almost entirely to degradation; chromatographic analysis revealed that the radioactive material released from diaphragms treated with myasthenic immunoglobulin consisted of iodotyrosine, a breakdown product of ^{125}I-α-BuTx. Similar findings have been reported from other laboratories using EAMG sera in vitro (Reiness et al 1978, Merlie et al 1979). These results suggest that accelerated degradation may be an important mechanism in bringing about the reduction of AChRs at neuromuscular junctions of patients with myasthenia gravis.

Synthesis and Incorporation of AChRs

In contrast to its pronounced influence on AChR degradation, myasthenic immunoglobulin does not appear to have an effect on the rates of synthesis and incorporation of AChR in muscle membranes. Synthesis and incorporation were measured in rat skeletal muscle cultures (Drachman et al 1978a). The cultures were first treated overnight with immunoglobulin from myasthenic or control patients. Surface AChRs were then blocked with non-labeled α-BuTx, and an interval of 6 hr was allowed for the synthesis and incorporation of new AChRs. At the end of that time ^{125}I-α-BuTx was added to measure the number of new surface AChRs. There was no difference in the amount of bound ^{125}I-α-BuTx in cultures treated with myasthenic, as compared with control, immunoglobulin. These findings indicate that the rates of synthesis and incorporation of new AChRs are not directly affected by myasthenic immunoglobulin. Whether a compensatory change in synthesis occurs in chronic MG has yet to be determined.

Blockade of AChRs

The hypothesis that a curare-like substance might be present in the serum of myasthenic patients was proposed as early as 1938 (Walker 1938), yet the role of blockade of AChRs in the pathogenesis of MG is still controversial. One would like to know whether antibodies from myasthenic patients can interfere with ACh transmission (a) by occupying the active site of the AChR itself, (b) by binding nearby and sterically hindering access to the site, or (c) by binding at a distance and altering the structure of the ACh molecule. Present evidence indicates that many myasthenic patients have antibodies capable of blocking AChRs, but the relative importance of this mechanism in impairing cholinergic transmission is not yet clear.

BUTX-BINDING STUDIES   Several studies have shown that the addition of sera from myasthenic patients to AChRs may interfere with the binding of α-BuTx, in a proportion of cases varying from 7 to 68% (Almon et al 1974, Bender et al 1975, Mittag et al 1976). These striking differences in results must be due to differences in the AChR preparations used, variation in the biological activity of individual patients' sera, and specific details of the experimental conditions. An even greater difficulty in interpretation arises because the antibody's ability to block α-BuTx-binding may not necessarily parallel its ability to interfere with ACh transmission at the neuromuscular junction. Although α-BuTx undoubtedly binds at or very near the active site of the receptor (Heidmann & Changeux 1978), there are important differences between α-BuTx and ACh (Lee 1972). α-BuTx is much larger than ACh (MW = 8,000 cf 182) and, therefore, could be excluded from a site only partially blocked by antibody, which might be accessible to the small ACh molecule. On the other hand, the affinity of α-BuTx for the receptor is much higher than that of ACh. Therefore, α-BuTx might displace the blocking antibody and bind to the receptor under circumstances in which ACh transmission could be impaired.

REVERSIBLE INHIBITION OF NEUROMUSCULAR TRANSMISSION Sera from some patients may have a reversible inhibitory effect on neuro-muscular transmission that is suggestive of receptor blocking. The addition of myasthenic sera to marginally affected EAMG animals in vivo (Heil-bronn et al 1979) or to neuromuscular junctions in vitro (Shibuya et al 1978) has been reported to interfere with ACh transmission, although no change was noted in another in vitro study (Albuquerque et al 1976a).

Plasmapheresis has been reported to produce very rapid clinical improvement within hours to days in occasional cases (Dau et al 1977), thus suggesting the removal of a blocking antibody. Similarly, thoracic duct drainage of lymph from myasthenic patients produced clinical improvements within 48 hr, whereas reinfusion of the lymph increased myasthenic weakness within hours (Matell et al 1976). The rapidity and reversibility of these changes have suggested that at least one action of the antibody might be functional blockade of AChRs, but such circumstantial evidence cannot rule out other possible explanations.

ANTIBODIES BINDING TO ACh R, BUT NOT BLOCKING   The large majority of MG patients have a significant proportion of antibodies that bind to the AChR at loci other than the active site. As described above, the commonly-used radioimmunoassay utilizes AChR with its active site already occupied by ^{125}I-α-BuTx, and can therefore measure only antibodies directed against non-site determinants (Appel et al 1975, Lindstrom et al 1976). This indicates that non-site-directed antibodies are present in myasthenic patients, and undoubtedly contribute to the pathogenesis of MG by other mechanisms detailed elsewhere in this review. However, the existence of non-site antibodies cannot be used as evidence for or against the possibility of an antibody population capable of blocking the receptor.

On balance, the evidence now available suggests that at least some myasthenic patients have antibodies capable of blocking AChRs under appropriate conditions; however, the relative role of receptor blockade in producing clinically significant impairment of neuromuscular transmission is as yet unresolved and is likely to differ in individual patients.

Structural Damage to Neuromuscular Junctions: Possible Role of Complement

Light and electron microscopic studies of neuromuscular junctions have revealed a variety of morphological abnormalities (Zacks et al 1962, Woolf 1966, Bergman et al 1971, Engel & Santa 1971) that may contribute to the impairment of neuromuscular transmission. The junctional folds are sparse and shallow, with a decreased area of contact with the nerve terminals. Undoubtedly, these structural changes result from the interaction of antibodies with AChRs. The antibodies, perhaps in collaboration with complement, may produce the fragmentation and shedding of postsynaptic membrane into the synaptic space that have been seen by electron microscopy (Rash et al 1976, Engel et al 1979). An alternative hypothesis, not yet tested, is that a sufficient amount of postsynaptic membrane may be endocytosed as a consequence of accelerated degradation of AChRs to result in the simplification of junctional folds. The altered geometry of the neuromuscular junction not only reflects the loss of AChRs, but may itself impair neuromuscular transmission as a result of the reduced contact between nerve and muscle membranes.

A possible role of complement in the pathogenesis of MG was first suggested by the observation that complement levels are lowered in myasthenic patients, especially during clinical exacerbations (Nastuk et al 1960). Moreover, Toyka et al (1977) found that the first part of the complement system (up to C3) enhanced the effect of passively transferred myasthenic immunoglobulin, although the latter part of the system (beyond C5) seemed to play no role. Both C3 and C9 have been detected at neuromuscular junctions of myasthenic patients, by immuno-electron microscopy (Engel et al 1979). It seems likely that complement may act in concert with IgG to produce local damage to the postsynaptic membrane. In vitro studies indicate that complement participates in neither the accelerated degradation of AChRs nor the receptor site blocking mediated by myasthenic IgG (Drachman et al 1980).

LYMPHOCYTES AND THE AUTOIMMUNE RESPONSE IN MG

Thus far, the discussion has focused on antibody-mediated mechanisms in the pathogenesis of MG; however, several recent studies have begun to reveal the important role played by lymphocytes in the autoimmune responses.

B-lymphocytes represent the "final common path" in the production of antibody. Both the peripheral blood lymphocytes and the thymic lymphocytes of myasthenic patients have been shown to produce anti-AChR antibody when cultured in vitro (Vincent et al 1979). These findings are not unexpected and identify two of the potential sites of production of anti-AChR antibody in myasthenic patients.

It has been suggested that some abnormality of the regulatory functions of lymphocytes could be involved in the development of the autoimmune response in at least some myasthenic patients (Penn 1979, Shore et al 1979). The coexistence of other autoimmune diseases such as rheumatoid arthritis, lupus erythematosus, and Hashimoto's thyroiditis in some myasthenic patients implies a loss of normal immunoregulatory control mechanisms (Simpson 1960). Detailed studies of subsets of lymphocytes have produced a bewildering variety of results. Suppressor cells have been studied, and conflicting reports describe either a decrease (Mischak et al 1979) or an increase (Birnbaum & Tsairis 1976) in suppressor function.

It has recently been shown that peripheral blood lymphocytes from myasthenic patients are stimulated to undergo increased cell division ("blast transformation") when incubated in the presence of AChR from electric eels or rays (Abramsky et al 1975b, Richman et al 1976, Conti-Tronconi et al 1979); however, the relationship between lymphocyte responsiveness and the clinical status of MG patients is as yet unresolved. In some studies, older males show the most consistent and pronounced responses (Richman et al 1979), although another study has failed to confirm these observations (Conti-Tronconi et al 1979). There are conflicting reports on whether the lymphocyte responsiveness reflects the clinical severity of myasthenia (Abramsky et al 1975b, Richman et al 1979) and whether the stimulation index returns toward normal in patients treated with immunosuppressive agents (Abramsky et al 1975a, Conti-Tronconi et al 1979). In general, stimulation of lymphocytes in response to a specific antigen indicates that the cells have been previously sensitized to that antigen (Greaves et al 1974); however, it is not yet known whether the positive responses in MG are derived from T- or B-cells, or what role the responding cells play in the immunopathogenesis of the disease.

The implications of these observations are by no means clear at present. The interactions between T-lymphocytes and B-lymphocytes are highly complex, and both classes of cells may be involved in humoral as well as cellular immune reactions (Greaves et al 1974). At present, there is little evidence for a cell-mediated effector mechanism in MG, but it seems likely that defects of immune regulation may exist in at least some myasthenic patients.

IS MYASTHENIA GRAVIS ONE DISEASE?

Although the target of the autoimmune attack in myasthenic patients is probably always the AChRs, it is doubtful that the immune mechanisms used are always the same; there are enough differences among patients to suggest the possibility of divergent origins and pathways of the immune processes.

Clinical differences between groups of myasthenic patients are well recognized. It has long been known that age and sex incidence rates are not random, younger females and older males being selectively affected (Osserman 1958). Furthermore, the distribution and severity of involvement of muscles may be so characteristic as to suggest different processes. Thus, in some patients weakness may be confined to the extraocular muscles, while in others widespread involvement of the somatic musculature may be the rule.

The thymic abnormality varies among groups of patients; approximately one-quarter have no histologic abnormalities, whereas the majority show germinal centers, and approximately 10% have thymic tumors (Castleman 1966, Namba et al 1976, Levine 1979). Patients may also be separated on the basis of histocompatibility typing. Young females with thymic hyperplasia have an excessively high prevalence (disequilibrium) of the HLA-B8 antigen (Feltkamp et al 1974, Pirskanen 1976). By contrast, HLA-B8 is uncommon in older males and patients with thymomas. Other clinical features that suggest possible unique pathogenetic mechanisms include familial occurrence in some cases (Herrmann 1971, Namba et al 1971) and the association with other autoimmune disorders in some but not all patients with MG (Simpson 1960, Oosterhuis & deHass 1968, Namba et al 1973).

As noted above, anti-AChR antibody levels vary widely from patient to patient, and correspond only approximately to the degree of weakness (Almon et al 1974, Appel et al 1975, Bender et al 1975, Lindstrom et al 1976, Mittag et al 1976, Toyka et al 1977). More than 10% of patients have no detectable antibody by present assay methods. Similarly, stimulation of peripheral lymphocytes in the presence of AChR occurs in some, but not all, myasthenic patients (Abramsky et al 1975b, Richman et al 1976, Conti-Tronconi et al 1979, Richman et al 1979).

It is possible that these observed differences merely represent minor variations of a single pathogenetic mechanism; however, it seems more likely that the autoimmune attack in MG may take a variety of pathways in different patients and still result in the same basic deficit of AChRs.

ORIGIN OF THE AUTIOMMUNE RESPONSE IN MG: POSSIBLE ROLE OF THE THYMUS

One of the unsolved problems in MG, as in other spontaneously occurring autoimmune diseases of man, concerns the origin of the autoimmune response. The pathologic changes in thymus glands of patients with MG and the favorable results of thymectomy first suggested the possibility that the process might originate intrathymically. As noted above, approximately 75% of myasthenic patients have thymic abnormalities. Of these, 85% show germinal center formation ("hyperplasia"), and gross or microscopic thymomas are found in approximately 15%. Subtle changes have been found in the lymphocytic cells of the thymus in myasthenic patients, including alterations in their antigenic properties, an increased proportion of B cells, and increased production of antibodies, especially those directed against AChR (Abdou et al 1974, Staber et al 1975, Richman et al 1976, Cook et al 1977, Levine 1979, Vincent et al 1979).

In addition to lymphocytes, the thymus gland is known to contain cells of other types, including muscle-like or "myoid" cells (Van de Velde & Friedman 1970). The possible relation of these cells to MG was suggested by the discovery that they cross-react with antimuscle antibody (van der Geld & Strauss 1966). More recently, typical skeletal muscle cells have been cultured from thymus glands of patients with MG and from thymuses of normal rats (Wekerle et al 1975, Kao & Drachman 1977a). These cells behave in all respects like normal mammalian skeletal muscle in tissue culture (Kao & Drachman 1977a). They are striated and multinucleated, contract in response to electrical stimulation, and accept innervation when co-cultured with cholinergic neuronal cells. Perhaps most intriguing was the finding that the cultured thymic muscle cells have surface AChRs, as demonstrated by autoradiographic and electrophysiological techniques (Kao & Drachman 1977a). They may represent the source of AChR that has been found in thymic extracts (Aharonov et al 1975, Vincent et al 1979). Because of their strategic location within the thymus, the AChR-bearing muscle cells may be particularly vulnerable to immune attack. Some alteration of the muscle cells or the lymphocytes of the thymus may serve to break tolerance and thereby initiate an autoimmune response directed against AChRs, as well as other components of skeletal muscles. The fact that antistriational antibodies that cross-react with the myoid cells are present in sera of a majority of MG patients suggests that the myoid cells or other intact skeletal muscle cells, rather than some alternative source of isolated AChR, may serve as the antigenic stimulus in MG.

The possibility that a viral infection of the thymus could trigger this process has been suggested (Datta & Schwartz 1974), and several studies have been undertaken to test this hypothesis. Antibodies against cytomegalovirus (CMV) were found in an increased proportion of myasthenic patients (Tindall et al 1978), but the relevance of raised antibody titers to such a ubiquitous agent as CMV is not clear. Culture and electron microscopy of thymuses from myasthenic patients have failed to yield evidence of the presence of CMV or other viruses (T. Aoki, D. B. Drachman, J. Wolinsky, C. Gibbs, in preparation). Gamboa et al (cited in Penn 1979) have apparently found an increase in prevalence of the herpes simplex viral antigen in myasthenic thymus glands, as compared with controls. Although the concept of a virus "trigger" in MG is attractive, there is as yet no firm evidence to support it. Many gaps remain in understanding the origin of the autoimmune response and the complex role of the thymus gland in MG.

THERAPY OF MYASTHENIA GRAVIS

Modern treatment has improved the outlook in MG so remarkably that most patients are able to return to a full productive life. The ultimate goal of a cure, however, has remained elusive; most patients must continue taking medications despite the risks of adverse side-effects. In this section, we review the principles of treatment of MG and discuss newer approaches.

In general, the following four methods of treatment are currently used: (*a*) enhancement of neuromuscular transmission, (*b*) immunosuppression, (*c*) surgical thymectomy, and (*d*) depletion of circulating antibodies.

Enhancement of Neuromuscular Transmission

The aim of this method of treatment is to increase the amount of ACh available to interact with AChRs at the neuromuscular junction. The most effective agents are the quaternary ammonium anticholinesterase drugs, such as pyridostigmine, which inhibit the enzymatic hydrolysis of ACh at the neuromuscular junction (Koelle 1975). The ACh released by the nerve can therefore interact repeatedly with the remaining AChRs, thus giving rise to larger epps. If the epp amplitude is raised above the threshold at many junctions, both the patient's strength and the electrical response to repetitive nerve stimulation will improve. The dosage of anticholinesterase drugs must be carefully regulated, since an excess can result in increased weakness, due to depolarization or desensitization block of the receptors (Osserman 1958, Flacke 1973). Moreover, very high doses of anticholinesterase agents have been shown experimentally to produce adverse effects similar to those of MG itself, including a reduction of AChRs at neuromuscular junctions (Chang et al 1973, Fambrough et al 1973), decreased mepp amplitudes (Roberts & Thesleff 1969), and alterations in ultrastructure of the post-synaptic membrane (Engel et al 1973); however, anticholinesterase agents continue to be used as the first line of treatment for most patients with MG.

Several drugs that are thought to enhance the release of ACh by cholinergic nerves have been used as adjuncts to more conventional treatment. Ephedrine was the first drug used in MG (Osserman 1958), and may act like other adrenergic agents to increase the release of ACh from cholinergic nerve terminals (MacIntosh & Collier 1977). Guanidine (see Osserman 1958, MacIntosh & Collier 1977), germine acetate (Flacke et al 1971), and 4-amino pyridine (Lundh et al 1979) all increase the impulse-dependent release of ACh but none of these are in general clinical use at present. As a rule, the presynaptic-acting drugs provide only marginal benefit for myasthenic patients.

Immunosuppression

Adrenal corticosteroids are the most commonly used immunosuppressive agents. Steroid treatment results in clinical improvement in 70 to 100% of patients, with good to excellent results in 63 to 100% (summarized in Drachman 1978). Steroid treatment is not without problems; to maintain the beneficial effects, medication must be continued indefinitely, thus resulting in a significant incidence of adverse side-effects, which may include osteoporosis, cataract formation, decreased resistance to infection, etc. Although differences of opinion remain about the indications for steroid therapy, it is generally agreed that steroid treatment may be considered for any myasthenic patient whose weakness is not satisfactorily controlled by anticholinesterase medication and/or thymectomy.

The precise mechanism of action of corticosteroid agents in MG has not yet been established. It is well known that they exert suppressive actions at many levels of the immune system (Claman 1972). They have been shown to decrease the size of the thymus, and to reduce the lymphycyte population in sensitive species. In MG, steroid treatment may produce a reduction of anti-AChR antibody levels (Tindall 1980) and a diminution of anti-AChR reactivity of peripheral lymphocytes (Abramsky et al 1975a). Whether these effects alone are sufficient to account for the benefit of steroid therapy is not yet known. Experimental studies have demonstrated that high concentrations of steroids exert certain direct influences on neuromuscular transmission (Wilson et al 1974, Wolters 1976), but the clinical relevance of such effects in MG has not been established.

Other immunosuppressive drugs, including azathioprine, 6-mercaptopurine, and cyclophosphamide, have been reported to benefit myasthenic patients, often after unsuccessful treatment with adrenal corticosteroids and thymectomy (Matell et al 1976, Hertel et al 1979). These agents are reported to decrease the anti-AChR antibody titers (Lefvert et al 1978, Reuther et al 1979), but little is known about their precise mechanisms of action. The disadvantages of prolonged immunusuppressive therapy include the risks of bone marrow depression, decreased resistance to infection, and the late occurrence of neoplasms.

Thymectomy

The early reports of Schumacher & Roth (1913) and Blalock et al (1939) first suggested that removal of the thymus gland might lead to clinical improvement in patients with MG. These observations have been confirmed by the published reports of large series of surgically treated cases, with improvement occurring in 57 to 86% of patients, and "remission" in 20 to 36% (reviewed in Drachman 1978, Keesey 1979). In spite of these favorable results, many unanswered questions remain regarding the indications for thymectomy, the prognostic influence of a variety of factors, and the mechanism of the beneficial effect.

The most widely accepted indication for thymectomy is the presence of a thymoma, since the tumor must usually be removed for its own sake, to prevent local spread. In patients without thymomas, the main advantage of thymectomy lies in the possibility of permanent remission or improvement in some cases. On the other hand, it entails the short-term risks inherent in any major surgical procedure undertaken on a weak patient. Moreover, improvement may be delayed for up to ten years after thymectomy, and many patients continue to require anticholinesterase medication or adrenal corticosteroids or both after thymectomy. As a matter of practice, most physicians recommend thymectomy for adult patients with generalized weakness of moderate or severe degree.

Little is known about the mechanism by which thymectomy produces its beneficial effect in MG. On theoretical grounds, the following possibilities may be suggested:

1. Removal of the thymus may eliminate a source of continuing antigenic stimulation. If the thymic "myoid" cells serve as the source of antigen in initiating the autoimmune response in MG (Kao & Drachman 1977a), then their removal might allow the immune response to subside.

2. Thymectomy may remove an important source of cells secreting anti-AChR antibody. As noted above, thymic lymphocytes are capable of producing anti-AChR antibody (Vincent et al 1979). If the thymus serves as a reservoir for a significant proportion of the patient's anti-AChR antibody secreting cells, its removal should result in decreased antibody production, and clinical improvement; however, there are conflicting reports of the effect of thymectomy on anti-AChR antibody levels, even in clinically successful cases (Scadding et al 1977, Lefvert et al 1978, Seybold & Lindstrom 1979).

3. The thymus may participate in a disturbance of immune regulation that permits the production of autoantibodies in MG. The central role of the thymus in immune regulation, and the alterations noted above in thymic lymphocytes in myasthenic patients, suggest that thymectomy may in some way correct the aberration of immune function.

It is not yet clear which if any of these effects is most important. Moreover, the fact that not all patients benefit from thymectomy suggests that the role of the thymus in the immunopathogenesis of MG may differ in different individuals.

Antibody Depletion Therapy

During the past several years, therapeutic measures aimed at reducing serum antibody levels have been shown to benefit some myasthenic patients, thus adding support to the concept of an antibody-mediated process and leading to clinical applications. The first antibody-depleting procedure used in MG was drainage of lymph via the thoracic duct (Matell et al 1976). Improvement began within 48 hr, and persisted throughout the period of drainage. More recently, the technically simpler methods of plasma exchange (Pinching et al 1976) and plasmapheresis (Dau et al 1977) have been applied to MG, with generally beneficial effects. Following each course of plasmapheresis, the serum anti-AChR antibody levels fall. This generally correlates with clinical improvement (Vincent et al 1977). In some cases, the improvement has been dramatic, giving rise to exaggerated publicity in the lay press; however, the benefit is temporary, unless the patient undergoes repeated courses of plasmapheresis or is treated concomitantly with immunosuppressive agents (Newsom-Davis et al 1979).

The role of plasmapheresis in the clinical therapy of MG has yet to be determined, but there is concensus (Dau 1979) that it is useful for the short-term treatment of severely involved patients, in order to get them through a difficult period.

Specific Immunotherapy—Experimental Approaches

Ideally, the goal of therapy in MG should be to eliminate the lymphocytes specifically sensitized to AChR and capable of producing anti-AChR antibody. Certain novel approaches that have been tested in the experimental animal model may ultimately find application in treating the human disease.

An interesting therapeutic strategy, based on the production of antiidiotype antibodies, has been suggested by Schwartz et al (1978). The antigen-combining sites of antibodies are structurally unique, and are themselves antigenic. According to the "network theory" of Jerne (1973), the immune system may normally produce anti-idiotype antibodies directed against these sites, which serve to regulate the lymphocytes' antibody production. If anti-idiotype antibodies can be made against the specific determinants of anti-AChR antibodies, they might be used to control the autoimmune reaction in MG. Thus far, small amounts of anti-idiotype antibodies against anti-AChR antibodies have been prepared by an in vitro method, but have not yet been tested in the intact animal. One of the problems inherent in the anti-idiotype approach is the heterogeneity of anti-AChR antibodies. There are many different antigenic determinants on the AChR molecule, and individual patients (or EAMG animals) undoubtedly have multiple antobodies with specificity for multiple different sites. It is rather surprising and encouraging that the anti-idiotypes prepared against mouse anti-AChR antibodies seem to have broad cross-reactivity with anti-AChR antibodies from a wide variety of species, including mouse, rabbit, rat, and monkey.

Another therapeutic strategy designed to eliminate the antibody-producing lymphocytes has been used successfully in rats with EAMG (Pestronk et al 1980). A single high dose of cyclophosphamide was given, because of its known effectiveness against B-lymphocytes. The rats were "rescued" from the otherwise lethal effect of the drug by bone marrow cell transplantation. This treatment produced a rapid and sustained fall of antibody titers against both the immunizing antigen (torpedo AChR) and the autoantigen (rat AChR). In treated animals with the most pronounced fall of antibody titers, mepp amplitudes and AChRs returned toward normal at neuromuscular junctions.

Recently, Bartfeld & Fuchs (1978) have reported that immunization of rabbits with a denatured AChR preparation can prevent the development of EAMG, or reduce its severity if already established. Rabbits immunized with this reduced carboxymethylated preparation did not show clinical signs of EAMG, although they developed high titers of antibody, which cross-reacted with native torpedo AChR. The mechanism by which immunization with an altered AChR preparation protects against EAMG is not yet certain. The pathogenic antibodies in EAMG rabbits appear to block the active site of the AChR, whereas the antibodies produced in response to denatured AChR may bind to other sites on the junctional AChRs and protect against the more harmful blocking antibody.

CONCLUSIONS

The basic abnormality in MG is a decrease of AChRs at neuromuscular junctions. This results in the typical clinical features of weakness and neuromuscular fatigue, the characteristic electrophysiological findings of low amplitude mepps and decremental responses, and the pharmacological responses to anticholinesterase drugs.

The reduction of receptors is brought about by an autoimmune reaction, mediated by anti-AChR antibodies. The mechanisms of antibody action include accelerated degradation of AChRs, blockade of the receptors, and focal damage to the postsynaptic membrane, probably enhanced by complement. Much has been learned during the past few years about the mechanism of accelerated AChR degradation, which involves cross-linking by divalent antibody molecules, aggregation and endocytosis of the receptors, and a final step of enzymatic degradation within the muscle cells.

The present understanding of the neurobiology and immunology of MG has provided rational explanations for therapeutic methods in current use. Based on knowledge of the pathogenesis of Mg, specific methods of immunotherapy are now being devised, with the ultimate goal of curing the underlying disorder.

ACKNOWLEDGMENTS

I am deeply indebted to the many colleagues who participated in the studies described here, including D. Fambrough, S. Satya-Murti, F. Slone, K. Toyka, A. Pestronk, I. Kao, D. Griffin, J. Winkelstein, C. Angus, R. Adams, E. Stanley, K. Fischbeck, A. Murphy, J. Michelson, and G. Hoffman. Ms. C. Barlow provided expert assistance with the manuscript. The original research carried out in the author's laboratory was supported in part by NIH grants #5 PO1 NS10920 and 5 RO1 HD04817, and grants from the Muscular Dystrophy Association and the Myasthenia Gravis Foundation.

Literature Cited

Abdou, N. I., Lisak, R. P., Zweiman, B., Abrahamsohn, I., Penn, A. S. 1974. The thymus in myasthenia gravis. Evidence for altered cell populations. *N. Engl. J. Med.* 291:1271–75

Abramsky, O., Aharonov, A., Teitelbaum, D., Fuchs, S. 1975a. Myasthenia gravis and acetylcholine receptor. *Arch. Neurol.* 32:684–87

Abramsky, O., Aharonov, A., Webb, C., Fuchs, S. 1975b. Cellular immune response to acetylcholine receptor-rich fraction in patients with myasthenia gravis. *Clin. Exp. Immunol.* 19:11–16

Aharonov, A., Tarrab-Hazdai, R., Abramsky, O., Fuchs, S. 1975. Immunological relationship between acetylcholine receptor and thymus: A possible significance in myasthenia gravis. *Proc. Natl. Acad. Sci. USA* 72:1456–60

Albuquerque, E. X., Lebeda, J. J., Appel, S. H., Almon, R., Kauffman, F. C., Mayer, R. F., Narahashi, T., Yeh, J. Z.

1976a. Effects of normal and myasthenic serum factors on innervated and chronically denervated mammalian muscles. *Ann. NY Acad. Sci.* 274:475–92

Albuquerque, E. X., Rash, J. E., Mayer, R. F., Satterfield, J. R. 1976b. An electrophysiological and morphological study of the neuromuscular junctions in patients with myasthenia gravis. *Exp. Neurol.* 51:536–63

Almon, R. R., Andrew, C. G., Appel, S. H. 1974. Serum globulin in myasthenia gravis: Inhibition of α-bungarotoxin binding to acetylcholine receptors. *Science* 186:55–57

Appel, S. H., Almon, R. R., Levy, N. 1975. Acetylcholine receptor antibodies in myasthenia gravis. *N. Engl. J. Med.* 293:760–61

Appel, S. H., Anwyl, R., McAdams, M. W., Elias, S. 1977. Accelerated degradation and acetylcholine receptor from cultured rat myotubes with myasthenia gravis sera and globulins. *Proc. Natl. Acad. Sci. USA* 74:2130–34

Barrett, E. F., Magleby, K. L. 1976. Physiology of cholinergic transmission. In *Biology of Cholinergic Function*, ed. A. M. Goldberg, I. Hanin, pp. 29–100. New York: Raven

Bartfeld, D., Fuchs, S. 1978. Specific immunosuppression of experimental autoimmune myasthenia gravis by denatured acetylcholine receptor. *Proc. Natl. Acad. Sci. USA* 75:4006–10

Bender, A. N., Engel, W. K., Ringel, S. P., Daniels, M. P., Vogel, Z. 1975. Myasthenia gravis: A serum factor blocking acetylcholine receptors of the human neuromuscular junction. *Lancet* 1:607–8

Berg, D. K., Hall, Z. W. 1975. Loss of α-bungarotoxin from junctional and extrajunctional receptors in rat diaphragm muscle in vivo and in organ culture. *J. Physiol. London* 252:771–89

Bergman, R. A., Johns, R. J., Afifi, A. K. 1971. Ultrastructural alterations in muscle from patients with myasthenia gravis and Eaton-Lambert syndrome. *Ann. NY Acad. Sci.* 183:88–122

Bevan, S., Kullberg, R. W., Heinemann, S. F. 1977. Human myasthenic sera reduce acetylcholine sensitivity of human muscle cells in tissue culture. *Nature* 267:263–65

Birnbaum, G., Tsairis, P. 1976. Suppressor lymphocytes in myasthenia gravis and effect of adult thymectomy. *Ann. NY Acad. Sci.* 274:527–35

Blalock, A., Mason, M. F., Morgan, H. J., Riven, S. S. 1939. Myasthenia gravis and tumors of the thymic region. *Ann. Surg.* 110:544–61

Brockes, J. P., Berg, D. K., Hall, Z. W. 1976. The biochemical properties and regulation of acetylcholine receptors in normal and denervated muscle. *Cold Spring Harbor Symp. Quant. Biol.* 30:253–62

Campbell, H., Bramwell, E. 1900. Myasthenia gravis. *Brain* 23:277–336

Castleman, B. 1966. The pathology of the thymus gland in myasthenia gravis. *Ann. NY Acad. Sci.* 135:496–503

Chang, C. C., Chen, T. F., Chuang, S.-T. 1973. Influence of chronic neostigmine treatment on the number of acetylcholine receptors and the release of acetylcholine from the rat diaphragm. *J. Physiol. London* 230:613–18

Chang, C. C., Huang, M. C. 1975. Turnover of junctional and extrajunctional acetyl-

choline receptors of the rat diaphragm. *Nature* 253:643–44

Chang, C. C., Lee, C. Y. 1966. Electrophysiological study of neuromuscular blocking action of cobra neurotoxin. *Br. J. Pharmacol. Chemother.* 28:172–81

Claman, H. N. 1972. Cortocosteroids and lymphoid cells. *N. Engl. J. Med.* 287:388–97

Cohen, S. A., Pumplin, D. W. 1979. Clusters of intramembrane particles associated with binding sites for α-bungarotoxin in cultured chick myotubes. *J. Cell Biol.* 82:494–516

Conti-Tronconi, B. M., Morgutti, M., Sghirlanzoni, A., Clementi, F. 1979. Cellular immune response against acetylcholine receptor in myasthenia gravis. *Neurology* 29:496–501

Cook, J. D., Trotter, J. L., Engel, W. K., McIntosh, C. L. 1977. Altered immunologic cell populations in thymuses from myasthenia gravis patients. *Neurology* 27:365

Cull-Candy, S. G., Miledi, R., Trautmann, A. 1979. End-plate currents at normal and myasthenic human end-plates. *J. Physiol. London* 287:247–65

Cull-Candy, S. G., Miledi, R., Trautmann, A., Uchitel, O. D. 1980. On the release of transmitter at normal, myasthenia gravis and myasthenic syndrome affected human endplates. *J. Physiol. London* 299:621–38

Datta, S. K., Schwartz, R. S. 1974. Infectious–myasthenia. *N. Engl. J. Med.* 291:1304–5

Dau, P. C., ed. 1979. *Plasmapheresis and The Immunobiology of Myasthenia Gravis.* Boston: Houghton Mifflin

Dau, P. C., Lindstrom, J. M., Cassel, C. K., Denys, E. H., Shev, E. E., Spitler, L. E. 1977. Plasmapheresis and immunosuppressive drug therapy in myasthenia gravis. *N. Engl. J. Med.* 297:1134–40

Desmedt, J. E. 1973. The neuromuscular disorder in myasthenia gravis. In *New Developments in EMG and Clinical Neurophysiology*, ed. J. E. Desmedt, pp. 241–304. Basel: Karger

Devreotes, P. N., Fambrough, D. M. 1975. Acetylcholine receptor turnover in membranes of developing muscle fibers. *J. Cell Biol.* 65:335–58

Devreotes, P. N., Fambrough, D. M. 1976. Turnover of acetylcholine receptors in skeletal muscle. *Cold Spring Harbor Symp. Quant. Biol.* 40:237–51

Drachman, D. B. 1978. Myasthenia gravis. *N. Engl. J. Med.* 298:136–42, 186–93

Drachman, D. B., Adams, R. N., Stanley, E. F., Pestronk, A. 1980. Mechanisms of acetylcholine receptor loss in myasthenia gravis. *J. Neurol. Neurosurg. Psychiatr.* 43:601

Drachman, D. B., Angus, C. W., Adams, R. N., Kao, I. 1978a. Effect of myasthenic patients' immunoglobulin on acetylcholine receptor turnover: Selectivity of degradation process. *Proc. Natl. Acad. Sci. USA* 75:3422–26

Drachman, D. B., Angus, C. W., Adams, R. N., Michelson, J., Hoffman, G. J. 1978b. Myasthenic antibodies crosslink acetylcholine receptors to accelerate degradation. *N. Engl. J. Med.* 298:1116–22

Drachman, D. B., Kao, I., Pestronk, A., Toyka, K. V. 1976. Myasthenia gravis as a receptor disorder. *Ann. NY Acad. Sci.* 274:226–34

Elmqvist, D., Hofmann, W. W., Kugelberg, J., Quastel, D. M. J. 1964. An electrophysiological investigation of neuromuscular transmission in myasthenia gravis. *J. Physiol. London* 174:417–34

Engel, A. G., Lambert, E. H., Santa, T. 1973. Study of long-term anticholinesterase therapy, effects on neuromuscular transmission and on motor end-plate fine structure. *Neurology* 23:1273–81

Engel, A. G., Lindstrom, J. M., Lambert, E. H., Lennon, V. A. 1977. Ultrastructural localization of the acetylcholine receptor in myasthenia gravis and in its experimental autoimmune model. *Neurology* 27:307–15

Engel, A. G., Sahashi, K., Lambert, E. H., Howard, F. M. 1979. The ultrastructural localization of the acetylcholine receptor, immunoglobulin G and the third and ninth complement components at the motor endplate and their implications for the pathogenesis of myasthenia gravis. In *Current Topics in Nerve and Muscle Research*, ed. A. J. Aguayo, G. Karpati, pp. 111–22. Amsterdam: Excerpta Medica

Engel, A. G., Santa, T. 1971. Histometric analysis of the ultrastructure of the neuromuscular junction in myasthenia gravis and in the myasthenic syndrome. *Ann. NY Acad. Sci.* 183:46–63

Engel, A. G., Tsujihata, M. T., Lambert, E. H., Lindstrom, J. M., Lennon, V. A. 1976. Experimental autoimmune myasthenia gravis. A sequential and quantitative study of the neuromuscular junction ultrastructure and electrophysiologic correlation. *J. Neuropathol. Exp. Neurol.* 35:563–87

Erb, W. 1879. Zur casuistik der bulbaren lahmungen. Uber einen neuen wahrschein-lich bulbaren symptom-complex. *Arch Psychiatr. Nervenkr.* 9:336–50

Fambrough, D. M., Drachman, D. B., Satyamurti, S. 1973. Neuromuscular junction in myasthenia gravis: Decreased acetylcholine receptors. *Science* 182:293–95

Feltkamp, T. E., van den Berg-Loonen, P. M., Nijenhuis, L. E., Engelfriet, C. P., van Rossom, A. L., van Lughem, J. J., Oosterhuis, H. J. 1974. Myasthenia gravis, autoantibodies and HL-A antigens. *Br. Med. J.* 1:131–33

Fields, W. S. 1971. Myasthenia gravis. *Ann. NY Acad. Sci.* 183:1–386

Flacke, W. E. 1973. Treatment of myasthenia gravis. *N. Engl. J. Med.* 288:27–31

Flacke, W. E., Blume, R. P., Scott, W. R., Foldes, F., Osserman, K. E. 1971. Germine mono and diacetate in myasthenia gravis. *Ann. NY Acad. Sci.* 183:316–33

Granato, D. A., Fulpius, B. W., Moody, J. F. 1976. Experimental myasthenia in Balb/C mice immunized with rat acetylcholine receptor from rat denervated muscle. *Proc. Natl. Acad. Sci. USA* 73:2872–76

Greaves, M. F., Owen, J. J. T., Raff, M. C. 1974. *T and B Lymphocytes: Origins, Properties and Roles in Immune Responses.* New York/Amsterdam: Am. Elsevier, Excerpta Medica

Green, D. P. L., Miledi, R., de la Mora, M. P., Vincent, A. 1975. Acetylcholine receptors. *Philos. Trans. R. Soc. London Ser. B* 270:551–59

Hartzell, H. C., Fambrough, D. M. 1972. Acetylcholine receptors. Distribution and extrajunctional density in rat diaphragms after denervation correlated with acetylcholine sensitivity. *J. Gen. Physiol.* 60:248–62

Harvey, A. M., Masland, R. L. 1941. The electromyogram in myasthenia gravis. *Bull. Johns Hopkins Hosp.* 69:1–13

Heidmann, T., Changeux, J.-P. 1978. Structural and functional properties of the acetylcholine receptor protein in its purified and membrane-bound states. *Ann. Rev. Biochem.* 47:317–57

Heilbronn, E., Hammarstrom, L., Lefvert, A. K., Mattsson, C., Smith, E., Stålberg, E., Thornell, L. E. 1979. Effects of acetylcholine receptor antibodies in mice and rabbits. See Dau 1979, Ch. 9, pp. 92–96

Heilbronn, E., Mattsson, C., Thornell, L. E., Sjöström, M., Stålberg, E., Hilton-Brown, P., Elmqvist, D. 1976. Experimental myasthenia in rabbits: Biochemical, immunological, electrophysiological and morphological aspects. *Ann. NY Acad. Sci.* 274:337–53

Herrmann, C. 1971. The familial occurrence of myasthenia gravis. *Ann. NY Acad. Sci.* 183:334–50

Hertel, G., Mertens, H. G., Reuther, P., Ricker, K. 1979. The treatment of myasthenia gravis with azathioprine. See Dau 1979, Ch. 29, pp. 315–28

Howard, J. F., Sanders, D. B. 1978. Passive transfer of human myasthenia gravis to rats. I. Electrophysiology of the developing neuromuscular block. *Neurology* 28:346

Ito, Y., Miledi, R., Vincent, A., Newsom-Davis, J. 1978. Acetylcholine receptors and endplate electrophysiology in myasthenia gravis. *Brain* 101:345–68

Jerne, N. K. 1973. The immune system. *Sci. Am.* 229:52–60

Kao, I., Drachman, D. B. 1977a. Thymic muscle cells bear acetylcholine receptors: Possible relation to myasthenia gravis. *Science* 195:74–75

Kao, I., Drachman, D. B. 1977b. Myasthenic immunoglobulin accelerates acetylcholine receptor degradation. *Science* 196:527–29

Katz, B., Miledi, R. 1972. The statistical nature of the acetylcholine potential and its molecular components. *J. Physiol. London* 224:665–99

Katz, B., Miledi, R. 1973. The binding of acetylcholine to receptors and its removal from the synaptic cleft. *J. Physiol. London* 231:549–74

Keesey, J. 1979. Indications for thymectomy in myasthenia gravis. See Dau 1979, Ch. 12, pp. 124–36

Koelle, G. B. 1975. In *The Pharmacological Basis of Therapeutics,* ed. L. S. Goodman, A. Gilman, Ch. 22, pp. 445–66. New York: Macmillan. 5th ed.

Kurtzke, J. F., Kurland, L. T. 1977. Epidemiology of neurologic disease in *Clinical Neurology,* ed. A. B. Baker, L. N. Baker, Vol. 3 Ch. 48. Hagerstown, Md: Harper & Row

Lambert, E. H., Lindstrom, J. M., Lennon, V. A. 1976. Endplate potentials in experimental autoimmune myasthenia gravis in rats. *Ann. NY Acad. Sci.* 274:300–18

Lee, C. Y. 1972. Chemistry and pharmacology of polypeptide toxins in snake venoms. *Ann. Rev. Pharmacol.* 12:265–86

Lefvert, A. K., Bergstrom, K., Matell, G., Osterman, P. O., Pirskanen, R. 1978. Determination of acetylcholine receptor antibody in myasthenia gravis: Clinical usefulness and pathogenetic implications. *J. Neurol. Neurosurg. Psychiatr.* 41:394–403

Lennon, V. A. 1978. Immunofluorescence analysis of surface acetylcholine receptors on muscle: Modulation by autoantibodies. In *Cholinergic Mechanisms and Psychopharmacology,* ed. D. J. Jenden, pp. 77–92. New York: Plenum

Levine, G. D. 1979. Pathology of the thymus in myasthenia gravis: Current concepts. See Dau 1979, Ch. 11, pp. 113–23

Libby, P., Bursztajn, S., Goldberg, A. J. 1980. Degradation of the acetylcholine receptor in cultured muscle cells: Selective inhibitors and the fate of undegraded receptors. *Cell.* 9:481–91

Lindstrom, J. M. 1979. The role of antibodies to the acetylcholine receptor protein and its component peptides in experimental autoimmune myasthenia gravis in rats. See Dau 1979, Ch. 1, pp. 3–19

Lindstrom, J. M., Lambert, E. H. 1978. Content of acetylcholine receptor and antibodies bound to receptor in myasthenia gravis, experimental autoimmune myasthenia gravis, and Eaton-Lambert syndrome. *Neurology* 28:130–38

Lindstrom, J. M., Seybold, M. E., Lennon, V. A., Whittingham, S., Duane, D. D. 1976. Antibody to acetylcholine receptor in myasthenia gravis: Prevalence, clinical correlates, and diagnostic value. *Neurology* 26:1054–59

Lundh, H., Nilsson, C., Rosen, J. 1979. Effects of 4-aminopyridine in myasthenia gravis. *J. Neurol. Neurosurg. Psychiatr.* 42:171–75

MacIntosh, F. C., Collier, B. 1977. Neurochemistry of cholinergic terminals. In *Neuromuscular Junction,* ed. E. Zaimis. Ch. 2, pp. 99–228. Heidelberg: Springer

McAdams, M. W., Roses, A. D. 1980. Comparison of antigenic sources for acetylcholine receptor antibody assays in myasthenia gravis. *Ann Neurol.* 8: 61–66

Matell, G., Bergstrom, K., Franksson, C., Hammarstrom, L., Lefvert, A. K., Moller, E., von Reis, G., Smith, E. 1976. Effect of some immunosuppressive procedures on myasthenia gravis. *Ann. NY Acad. Sci.* 274:659–76

Mendell, J. R., Whitaker, J. N., Engel, W. K. 1973. The skeletal muscle binding site of antistriated muscle antibody in myasthenia gravis: An electron microscopic immunohistochemical study using peroxidase conjugated antibody fragments. *J. Immunol.* 111:847–56

Merlie, J. P., Heinemann, S., Lindstrom, J. M. 1979. Acetylcholine receptor degradation in adult rat diaphragms in organ culture and the effect of anti-acetylcholine receptor antibodies. *J. Biol. Chem.* 254:6300–27

Mischak, R. P., Dau, P. C., Gonzalez, R. L., Spitler, L. E. 1979. In vitro testing of suppressor cell activity in myasthenia Gravis. See Dau 1979, Ch. 7, pp. 72–78

Mittag, T., Kornfeld, P., Tormay, A., Woo, C. 1976. Detection of antiacetylcholine receptor factors in serum and thymus from patients with myasthenia gravis. *N. Engl. J. Med.* 294:691–94

Molenaar, P. C., Polak, R. L., Miledi, R., Alema, S., Vincent, A., Newsom-Davis, J. 1979. The cholinergic synapse. *Progr. Brain Res.,* 49:449–58

Namba, T., Brown, S. B., Grob, D. 1970. Neonatal myasthenia gravis: Report of two cases and review of the literature. *Pediatrics* 45:488–504

Namba, T., Brunner, N. G., Brown, S. B., Muguruma, M., Grob, D. 1971. Familial myasthenia gravis. Report of 27 patients in 12 families and review of 164 patients in 73 families. *Arch. Neurol.* 25:49–60

Namba, T., Brunner, N. G., Grob, D. 1973. Association of myasthenia gravis with pemphigus vulgaris, candida albicans infection, polymyositis and myocarditis. *J. Neurol. Sci.* 20:231–42

Namba, T., Nakata, Y., Grob, D. 1976. The role of cellular and humoral immune factors in the neuromuscular block of myasthenia gravis. *Ann. NY Acad. Sci.* 274:493–513

Nastuk, W. L., Niemi, W. D., Alexander, J. T., Chang, H. W., Nastuk, M. A. 1979. Myasthenia in frogs immunized against cholinergic-receptor protein. *Ann. J. Physiol.* 236:C53–57

Nastuk, W. L., Plescia, O. J., Osserman, K. E. 1960. Changes in serum complement activity in patients with myasthenia gravis. *Proc. Soc. Exp. Biol.* 105:177–84

Nastuk, W. L., Strauss, A. J. L. 1961. Further developments in the search for a neuromuscular blocking agent in the blood of patients with myasthenia gravis. In *Myasthenia Gravis,* ed. H. R. Viets, Ch. 3, pp. 229–237. Springfield Ill: Thomas

Nastuk, W. L., Strauss, A. J. L., Osserman, K. E. 1959. Search for a neuromuscular blocking agent in the blood of patients with myasthenia gravis. *Am. J. Med.* 26:394–409

Newsom-Davis, J., Ward, C. D., Wilson, S. G., Pinching, A. J., Vincent, A. 1979. plasmapheresis: short and long term benefits. See Dau 1979, ch. 19, pp. 199–208

Oda, K., Goto I., Kuroiwa, Y., Onoue, K., Ito, Y. 1980. Myasthenia gravis: Antibodies to acetylcholine receptor with human and rat antigens. *Neurology* 30:543–46

Oosterhuis, H. J., deHass, W. H. D. 1968. Rheumatic diseases in patients with myasthenia gravis. An epidemiological and clinical investigation. *Acta Neurol. Scand.* 44:219–27

Oppenheim, H. 1901. *Zur Myasthenische Paralyse.* Berlin: Karger

Orci, L., Carpentier, J.-L., Perrelet, A., Anderson, R. G. W., Goldstein, J. L. 1978. Occurrence of low density lipoprotein receptors within large pits on the surface of human fibroblasts as demonstrated by freeze-etching. *Exp. Cell Res.* 113:1–13

Osserman, K. E. 1958. *Myasthenia Gravis.* New York: Grune & Stratton

Ozdemir, C., Young, R. R. 1976. The results to be expected from electrical testing in the diagnosis of myasthenia gravis. *Ann. NY Acad. Sci.* 274:203–22

Pagala, M. K. D., Tada, S., Namba, T., Grob, D. 1977. Effect of repeated injections of serum immunoglobulins from patients with myasthenia gravis on neuromuscular transmission in neonatal mice. *Ann Sci. Meet. Myasthenia Gravis Found.,* New York, NY, 1977 (Abstr.)

Patrick, J., Lindstrom, J. 1973. Autoimmune response to acetylcholine receptor. *Science* 180:871–72

Penn, A. S. 1979. Immunological features of myasthenia gravis. See Engel et al 1979, pp. 123–32

Penn, A. S., Chang, H. W., Lovelace, R. E., Niemi, W., Miranda, A. 1976. Antibodies to acetylcholine receptors in rabbits: Immunological and electrophysiological studies *Ann. NY Acad Sci.* 274:354–76

Pestronk, A., Drachman, D. B., Stanley, E. F., Adams, R. N. 1980. Treatment of experimental autoimmune myasthenia gravis (EAMG) using single, high-dose cyclophosphamide. *Abstr. Neurol.* 30: 388

Pinching, A. J., Peters, D. K., Newsom-Davis, J. 1976. Remission of myasthenia gravis following plasma-exchange. *Lancet* 2:1373–76

Pirskanen, R. 1976. On the significance of HL-A and LD antigens in myasthenia gravis. *Ann. NY Acad. Sci.* 274:451–60

Rash, J. E., Albuquerque, E. X., Hudson, C. S., Mayer, R. F., Satterfield, J. R. 1976. Studies on human myasthenia gravis. Electrophysiological and ultrastructural evidence compatible with antibody labeling of the acetylcholine receptor complex. *Proc. Natl. Acad. Sci.* 73:4584–88

Reiness, C. G., Weinberg, C. B., Hall, Z. W. 1978. Antibody to acetylcholine receptor increases degradation of junctional and extrajunctional receptors in adult muscle. *Nature* 274:68–70

Remen, L. 1932. Zur pathogenese und therapie der myasthenia gravis pseudoparalytica. *Dtsch. Z. Nervenheilkd.* 128:66–78

Reuther, P., Fulpius, B. W., Mertens, H. G., Hertel, G. 1979. Antiacetylcholine receptor antibody under long-term azathioprine treatment in myasthenia gravis. See Dau 1979, Ch. 30, pp. 329–42

Richman, D. P., Antel, J. P., Patrick, J. W., Arnason, B. G. W. 1979. Cellular immunity to acetylcholine receptor in mayasthenia gravis: Relationship to histocompatibility type and antigenic site. *Neurology* 29:291–96

Richman, D. P., Patrick, J., Arnason, B. G. W. 1976. Cellular immunity in myasthenia gravis. *N. Engl. J. Med.* 294:694–98

Roberts, D. V., Thesleff, S. 1969. Acetylcholine release from motor-nerve endings in rats treated with neostigmine. *Eur. J. Pharmacol.* 6:281–85

Rowland, L. P., Aranow, H. Jr., Hoefer, P. F. A. 1961. Observations on the curare test in the differential diagnosis of myasthenia gravis. See Nastuk et al 1961, Ch. 5, pp. 411–34

Sanders, D. B., Schleifer, L. S., Eldefrawi, M. E., Norcross, N. L., Cobb, E. E. 1976. An immunologically induced defect of neuromuscular transmission in rats and rabbits. *Ann. NY Acad. Sci.* 274:319–36

Satyamurti, S., Drachman, D. B., Slone, F. 1975. Blockade of acetylcholine receptors: A model of myasthenia gravis. *Science* 187:955–57

Scadding, G. K., Thomas, H. C., Havard, C. W. 1977. Myasthenia gravis: Acetylcholine receptor antibody titres after thymectomy. *Br. Med. J.* 1:1512

Schumacher, M., Roth, J. 1913. Thymektomie bei einem fall von morbus basedowi mit myasthenie mitteilungen aus den grenzgebieten der medizin und chirurgie. *Mitt. Grenzgeb. Med. Chir.* 25:746

Schwartz, M., Novick, D., Givol, D., Fuchs, S., 1978. Induction of anti-idiotypic antibodies by immunisation with syngeneic spleen cells educated with acetylcholine receptor. *Nature* 273:343–45

Seybold, M. E., Lindstrom, J. M. 1979. Serial anti-acetylcholine receptor antibody tit-

ers in patients with myasthenia gravis: Effects of steroid therapy. See Dau 1979, Ch. 28

Shibuya, N., Mori, K., Nakazawa, Y. 1978. Serum factor blocks neuromuscular transmission in myasthenia gravis: Electrophysiological study with intracellular microelectrodes. *Neurology* 28:804–11

Shore, A., Limatibul, S., Dosch, H.-M, Gelfand, E. W. 1979. Identification of two serum components regulating the expression of T-lymphocyte function in childhood myasthenia gravis. *N. Engl. J. Med.* 301:625–29

Simpson, J. A. 1960. Myasthenia gravis: A new hypothesis. *Scott. Med. J.* 4:419–36

Staber, F. G., Fink, J., Sack, W. 1975. B lymphocytes in the thymus of patients with myasthenia gravis. *N. Engl. J. Med.* 292:1032–33

Stålberg, E., Trontelj, J. V., Schwartz, M. S. 1976. Single muscle-fiber recording of the jitter phenomenon in patients with myasthenia gravis and in members of their families. *Ann. NY Acad. Sci.* 274:189–202

Stanley, E. F., Drachman, D. B. 1978. Effect of myasthenic immunoglobulin on acetylcholine receptors of intact mammalian neuromuscular junctions. *Science* 200:1285–87

Strauss, A. J. L., Segal, B. C., Hsu, K. C., Burkholder, P. M., Nastuk, W. L., Osserman, K. E. 1960. Immunofluorescence demonstration of a muscle binding complement-fixing serum globulin fraction in myasthenia gravis. *Proc. Soc. Exp. Biol. Med.* 105:184–91

Sugiyama, H., Benda, P., Meunier J.-C., Changeux, J.-P. 1973. Immunological characterization of the cholinergic receptor protein from electrophorus electricus. *FEBS Lett.* 35:124–28

Tarrab-Hazdai, R., Aharonov, A., Silman, I., Fuchs, S. 1975. Experimental autoimmune myasthenia induced in monkeys by purified acetylcholine receptor. *Nature* 256:128–30

Tarrab-Hazdai, R., Yaffe, D., Prives, Y., Amsterdam, A., Fuchs, S. 1979. Effects of macrophages and antibodies from myasthenic animals on muscle cells in culture. See Dau 1979, Ch. 3

Taylor, R. B., Duffus, W. P. H., Raff, M. C., dePetris, S. 1971. Redistribution and pinocytosis of lymphocyte surface immunoglobulin antibody. *Nature New Biol.* 233:225–29

Thies, R. E. 1965. Neuromuscular depression and the apparent depletion of transmitter in mammalian muscle. *J. Neurophysiol.* 28:427–42

Tindall, R. S. A. 1980. Humoral immunity in myasthenia gravis: Effect of steroids and thymectomy. *Neurology* 30:554–57

Tindall, R. S. A., Cloud, R., Luby, J., Rosenberg, R. N. 1978. Serum antibodies to cytomegalovirus in myasthenia gravis: Effects of thymectomy and steroids. *Neurology* 28:273–77

Toyka, K. V., Drachman, D. B., Griffin, D. E., Pestronk, A., Winkelstein, J. A., Fischbeck, K. H., Kao, I. 1977. Myasthenia gravis: Study of humoral immune mechanisms by passive transfer to mice. *N. Engl. J. Med.* 296:125–31

Toyka, K. V., Drachman, D. B., Pestronk, A., Kao, I. 1975. Myasthenia gravis: Passive transfer from man to mouse. *Science* 190:397–99

van der Geld, H. W. R., Strauss, A. J. L. 1966. Myasthenia gravis. Immunological relationship between striated muscle and thymus. *Lancet* 1:57–60

Van de Velde, R. L., Friedman, N. B. 1970. Thymic myoid cells and myasthenia gravis. *Am. J. Pathol.* 59:347–68

Vincent, A., Pinching, A. J., Newsom-Davis, J. 1977. Circulating anti-acetylcholine receptor antibody in myasthenia gravis treated by plasma exchange. *Neurology* 27:364

Vincent, A., Scadding, G. K., Clarke, C., Newsom-Davis, J. 1979. Anti-acetylcholine receptor antibody synthesis in culture. See Dau 1979, Chap. 6, pp. 59–71

Walker, M. B. 1934. Treatment of myasthenia gravis with physostigmine. *Lancet* 1:1200–1

Walker, M. B. 1938. Myasthenia gravis: Case in which fatigue of forearm muscles could induce paralysis of extraocular muscle. *Proc. R. Soc. Med.* 31:722

Waud, D. R. 1971. A review of pharmacological approaches to the acetylcholine receptors at the neuromuscular junction. *Ann. NY Acad. Sci.* 183:147–57

Wekerle, H., Paterson, B., Ketelsen, U. P., Feldman, M. 1975. Striated muscle fibres differentiate in monolayer cultures of adult thymus reticulum. *Nature* 256:493–94

Willis, T. 1672. *De Anima Brutorum,* pp. 404–6. Oxford: Theatro Sheldoniano

Wilson, R. W., Ward, M. D., Johns, T. R. 1974. Corticosteroids: A direct effect on the neuromuscular junction. *Neurology* 24:1091–95

Wolters, E. C. M. J. 1976. *Corticosteroids and neuromuscular transmission,* pp. 1–104. Neth: Drukkerij en Uitegeverij P.E.T.

Woolf, A. L. 1966. Morphology of the myasthenic neuromuscular junction. *Ann. NY Acad. Sci.* 135:35–58

Zacks, S. I., Bauer, W. C., Blumberg, J. M. 1962. The fine structure of the myasthenic neuromuscular junction. *J. Neuropathol. Exp. Neurol.* 21:335–47

Ann. Rev. Neurosci. 1981. 4:227–72
Copyright © 1981 by Annual Reviews Inc. All rights reserved

THE ENTERIC NERVOUS SYSTEM

♦11554

Michael D. Gershon

Department of Anatomy, College of Physicians & Surgeons of Columbia
University, New York, New York 10032

DEFINITION

The enteric nervous system can be defined as the intrinsic innervation of the gastrointestinal tract. It is comprised of two major plexuses of ganglion cells and interconnecting fibers as well as several subsidiary groupings of fibers (Schofield 1968, Gabella 1976, Furness & Costa 1980). The two ganglionated plexuses are the submucosal (or Meissner's plexus) and the myenteric (or Auerbach's plexus). The submucosal plexus lies within the dense, irregularly arranged connective tissue of the submucosa, whereas the myenteric plexus is characteristically found between the circular and longitudinal layers of smooth muscle that form the gut's muscularis externa. Subsidiary aggregations of nerve fibers without ganglion cells are found (*a*) within the circular layer of smooth muscle, (*b*) as a deep muscular plexus in the submucosa just below the circular muscle, (*c*) within the muscularis mucosae, and (*d*) within the loose areolar connective tissue of the lamina propria of the mucosa.

The enteric nervous system is a part of the autonomic nervous system (ANS) and, in fact, was originally classified by Langley (1921) as the third division of the ANS. Unfortunately, Langley's classification is not in vogue today (see Gershon 1979a). The ANS now is usually considered to contain only two divisions, the sympathetic and the parasympathetic (Kuntz 1953), which are defined primarily by their connections to the central nervous system (CNS). The sympathetic division consists of a thoracicolumbar preganglionic outflow of axons arising from the cell bodies of neurons located mainly, if not entirely, in the intermedio-lateral column of the spinal cord and synapsing with postganglionic neurons in the paravertebral

227

0147-006X/81/0301-0227$01.00

or prevertebral ganglia. The parasympathetic division involves a craniosacral preganglionic outflow from the CNS and post-ganglionic cell bodies situated close to, or within, the innervated organs. Both divisions thus represent efferent pathways leading to effector organs from the CNS. Langley recognized that the enteric nervous system receives sympathetic and parasympathetic inputs. He believed, however, that many of the neurons of the enteric nervous system are probably not connected to the CNS at all. There are, for example, fewer than 20,000 fibers (afferent and efferent) in the vagus nerves as these nerves pass through the diaphragm, whereas there are over 6×10^6 ganglion cells in the small intestine of the guinea pig (Irwin 1931). In humans, there may be more than 10^8 ganglion cells, a number that is of the same order of magnitude as the number of neurons in the spinal cord (Furness & Costa 1980). The human vagus nerves, however, contain fewer than 2×10^3 efferent fibers (Hoffman & Schnitzlein 1969). Although arguments that depend on differences between numbers of fibers and numbers of cells are weakened by the fact that a single nerve fiber can innervate as many as 1000 cells, it is difficult to imagine how fewer than 2×10^3 fibers could usefully innervate over 10^8 neurons. If, therefore, as this discrepancy between numbers of fibers and numbers of cells suggests, many enteric neurons do not receive an input from the CNS, the enteric nervous system cannot be included within the definitions of either the sympathetic or parasympathetic divisions of the ANS. This distinction is of more than just semantic significance. The anatomical independence of the enteric nervous system from the CNS is matched by a similar physiological independence.

The autonomy of the enteric nervous system was recognized as early as 1899 by Bayliss & Starling. It is worth quoting some of their prescient conclusions (Bayliss & Starling 1899):

> The peristaltic contractions are true coordinated reflexes, started by mechanical stimulation of the intestine, and carried out by the local nervous mechanism (Auerbach's plexus). They are independent of the connection of the gut with the central nervous system. They travel only in one direction, from above downwards, and are abolished on paralysing the local nervous apparatus by means of nicotin or cocaine.

> The production of the true peristaltic wave is dependent on the unvarying response of the intestinal nervous mechanism to local stimulation, the law of the intestine. This law is as follows: local stimulation of the gut produces excitation above and inhibition below the excited spot. These effects are dependent on the activity of the local nervous mechanism.

This "local nervous mechanism" can be excited in-vitro to produce the peristaltic reflex (Trendelenburg, 1917). This ability to manifest reflex activity in vitro, cut off from the CNS, suggests that the gut may have sensory receptors, intrinsic primary afferent neurons (as dorsal root and cranial nerve ganglia are not present in in vitro preparations of gut), interneurons, and motor neurons. The gut is thus quite different in the completeness of its innervation from other peripheral organs and, as a consequence, the extrinsic innervation of the gut can be interrupted with relatively little adverse effect on its function. In fact, the enteric nervous system can apparently actually "talk back," at least to ganglia that relay CNS input. Centripetal fibers originating in the colon, for example, project to the inferior mesenteric ganglion and there influence synaptic transmission (Szurszewski & Weems 1976).

Actually, the gut is capable of more complex patterns of activity than simply the peristaltic reflex (Code & Carlson 1968, Hightower 1968). Not all intestinal motility is propulsive. The gut also exhibits pendular non-propulsive movements that may have a mixing function and these have to be appropriately coordinated with propulsive activity. In addition, the prominent myogenic activity of the gut (Bortoff 1972) must be integrated with neurogenic control. The importance of the innervation of the gut to its motor function is illustrated by the obstruction brought about when a segment of bowel is congenitally deprived of ganglion cells, as occurs in Hirschsprung's disease (Bodian et al 1949), or in the piebald and lethal spotted strains of mutant mice (Bolande 1975). The aganglionic segments are not denervated, as was once supposed (Penninckx & Kerremans 1975). They receive an adrenergic (Gannon et al 1969, Garrett et al 1969, Webster 1974) and a cholinergic (Kamijo et al 1953, Niemi et al 1961, Webster 1974) innervation; however, the absence of ganglion cells and even the axons of intrinsic enteric neurons (Rogawski et al 1978) eliminates the local coordination and integrative activity associated with the intrinsic elements of the enteric nervous system that lead to intestinal obstruction. The enteric nervous system, therefore, is an independent division of the ANS, is capable of mediating rather complex behavior, and is essential for the motor control of the gut.

The uniquely independent nature of the enteric nervous system and the relative complexity of the behaviors it controls are reflected in the structure and neurochemistry of the system. As might have been expected, these appear to be different from, and more complex than, other regions of the peripheral nervous system. The neurophysiology of the enteric nervous system is probably similarly complicated, but, perhaps because the system was thought of for many years as an array of simple parasympathetic relays, relatively little has been learned about it until recently.

ANATOMY

The anatomy of the enteric plexuses had been extensively studied by light microscopy prior to the introduction of the electron microscope. It was recognized, for example, that the submucosal plexus contains pseudounipolar or bipolar neurons, a morphology characteristic of primary affer- ent cells (Schofield 1968). The myenteric plexus, particularly, was found to contain a number of different types of neuron that could be distinguished one from another morphologically (Gunn 1959, 1968, Schofield 1968, Feher & Vajda 1972), by their histochemical reactivity for cholinesterase (Coup- land & Holmes 1958, Leaming & Cauna, 1961, Lassmann 1962, Sutherland 1967) or monoamine oxidase (Furness & Costa 1971), or by differential susceptibility to staining with silver solutions (Schofield 1968). When the myenteric plexus was first examined ultrastructurally, therefore, no sur- prises were anticipated; however, a novel structure was seen (Figure 1). The plexus was found to have characteristics that are dissimilar to all other regions of the peripheral nervous system and to resemble the CNS (Baum- garten et al 1970, Gabella 1972a). These characteristics include a paucity of extracellular space and the absence of perineurial or endoneurial sheaths surrounding myenteric axons. In fact, the myenteric plexus contains no collagen at all. The supporting cells of the myenteric plexus resemble the astroglia of the CNS more than they do other Schwann cells of the periph- eral nervous system (PNS) (Gabella 1971, Cook & Burnstock 1976b). These supporting cells, moreover, form an enclosing tube that almost completely separates the neuronal elements of the myenteric plexus from the surround- ing connective tissue space (Gershon & Bursztajn 1978). This sheath does part at intervals to permit axonal varicosities to reach the surface. These sites, where axonal varicosities are bare and uncovered by supporting cell processes, are probably regions where axons can release transmitter for the control of the enteric smooth muscle.

The interior of the myenteric plexus is also unusual in that it is totally avascular (Gabella 1972a). All of the capillaries that supply the neural tissue lie outside of the enclosing glial sheath of the plexus. These myenteric capillaries, moreover, are quite different in their fine structure from the capillaries found elsewhere in the gut. As is the case for most visceral capillaries, ordinary enteric capillaries are fenestrated and are exceedingly permeable (Clementi & Palade 1969, Simionescu et al 1971, 1974). Conse- quently, the intestinal lymph is very rich in protein that has leaked out of the vasculature (Mayerson et al 1960); however, the capillaries in the muscularis externa just outside the sheath of the myenteric plexus are much thicker walled than other enteric capillaries and they are non-fenestrated (Gershon & Bursztajn 1978). In this sense the myenteric capillaries resem- ble cerebral capillaries. These too are non-fenestrated and, in fact, are an important component of the blood-brain barrier (Reese & Karnovsky 1967, Brightman & Reese 1969). The cerebral and myenteric capillaries are also similar in that both have impermeable junctions that prevent the passage of intravascular tracer molecules, such as Evans blue-labeled albumin or

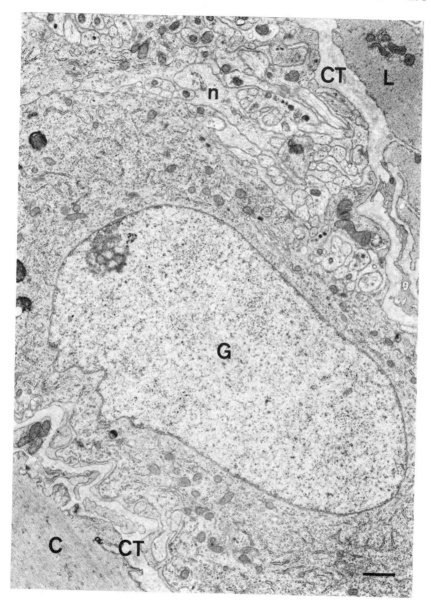

Figure 1   An electron micrograph of a portion of the myenteric plexus of the small intestine of a mouse. A ganglion cell (*G*) fills the center of the field. A connective tissue space (*CT*) separates the longitudinal (*L*) and (*C*) layers of smooth muscle from the neuropil (*n*) of the plexus. The marker = 1.0 μm.

horseradish peroxidase, between endothelial cells (Gershon & Bursztajn 1978). There is a slow leakage of macromolecular tracers out of myenteric capillaries that is probably accounted for by transport through the endothelial cells within plasmalemmal vesicles. A backup system of phagocytic cells, however, removes this material and prevents the tracers leaking out of vessels from reaching detectable concentrations in the extravascular space or within the myenteric plexus. The backup system of phagocytes in the gut is similar to the macrophages found in the cortex of the thymus that function as a part of the blood-thymic barrier (Raviola & Karnovsky 1972). There is, therefore, a blood-myenteric plexus barrier to macromolecules that resembles the blood-thymic barrier and may be functionally analogous to the blood-brain barrier. As is true also of the brain, the glial sheath cells do not appear to constitute any barrier to the entrance of tracers such as lathanum ions (Bursztajn & Gershon 1977) or macromolecules (Jacobs 1977, Gershon & Bursztajn 1978) into the plexus.

Perhaps the most striking feature of the ultrastructure of the myenteric plexus that resembles the CNS is the diversity of the types of neuron and axon terminal found in the plexus (Cook & Burnstock 1976a). Neuronal cell bodies have been classified into nine different types using neuronal size, distribution of organelles, location of the neurons, and their relationship to satellite cells as criteria. Axon terminals have been classified into eight (Cook & Burnstock 1976a) or ten (Furness & Costa 1980) morphologically distinct types using the size, shape, and content of the synaptic vesicles seen in the profiles of axonal varicosities as the identifying criteria. A correlation of the types of nerve cell body identified by electron microscopy with the types previously identified by light microscopy has not yet been done; no correlation, furthermore, has been made between the types of neuron identified morphologically and the types identified histochemically or physiologically. Moreover, it is still not possible to identify the neurotransmitter of most axons from the morphology of the axon terminals. The multiplicity of types of neuron and varicosity does, however, highlight the fact that there are not only many neurons in the enteric nervous system, but many types of neuron as well.

ENTERIC NEURONAL DIVERSITY

The morphological evidence for a multiplicity of enteric neuronal types has been confirmed by recent neurochemical and histochemical studies that have provided evidence for a similar multiplicity of enteric neurotransmitters. This evidence varies in strength for different neurotransmitter substances. It includes small molecules such as acetylcholine (ACh), norepinephrine (NE), 5-hydroxytryptamine (5-HT), and dopamine (DA);

a nucleotide (ATP); and larger molecules, neuropeptides, such as vasoactive intestinal polypeptide (VIP), substance P, somatostatin, enkephalin, a gastrin-cholecystokinin-like substance, bombesin, and neurotensin. The rapid expansion of the list of proposed enteric neurotramsmitters in recent years makes it unlikely that the list as it appears today is at all complete. It is, moreover, extremely difficult actually to prove that a given neurotransmitter candidate is, in fact, a neurotransmitter. Of the substances listed here as putative enteric neurotransmitters, only ACh, NE, and probably 5-HT can be said to have satisfied all of the criteria (Iversen 1979) necessary for establishing a neurotransmitter. The evidence for the various neurotransmitters and potential neurotransmitters is reviewed below. 5-HT has been singled out for special attention to serve as a model. More is known about it than about the other newly implicated substances and it illustrates the problems involved in neurotransmitter identification very well. ACh and NE are considered first because they have long served to focus research on the ENS and basic principles of enteric physiology have been derived from their study.

Acetylcholine

ACh was the first neurotransmitter to be identified in the gut (Dale 1937a,b). It gained acceptance as a neurotransmitter substance, however, at a time when relatively few sophisticated methods of analysis were available, and it is problematic as to whether the evidence that sufficed initially to establish the existence of cholinergic neurons in the gut would suffice to do so today (see Gershon 1970 for a critical review of the early evidence for ACh). For example, although it was known that the concentration of ACh in the myenteric plexus was extremely high (Welsh & Hyde 1944), it was not until 1968 that ACh was unequivocally shown to be synthesized and released from the myenteric plexus (Paton & Zar 1968). To be sure, it had been demonstrated earlier that ACh was released from the gut spontaneously (Feldberg & Lin 1949, Chujyo 1953) and when enteric nerves were stimulated (Schaumann 1957, Paton 1957); however, the source of the released ACh within the gut had not previously been located. In 1968 Paton & Zar finally showed that ACh was released by electrical stimulation from dissected strips of longitudinal muscle with adherent myenteric plexus (LM-MP), but that no ACh was contained in, or was released by, similar strips of longitudinal muscle from which the myenteric plexus had been removed mechanically. ACh, therefore, was established as a neurotransmitter before it had satisfied all of the criteria currently used to identify neurotransmitter substances (Iversen 1979); nevertheless, the supposition that ACh is an enteric neurotransmitter has shaped research on the enteric nervous system for over half a century. During that period no contrary evidence has emerged to indicate that ACh might not be an enteric neuro-transmitter. Consequently, ACh is now thought of as the best established and least controversial of the proposed enteric neurotransmitter substances.

RELEASE OF ACh   The characteristics of neuronal release of ACh from enteric neurons have been carefully studied using biossay to measure ACh (Paton et al 1971). The guinea pig's myenteric plexus in-vitro spontane-ously releases large quantities of ACh, which makes this preparation a convenient one for the study of the ACh release mechanism. The guinea pig LM-MP preparation also takes up choline by a high affinity, NA^+-depend-ent transport process (Pert & Snyder 1974). The uptake of choline by enteric nerves is closely coupled to ACh biosynthesis (Pert & Snyder 1974, Szerb 1975). This property has led more recent investigators, interested in study-ing the stimulated release of ACh, to preload LM-MP strips or whole gut with radioactive ACh by incubating tissues with [^{3}H]-choline (Szerb 1975, 1976, Wikberg 1977a, Manber & Gershon 1979). Stimulation of the LM-MP strips then releases [^{3}H]-ACh. Since the [^{3}H]-ACh synthesized from exogenous [^{3}H]-choline appears to mix with endogenous stores of ACh (Szerb 1976, Wikberg 1977a), the technique of preloading the neurons with radioactive ACh seems to be valid and greatly improves the specificity and sensitivity of detection of released ACh (Wikberg 1977a).

The spontaneous (nonstimulated) release of ACh from the myenteric plexus can be reduced but not abolished by ganglion-blocking agents, such as hexamethonium, or by agents, such as tetrodotoxin, that abolish Na^+-dependent nerve conduction (Paton & Zar 1968, Paton et al 1971). About 40% of the control level of ACh release persists in the presence of these agents. This suggests that there are cholinergic interneurons in the myen-teric plexus that are spontaneously active even when the gut is apparently at rest in vitro and that this activity involves release of ACh. Indeed, almost all the ACh spontaneously released in vitro originates from intrinsic neu-rons (Paton et al 1971); the extrinsic, preganglionic vagal fibers make a negligible contribution to the measured release of ACh. The release of ACh that persists after impulse activity in neurons has been blocked may be analogous to the tetrodotoxin-resistant spontaneous quantal release of ACh from the neuromuscular junction (Katz & Miledi 1967). In fact, the total spontaneous release can be decreased by omitting Ca^{2+} from the incubating medium (Paton et al 1971); therefore, the spontaneously released ACh can be attributed to two sources: (a) impulse-evoked release and (b) spontane-ous quantal release analogous to that described at the neuromuscular junction, which is largely independent of external Ca^{2+}. The electrically stimulated release of ACh is similarly Ca^{2+}-dependent (Cowie et al 1978) and is abolished by tetrodotoxin (Paton et al 1971). The myenteric cholinergic nerves thus behave in an orthodox manner with respect to the Ca^{2+} hypothesis of neurotransmitter release (see Llinas & Steinberg 1977 for references).

MORPHOLOGY OF CHOLINERGIC TERMINALS   Unfortunately, there are no good histochemical methods for the detection of cholinergic neurites. Enteric neurites contain acetylcholinesterase (see above), but this enzyme cannot be relied upon as a cholinergic marker. It is found, for example, in the adrenergic nerves to the pineal gland (Eränkö et al 1970) and the kidney (Barajas & Wang 1975). Some axon terminals known to be cholinergic, such as those found at neuromuscular junctions (Birks et al 1960, Couteaux 1960) or in the electric organs of fish (Boyne et al 1975), contain a preponderance of small (about 50 nm in diameter) electron-lucent synaptic vesicles together with occasional larger dense-cored vesicles. Many of the axon terminals in the enteric nerve plexuses fit this description (Baumgarten et al 1970, Gabella 1972a,b, Cook & Burnstock 1976a) and it is probable that some of these are cholinergic; however, not all terminals with a preponderance of small clear vesicles appear to be cholinergic. Some serotonergic axon terminals in the brain also have this appearance (Chan-Palay 1975, 1978, Calas et al 1976), and some terminals with a preponderance of small electron-lucent vesicles in the gut specifically take up $[^3H]$-5-HT (Rothman et al 1976, Dreyfus et al 1977b, Jonakait et al 1979a,b); therefore, even if it should prove true that cholinergic terminals all have the same morphology, morphology alone cannot be used to establish that any given enteric axonal varicosity is cholinergic.

FUNCTION OF ENTERIC CHOLINERGIC NEURONS   Cholinergic neurons are clearly involved in the control of intestinal motility by the enteric nervous system (see review by Kosterlitz 1968). Many, if not all, of the final excitatory neurons to both the circular and longitudinal muscle coats are cholinergic (Dale 1937a,b, Kosterlitz & Lees 1964, Kosterlitz 1968). In addition, cholinergic ganglionic transmission is critically involved in the mediation of the peristaltic reflex. Thus, the peristaltic reflex can be interrupted either by muscarinic antagonists, such as atropine or hyoscine, that interfere with cholinergic neuro-smooth muscular transmission, or by nicotinic antagonists, such as hexamethonium, that block ganglionic transmission (Feldberg 1951, Kosterlitz & Lees 1964). The ganglionic site of action of acetylcholine within the myenteric plexus that is critical to peristalsis has not yet been localized on a cellular level. Both the final intrinsic excitatory and inhibitory myenteric neurons, however, are stimulated by ACh (Kosterlitz & Lees 1964, Burnstock et al 1966, Campbell 1966, Bülbring & Gershon 1967, Gershon 1967a).

Neurophysiological Studies

Electrophysiological exploration of the myenteric plexus has been carried out using either extracellular or intracellular electrodes (see reviews by Wood 1975a, 1979); however, relatively little correlation has been made between the results obtained with the two techniques (Wood & Mayer 1978c). Although six different types of neuron have been described from the patterns of activity of single units recorded extracellularly (Wood 1973), only two types of cell clearly identified as neuronal have been encountered with intracellular electrodes (Nishi & North 1973a, Hirst et al 1974). A third type of cell has been recorded intracellularly, but it usually appears to be inexcitable and so has been presumed to be glial (Nishi & North 1973a, Hirst et al 1974). This latter identification may be questioned, however, because many of the cells that seem to be inexcitable when impaled initially, become excitable 20 to 30 min after impalement (Hodgkiss & Lees 1978).

EXTRACELLULAR RECORDING   The types of neuron recognized with extracellular electrodes can be conveniently classified into *burst-type* units, *mechanosensitive* units, and *single-spike* units (Wood, 1975a, Wood 1979). Burst-type units discharge trains of spikes episodically and are silent during the intervals between the episodes, or bursts, of activity (Wood 1970, Wood 1973). It seems likely that the recorded activity of these cells reflects the spontaneous activity of enteric neurons that was also detected by analyzing the release of ACh (see discussion above) and is not due to an artifact introduced by mechanical irritation of the gut by the recording electrodes. The possibility of an artifact has been raised to explain why burst-type activity is not encountered when intracellular electrodes are used (Nishi & North 1973a). There is a considerable amount of evidence against such an artifact. Electrical stimulation of synaptic inputs to single units, for example, can elicit trains of spikes that mimic the spontaneous "bursts" (Cooke et al 1979). Furthermore, the spontaneous burst-type activity continues for hours in-vitro, mechanical distortion of ganglia does not alter the pattern of burst-type activity, and the activity is recorded even with fine-tipped micropipettes filled with NaCl (Wood & Mayer 1978c).

The set of burst-type units has been subdivided into subsets of *steady-bursters* and *erratic-bursters* on the basis of the regularity, or lack thereof, of the interburst intervals. The erratic pattern of activity seems to be generated by a synaptic input to the neurons, whereas the more regular pattern persists in high Mg^{2+}-containing solutions, and thus seems to involve an intrinsic cellular pacemaker mechanism (Wood 1975b). It has been postulated that the steady-burster neurons can synaptically drive the erratic-bursters (Wood 1975a), although the neurotransmitter involved is unknown (Cooke et al 1979).

Mechanosensitive units are defined as cells whose pattern of discharge is altered by mechanical distortion of the ganglion in which the unit resides (Wood 1975a, 1979). There are three subsets of mechanosensitive units. These subsets are differentiated from one another by their pattern of adaptation to mechanical perturbation; thus, there are slowly-adapting units, rapidly adapting units, and non-adapting units that discharge trains of spikes without adaptation for long periods of time (Mayer & Wood 1975). This last mechanosensitive unit has been called a *tonic-type* neuron (Mayer & Wood 1975). The tonic-type neuron has been postulated to be an interneuron activated by the other "mechanosensitive" units. The sensory receptors responsible for activation of the mechanosensitive units have not been identified. The receptive field of mechanoreceptor units is limited to the ganglion containing the recording electrode; fiber tracts do not contain mechanoreceptors (Ohkawa & Prosser 1972). This limited receptive field indicates that they are not the pressure receptors responsible for activating the peristaltic reflex. Those receptors are located, not in myenteric ganglia, but in the intestinal mucosa (Bülbring & Crema 1958, Bülbring & Lin 1958, Kosterlitz & Lees 1964). The peristaltic reflex can be initiated by selectively applying pressure or agents, such as 5-HT, to the mucosa and the reflex can be blocked by selectively anaesthetizing or removing the mucosa.

Single-spike units characteristically discharge, not trains, but individual spikes at a relatively low frequency (Wood 1975a, 1979). The discharge of spikes by these cells is not blocked by adding a high concentration of Mg^{2+} to the incubating solution (Wood 1975b); therefore, the activity of these cells does not depend on their receiving a synaptic input. Elevating the concentration of Mg^{2+} may, however, change the frequency or pattern of discharge of the cells, which indicates that their activity is probably altered synaptically.

INTRACELLULAR RECORDING   The two types of neuron that have been described as a result of intracellular recording have been called the type 1 (or type S) and type 2 (or type AH) cells by Nishi & North (1973a) and by Hirst, Holman & Spence (1974). The type 1(S) cells have a lower resting membrane potential than type 2(AH) cells and a higher input resistance. The type 1(S) cells are slowly adapting and respond to the application of a prolonged depolarizing current with a continuous discharge of spikes; the frequency of this discharge is a direct function of the intensity of the applied current. The type 2(AH) cells are rapidly adapting and respond to depolarizing current with only one or two action potentials followed by a prominent and prolonged hyperpolarizing after-potential during which the cell is relatively inexcitable. No such hyperpolarizing after-potentials are seen in type 1(S) cells. Action potentials, recorded in type 1(S) cells, are, moreover, abolished by tetrodotoxin, whereas the action potentials of the type 2(AH) cell are tetrodotoxin-resistant. The rising phase of the action potenial in type 2(AH) neurons is associated with an inward movement of Ca^{2+} (Hirst & Spence 1973, North 1973, North & Nishi 1976). The entry of Ca^{2+} appears to activate a calcium-dependent K^+ conductance (Meech & Standen 1975) that is responsible for the after-hyperpolarization that characterizes these cells (North 1973, North & Nishi 1976).

When the type 2(AH) neurons were first encountered, no evidence of synaptic input to these cells was found (Nishi & North 1973a, Hirst et al 1974). It was therefore claimed that the somata of these cells, at least, are not innervated and it was proposed that the type 2(AH) cells are primary afferent neurons. Recently, however, synaptic input to type 2(AH) neurons has been discovered (Katayama & North 1978, Wood & Mayer 1978a,b, Grafe et al 1979a,b, Wood & Mayer 1979a,b, Johnson et al 1980a) and the hypothesis that these cells might be non-innervated sensory neurons is no longer tenable. Both type 1(S) and type 2(AH) neurons thus receive synaptic inputs. In type 1(S) cells, fast (less than 50 msec in duration) excitatory postsynaptic potentials (epsps) are more readily seen than in type 2(AH) cells (Nishi & North 1973a, Hirst et al 1974), but fast epsps do occur in both types of physiologically classified enteric neuron. The amplitude of fast epsps is usually smaller, however, in type 2(AH) neurons (Grafe et al 1979a). The fast epsps are mediated by nicotinic receptors (Nishi & North 1973a, Hirst et al 1974). There is some pharmacological evidence that has been interpreted to indicate that myenteric ganglion cells also have muscarinic receptors. Muscarine restores the peristaltic reflex in preparations in which the reflex has been blocked with any one of a large number of unrelated pharmacological agents (Beleslin & Samardzic 1978). This type of experiment is difficult to interpret, however, and no direct physiological evidence is available to indicate that cholinoceptive sites in enteric ganglia are anything but nicotinic. In fact, assays of enteric muscarinic receptors indicate that the bulk of the muscarinic receptors are in the smooth muscle (Yamamura & Snyder 1974). The longitudinal muscle is particularly rich in these receptors. There are, however, some muscarinic receptors on enteric axon terminals; these appear to modulate release of ACh (Kilbinger 1977) and NE (Manber & Gershon 1979) (see discussion below).

The observation that all fast epsps in myenteric neurons are mediated by ACh raises the question of what the role might be of other substances postulated to be enteric neurotransmitters. One possible role is the mediation of presynaptic inhibition or other effects at axo-axonic enteric synapses (discussed below). In addition, non-cholinergic, slow epsps have been recorded from enteric neurons (Wood & Mayer 1978a,b, 1979a,b, Johnson et al 1980a), as have inhibitory postsynaptic potentials (ipsps) (Hirst & McKirdy 1975, Hirst & Silinsky 1975, Johnson et al 1980a).

Slow epsps are seen most frequently in type 2(AH) neurons but they also occur in type 1(S) cells. They can be evoked by application of single shocks to interganglionic connectives (Wood & Mayer 1979a) or by focal stimulation of the surface of a ganglion (Johnson et al 1980a). Repetitive stimulation evokes slow epsps in a greater proportion of cells than does the application of single shocks (Wood & Mayer 1978a,b, 1979a, Johnson et al 1980a). The time course of the slow epsp is about one thousand times slower than that of the fast epsp. The slow epsp is associated with increased membrane resistance and therefore an enhanced excitability, particularly of type 2(AH) neurons. In fact, during the slow epsps these neurons respond to the intracellular injection of depolarizing current, not with just one or two action potentials as they normally do, but with trains of spikes (Wood & Mayer 1979a). These trains are reminiscent of the trains of spikes discharged by tonic-type mechanosensitive neurons recorded extracellularly (Mayer & Wood 1975). Therefore, it has been suggested that the type 2(AH) neurons may be the intracellularly-recorded equivalent of the extracellularly-recorded tonic-type unit (Wood & Mayer 1979a). In addition to the enhanced excitability of cells during the slow epsps, the after-hyperpolarization of the type 2(AH) neurons is also diminished (Wood & Mayer 1979a). This and additional electrophysiological data indicate that the slow epsp results from a depression of the Ca^{2+}-activated K^+ conductance of the type 2(AH) cell membrane (Wood & Mayer 1979a, Johnson et al 1980a). Cations, such as Mn^{2+} or Mg^{2+}, that antagonize the movement of Ca^{2+} into cells, reduce the K^+ conductance of type 2(AH) cells and mimic the slow epsp (Grafe et al 1980). The slow epsp also persists in the presence of atropine and hexamethonium, thus confirming that ACh is not the transmitter responsible for mediating this response.

Ipsps have been seen in a higher proportion of cells in the submucosal (Hirst & McKirdy 1975, Hirst & Silinsky 1975) than in the myenteric plexus (Wood & Mayer 1978a, Johnson et al 1980a). As in the case of the slow epsp, an ipsp can be evoked by single shocks but is seen in a greater proportion of cells when repetitive stimulation is used. The ipsps recorded from submucosal neurons are of shorter duration than the ipsps recorded from myenteric neurons (1 to 5 sec vs 2 to 40 sec). The membrane mechanism of both appears to be an activation of a K^+ conductance (Johnson et al 1980a). The adrenergic-neuron-blocking agent, guanethidine, blocks the ipsps of submucosal neurons and the potentials are mimicked by NE or DA, although the ipsps persist in the face of extrinsic denervation of the gut, which indicates that they are not sympathetically mediated (Hirst & McKirdy 1975, Hirst & Silinsky 1975). The iontophoretic application of

NE to a proportion of type 2 cells does produce a hyperpolarization similar in mechanism to the ipsp (North & Henderson 1975, Johnson et al 1980a). It has therefore been proposed that the slow ipsp results from the activation of sympathetic, noradrenergic terminals in the enteric plexuses; however, this possibility seems remote. It cannot, for example, explain the existence of ipsps in type 1(S) neurons that are not responsive to NE (Nishi & North 1973b), nor can it explain the persistence of the potentials following extrinsic denervation of the gut (see below).

Norepinephrine

Sympathetic nerve stimulation has long been known to relax non-sphincteric regions of the mammalian small intestine (Finkleman 1930). This effect was attributed to the release of an adrenergic substance (Finkleman 1930) that has been presumed to be NE since the work of von Euler (von Euler 1951). Intestinal relaxation was for some time thought to be due to a direct effect of NE, released from sympathetic post-ganglionic axon terminals, on the enteric smooth muscle itself (Gillespie & McKenna 1961). The application to the gut of histochemical techniques for the localization of NE, especially formaldehyde-induced fluorescence, however, changed this view of enteric adrenergic mechanisms. Noradrenergic sympathetic nerves were found to be quite sparse in the muscularis externa of the gut and to be virtually absent from the longitudinal layer of smooth muscle (Norberg 1964, Jacobowitz 1965). Instead, adrenergic nerves were found to be concentrated in the submucosal and myenteric plexuses. These findings led to the proposal by Norberg & Sjöqvist (1966) that the action of sympathetic nerves to the gut was not to relax smooth muscle directly, as had previously been thought, but rather to inhibit ganglionic transmission in the excitatory neural pathway. According to this hypothesis, the relaxation that follows stimulation of sympathetic nerves is due to a removal of an excitatory cholinergic tone.

THE ENTERIC SITE(S) OF ACTION OF THE SYMPATHETIC TRANSMITTER   Since its formulation, the hypothesis of Norberg & Sjöqvist (1966) has received experimental support. Exogenous NE powerfully inhibits the release of ACh from the gut (Schaumann 1958, Paton & Vizi 1969). Moreover, sympathetic nerve stimulation is more effective in antagonizing contractions of the gut brought about by stimulation of the vagus nerves than it is in blocking the contractions produced by the direct application of ACh (Kewenter 1965). This observation suggests that sympathetic nerves, as well as exogenous NE, inhibit the release of ACh. Indirect pharmacological evidence has been marshalled to support this action of sympathetic nerves (Furness & Costa 1974, Cowie et al 1978). Although these experiments do not prove that the action of NE is to inhibit ganglionic transmission (an action on the terminals of the final excitatory cholinergic neurons is equally likely), the idea that sympathetic nerves act on ganglion cells gained wide acceptance (Burnstock & Costa 1973).

Not all of the evidence derived from studies of the targets of the enteric sympathetic innervation is compatible with an explanation of sympathetic action that rests entirely on inhibition of ganglionic transmission. NE has a direct action on enteric smooth muscle mediated, in the guinea pig, through β-adrenoceptors (Bülbring 1972, Wikberg 1977b), and in the rabbit, through both α- and β-adrenoceptors (Wikberg 1977b). Sympathetic nerve stimulation, moreover, will readily relax the small intestine in vitro even after receptors for ACh have been entirely blocked with hyoscine (Gershon 1967b). Furthermore, the concept that there is a resting excitatory tone that can be removed by sympathetic stimulation, cannot be supported (Gershon 1976b). In fact, under appropriate conditions, atropine, local anaesthetics, or tetrodotoxin will cause a contraction, not a relaxation, of the intestinal smooth muscle (Wood 1972, Bortoff & Miller 1975). The predominant neural action on the intestinal smooth muscle, therefore, may be inhibitory rather than excitatory. Finally, sympathetic nerve stimulation relaxes the gut without altering the electrical activity of enteric ganglia (Takayanagi et al 1977). Under some circumstances, NE released from sympathetic nerves probably can reach and effect receptors on the enteric smooth muscle. It should be noted that electrical coupling and gap junctions between enteric smooth muscle cells (Barr et al 1968, Daniel et al 1976, Gabella & Blundell 1979) make it possible for a transmitter released only at the surface of the myenteric plexus to affect an entire smooth muscle layer.

A closer analysis of the actions of NE on neurons of the myenteric plexus indicates that the primary site of action of the amine is not on ganglion cell bodies. Intracellular records from myenteric ganglion cells usually, as noted above, reveal no change in membrane potential in response to the application of NE (Holman et al 1972, North & Nishi 1974). Instead, the dominant effect of NE seems to be a presynaptic inhibition of ACh release; NE reduces the size of the cholinergically mediated fast epsp (Nishi & North 1973b). [This is in contradistinction to the submucosal plexus or some myenteric type 2(AH) cells where a substance that resembles a catecholamine pharmacologically may be the mediator of the ipsps recorded from ganglion cells (Hirst & Silinsky 1975, Johnson et al 1980a).] Exogenous NE, moreover, is able to inhibit the release of ACh even when postganglionic cholinergic axons are stimulated directly (Kosterlitz et al 1970), thus indicating an action of the compound on a portion of the cholinergic neuron distal to its cell body.

Recently, evidence has been presented indicating that noradrenergic sympathetic terminals make reciprocal axo-axonic synapses with cholinergic terminals in the gut (Manber & Gershon 1979). When sympathetic nerves in the perivascular supply to a loop of rabbit jejunum are stimulated simultaneously with the intrinsic cholinergic nerves in the loop, the release of ACh is inhibited. The effect of NE on release of ACh is mediated by α-adrenoceptors (Kosterlitz et al 1970, Wikberg 1977b, Cowie et al 1978). Prior chemical sympathectomy with 6-hydroxydopamine (6-HD) prevents the reduction of stimulated ACh release by perivascular nerve stimulation (Manber & Gershon 1979). A reciprocity of connections is indicated by the effect of ACh. ACh inhibits the release of NE from stimulated enteric sympathetic nerves (Manber & Gershon 1979). This action of ACh is muscarinic and is blocked by atropine. Atropine, in turn, enhances the release of NE when both adrenergic and cholinergic nerves are stimulated together. ACh, it should be noted, has a similar muscarinic action to inhibit its own release (Kilbinger 1977). It therefore seems likely that cholinergic nerves inhibit the release of NE (and ACh) as adrenergic nerves do the release of ACh.

LOCALIZATION OF NORADRENERGIC AXON TERMINALS AND ITS SIGNIFICANCE   Early anatomical evidence suggesting the concentration of noradrenergic nerves within the enteric plexuses was ambiguous as to the localization within the myenteric plexus of the noradrenergic axon terminals (Norberg 1964, Jacobowitz 1965). Some histofluorescent images were interpreted to indicate that a ring of noradrenergic varicosities surround ganglion cells, but the histochemical method lacks the resolution necessary to distinguish axo-somatic or axo-dendritic from axo-axonic endings. Actually, a wider distribution of noradrenergic terminals has become apparent from more recent histochemical work. The terminals are not confined to the enteric plexuses, but are also found in the circular muscle, especially in sphincters (Gillespie & Maxwell 1971, Furness & Costa 1974), in the taenia coli (Åberg & Eränkö 1967), and in the mucosa where they innervate enteroendocrine cells, and can cause the discharge of 5-HT-containing enterochromaffin cell granules (Ahlman et al 1976). The ultrastructural identification of noradrenergic terminals within the gut is complicated by the difficulty of preserving the cores of the small (about 50 nm diameter) granular vesicles (SGV) (Gabella 1972b, Furness & Costa 1974, Wong 1977), upon which the identification of these terminals depends (Geffen & Livett 1971). This difficulty can be overcome by using specific cytochemical markers including 6-HD, 5-hydroxydopamine (5-HD) (Baumgarten et al 1970, Howard & Garrett 1973, Wong et al 1974, Wong 1975, 1977, Gordon-Weeks & Hobbs 1979), and fixation with $KMnO_4$ (Hökfelt 1971, Feher

1974, 1975, Manber & Gershon 1979). A complication has been introduced by the recent finding of a type of terminal in the gut that, after aldehyde fixation, contains SGV but does not take up 5-HD or 6-HD (Gordon-Weeks & Hobbs 1979). These latter terminals are presumably not adrenergic, or if they are adrenergic, they are certainly atypical. Ultrastructural studies that do not use specific markers thus cannot be relied upon.

Those studies that have used specific cytochemical markers for noradrenergic terminals have not resulted in the publication of many pictures of noradrenergic axo-dendritic or axo-somatic synapses in the myenteric plexus (Baumgarten et al 1970, Gershon et al 1980b). If these exist at all, they must be extremely rare. In fact, radioautographic studies have shown that the overwhelming majority of noradrenergic fibers are not located deep within the myenteric plexus in close proximity to ganglion cells (Manber & Gershon 1979, Gershon 1979c). Instead, they are strikingly concentrated at the periphery of the myenteric plexus at the interface with the connective tissue space that lies between the myenteric plexus and the surrounding layers of smooth muscle. Ultrastructural examination of the myenteric plexus fixed with $KMnO_4$ (Manber & Gershon 1979) has revealed that noradrenergic terminals are a minor component of the myenteric plexus, comprising far less than 1% of all varicosities in the plexus. At the periphery of the plexus they form complexes with other varicosities that are non-adrenergic. These complexes consist of two or more varicosities filled with synaptic vesicles abutting on one another without an intervening supporting cell process. These complexes often protrude through the supporting cell sheath that surrounds the plexus and expose a bare surface to the connective tissue space. These complexes have also been seen with aldehyde fixation and are revealed by scanning electron microscopy to be a prominent feature of the surface of the myenteric plexus (Gershon & Bursztajn 1978, Gershon 1979c). No synaptic membrane specializations have been seen within or between the abutting varicosities; however, most myenteric varicosities (Jonakait et al 1979a), and autonomic postganglionic terminals in general, do not show such specializations (Richardson 1964, Burnstock, 1979a). These morphological observations should be interpreted in the light of the physiological and biochemical data, outlined above, on adrenergic-cholinergic interactions. It seems likely that the complexes of varicosities identified at the periphery of the myenteric plexus provide a morphological basis for the existence of an adrenergic-cholinergic axo-axonic synapse in the gut. Such synapses appear to be widely distributed in the ANS (Vanhoutte 1977).

EXTRINSIC NATURE OF THE SYMPATHETIC INNERVATION   Unlike ACh, NE is the neurotransmitter of neurons whose cell bodies lie [with rare exceptions, such as the guinea pig's proximal colon (Costa et al 1971, Gabella & Juorio 1975)] entirely outside the gut itself. Extrinsic denervation, therefore, depletes NE, detected biochemically or histochemically, from the wall of the gut (Hamberger & Norberg 1965, Jacobowitz 1965, Ahlman et al 1973, Furness & Costa 1974, 1978, Juorio & Gabella 1974). It also eliminates the specific uptake of NE (Gabella & Juorio 1975) and results in the complete disappearance of all biochemically detectable tyrosine hydroxylase activity and immunohistochemically demonstrable dopamine β-hydroxylase (Furness et al 1979). Long-term organotypic tissue cultures of gut, moreover, contain no neurons or neurites showing catecholamine histofluorescence, that specifically take up NE (Dreyfus et al 1977a,b), or that contain dopamine β-hydroxylase immunoreactivity (Schultzberg et al 1978). The gut, therefore, except for oddities such as the proximal colon of the guinea-pig, contains no intrinsic epinephrine-containing, noradrenergic, or dopaminergic neurons. This absence would seem to rule out the possibility that a catecholamine could be the mediator of the ipsps recorded from enteric neurons (Hirst & Silinsky 1975, Johnson et al 1980a); the potentials persist, whereas catecholamines do not, following extrinsic denervation of the gut. The extrinsically denervated gut does, however, contain intrinsic neurons that are able to decarboxylate aromatic amino acids such as L-dihydroxyphenylalanine and 5-hydroxytryptophan (5-HTP) to their respective amines, DA or 5-HT (Costa et al 1976, Furness & Costa 1978). These intrinsic neurons are probably not catecholaminergic but indoleaminergic (Furness & Costa 1978, Costa & Furness 1979a, Furness et al 1979, Furness & Costa 1980). It is probable that the transmitter of these "amine-handling" neurons is 5-HT (see below).

5-Hydroxytryptamine

ENTEROCHROMAFFIN CELLS   The major storage depot for 5-HT in the body is the gut. The bulk of this material is present in enteroendocrine (enterochromaffin, EC) cells of the mucosa (Erspamer 1966). These cells continuously release 5-HT into the circulation so that the concentration of 5-HT in portal venous blood far exceeds that of the general circulation (Toh 1954). 5-HT, applied to the mucosal surface of the gut, triggers the peristaltic reflex and pressure releases mucosal 5-HT (Bülbring & Crema 1958, Bülbring & Lin 1958). Since pressure also triggers the peristaltic reflex, these observations led to the hypotheses that mucosal EC cells are pressure receptors and that their pressure-stimulated release of 5-HT activates mucosal afferent nerve endings, thus initiating the peristaltic reflex (Bülbring & Crema 1958, Bülbring & Lin 1958). When it became apparent, however, that the mucosa could be depleted of 5-HT without blocking the peristaltic reflex, these hypotheses had to be abandoned (Boullin 1964). The function of 5-HT in EC cells is not yet known.

THE NEUROTRANSMITTER HYPOTHESIS   A new proposal for a function of enteric 5-HT has emerged more recently. This proposal is that 5-HT is the neurotransmitter of a population of intrinsic enteric neurons. This idea was based initially on pharmacological and radioautographic investigations. Pharmacological studies revealed that the action of 5-HT on the gut is mediated through myenteric neurons and that these neurons have receptors for 5-HT that are distinct from the nicotinic receptors for ACh (Gaddum & Picarelli 1957, Brownlee & Johnson 1963, Gershon 1967a, Drakontides & Gershon 1968, Costa & Furness 1979a,b, Johnson et al 1980b, North et al 1980). 5-HT can, in fact, like ACh, activate enteric neurons that are either excitatory or inhibitory to the gut's smooth muscle (Bülbring & Gershon 1967, Gershon 1967a, Costa & Furness 1979b). Axons of the myenteric and submucosal plexuses, furthermore, were found to synthesize [^{3}H]-5-HT from its immediate precursor, [^{3}H]-5-hydroxytryptophan $\{[^3H]$-5-HTP$\}$ and, after doing so, to become labeled and radioautographically detectable (Gershon et al 1965, Gershon & Ross 1966a,b). Since noradrenergic nerves did not label under these conditions, the radioautographic observations were taken as evidence suggesting the presence of serotonergic axons in the gut. Initiation of the peristaltic reflex by mucosal stimulation, moreover, was shown to release the labeled 5-HT from the myenteric plexus (Gershon et al 1965). Additional pharmacological evidence was obtained to implicate 5-HT in vagal relaxation of the stomach: after muscarinic blockade 5-HT mimicked vagal relaxation, desensitization of neural receptors to 5-HT inhibited the response, and 5-HT was released from the wall of the stomach by electrical stimulation (Bülbring & Gershon 1967). In the years since these early observations first suggested a neurotransmitter role for 5-HT in the enteric nervous system, a great deal of supporting evidence has accumulated. In fact, all of the criteria generally required for establishing a substance as a neurotransmitter (Iversen 1979) have been satisfied for 5-HT in the enteric nervous system (Gershon 1979b); nevertheless, the amine still evokes controversy. Although there seems to be general agreement that some indoleamine is an enteric neurotransmitter, some investigators have speculated that the actual substance may be 5-HT-like, rather than 5-HT itself (Costa & Furness 1979a, Furness & Costa 1980.) It is therefore worthwhile to review the evidence for 5-HT following the usual criteria for transmitter identification.

Presence in enteric neurons   A relatively small amount of 5-HT is found in the wall of the gut after the mucosa has been removed (Feldberg & Toh

1953, Robinson & Gershon 1971, Juorio & Gabella 1974, Gershon & Tamir 1979, 1980). Although the existence of this non-mucosal store of 5-HT has been known for many years (Feldberg & Toh 1953), its significance has often been discounted because of the much larger store of 5-HT in the mucosa (Costa & Furness 1979a). The concentration of 5-HT in the guinea-pig LM-MP (longitudinal muscle with adherent myenteric plexus) preparation, for example, is 80 to 110 ng/g (Juorio & Gabella 1974, Robinson & Gershon 1971, Gershon & Tamir 1979, 1980); however, this figure underestimates the actual 5-HT concentration in the neural tissue. The myenteric plexus has not been obtained in pure form and constitutes only a small part of the LM-MP strip, the bulk of this preparation being muscle. The turnover of enteric neural 5-HT, moreover, is more rapid than that of the EC cells (Gershon & Sleisenger 1966) and consequently there is a rapid accumulation of 5-HT in the LM-MP preparation following inhibition of monoamine oxidase (MAO) (Robinson & Gershon 1971, Feher 1974, 1975).

Probably the main source of skepticism toward 5-HT as an enteric neurotransmitter is the difficulty that has been encountered by several investigators in demonstrating 5-HT within mammalian enteric neurons by histofluorescence (Baumgarten et al 1973, Ahlman & Enerback 1974, Dubois & Jacobowitz 1974, Costa & Furness 1979a). It is also, however, difficult to demonstrate 5-HT in central serotonergic neurons (Fuxe & Jonsson 1967) and, unless the endogenous 5-HT concentration is augmented, many central serotonergic neurons and their terminals go undetected (Aghajanian & Asher 1971). 5-HT has been detected histochemically in the myenteric plexus following inhibition of MAO (Robinson & Gershon 1971, Feher 1974, 1975). The same procedures, involving administration of L-tryptophan, with or without parachlorophenylalanine (PCPA), as used for the selective enhancement and detection of the histofluorescence of the cell bodies and terminals of central serotonergic raphe neurons (Aghajanian & Asher 1971, Aghajanian et al 1973), also demonstrate the histofluorescence of 5-HT in enteric neurons (Dreyfus et al 1977a). The fluorescent material shown in enteric neurons with these techniques exhibits the same activation and emission spectra by microspectrofluorophotometry as the authentic 5-HT-formaldehyde fluorophore (Dreyfus et al 1977a). No non-neuronal cells other than EC cells (and mast cells of rats and mice only) have been found to contain 5-HT. 5-HT can readily be demonstrated histochemically in enteric neurons when the wall of the gut is grown in organotypic tissue culture (Dreyfus et al 1977a,b). These cultures contain virtually no EC cells. Since enteric 5-HT-containing neurons persist in culture, they must be intrinsic to the enteric nervous system. Enteric neurons, moreover, both in culture and in situ, convert

[^{3}H]-L-tryptophan to [^{3}H]-5-HT (Dreyfus et al 1977a). They also contain material that cross-reacts with, and enables the neurons to be demonstrated immunocytochemically by, an antibody prepared against tryptophan hydroxylase purified from central serotonergic neurons (Gershon et al 1977). Enteric neurons thus resemble central serotonergic neurons in being able to synthesize 5-HT from the dietary amino acid L-tryptophan. This property is a marker for central serotonergic neurons (Kuhar et al 1972). The nonmucosal 5-HT detected biochemically in the wall of the gut, therefore, is undoubtedly present in neurons in the enteric nervous system.

Tryptophan uptake   Enteric serotonergic neurons resemble their central counterparts in additional ways. The enteric neurons have a relatively high affinity, saturable, uptake mechanism for L-tryptophan as well as 5-HT (Dreyfus et al 1977a). The Km for L-tryptophan uptake is about 50 uM and the transport system appears to be associated with terminal varicosities (Jonakait et al 1979a). Tryptophan uptake by enteric neurons seems also to be regulated by neuronal activity (Gershon & Dreyfus 1980). It is stimulated by high K^+-containing solutions and by low concentrations (0.1 uM) of 5-HT itself. This effect of 5-HT on L-tryptophan uptake is indirect; it is blocked by tetrodotoxin and by elevating the Mg^{2+} or removing the Ca^{2+} from the gut's suspending medium. The stimulatory action of 5-HT, therefore, probably depends on the release of another, as yet unidentified, neurotransmitter.

Serotonin-binding protein   Enteric neurons, like central serotonergic neurons, also contain a highly specific serotonin-binding protein (SBP) (Jonakait et al 1977). This protein has a very high affinity for 5-HT; there are two dissociation constants for the gut protein, 0.7 nM and 0.5 uM. SBP is confined in the guinea pig LM-MP preparation to the myenteric plexus and it appears in the gut during ontogeny coincidentally with the appearance of enteric serotonergic neurons detected by other means. SBP is also released by electrical stimulation from the gut along with [^{3}H]-5-HT (Jonakait et al 1979b). This release is Ca^{2+}-dependent. Since the binding of 5-HT by SBP in both brain and gut is inhibited by reserpine, it has been proposed that SBP may be a 5-HT storage protein (Tamir & Gershon 1979, Gershon & Tamir 1980). Its role, according to this hypothesis, is to prevent swelling of 5-HT storage organelles by binding intravesicular 5-HT, thus reducing the intravesicular osmotic pressure. In support of this hypothesis, the Ca^{2+}-dependent release of SBP from stimulated enteric neurons is apparently by exocytosis. The cytosol marker protein, lactic dehydrogenase, is not simultaneously released with SBP by nerve stimulation (Jonakait et al 1979b). Exocytosis implies a vesicular means of storage. The SBP in the

CNS also seems to be confined to serotonergic neurites in both brain (Tamir & Kuhar 1975) and spinal cord (Tamir & Gershon 1979). The SBP of central origin is transported proximodistally by fast axonal transport and is enriched in purified preparations of synaptic vesicles (Tamir & Gershon 1979). SBP, therefore, is probably a component of the neuronal 5-HT vesicular storage mechanism.

Release of 5-HT   Release of 5-HT from enteric neurons is made difficult to demonstrate by the disparity between the sizes of the large EC cell and the smaller neuronal pool of the amine. Paton & Vane (1963), for example, found that stimulation of the vagus nerves caused the release of 5-HT from the guinea pig stomach. They did not, however, identify the source of the released amine. Bülbring & Gershon (1967) attempted to eliminate the uncertainty introduced by EC cells by destroying the gastric mucosa through selective asphyxiation prior to attempting to evoke the release of 5-HT. They found that electrical stimulation released 5-HT from the mouse stomach even after mucosal asphyxiation and the release, moreover, was blocked by tetrodotoxin. They did not, however, show Ca^{2+}-dependence of the release, nor could they eliminate the possibility that nerves in the stomach took up 5-HT released from the mucosa during its asphyxiation. It is thus conceivable, if not probable, that although the stomach could be demonstrated histologically to have no EC cells remaining at the time of stimulation, the EC cells were nevertheless the source of the 5-HT ulti-mately released from enteric nerves (Costa & Furness 1979a); however, evidence for neural release of 5-HT has also been obtained from studies of other preparations of gut. The transmurally stimulated distal colon of the guinea pig is one such example (Furness & Costa 1973). When this prepara-tion is stimulated at high frequency, there is a tetrodotoxin-sensitive release of a substance that acts on smooth muscle through the same receptors as 5-HT. High frequency stimulation of the guinea pig LM-MP preparation similarly releases a material in the myenteric plexus that causes a prolonged synaptic activation of neurons that can be recorded with suction electrodes for 10 to 30 sec (Dingledine et al 1974). This activation is blocked by desensitization of receptors for 5-HT. All of these experiments suggest that 5-HT is released, but all essentially rely on bioassay for identification of the released material. Obviously, if there is another substance that interacts with 5-HT receptors in these bioassays, as does 5-HT, then that material would be subject to misidentification.

Another approach has been to take advantage of the difference that exists in serotonergic mechanisms between EC cells and enteric neurites. The EC cells lack the specific uptake mechanism (Rubin et al 1971) for 5-HT that characterizes enteric serotonergic neurites (Gershon & Altman 1971, Robinson & Gershon 1971, Gershon et al 1976). The gut can thus be incubated with [^{3}H]-5-HT at a low concentration in vitro and load enteric serotonergic neurites but not EC cells with the labeled amine (Jonakait et al 1979b). A convenient way of doing this is to evert (turn inside out) a loop of intestine and to perfuse it through the newly created serosal lumen (Jonakait et al 1979a,b, Gershon & Tamir 1979, 1980). Light and electron microscopic radioautography of tissues from preparations perfused through the serosal lumen with [^{3}H]-5-HT reveal no barrier to the free passage of the amine between the perfusing fluid and the submucosal and myenteric plexuses. Axons become labeled in both; however, very little radioactive material traverses the wall of the gut to reach the medium bathing the mucosal surface of the gut. Conversely, if radioactive 5-HT is added to the mucosal side of the everted preparation, very little radioactivity reaches the serosal perfusate for up to 15 to 20 min (Gershon & Tamir 1979, 1980). There thus appears to be a barrier to the transmural passage of 5-HT. The radioauto-graphic results, mentioned above, showing that 5-HT added to the serosal surface of the intestine penetrates as far as the submucosa, indicate that the barrier must be located between the submocosa and the mucosa; however, no anatomical structure constituting the barrier has yet been identified. The barrier may serve to protect the neural plexuses from the continuous release of 5-HT from EC cells.

Transmural stimulation of perfused everted preparations of gut, pre-loaded with [^{3}H]-5-HT, releases the labeled amine into the serosal perfusate (Jonakait et al 1979b). This release appears to be synaptic as it is blocked by tetrodotoxin, removal of Ca^{2+}, or elevation of the Mg^{2+} concentration of the medium. [^{3}H]-5-HT is also released from LM-MP preparations of guinea pig small intestine by high K^+ solutions or the Ca^{2+} ionophore X-537A (Jonakait et al 1979b). In these instances, however, the dependence of the release on Ca^{2+} is slight or not demonstrable, respectively. Internal tissue Ca^{2+} could be mobilized by these agents or they could be acting through a Ca^{2+}-independent mechanism. [^{3}H]-5-HT is also released from LM-MP strips by electrical stimulation (Schulz & Cartwright 1974a).

The barrier to the passage of 5-HT from the mucosa to the serosa of the gut makes the perfused everted intestine a good preparation to use to investigate the release of endogenous 5-HT. Since mucosal 5-HT does not reach the serosal perfusate, 5-HT appearing in this perfusate must be of neural origin. The serosal perfusate from these preparations has been found to contain 5-HT (Gershon & Tamir 1979, 1980). 5-HT was detected by a sensitive and specific radioenzymatic assay (Saavedra et al 1973, Boireau et al 1976). This identification of 5-HT, therefore, is not subject to the uncer-tainty that was true of earlier studies involving bioassays. The concentration of 5-HT in the serosal perfusate shows a Ca^{2+}-dependent rise upon electrical stimulation (Gershon & Tamir 1979, 1980). 5-HT, therefore, is released from stimulated enteric neurites in a manner expected for a neurotransmitter. Since there is also a vast increase in the release of 5-HT from the serosal surface of the gut that accompanies peristalsis (Gwee & Yeoh 1968), it is likely that enteric serotonergic neurons are involved in that activity. Evidence to support this involvement has also been obtained from pharmacological analyses of the neural pathways involved in the mediation of the peristaltic reflex (Costa & Furness 1976).

Target response Pharmacological analyses of the actions of 5-HT have revealed that the amine affects many sites in the gut. The smooth muscle is variably responsive to 5-HT, depending on the region of the gut and the species examined (Gershon 1967a, Drakontides & Gershon 1968, Costa & Furness 1979b). For example, the longitudinal muscle of the fundus of the rat stomach (Vane 1957) is very sensitive to 5-HT, but the longitudinal muscle of the small intestine is insensitive to the amine (Gershon 1967a). The circular muscle of the guinea pig stomach (Bülbring & Gershon 1967), ileum, and colon (Costa & Furness 1979b), moreover, is virtually unresponsive to 5-HT. The neural effects of 5-HT, however, are much more consistent than those on smooth muscle. The amine clearly acts on enteric ganglia (Gaddum & Picarelli 1957, Brownlee & Johnson 1963) to activate neurons that release ACh and cause the contraction of the longitudinal (Gershon 1967a, Drakontides & Gershon 1968, Vizi & Vizi 1978) and circular (Harry 1963, Costa & Furness 1979b) layers of smooth muscle. 5-HT also activates non-adrenergic non-cholinergic intrinsic inhibitory neurons (Bülbring & Gershon 1967, Gershon 1967a, Drakontides & Gershon 1968, Furness & Costa 1973, Costa & Furness 1979b) and can therefore relax the gut if excitatory neuromuscular transmission is blocked. These results imply that the role of 5-HT in the enteric nervous system is probably to serve as the transmitter of interneurons. This conclusion is consistent with the presence of 5-HT in intrinsic enteric neurons and its absence from extrinsic fibers in the vagus nerves (Dreyfus et al 1977a,b).

One difficulty that has plagued attempts to analyze the enteric actions of 5-HT has been the absence of a reliable pharmacological antagonist, active at neural 5-HT receptors. Receptors for 5-HT on the smooth muscle are consistently inhibited by methysergide at concentrations of the drug low enough to be specific (Drakontides & Gershon 1968, Costa & Furness 1979a,b). Neural actions of 5-HT are also blocked by this drug (Costa & Furness 1976, 1979a,b, Wood & Mayer 1979b), but at higher concentrations. In the mouse duodenum, separate neural and muscular receptors for 5-HT have been distinguished (Drakontides & Gershon 1968). Neural receptors are insensitive to tryptamine but are first excited and then blocked by phenylbiguanide. Muscular 5-HT receptors are insensitive to phenylbiguanide but are activated by tryptamine and blocked by methysergide. Unfortunately, phenylbiguanide does not act on guinea-pig enteric neurons as it does on those of mice (Costa & Furness 1979b). Neural receptors for 5-HT are distinct from those for ACh, and, in general, antagonists of nicotinic receptors for ACh, such as hexamethonium or pentolinium, do not antagonize 5-HT's stimulation of neurons (Bülbring & Gershon 1967, Drakontides & Gershon 1968, Costa & Furness 1979b). D-tubocurarine, however, is an exception. The ileal contraction caused by stimulation of cholinergic neurons by 5-HT in the guinea pig is blocked by D-tubocurarine (Costa & Furness 1979b), as is the excitatory effect of the iontophoretic application of 5-HT to submucosal ganglion cells (Hirst & Silinsky 1975). It is interesting, in this regard, that D-tubocurarine is also an antagonist at "A" receptors for 5-HT in ganglia of *Aplysia* (Gerschenfeld & Paupardin-Tritsch 1974). There are several different types of 5-HT receptor in *Aplysia* ganglia and it is certainly conceivable that such a multiplicity of receptors exists in the enteric nervous system as well.

In the absence of a specific drug, desensitization of 5-HT appears to be the most specific antagonistic treatment available. Desensitization is rapid when 5-HT is added to the fluid bathing preparations of gut (Wood & Mayer 1979b), while responses to other transmitters, such as ACh, persist (Bülbring & Gershon 1967, Furness & Costa 1973, Costa & Furness 1976, 1979a,b).

Electrophysiological studies have confirmed the excitatory effects of 5-HT on enteric neurons (Sato et al 1974, Dingledine et al 1974, Dingledine & Goldstein 1976). With intracellular electrodes, 5-HT has been found to depolarize both type 1(S) and type 2(AH) cells in the myenteric plexus (Wood & Mayer 1979b, Johnson et al 1980b). The most striking effect is that of iontophoretic application of 5-HT to type 2(AH) cells (Wood & Mayer 1979b). 5-HT mimics the effect on these cells of the natural transmitter released by fiber tract stimulation to induce slow epsps (Wood & Mayer 1979a). Both the endogenous transmitter and the exogenous 5-HT produce an extremely long-lasting membrane depolarization (up to 1 min); both increase neuronal input resistance; both abolish the hyperpolarization that follows action potentials in type 2(AH) cells; both augment membrane excitability (Wood & Mayer 1979b). The slow epsp and the depolarizing response to 5-HT are both probably due to depression of Ca^{2+}-dependent K^+ conductance that characterizes type 2(AH) cells (Grafe et al 1980, Johnson et al 1980b). Cholinergic agonists and antagonists do not alter slow epsps or the response to iontophoretic application of 5-HT (Wood & Mayer 1979b). The response to fiber tract stimulation and to exogenous 5-HT is blocked by methysergide and by desensitization of 5-HT receptors with excess 5-HT in the perfusion solution. Other substances, when applied iontophoretically, also act on type 2(AH) cells as do fiber tract stimulation and 5-HT; substance P is one such example (Katayama & North 1978, Katayama et al 1979). It is unlikely, however, that substance P is the mediator of slow epsps (Grafe et al 1979b). The pharmacology is not correct. Desensitization to 5-HT and methysergide both block the slow epsp (and the effect of 5-HT) but neither interferes with the action of substance P.

The action of NE on fiber tract stimulation is interesting. NE antagonizes the generation of slow epsps in type 2(AH) neurons following fiber tract stimulation (Wood & Mayer 1979c). NE does not, however, antagonize the action of iontophoretically applied 5-HT. This has led to the hypothesis that NE presynaptically inhibits the axonal release of 5-HT. Exogenous NE has also been found to inhibit the stimulated release of [^{3}H]-5-HT from myenteric axons preloaded with that radioactive amine (Gershon 1980), thus confirming the conclusions drawn from electrophysiological experiments with NE (Wood & Mayer 1979c). The effect of sympathetic nerve stimulation, however, is not identical to that of exogenous NE (Gershon 1980). The interaction between NE and 5-HT, therefore, needs further study.

5-HT has additional actions in the myenteric plexus besides its apparent mediation of slow epsps in type 2(AH) neurons (North et al 1980, Johnson et al 1980b). 5-HT also depresses the cholinergic fast epsp in all myenteric neurons. Since 5-HT does not affect the amplitude or time course of the depolarization produced by iontophoretic application of ACh, it has been concluded that the action of 5-HT on the fast epsp is due to presynaptic inhibition of ACh release. A final, though rare action of 5-HT, following its iontophoretic application, is to hyperpolarize occasional type 2(AH) cells (Johnson et al 1980b). These responses are abolished in Ca^{2+}-free solutions and are therefore probably due to the release of another transmitter by 5-HT. Exogenous 5-HT has been shown to be taken up nonspecifically by enteric noradrenergic axons and, if present in high enough concentration, 5-HT can displace endogenous NE (Drakontides & Gershon 1972). It is thus possible that the rare indirect hyperpolarizing responses to iontophoretic application of 5-HT are the result of the liberation of endogenous NE.

Pharmacological identity   A number of neural effects, such as the elicitation of slow epsps in type (AH) or tonic-type neurons (Wood & Mayer 1979b), vagal relaxation of the stomach (Bülbring & Gershon 1967), vagal relaxation of the lower esophageal sphincter of the oppossum (Rattan & Goyal 1978), post-train synaptic excitation in the myenteric plexus (Dingledine & Goldstein 1976), and ascending peristaltic excitation (Furness &

Costa 1973, Costa & Furness 1976) are inhibited by antagonism of the effects of 5-HT. Chronic depletion of 5-HT, with the tryptophan hydroxylase inhibitor, parachlorophenylalanine (PCPA), interferes with intestinal motility (Welch & Welch 1968, Weber 1970, Breisch et al 1976, Saller & Stricker 1978) and makes the gut supersensitive to 5-HT (Schulz & Cartwright 1974b). Supersensitivity would be expected from chronic depletion of a neurotransmitter. PCPA also blocks the neural inhibition of vagally evoked excitatory junction potentials that accompanies descending inhibition in the colon (Julé 1980). The PCPA block can be overcome by L-tryptophan and is mimicked by inhibition of synthesis of 5-HT from 5-HTP. Inhibition of 5-HT re-uptake potentiates and prolongs the duration of inhibition of vagally evoked responses. These observations indicate that the phase of descending inhibition in peristalsis has at least two neural components. One is the relaxation of smooth muscle by non-adrenergic intrinsic inhibitory neurons (Costa & Furness 1976, Julé 1980); the other is a corresponding inhibition of cholinergic excitation (Julé 1980). 5-HT may be involved in the mediation of this latter pathway.

Inactivation   An inactivating mechanism exists for 5-HT in the enteric nervous system. Axons in the myenteric and submucosal plexuses take up 5-HT (Gershon & Altman 1971, Robinson & Gershon 1971, Gershon et al 1976, Furness & Costa 1978). The uptake is largely a feature of axons; varicosities isolated from the gut that has been exposed to [^{3}H]-5-HT contain and are enriched with the labeled amine, and isolated varicosities obtained from the myenteric plexus actively take up [^{3}H]-5-HT (Jonakait et al 1979a). The uptake mechanism for 5-HT is distinct from the uptake mechanism that also exists in the enteric plexuses for NE. The uptake of NE is lost after chemical sympathectomy with 6-HD, whereas the uptake of 5-HT is unaffected by this treatment (Gershon & Altman 1971, Gershon et al 1976, Gershon et al 1980b). Unlike the uptake of NE, the 5-HT uptake mechanism also persists after extrinsic denervation of the gut (Furness & Costa 1978). During ontogeny, moreover, the uptake mechanism for 5-HT in the enteric nervous system precedes that for NE by a wide margin in rabbits (Rothman et al 1976), mice (Rothman et al 1979), guinea pigs (Gintzler et al 1980), and chicks (Epstein et al 1980, Gershon et al 1980a). NE, furthermore, fails to inhibit the uptake of 5-HT even when present at 1000 times the 5-HT concentration (Rothman et al 1976).

Elements responsible for 5-HT uptake   The neurons responsible for uptake of 5-HT are intrinsic enteric neurons and survive in organotypic tissue culture for long periods of time (Dreyfus et al 1977b). The neurons also develop in cultures of gut that has been removed from fetal mice prior to the appearance of cells recognizable as neurons (Rothman et al 1979). A well-organized myenteric plexus forms in these cultures and, in addition to cholinergic neurons, the plexus in culture contains neurons that display 5-HT histofluorescence and the 5-HT uptake mechanism. Similarly, neurons that take up 5-HT develop in explants of embryonic chick gut removed prior to the appearance in the gut of cells identifiable as neurons and grown on the chorio-allantoic membrane of chick embryo hosts (Gershon et al 1980a). These results indicate that intrinsic serotonergic neurons must be derived from precursors that cannot be recognized morphologically when they colonize the gut. Definitive phenotypic characteristics, such as acquisition of the specific 5-HT uptake mechanism and stores of the amine, are acquired within the enteric microenvironment.

Properties of the 5-HT uptake mechanism   The 5-HT uptake mechanism in enteric neurons has been extensively characterized (Gershon & Altman 1971, Gershon et al 1976, Gershon & Jonakait 1979). The uptake is a saturable, energy-dependent mechanism; the Km is about 0.7 μM. The uptake of 5-HT has specific ionic requirements as well. Uptake requires NA^+, is antagonized by K^+ in a direct proportion to the logarithm of the K^+ concentration, and is dependent upon the normal Ca^{2+} concentration (about 2.5 mM). Uptake of 5-HT is inhibited in the absence of Ca^{2+} or in the presence of an elevated Ca^{2+} concentration. Uptake of 5-HT is also inhibited by ouabain.

One of the most striking features of the enteric 5-HT uptake mechanism is its narrow molecular structural requirements (Gershon et al 1976). Even very slight changes in the 5-HT molecule greatly decrease the affinity of the compound for the uptake site. Methylation of the terminal amino group of 5-HT reduces affinity a hundred-fold, as does removal of the ring hydroxyl group or movement of the hydroxyl group to the number six position of the indole ring. These analogs do, however, competitively inhibit (albeit poorly) uptake of [^{3}H]-5-HT. Removal of the aliphatic side chain eliminates the ability of the compounds to compete with 5-HT for uptake.

5,6-Dihydroxytryptamine (5,6-DHT) and 5,7-dihydroxytryptamine (5,7-DHT) have about 25% of 5-HT's affinity for the uptake site (Gershon et al 1976). These compounds, and 6-HT as well, are very useful, however, for the histochemical and cytochemical study of enteric serotonergic neurons. 5,6-DHT and 5,7-DHT are cytotoxic and can be used to identify enteric serotonergic axons ultrastructurally (Gershon et al 1980b). The serotonergic varicosities have not been found to be very distinctive in their appearance. They contain a mixed population of small electron-lucent and larger (about 80 to 100 nm in diameter) dense-cored synaptic vesicles. 6-HT is useful for the study of serotonergic terminals by histofluorescence because it has a better fluorescent yield than 5-HT (Gershon et al 1976, Furness & Costa 1978).

The susceptibility of the enteric 5-HT uptake mechanism to inhibition by drugs is very similar to the drug susceptibility of the uptake of 5-HT by serotonergic neurons of the CNS (Gershon et al 1976). Tricyclic antidepressants inhibit the uptake of 5-HT and NE; however, chlorimipramine is most effective against 5-HT uptake, whereas desmethylimipramine is most potent against uptake of NE. Chlorinated amphetamines also inhibit 5-HT uptake. The tricyclics are competitive uptake inhibitors, whereas the chlorinated amphetamines behave in a manner that is neither purely competitive nor non-competitive. Chlorimipramine, moreover, appears to release 5-HT to some extent as well as to inhibit its uptake (Gershon & Jonakait 1979). The most specific and potent inhibitor of uptake of 5-HT by enteric neurons found thus far is fluoxetine (Gershon & Jonakait 1979). At concentrations that do not inhibit uptake of NE, this drug can almost abolish 5-HT uptake.

The specific uptake of 5-HT is a valuable marker for studies of enteric serotonergic neurons. The uptake of $[^3H]$-5-HT can be accurately measured and it also makes enteric serotonergic neurites demonstrable by radioautography (Gershon & Altman 1971, Robinson & Gershon 1971). $[^3H]$-5-HT can be fixed in enteric neurites with aldehyde-containing fixatives without causing detrimental movement of the labeled amine (Fischman & Gershon 1964, Gershon & Ross 1966a,b). The uptake of $[^3H]$-5-HT thus makes possible the untrastructural radioautographic identification of serotonergic terminals (Rothman et al 1976, Dreyfus et al 1977b, Jonakait et al 1979a,b). Terminals that take up $[^3H]$-5-HT have the same appearance as those marked by 5, 7-DHT (Gershon et al 1980b). Except during early development, labeling with $[^3H]$-5-HT is restricted to axons (Epstein et al 1980, Gershon et al 1980a). Serotonergic neurons thus seem to resemble cholinergic neurons; in cholinergic neurons the uptake mechanism for choline is similarly restricted to the axons of these cells (Suszkiw & Pilar 1976).

Phylogeny of enteric serotonergic neurons 5-HT is present in the bowel of all vertebrates (Erspamer 1966). It is difficult to generalize from this to the enteric nervous system, however, because of the large quantity of 5-HT in mucosal EC cells. Some lower vertebrates, including cyclostomes and teleosts, do not have EC cells (Erspamer 1966, Watson 1979) but do have 5-HT in their intestines (Baumgarten et al 1973, Watson 1979, Goodrich et al 1980). Dissection of hagfish gut and measurement of the concentration of 5-HT in layers of the intestinal wall reveal a distribution of the amine that is consistent with its being located in the myenteric plexus and 5-HT can be shown histochemically in hagfish (Goodrich et al 1980), lamprey (Baumgarten et al 1973), and teleost (Watson 1979) enteric neurons. The hagfish gut, moreover, takes up [^{3}H]-5-HT by a mechanism that appears to be identical to that of mammals (Goodrich et al 1980). Electron microscopic radioautography in hagfish as in mammals reveals an uptake restricted to neuronal elements, especially axonal varicosities containing many small electron-lucent synaptic vesicles.

Enteric serotonergic neurons, all with the specific uptake mechanism for 5-HT, have now been reported in a substantial number of vertebrates. In addition to cyclostomes they have been found in teleosts and amphibia (Goodrich et al 1980), birds (Epstein et al 1980, Gershon et al 1980a), rabbits (Rothman et al 1976), guinea pigs (Gershon & Altman 1971, Gershon et al 1976), mice (Rothman et al 1979), rats (Gershon et al 1977), humans, and nonhuman primates (Goodrich & Gershon 1977, Rogawski et al 1978). They do not, however, seem to exist in tunicates or echinoderms, groups that may have given rise to the vertebrates (Goodrich et al 1980). Consequently, the idea that enteric serotonergic neurons are present in cyclostomes because they are an invertebrate vestige in an invertebrate-vertebrate transition group (Baumgarten et al 1973) is probably not correct. Actually, yellow (5-HT-like) formaldehyde-induced fluorescence has been reported in enteric neurons in the endostyle and gut of amphioxus (Salimova 1978). This suggests that vertebrate enteric serotonergic neurons may first have arisen in this cephalochordate.

Although not every vertebrate class has been examined, enteric serotonergic neurons have been found in both early and late-evolving species. Intermediate classes also have the neurons. Since both birds and mammals have retained these neurons, it is probable that they are a general feature of the vertebrate bowel (Goodrich et al 1980). Enteric serotonergic neurons thus seem to have arisen early and to have been retained throughout vertebrate evolution. Consequently, although the precise role of these neurons in gastrointestinal physiology remains to be determined, it seems likely that they are important. The early origin of enteric serotonergic neurons in phylogeny parallels their similarly early appearance in mammalian (Rothman et al 1976, 1979, Gintzler et al 1980) and avian (Epstein et al 1980, Gershon et al 1980a) ontogeny.

Since 5-HT satisfies all of the criteria necessary to identify it as a neurotransmitter, it seems reasonable to conclude that 5-HT is a neurotransmitter in the enteric nervous system. The enteric serotonergic neurons are intrinsic to the gut. Their role will have to be defined in future studies.

Purinergic Neurons

Non-adrenergic, non-cholinergic neurally mediated inhibition of intestinal muscle was first discovered in the taenia of the guinea pig caecum (Burnstock et al 1964, 1966). This effect differs from sympathetically mediated inhibition in several ways:

1. It is more effective than sympathetic relaxation at low frequencies of nerve stimulation.
2. It is not inhibited by adrenergic neuron-blocking drugs.
3. It is mediated by intrinsic neurons (Rikimaru 1971).
4. It is accompanied by inhibitory junction potentials in smooth muscle cells (Bennett et al 1966) that are not comparable to the effects on smooth muscle of sympathetic nerve stimulation. It was soon found that intrinsic inhibitory neurons are widely distributed in the gastrointestinal tract and, in the stomach and rectum, they receive an input from the CNS via the vagus nerves (Martinson 1965, Campbell 1966, Bülbring & Gershon 1967) or pelvic nerves (Julé & Gonella 1972, Costa & Furness 1973), respectively. It is probable that the intrinsic inhibitory neurons between the stomach and rectum are not under CNS control but are involved in the mediation of the descending inhibitory phase of the peristaltic reflex (Burnstock & Costa 1973, Costa & Furness 1976, Julé 1980).

The non-adrenergic nature of intrinsic inhibitory neurons was confirmed when it was found that they develop before the adrenergic innervation reaches the gut during ontogeny (Gershon & Thomson 1973). In 1970 it was proposed that the transmitter of the intrinsic inhibitory neurons is either ATP or a related purine nucleotide (Burnstock et al 1970). Burnstock (1971) suggested that the nerves could appropriately be called purinergic. Since the original proposal of ATP as an enteric neurotransmitter, Burnstock (1971, 1979b) has reviewed the extensive evidence that has accumulated in support of this proposal. This evidence includes indications that (a) enteric nerve terminals contain ATP and release ATP on stimulation, (b) ATP mimics the action on smooth muscle of the intrinsic inhibitory neurotransmitter, (c) the gut has enzymes to catalyze the breakdown of ATP and enteric nerves have a high affinity uptake mechanism for adenosine but not ATP, and (d) drugs (albeit not entirely specific) that block the action of ATP on smooth muscle also antagonize the effects of intrinsic inhibitory neurons. Some enteric axons take up quinacrine which can be used to demonstrate these axons histofluorometrically (Olson et al 1976). Burnstock (1979b) has suggested that this quinacrine histofluorescence is a marker for purinergic nerves in tissues. Quinacrine does appear to be released from nerves that have taken it up by high K^+ or by treatment with veratridine (Ålund & Olson 1979), but it has yet to be shown that these nerves take up and release the compound because it binds to ATP (Burnstock 1979b).

Despite the impressive array of supporting observations, a number of recent studies have reported instances in which the responses of various gastrointestinal preparations to stimulation of intrinsic inhibitory neurons have been different from the responses of these preparations to ATP or other purine nucleotides (Szurszewski 1979, Campbell & Gibbons 1979). Doubt has thus been expressed about the validity of the supposition that ATP is a transmitter (Campbell & Gibbins 1979). The recent doubt about ATP as the transmitter of intrinsic inhibitory neurons has coincided with a recent burst of enthusiasm for enteric peptidergic neurotransmission. One such peptide, VIP, has emerged as a new candidate for the inhibitory neurotransmitter (Campbell & Gibbins 1979, Fox et al 1979, Furness & Costa 1979a,b, 1980, Goyal et al 1979).

Peptidergic Neurons

The first serious suggestion that the enteric nervous system contains peptidergic axons was made by Baumgarten, Holstein & Owman (1970), who noted that one of the three types of enteric terminal they studied contained large dense-cored vesicles (90 to 150 nm in diameter). Because of its similarity in appearance to the terminals of neurosecretory axons in the posterior pituitary gland, they called the terminals *p-type* for peptide-containing. Subsequently, VIP-like immunoreactivity was indeed found in p-type terminals (Larsson 1977). Since 1970, moreover, a variety of antisera have become available and these have made possible the immunocytochemical detection of a number of neuropeptides in neurons in the gut. Because the identification of most of these neuropeptides is heavily dependent upon immunocytochemical evidence, their identification is only as selective as the particular antibody being used. The possibility of cross reactivity with other peptides is therefore always to be considered, and peptide-like immunoreactivity should be understood to be implied by most of these identifications. Enteric peptide-containing neurons have recently been reviewed (Hökfelt 1979, Furness & Costa 1980, Furness et al 1980). This topic is therefore covered only briefly here.

VIP   This peptide has been found in many regions of the gut in a wide variety of mammals (see Hökfelt 1979, Furness & Costa 1980). Since VIP immunoreactivity survives when the gut is grown for extended periods of time in organotypic tissue culture, at least some of the VIP neurons are intrinsic to the gut (Schultzberg et al 1978). In the guinea pig ileum 42% of the neurons in the submucosal plexus and 2.4% of those in the myenteric plexus display VIP-like immunoreactivity. The projections of the VIP immunoreactive neurons have been mapped by combining a series of lesions with immunocytochemistry (Furness & Costa 1979a, Furness et al 1980). The axons of the VIP neurons project in an oral-to-anal direction. Myenteric VIP neurons send extensive terminal ramifications to the circular muscle and additional terminals to other neurons in the myenteric plexus. The projection to the circular muscle and the anally running direction of the fibers is consistent with VIP serving as the intrinsic inhibitory neurotransmitter, which, it should be recalled, is involved in the mediation of descending inhibition (Hirst & McKirdy 1974, Costa & Furness 1976, Furness & Costa 1979a). VIP relaxes smooth muscle (Piper et al 1970, Uddman et al 1978, Campbell & Gibbins 1979, Furness & Costa 1980) and is released from the gut upon vagal stimulation Fahrenkrug et al 1978), an effect that is antagonized by splanchnic nerve stimulation or by administration of hexamethonium. These observations imply that VIP neurons are innervated by the vagus, as are the inhibitory neurons of the stomach. Although all of these properties suggest that VIP is a strong candidate for the intrinsic inhibitory neuromuscular transmitter, it is unlikely that this could be the peptide's only role in the enteric nervous system. VIP neurons project to myenteric neurons, as well as to muscle, they are prominent in the submucosal plexus, and they project to villi (Hökfelt 1979). VIP increases the firing rate of many myenteric neurons (Williams & North 1979). VIP neurons also appear to project centripetally out of the gut to the inferior mesenteric ganglion (Hökfelt 1979). Cocks & Burnstock (1979), moreover, have reported that VIP does not exactly mimic the intrinsic inhibitory neuron's effect on the gut and, in fact, they argue that the peptide mimics the natural transmitter less well than ATP. The role of enteric VIP neurons thus remains to be established and the intrinsic inhibitory transmitter has still to be identified with certainty.

SUBSTANCE P   The myenteric and submucosal plexuses are rich in axons that display substance P-like immunoreactivity (Hökfelt 1979, Furness & Costa 1980). Most of these substance P terminals arise from intrinsic enteric neurons (Schultzberg et al 1978, Franco et al 1979b). There are, however, also numerous fibers containing substance P-like immunoreactivity in the vagus nerves (Gamse et al 1979, Hökfelt 1979). Substance P seems to be released from enteric nerves upon stimulation (Franco et al 1979a). The compound also, as noted earlier, mimics the slow epsp in type 2(AH) myenteric neurons (Katayama & North 1978, Katayama et al 1979), although substance P does not appear to be the mediator of that response (Grafe et al 1979b). Substance P neurons have been estimated to comprise 11% of the neurons in the submucosal plexus and about 3.5% of the neurons in the myenteric plexus (Furness & Costa 1980, Furness et al 1980). As with all of the neuropeptides, the role of substance P in enteric function is unknown.

SOMATOSTATIN   Most of the neurites in the gut that contain somatostatin-like immunoreactivity are, like VIP and substance P, of intrinsic origin (Schultzberg et al 1978, Furness & Costa 1980, Furness et al 1980). About 17% of submucosal neurons and 3% of myenteric neurons display somatostatin immunoreactivity (Furness et al 1980). In the myenteric plexus, somatostatin axons appear to project in an oral-to-anal direction. The majority of these terminals appear to end within the enteric nervous system on other somatostatin immunoreactive neurons and on other unidentified neurons. Somatostatin seems to have little direct effect on the smooth muscle. It does, however, decrease release of ACh and activate intrinsic inhibitory neurons (Guillemin 1976, Cohen et al 1978, Furness & Costa 1979b). Within the myenteric plexus, somatostatin has been found, using intracellular electrodes, to depolarize some neurons while hyperpolarizing others (Katayama & North 1980). The effect of the peptide differed in some cells, depending on whether it was introduced into the bathing medium or applied locally by iontophoresis. Depolarizing responses were associated with a fall and hyperpolarizing responses with a rise in cell input resistance. It is possible that the cells depolarized by somatostatin are intrinsic inhibitory neurons and those hyperpolarized by the peptide are cholinergic excitatory neurons. If this were true, and no evidence to support the possibility yet exists, then somatostatin would be a good candidate for a neurotransmitter of interneurons involved in mediating the phenomenon of descending inhibition. The anal-going projection of somatostatin neurites and their restriction to the enteric plexuses are consistent with this possibility (Furness & Costa 1980).

ENKEPHALIN   Both leu- and met-enkephalin have been detected immunocytochemically in neurites in the gut (Hökfelt 1979, Furness & Costa 1980). Neurons that display enkephalin immunoreactivity survive in culture and so at least some are intrinsic to the gut. Cell bodies of these neurons have been found in the myenteric plexus where they have been estimated to constitute approximately 25% of the total neuronal population (Furness & Costa 1980, Furness et al 1980). No cell bodies and few enkephalin axons are found in the submucosal plexus or the mucosa. A fairly dense projection of enkephalin-immunoreactive fibers extends to the circular muscle. Enkephalins are apparently released from the enteric nervous system by electrical stimulation (Schulz et al 1977, Puig et al 1977). Synthesis of enkephalin has also been demonstrated, probably involving derivation of enkephalin from an initially formed, but unidentified, precursor (Sosa et al 1977). Opiates have long been known to inhibit the stimulated and resting release of ACh from the guinea pig gut (Schaumann 1957, Paton 1957). Enkephalins have a similar effect, although there appears to be more than one kind of opiate receptor in the gut (Lord et al 1977, Waterfield et al 1977, Kosterlitz 1979). Enkephalins also have a direct contractile action on the longitudinal muscle (Kosterlitz 1979). Enkephalins hyperpolarize myenteric neurons when present in the fluid perfusing the gut but not when the peptides are applied by iontophoresis to the membranes of cell bodies

(North et al 1979). It may be that receptors are located on processes not cell bodies, but the possibility that inhibition of ACh release is a presynaptic action must also be considered. If so, then the enkephalin fibers in the circular muscle may be there to inhibit axo-axonically the release of ACh from neighboring excitatory cholinergic axon terminals.

Other Neuronal Types

Other neuropeptides found in the enteric nervous system include immunoreactive pancreatic polypeptide (Loren et al 1979), gastrin or cholecystokinin (Larsson & Rehfeld 1979), neurotensin (see Hökfelt 1979), and bombesin (see Furness & Costa 1980). The projections of neurons containing these peptides and their function have not been worked out. Some pharmacological evidence has been advanced to support other neurotransmitters such as dopamine (Valenzuela 1976, Szurszewski 1979); however, the gut contains no intrinsic dopaminergic neurons (Furness et al 1979). Nevertheless, estimates of proportions of various types of enteric neuron (Furness & Costa 1980, Furness et al 1980) in the enteric plexuses leave 63.5% of neurons in the myenteric plexus and 18% of neurons in the submucosal plexus unaccounted for. Many of these are probably cholinergic but there is still room for more neuronal types to be discovered.

CONCLUSIONS

Although a great deal of ganglionic transmission in the enteric plexuses is cholinergic and nicotinic, other transmitters and receptors are certainly also involved. Electrophysiological studies have not yet clarified this issue. There is a major discrepancy between a relatively large number of types of neuron that have been identified with extracellular electrodes and a smaller number identifiable when intracellular microelectrodes have been used. This discrepancy may be due to sampling artifacts, mechanical artifacts, or to a restriction of some electrical events in enteric neurons to their processes. The description of neurons with either type of electrical recording fails to match the abundance of neuronal types suggested by histochemical, immunocytochemical, and anatomical evidence. The physiology of the enteric nerve plexuses, therefore, is still at an early stage of understanding and an exciting frontier of research.

The enteric nervous system represents both a problem and an opportunity for neurobiologists. The system is attractive to study because it seems to be a relatively simple nervous system (decreasingly so in recent years, but still simpler than the brain). It survives and functions well in-vitro and is particularly accessible. These features facilitate the study of the properties of those neurons that have been identified within the gut. The guinea pig ileum has been a special boon to pharmacologists in this regard. For example, this preparation contains opiate receptors and was invaluable in the isolation of the enkephalins (see Kosterlitz 1979 for references). The enteric nervous system, therefore, is a good model system, easier in many ways to study than the brain. On the other hand, the enteric nervous system itself is not well understood. It contains many types of neuron but the relationship of this abundance to function remains unresolved. The interconnections of the various neurons with one another and the possible presence of definable pathways in the myenteric plexus are still unknown. Whether the cohort of cell types in each enteric ganglion is the same or different or, if different, whether there is a modular repetition of a limited number of ganglionic aggregations remains to be established. In fact, more is known of the "wiring diagram" of the CNS than the enteric nervous system. Nevertheless understanding the function of the enteric nervous system would be valuable, not only because of the obvious benefits to public health that would accrue from learning how gastrointestinal motility is neurally controlled, but also because a system like the gut, with its limited repertory of behavior, should ultimately be useful in bridging the gap between the study of neuronal events and the resultant behavioral output.

ACKNOWLEDGMENTS

The author wishes to thank Diane Sherman, Laraine Field, Christine Wade, and Steven Erde for help in the preparation of this manuscript.

Literature Cited

Åberg, G., Eränkö, O. 1967. Localization of noradrenaline and acetylcholinesterase in the taenia of the guinea pig cecum. *Acta Physiol. Scand.* 69:383–84

Aghajanian, G. K., Asher, I. M. 1971. Histochemical fluorescence of raphe neurons: Selective enhancement by tryptophan. *Science* 172:1159–61

Aghajanian, G. K., Kuhar, M. J., Roth, R. H. 1973. Serotonin-containing neuronal perikarya and terminals: Differential effects of p-chlorophenylalanine. *Brain Res.* 54:85–101

Ahlman, H., Enerback, L. 1974. A cytofluorometric study of the myenteric plexus in the guinea-pig. *Cell Tissue Res.* 153:419–34

Ahlman, H., Enerback, L., Kewenter, J., Storm, B. 1973. Effects of extrinsic denervation on the fluorescence of monoamines in the small intestine of the cat. *Acta Physiol. Scand.* 89:429–35

Ahlman, H., Lundberg, J., Dahlstrom, A., Kewenter, J. 1976. A possible vagal adrenergic release of serotonin from en-

terochromaffin cells in the cat. *Acta Physiol. Scand.* 98:366–75

Ålund, M., Olson, L. 1979. Depolarization-induced decreases in fluorescence intensity of gastrointestinal quinacrine-binding nerves. *Brain Res.* 166:121–37

Barajas, L., Wang, P. 1975. Demonstration of acetylcholinesterase in the adrenergic nerves of the renal glomerular arterioles. *J. Ultrastruct. Res.* 53:244–53

Barr, L., Berger, W., Dewey, M. M. 1968. Electrical transmission at the nexus between smooth muscle cells. *J. Gen. Physiol.* 51:347–68

Baumgarten, H. G., Bjorklund, A., Lachenmayer, L., Nobin, A., Rosengren, E. 1973. Evidence for existence of serotonin-, dopamine- and noradrenaline-containing neurons in the gut of *Lampetra fluviatilis. Z. Zellforsch.* 141:33–54

Baumgarten, H. G., Holstein, A. F., Owman, C. 1970. Auerbach's plexus of mammals and man: Electron microscopic identification of three different types of neuronal processes in myenteric ganglia of the large intestine from Rhesus monkeys, guinea-pigs and man. *Z. Zell-forsch. Mikros. Anat.* 106:376–97

Bayliss, W. M., Starling, E. H. 1899. The movements and innervation of the small intestine. *J. Physiol. London* 24:99–143

Beleslin, D. B., Samardzic, R. 1978. The effect of muscarine on cholinoceptive neurons subserving the peristaltic reflex of guinea-pig isolated ileum. *Neuropharmacology* 17:793–98

Bennett, M. R., Burnstock, G., Holman, M. E. 1966. Transmission from intramural inhibitory nerves in the smooth muscle of the guinea-pig taenia coli. *J. Physiol. London* 182:541–58

Birks, R., Huxley, H. E., Katz, B. 1960. The fine structure of the neuromuscular junction of the frog. *J. Physiol. London* 150:134–44

Bodian, M., Stephens, F. D., Ward, B. C. H. 1949. Hirschsprung's disease and idiopathic megacolon. *Lancet* 1:6–11

Boireau, A., Ternaux, J. P., Bourgoin, S., Hery, F., Glowinski, J., Hamon, M. 1976. The determination of picogram levels of 5-HT in biological fluids. *J. Neurochem.* 26:201–4

Bolande, R. P. 1975. Animal model of human disease. Aganglionic or hypoganglionic megacolon; animal model: Aganglionic megacolon in piebald and spotted mutant mouse strains. *Am. J. Pathol.* 79:189–92

Bortoff, A. 1972. Digestion: Motility. *Ann. Rev. Physiol.* 34:261–90

Bortoff, A., Miller, R. 1975. Stimulation of intestinal muscle by atropine, procaine, and tetrodotoxin. *Am. J. Physiol.* 229:1609–13

Boullin, D. J. 1964. Observations on the significance of 5-hydroxytryptamine in relation to the peristaltic reflex of the rat. *Br. J. Pharmacol.* 23:14–33

Boyne, A. F., Bohan, T. P., Williams, T. H. 1975. Changes in cholinergic synaptic vesicle populations and the ultrastructure of the nerve terminal membranes of *Narcine Brasiliensis* electric organ stimulated to fatigue in vivo. *J. Cell Biol.* 67:814–25

Breisch, S. T., Zemlan, F. P., Hoebel, B. G. 1976. Hyperphagia and obesity following serotonin depletion by intravesicular p-chlorophenylalanine. *Science* 192:382–85

Brightman, M. W., Reese, T. S. 1969. Junction between intimately apposed cell membranes in the vertebrate brain. *J. Cell Biol.* 40:648–77

Brownlee, G., Johnson, E. S. 1963. The site of the 5-hydroxytryptamine receptor on the intramural nervous plexus of the guinea-pig isolated ileum. *Br. J. Pharmacol.* 21:306–22

Bülbring, E. 1972. Action of catecholamines on the smooth muscle cell membrane. In *Drug Receptors,* ed. H. P. Rang, pp. 1–13. Baltimore: Univ. Park

Bülbring, E., Crema, A. 1958. Observations concerning the action of 5-hydroxytryptamine on the peristaltic reflex. *Br. J. Pharmacol.* 13:444–57

Bülbring, E., Gershon, M. D. 1967. 5-Hydroxytryptamine participation in the vagal inhibitory innervation of the stomach. *J. Physiol. London* 192:823–46

Bülbring, E., Lin, R. C. Y. 1958. The effect of intraluminal application of 5-hydroxytryptamine and 5-hydroxytryptophan on peristalsis, the local production of 5-hydroxytryptamine and its release in relation to intraluminal pressure and propulsive activity. *J. Physiol. London* 140:381–407

Burnstock, G. 1971. Neural nomenclature. *Nature* 229:282–83

Burnstock, G. 1979a. Autonomic neuroeffector junction. In *Neurosciences Research Program Bulletin.* Vol. 17, *Non-adrenergic, Non-cholinergic Autonomic Neurotransmission Mechanisms,* ed. G. Burnstock, M. D. Gershon, T. Hökfelt, L. L. Iversen, H. W. Kosterlitz, J. H. Szurszewski, pp. 388–91. Cambridge: MIT Press

Burnstock, G. 1979b. Adenosine triphosphate. See Burnstock 1979a, pp. 406–14

Burnstock, G., Campbell, G., Bennett, M., Holman, M. E. 1964. Innervation of the guinea-pig taenia coli: Are there intrinsic inhibitory nerves which are distinct from sympathetic nerves? *Int. J. Neuropharmacol.* 3:163–66

Burnstock, G., Campbell, G., Rand, M. J. 1966. The inhibitory innervation of the taenia of the guinea pig caecum. *J. Physiol. London* 182:504–26

Burnstock, G., Campbell, G., Satchell, D., Smythe, A. 1970. Evidence that adenosine triphosphate or a related nucleotide is the transmitter substance released by non-adrenergic inhibitory nerves in the gut. *Br. J. Pharmacol.* 40:668–88

Burnstock, G., Costa, M. 1973. Inhibitory innervation of the gut. *Gastroenterology* 64:141–43

Bursztajn, S., Gershon, M. D. 1977. Discrimination beteen nicotinic receptors in vertebrate ganglia and skeletal muscle by alphabungarotoxin and cobra venoms. *J. Physiol. London* 269:17–31

Calas, A., Besson, M. J., Gaughy, G., Alonso, G., Glowinski, J., Cheramy, A.

1976. Radioautographic study of in vivo incorporation of ^{3}H-monoamines in the cat caudate nucleus: Identification of serotoninergic fibres. *Brain Res.* 118:1–13

Campbell, G. 1966. The inhibitory nerve fibers in the vagal supply to the guinea-pig stomach. *J. Physiol. London* 185:600–12

Campbell, G., Gibbins, I. L. 1979. Nonadrenergic, noncholinergic transmission in the autonomic nervous system: Purinergic nerves. In *Trends in Autonomic Pharmacology,* ed. S. Kalsner, 1:103–44. Balitmore/Munich: Urban & Schwarzenberg

Chan-Palay, V. 1975. Fine structure of labelled axons in the cerebellar cortex and nuclei of rodents and primates after intraventricular infusions with tritiated serotonin. *Anat. Embryol.* 148:235–65

Chan-Palay, V. 1978. The paratrigeminal nucleus. II. Identification and interrelations of catecholamine axons, indoleamine axons, and Substance P immunoreactive cells in the neuropil. *J. Neurocytol.* 7:419–42

Chujyo, N. 1953. Site of acetylcholine production in the wall of intestine. *Am. J. Physiol.* 174:196–98

Clementi, F., Palade, G. E. 1969. Intestinal capillaries. I. Permeability to peroxidase and ferritin. *J. Cell Biol.* 41:33–58

Cocks, T., Burnstock, G. 1979. Effects of neuronal polypeptides on intestinal smooth muscle: A comparison with non-adrenergic, non-cholinergic nerve stimulation and ATP. *Eur. J. Pharmacol.* 54:251–59

Code, C. F., Carlson, H. C. 1968. Motor activity of the stomach. In *Handbook of Physiology.* Sect. 6, *Alimentary Canal,* Vol. 4, *Motility,* ed. C. F. Code, pp. 1903–16. Washington DC: Am. Physiol. Soc.

Cohen, M. L., Rosing, E., Wiley, K. S., Slater, I. H. 1978. Somatostatin inhibits adrenergic and cholinergic neurotransmission in smooth muscle. *Life Sci.* 23:1659–64

Cook, R. D., Burnstock, G. 1976a. The ultrastructure of Auerbach's plexus in the guinea-pig. I. Neuronal elements. *J. Neurocytol.* 5:171–94

Cook, R. D., Burnstock, G. 1976b. The ultrastructure of Auerbach's plexus in the guinea-pig. II. Non-neuronal elements. *J. Neurocytol.* 5:195–206

Cooke, A. R., Athey, G. F., Wood, J. D. 1979. synaptic activation of burst-type myenteric neurons in cat small intestine. *Fed. Proc.* 28:959

Costa, M., Furness, J. B. 1973. The innervation of the internal anal sphincter in the guinea-pig. *Rend. Gastroenterologica* 5:37–38

Costa, M., Furness, J. B. 1976. The peristaltic reflex: An analysis of the nerve pathways and their pharmacology. *Naunyn Schmiedebergs Arch. Pharmacol.* 294:47–60

Costa, M., Furness, J. B. 1979a. Commentary: On the possibility that an indoleamine is a neurotransmitter in the gastrointestinal tract. *Biochem. Pharmacol.* 28:565–71

Costa, M., Furness, J. B. 1979b. The sites of action of 5-HT in nerve muscle preparations from guinea-pig small intestine and colon. *Br. J. Pharmacol.* 65:237–48

Costa, M., Furness, J. B., Gabella, G. 1971. Catecholamine containing nerve cells in the mammalian myenteric plexus. *Histochem.* 25:103–6

Costa, M., Furness, J. B., McLean, J. R. 1976. The presence of aromatic l-amino acid decarboxylase in certain intestinal nerve cells. *Histochemistry* 48:129–43

Coupland, R. E., Holmes, R. L. 1958. Auerbach's plexus in the rabbit. *J. Anat. London* 92:651

Couteaux, R. 1960. Motor endplate structure. In *The Structure and Function of Muscle,* ed. G. H. Bourne 1:337–80. New York: Academic

Cowie, L. A., Kosterlitz, H. W., Waterfield, A. A. 1978. Factors influencing the release of acetylcholine from the myenteric plexus of the ileum of the guinea-pig and rabbit. *Br. J. Pharmacol.* 64:565–80

Dale, H. H. 1937a. Acetylcholine as a chemical transmitter of the effects of nerve impulses. I. History of ideas and evidence. Peripheral autonomic actions. Functional nomenclature of nerve fibers. *J. Mt. Sinai Hosp.* 4:401–15

Dale, H. H. 1937b. Acetylcholine as a chemical transmitter of the effects of nerve impulses. II. Chemical transmission at ganglionic synapses and voluntary motor nerve endings, some general considerations. *J. Mt. Sinai Hosp.* 4:416–29

Daniel, E. E., Daniel, V. P., Duchon, G., Garfield, R. E., Nichols, M., Malhotra, S. K., Oki, M. 1976. Is the nexus necessary for cell-to-cell coupling in smooth muscle? *J. Membr. Biol.* 28:207–39

Dingledine, R., Goldstein, A. 1976. Effect of synaptic transmission blockade on morphine action in the guinea pig myenteric plexus. *J. Pharmacol. Exp. Ther.* 196:97–106

Dingledine, R., Goldstein, A., Kendig, J. 1974. Effects of narcotic opiates and serotonin on the electrical behavior of neurons in the guinea pig myenteric plexus. *Life Sci.* 14:2299–2309

Drakontides, A. B., Gershon, M. D. 1968. 5-HT receptors in the mouse duodenum. *Br. J. Pharmacol.* 33:480–92

Drakontides, A. B., Gershon, M. D. 1972. Studies of the interaction of 5-hydroxytryptamine (5-HT) and the perivascular innervation of the guinea-pig caecum. *Br. J. Pharmacol.* 45:417–34

Dreyfus, C. F., Bornstein, M. B., Gershon, M. D. 1977a. Synthesis of serotonin by neurons of the myenteric plexus in situ and in organotypic tissue culture. *Brain Res.* 128:125–39

Dreyfus, C. F., Sherman, D., Gershon, M. D. 1977b. Uptake of serotonin by intrinsic neurons of the myenteric plexus grown in organotypic tissue culture *Brain Res.* 128:109–23

Dubois, A., Jacobowitz, D. M. 1974. Failure to demonstrate serotonergic neurons in the myenteric plexus of the rat. *Cell Tissue Res.* 150:493–96

Epstein, M. L., Sherman, D. L., Gershon, M. D. 1980. Development of serotonergic neurons in the chick duodenum. *Dev. Biol.* 77:20–40

Eränkö, O., Rechardt, L., Eränkö, L., Cunningham, A. 1970. Light and electron microscopic histochemical observations on cholinesterase-containing sympathetic nerve fibers in the pineal body of the rat. *Histochem. J.* 2:479–89

Erspamer, V. 1966. Occurrence of indolealkylamines in nature. In *Handbook of Experimental Pharmacology.* Vol. 19, *5-Hydroxytryptamine and Related Indoleakylamines,* ed. V. Erspamer, pp. 132–81. New York: Springer

Fahrenkrug, J., Haglund, U., Jodal, M., Lundgren, O., Olbe, L., Schaffalitzky de Muckadell, O. B. 1978. Nervous release of vasoactive intestinal polypeptide in the gastrointestinal tract of cats: Possible physiological implications. *J. Physiol.* 284:291–305

Feher, E. 1974. Effect of monoamine oxidase inhibitor on the nerve elements of the isolated cat ileum. *Acta Morphol. Acad. Sci. Hung.* 22:249–63

Feher, E. 1975. Effects of monoamine inhibitor on the nerve elements of the isolated cat's ileum. *Verh. Anat. Ges.* 69:477–82

Feher, E., Vajda, J. 1972. Cell types in the nerve plexus of the small intestine. *Acta Morph. Acad. Sci. Hung.* 20:13–25

Feldberg, W. 1951. Effects of ganglion-blocking substances on the small intestine. *J. Physiol. London* 113:483–585

Feldberg, W., Lin, R. C. Y. 1949. The effect of cocaine on the acetylcholine output of the intestinal wall. *J. Physiol. London* 109:475–87

Feldberg, W., Toh, C. C. 1953. Distribution of 5-hydroxytryptamine (serotonin, enteramine) in the wall of the digestive tract. *J. Physiol. London* 119:352–62

Finkleman, B. 1930. On the nature of inhibition in the intestine. *J. Physiol. London* 70:145–57

Fischman, D. A., Gershon, M. D. 1964. A method for studying intracellular movement of water-soluble isotopes prior to radioautography. *J. Cell Biol.* 21:139–43

Fox, J., Said, S. I., Daniel, E. E. 1979. Is vasoactive intestinal polypeptide (VIP) an inhibitory neurotransmitter in the lower oesophageal sphincter (LES) in the North American opossum. *Gastroenterology* 76:1134

Franco, R., Costa, M., Furness, J. B. 1979a. Evidence for the release of endogenous substance P from intestinal nerves. *Naunyn Schmiedebergs Arch. Pharmacol.* 306:185–201

Franco, R., Costa, M., Furness, J. B. 1979b. Evidence that axons containing substance P in the guinea-pig ileum are of intrinsic origin. *Naunyn Schmiedebergs Arch. Pharmacol.* 307:57–63

Furness, J. B., Costa, M. 1971. Monoamine oxidase histochemistry of enteric neurons in the guinea-pig. *Histochemie* 28:324–36

Furness, J. B., Costa, M. 1973. The nervous release and the action of substances which affect intestinal muscle through neither adrenoreceptors nor cholinoreceptors. *Philos. Trans. R. Soc. London Ser. B* 265:123–33

Furness, J. B., Costa, M. 1974. The adrenergic innervation of the gastrointestinal tract. *Ergeb. Physiol. Biol. Chem. Exp. Pharmakol.* 69:1–51

Furness, J. B., Costa, M. 1978. Distribution of intrinsic nerve cell bodies and axons which take up aromatic amines and their precursors in the small intestine of the guinea-pig. *Cell Tissue Res.* 188:527–43

Furness, J. B., Costa, M. 1979a. Projections of intestinal neurons showing immunoreactivity for vasoactive intestinal polypeptide are consistent with these neurons being the inhibitory neurons. *Neurosci. Lett.* 15:199–204

Furness, J. B., Costa, M. 1979b. Actions of somatostatin on excitatory and inhibitory nerves in the intestine. *Eur. J. Pharmacol.* 56:69–74

Furness, J. B., Costa, M. 1980. Types of nerves in the enteric nervous system. *Neuroscience* 5:1–20

Furness, J. B., Costa, M., Franco, R., Llewellyn-Smith, J. J. 1980. Neuronal peptides in the intestine: Distribution and possible functions. In *Advances in Biochemical Psychopharmacology Neural Peptides and Neuronal Communications,* ed. E. Costa, M. Trabucchi, 22:608–18. New York: Raven

Furness, J. B., Costa, M., Freeman, C. G. 1979. Absence of tyrosine hydroxylase activity and dopamine B-hydroxylase immunoreactivity in intrinsic nerves of guinea-pig ileum. *Neuroscience* 4:305–10

Fuxe, K., Jonsson, G. 1967. A modification of the histochemical fluorescence method for the improved localization of 5-hydroxytryptamine. *Histochemie* 11:161–66

Gabella, G. 1971. Glial cells in the myenteric plexus. *Z. Naturforsch.* 26B:244–45

Gabella, G. 1972a. Fine structure of the myenteric plexus in the guinea-pig ileum. *J. Anat. London* 111:69–97

Gabella, G. 1972b. Innervation of the intestinal muscular coat. *J. Neurocytol.* 1:341–62

Gabella, G. 1976. Structure of the autonomic nervous system. London: Chapman & Hall

Gabella, G., Blundell, D. 1979. Nexuses between the smooth muscle cells of the guinea-pig ileum. *J. Cell Biol.* 82:239–47

Gabella, G., Juorio, A. V. 1975. Effect of extrinsic denervation on exogenous noradrenaline uptake in the guinea-pig colon. *J. Neurochem.* 25:631–34

Gaddum, J. H., Picarelli, Z. P. 1957. Two kinds of tryptamine receptor. *Br. J. Pharmacol.* 12:323–28

Gamse, R., Lembeck, F., Cuello, A. C. 1979. Substance P in the vagus nerve. Immunochemical and immunohistochemical eivdence for axoplasmic transport. *Naunyn Schmiedebergs Arch Pharmacol.* 306:37–44

Gannon, B. J., Noblett, H. R., Burnstock, G. 1969. Adrenergic innervation of bowel in Hirschsprung's disease. *Brit. Med. J.* 3:338–40

Garrett, J. R., Howard, E. R., Nixon, H. H. 1969. Autonomic nerves in rectum and colon in Hirschsprung's disease. *Arch. Dis. Child.* 44:406–17

Geffen, L. B., Livett, B. G. 1971. Synaptic vesicles in sympathetic neurons. *Physiol. Rev.* 51:98–157

Gerschenfeld, H. M., Paupardin-Tritsch, D. 1974. Ionic mechanisms and receptor properties underlying the responses of molluscan neurones to 5-hydroxytryptamine. *J. Physiol. London* 243:427–56

Gershon, M. D. 1967a. Effects of tetrodotoxin on innervated smooth muscle preparations. *Br. J. Pharmacol.* 29:259–79

Gershon, M. D. 1967b. Inhibition of gastrointestinal movement by sympathetic nerve stimulation: The site of action. *J. Physiol. London* 189:317–29

Gershon, M. D. 1970. The identification of neurotransmitters to smooth muscle. In *Smooth Muscle,* ed. E. Bülbring, A. F. Brading, A. W. Jones, T. Tomita, pp. 496–524. London: Edward Arnold

Gershon, M. D. 1979a. The autonomic nervous system. See Burnstock 1979a, pp. 384–88

Gershon, M. D. 1979b. Putative neurotransmitters: Serotonin. See Burnstock 1979a, pp. 414–24

Gershon, M. D. 1979c. Modulation in the enteric nervous system: A reciprocally inhibitory adrenergic-cholinergic axoaxonic junction in the gut. See Burnstock 1979a, pp. 474–79

Gershon, M. D. 1980. Storage and release of serotonin and serotonin binding protein by serotonergic neurons. *UCLA Forum.* In press

Gershon, M. D., Altman, R. F. 1971. An analysis of the uptake of 5-hydroxytryptamine by the myenteric plexus of the small intestine of the guinea pig. *J. Pharmacol. Exp. Ther.* 179:29–41

Gershon, M. D., Bursztajn, S. 1978. Properties of the enteric nervous system: Limitation of access of intravascular macromolecules to the myenteric plexus and muscularis externa. *J. Comp. Neurol.* 180:467–88

Gershon, M. D., Drakontides, A. B., Ross, L. L. 1965. Serotonin: Synthesis and release from the myenteric plexus of the mouse intestine. *Science* 149:197–99

Gershon, M. D., Dreyfus, C. F. 1980. Stimulation of tryptophan uptake into enteric neurons by 5-hydroxytryptamine: A novel form of neuromodulation. *Brain Res.* 184:229–33

Gershon, M. D., Dreyfus, C. F., Pickel, V. M., Joh, T. H., Reis, D. J. 1977. Serotonergic neurons in the peripheral nervous system: Identification in gut by immunohistochemical localization of tryptophan hydroxylase. *Proc. Natl. Acad. Sci. USA* 74:3086–89

Gershon, M. D., Epstein, M. L. Hegstrand, L. 1980a. Colonization of the chick gut by progenitors of enteric serotonergic neurons: Distribution, differentiation, and maturation within the gut. *Dev. Biol.*, 77:41–51

Gershon, M. D., Jonakait, G. M. 1979. Uptake and release of 5-hydroxytryptamine by enteric serotonergic neurons: Effects of fluoxetine (Lilly 110140) and chlorimipramine. *Br. J. Pharmacol.* 66:7–9

Gershon, M. D., Robinson, R. G., Ross, L. L. 1976. Serotonin accumulation in the guinea pig's myenteric plexus: Ion dependence, structure activity relationship, and the effect of drugs. *J. Pharmacol. Exp. Ther.* 198:548–61

Gershon, M. D., Ross, L. L. 1966a. Radioisotopic studies of the binding, exchange and distribution of 5-hydroxytryptamine synthesized from its radioactive precursor. *J. Physiol. London* 186:451–76

Gershon, M. D., Ross, L. L. 1966b. Localization of sites of 5-hydroxytryptamine storage and metabolism by radioautography. *J. Physiol. London* 186:477–92

Gershon, M. D., Sherman, D., Dreyfus, C. F. 1980b. Effects of indolic neurotoxins on enteric serotonergic neurons. *J. Comp. Neurol.* 190:581–96

Gershon, M. D., Sleisenger, M. 1966. Anaphylaxis and serotonin pools in the mouse gut. *Clin. Res.* 13:286

Gershon, M. D., Tamir, H. 1979. Serotonin (5-HT) release from stimulated peripheral neurons. *Proc. 7th Ann. Meet. Int. Soc. Neurochem., Jerusalem, 1979,* p. 349

Gershon, M. D., Tamir, H. 1980. Serotonin binding protein: Role in transmitter storage in central and peripheral serotonergic neurons. In *Serotonin,* ed. S. Haber, S. Gabay, M. Isidorides, S. Alivesatos. New York: Plenum. In press

Gershon, M. D., Thompson, E. B. 1973. The maturation of neuromuscular function in a multiply innervated structure: Development of the longitudinal smooth muscle of the foetal mammalian gut and its cholinergic excitatory, adrenergic inhibitory, and non-adrenergic inhibitory innervation. *J. Physiol. London* 234:257–77

Gillespie, J. S., Maxwell, J. D. 1971. Adrenergic innervation of sphincteric and nonsphincteric smooth muscle in the rat intestine. *J. Histochem. Cytochem.* 19:676–81

Gillespie, J. S., McKenna, B. R. 1961. The inhibitory action of the sympathetic nerves on the smooth muscle of the rabbit gut, its reversal by reserpine and restoration by catecholamines and by dopa. *J. Physiol. London* 156:17–34

Gintzler, A. R., Rothman, T. P., Gershon, M. D. 1980. Ontogeny of opiate mechanisms in relation to the sequential development of neurons known to be components of the guinea pig's enteric nervous system. *Brain Res.* 189:31–48

Goodrich, J. T., Bernd, P., Sherman, D. L., Gershon, M. D. 1980. Phylogeny of enteric serotonergic neurons. *J. Comp. Neurol.* 190:15–28

Goodrich, J. T., Gershon, M. D. 1977. Serotonergic neurons in the enteric nervous system of human and sub-human primates. *Proc. Int. Union Physiol. Sci. Paris,* 1977, 13:2173

Gordon-Weeks, P. R., Hobbs, M. J. 1979. A non-adrenergic nerve ending containing small granular vesicles in the guinea-pig gut. *Neurosci. Lett.* 12:81–86

Goyal, R. K., Said, S. I., Rattan, S. 1979. Influence of VIP antiserum on lower esophageal sphincter relaxation: Possible evidence for VIP as the inhibitory neurotransmitter. *Gastroenterology* 76:1142

Grafe, P., Mayer, C. J., Wood, J. D. 1979b. Evidence that substance P does not mediate slow synaptic excitation within the myenteric plexus. *Nature London* 279:720–21

Grafe, P., Mayer, C. J., Wood, J. D. 1980. Synaptic modulation of calcium-dependent potassium conductance in myenteric neurons. *J. Physiol. London.* In press

Grafe, P., Wood, J. D., Mayer, C. J. 1979a. Fast excitatory postsynaptic potentials in AH (type 2) neurons of guinea-pig myenteric plexus. *Brain Res.* 163:349–52

Guillemin, R. 1976. Somatostatin inhibits the release of acetylcholine induced electrically in the myenteric plexus. *Endocrinology* 991:1653–54

Gunn, M. 1959. Cell types in the myenteric plexus of the cat. *J. Comp. Neurol.* 111:83–93

Gunn, M. 1968. Histological and histochemical observations on the myenteric and submucosal plexuses of mammals. *J. Anat. London* 102:223–39

Gwee, M. C. E., Yeoh, T. S. 1968. The release of 5-hydroxytryptamine from rabbit small intestine in vitro. *J. Physiol. London* 194:817–25

Hamberger, B., Norberg, K. A. 1965. Studies of some systems of adrenergic synaptic terminals in the abdominal ganglia of the cat. *Acta Physiol. Scand.* 65:235–42

Harry, J. 1963. The action of drugs on the circular muscle strip from the guinea-pig isolated ileum. *Br. J. Pharmacol. Chemother.* 20:399–417

Hightower, N. C. Jr. 1968. Motor action in the small bowel. See Code & Carlson 1968, pp. 2001–24

Hirst, G. D. S., Holman, M. E., Spence, I. 1974. Two types of neurones in the myenteric plexus of duodenum in the guinea-pig. *J. Physiol. London* 236:303–26

Hirst, G. D. S., McKirdy, H. C. 1974. A nervous mechanism for descending inhibition in guinea-pig small intestine. *J. Physiol. London* 238:129–44

Hirst, G. D. S., McKirdy, H. C. 1975. Synaptic potentials recorded from neurones of the submucosa plexus of guinea-pig small intestine. *J. Physiol. London* 249:369–86

Hirst, G. D. S., Silinsky, E. M. 1975. Some effects of 5-hydroxytryptamine, dopamine and noradrenaline on neurones in the submucous plexus of guinea-pig small intestine. *J. Physiol. London* 251:817–32

Hirst, G. D. S., Spence, I. 1973. Calcium action potentials in mammalian peripheral neurones. *Nature London* 243:54–56

Hodgkiss, J. P., Lees, G. M. 1978. Correlated electrophysiological and morphological characteristics of myenteric neurons. *J. Physiol. London* 285:P19–20

Hoffman, H. H., Schnitzlein, H. N. 1969. The number of vagus nerves in man. *Anat. Rec.* 139:429–35

Hökfelt, T. 1971. Ultrastructural localization of intraneuronal monoamines. Some aspects on methodology. In *Progress in Brain Research. Histochemistry of Nervous Transmission,* ed. O. Eränkö, 34:213–22. New York: Elsevier

Hökfelt, T. 1979. Polypeptides: Localization. See Burnstock 1979a, pp. 424–43

Holman, M. E., Hirst, G. D. S., Spence, I. 1972. Preliminary studies of the neurons of Auerbach's plexus using intracellular microelectrodes. *Aust. J. Exp. Biol. Med. Sci.* 50:795–801

Howard, E. R., Garrett, J. R. 1973. The intrinsic myenteric innervation of the hind-gut and accessory muscles of defecation in the cat. *Z. Zellforsch. Mikrosk. Anat.* 136:31–44

Irwin, D. A. 1931. The anatomy of the Auerbach's plexus. *Am. J. Anat.* 49:141–66

Iversen, L. L. 1979. Criteria for establishing a neurotransmitter. See Burnstock 1979a, p. 406

Jacobowitz, D. 1965. Histochemical studies of the autonomic innervation of the gut. *J. Pharmacol. Exp. Ther.* 149:358–64

Jacobs, J. M. 1977. Penetration of systemically injected horseradish peroxidase into ganglia and nerves of the autonomic nervous system. *J. Neurocytol.* 6:607–18

Johnson, S. M., Katayama, Y., North, R. A. 1980a. Slow synaptic potentials in neurones of the myenteric plexus. *J. Physiol. London* 301:505–16

Johnson, S. M., Katayama, Y., North, R. A. 1980b. Multiple actions of 5-hydroxytryptamine on myenteric neurones of the guinea-pig ileum. *J. Physiol. London* 304:459–70

Jonakait, G. M., Gintzler, A. R., Gershon, M. D. 1979a. Isolation of axonal varicosities (autonomic synaptosomes) from the enteric nervous system. *J. Neurochem.* 32:1387–1400

Jonakait, G. M., Tamir, H., Gintzler, A. R., Gershon, M. D. 1979b. Release of [^{3}H] serotonin and its binding protein from enteric neurones. *Brain Res.* 174:55–69

Jonakait, G. M., Tamir, H., Rapport, M. M., Gershon, M. D. 1977. Detection of a soluble serotonin binding protein in the mammalian myenteric plexus and other peripheral sites of serotonin storage. *J. Neurochem.* 28:277–84

Julé, Y. 1980. Nerve-mediated descending inhibition in the proximal colon of the rabbit. *J. Physiol. London.* In press

Julé, Y., Gonella, J. 1972. Modifications de l'activite electrique du colon terminal de lapin par stimulation des fibres nerveuses pelviennes et sympathiques. *J. Physiol. Paris* 64:599–621

Juorio, A. V., Gabella, G. 1974. Noradrenaline in the guinea-pig alimentary canal: Regional distribution and sensitivity to denervation and reserpine. *J. Neurochem.* 221:851–58

Kamijo, K., Hiatt, R. B., Koelle, G. B. 1953. Congenital megacolon. A comparison of the spastic and hypertrophied segments with respect to cholinesterase activities and sensitivities to acetylcholine, DFP, and barium ion. *Gastroenterology* 24:173–85

Katayama, Y., North, R. A. 1978. Does substance P mediate slow synaptic excitation within the myenteric plexus? *Nature* 274:387–88

Katayama, Y., North, R. A. 1980. The action of somatostatin on neurones of the myenteric plexus of the guinea-pig ileum. *J. Physiol. London* 303:315–23

Katayama, Y., North, R. A., Williams, J. T. 1979. The action of substance P on neurons of the myenteric plexus of the guinea-pig small intestine. *Proc. R. Soc. London Ser. B* 206:191–208

Katz, B., Miledi, R. 1967. A study of synaptic transmission in the absence of nerve impulses. *J. Physiol. London* 192:407–36

Kewenter, J. 1965. The vagal control of the jejunal and ileal motility and blood flow. *Acta Physiol. Scand. Suppl.* 251:1–68

Kilbinger, H. 1977. Modulation by oxotremorine and atropine of acetylcholine release evoked by electrical stimulation of the myenteric plexus of the guinea-pig ileum. *Naunyn Schmiedebergs Arch. Pharmacol.* 300:145–51

Kosterlitz, H. W. 1968. Intrinsic and extrinsic nervous control of motility of the stomach and the intestine. See Code & Carlson 1968, pp. 2147–71

Kosterlitz, H. W. 1979. Enkephalins. See Burnstock 1979a, pp. 449–58

Kosterlitz, H. W., Lees, G. M. 1964. Pharmacological analysis of intrinsic intestinal reflexes. *Pharmacol. Rev.* 16:301–39

Kosterlitz, H. W., Lydon, R. J., Watt, A. J. 1970. The effects of adrenaline, noradrenaline and isoprenaline on inhibitory —and β-adrenoceptors in the longitudinal muscle of the guinea-pig ileum. *Br. J. Pharmacol.* 39:398–413

Kuhar, M. J., Aghajanian, G. K., Roth, R. H. 1972. Tryptophan hydroxylase activity and synaptosomal uptake of serotonin in discrete brain regions after midbrain raphe lesions: Correlations with serotonin levels and histochemical fluorescence. *Brain Res.* 44:165–76

Kuntz, A. 1953. *The Autonomic Nervous System.* Philadelphia: Lea & Febiger. 4th ed.

Langley, J. N. 1921. *The Autonomic Nervous System, Pt.* 1. Cambridge: Heffer

Larsson, L. I. 1977. Ultrastructural localization of a new neuronal peptide (VIP). *Histochemistry* 54:173–76

Larsson, L. I., Rehfeld, J. F. 1979. Localization and molecular heterogeneity of cholecystokinin in the central and peripheral nervous system. *Brain Res.* 165:201–18

Lassmann, G. 1962. The demonstration of specific cholinesterase in the nervous formations of the human appendix. *Acta Histochem.* 13:113–22

Leaming, D. B., Cauna, N. 1961. A qualitative and quantitative study of the my-

enteric plexus of the small intestine of the cat. *J. Anat. London* 95:160–69

Llinas, R. R., Steinberg, I. Z. 1977. The place of a calcium hypothesis in synaptic transmission In *Neuroscience Research Program Bulletin.* Vol. 15, *Depolarization-Release Coupling Systems in Neurons,* ed. R. R. Llinas, J. E. Heuser, pp. 565–74. Cambridge: MIT Press

Lord, J. A. H., Waterfield, A. A., Hughes, J., Kosterlitz, H. W. 1977. Endogenous opioid peptides: Multiple agonists and receptors. *Nature* 267:494–99

Loren, I., Alumets, J., Hakanson, R., Sundler, F. 1979. Immunoreactive pancreatic polypeptide (PP) occurs in the central and peripheral nervous system: Preliminary immunocytochemical observations. *Cell Tissue Res.* 200:179–86

Manber, L., Gershon, M. D. 1979. A reciprocal adrenergic-cholinergic axo-axonic synapse in the mammalian gut. *Am. J. Physiol.* 236(6):E738–45

Martinson, J. 1965. Vagal relaxation of the stomach. Experimental re-investigation of the concept of the transmission mechanism. *Acta Physiol. Scand.* 64:453–62

Mayer, C. J., Wood, J. D. 1975. Properties of mechanosensitive neurons within Auerbach's plexus of the small intestine of the cat. *Pflugers Arch.* 357:35–49

Mayerson, H. S., Wolfram, C. G., Shirley, H. H. Jr., Wasserman, K. 1960. Regional differences in capillary permeability. *Am. J. Physiol.* 198:155–60

Meech, R. W., Standen, N. B. 1975. Potassium activation in *Helix aspersa* neurons under voltage lamp: A component mediated by calcium influx. *J. Physiol. London* 249:211–39

Niemi, M., Kouvalainen, K., Hjelt, L. 1961. Cholinesterase and monoamine oxidase in congenital megacolon. *J. Pathol. Bacteriol.* 82:363–66

Nishi, S., North, R. A. 1973a. Intracellular recording from the myenteric plexus of the guinea-pig ileum. *J. Physiol. London* 231:471–91

Nishi, S., North, R. A. 1973b. Presynaptic action of noradrenaline in the myenteric plexus. *J. Physiol. London* 231:24–30P

Norberg, K. A. 1964. Adrenergic innervation of the intestinal wall studied by fluorescence microscopy. *Int. J. Neuropharmacol.* 3:379–82

Norberg, K. A., Sjöqvist, F. 1966. New possibilities for adrenergic modulation of ganglionic transmission. *Pharmacol. Rev.* 18:743–51

North, R. A. 1973. The calcium-dependent slow after-hyperpolarization in myenteric plexus neurones with tetrodotoxin-resistant action potentials. *Br. J. Pharmacol.* 49:709–11

North, R. A., Henderson, G. 1975. Action of morphine on guinea-pig myenteric plexus neurons studied by intracellular recording. *Life Sci.* 17:63–66

North, R. A., Henderson, G., Katayama, Y., Johnson, S. M. 1980. Electrophysiological evidence of presynaptic inhibition of acetylcholine release by 5-hydroxytryptamine in the enteric nervous system. *Neuroscience* 5:581–86

North, R. A., Katayama, Y., Williams, J. T. 1979. On mechanism and site of action of enkephalin on single myenteric neurons. *Brain Res.* 165:67–77

North, R. A., Nishi, S. 1974. Properties of the ganglion cells of the myenteric plexus of the guinea pig ileum determined by intracellular recording. In *Proc. 4th Int. Symp. Gastrointestinal Motility,* pp. 667–76. Vancouver: Mitchell

North, R. A., Nishi, S. 1976. The soma spike in myenteric plexus neurons with a calcium-dependent after-hyperpolarization. In *Physiology of Smooth Muscle,* ed. E. Bülbring, M. F., Shuba, pp. 303–7. New York: Raven

Ohkawa, H., Prosser, C. L. 1972. Electrical activity in myenteric and submucous plexuses of cat intestine. *Am. J. Physiol.* 222:1412–19

Olson, L., Alund, M., Norberg, K. 1976. Fluorescence microspinal demonstration of a population of gastrointestinal nerve fibers with a selective affinity for quinacrine. *Cell Tissue Res.* 171:407–23

Paton, W. D. M. 1957. The action of morphine and related substances on contraction and on acetylcholine output of coaxially stimulated guinea-pig ileum. *Br. J. Pharmacol.* 12:119–27

Paton, W. D. M., Vane, J. R. 1963. An analysis of the response of the isolated stomach to electrical stimulation and to drugs. *J. Physiol. London* 165:10–46

Paton, W. D. M., Vizi, E. S. 1969. The inhibitory action of noradrenaline and adrenaline on acetylcholine output by guinea-pig longitudinal muscle strip. *Br. J. Pharmacol.* 35:10–28

Paton, W. D. M., Vizi, E. S., Zar, M. A. 1971. The mechanism of acetylcholine release from parasympathetic nerves. *J. Physiol. London* 215:819–48

Paton, W. D. M., Zar, M. A. 1968. The origin of acetylcholine released from guinea-pig intestine and longitudinal muscle strips. *J. Physiol. London* 194:13–33

Penninckx, F., Kerremans, R. 1975. Pharmacological characteristics of the gan-glionic and aganglionic colon in Hirschsprung's disease. *Life Sci.* 17:1387–94

Pert, C. B., Snyder, S. H. 1974. High affinity transport of choline into the myenteric plexus of guinea-pig intestine. *J. Pharmacol. Exp. Ther.* 191:102–8

Piper, P. J., Said, S. I., Vane, J. R. 1970. Effects on smooth muscle preparations of unidentified vasodilator peptide from intestine and lung. *Nature* 225:1144–46

Puig, M. M., Gascon, P., Graviso, G. L., Musacchio, J. M. 1977. Endogenous opiate receptor ligand: Electrically induced release in the guinea-pig ileum. *Science* 195:419–20

Rattan, S., Goyal, R. K. 1978. Evidence of 5-HT participation in vagal inhibitory pathway to opossum LES. *Am. J. Physiol.* 234:E273–76

Raviola, E., Karnovsky, M. J. 1972. Evidence for a blood-thymus barrier using electron-opaque tracers. *J. Exp. Med.* 136:466–98

Reese, T. C., Karnovsky, M. J. 1967. Fine structural localization of a blood-brain barrier to exogenous peroxidase. *J. Cell Biol.* 34:207–17

Richardson, K. C. 1964. The fine structure of the albino rabbit iris with special reference to the identification of adrenergic and cholinergic nerves and nerve endings in its intrinsic muscles. *Am. J. Anat.* 114:173–205

Rikimaru, A. 1971. Contractile properties of organ-cultured intestinal smooth muscle. *Tohoku J. Exp. Med.* 105:199–200

Robertson, J. D. 1956. The ultrastructure of a reptilian myoneural junction. *J. Biophys. Biochem. Cytol.* 2:381–94

Robinson, R., Gershon, M. D. 1971. Synthesis and uptake of 5-hydroxytryptamine by the myenteric plexus of the small intestine of the guinea pig. *J. Pharmacol. Exp. Ther.* 179:29–41

Rogawski, M. A., Goodrich, J. T., Gershon, M. D., Touloukian, R. J. 1978. Hirschsprung's disease: Absence of serotonergic neurons in the aganglionic colon. *J. Pediat. Surg.* 13:608–15

Rothman, T. P., Dreyfus, C. F., Gershon, M. D. 1979. Differentiation of enteric neurons from unrecognizable precursors within the microenvironment of cultured fetal mouse gut. *Neurosci. Abstr.* 5:176

Rothman, T. P., Ross, L. L., Gershon, M. D. 1976. Separately developing axonal uptake of 5-hydroxytryptamine and norepinephrine in the fetal ileum of the rabbit. *Brain Res.* 115:437–56

Rubin, W., Gershon, M. D., Ross, L. L. 1971. Electron microscopic radioautographic identification of serotonin-synthesizing cells in the mouse gastric mucosa. *J. Cell Biol.* 50:399–415

Saavedra, J. M., Brownstein, M., Axelrod, J. 1973. A specific and sensitive enzymatic-isotopic microassay for serotonin in tissues. *J. Pharmacol. Exp. Ther.* 186:508–15

Salimova, N. 1978. Localization of biogenic monoamines in *Amphioxus Brachiostoma Lanceolatum. Dokl. Acad. Sci. USSR* 242:939–41

Saller, C. F., Stricker, E. M. 1978. Decreased gastrointestinal motility in rats after parenteral injection of p-chlorophenyl alanine. *J. Pharm. Pharmacol.* 30:646–47

Sato, T., Takayanagi, I., Takagi, K. 1974. Effects of acetylcholine releasing drugs on electrical activities obtained from Auerbach's plexus in the guinea-pig ileum. *Jpn. J. Pharmacol.* 24:447–51

Schaumann, W. 1957. Inhibition by morphine of the release of acetylcholine from the intestine of the guinea-pig. *Br. J. Pharmacol.* 12:115–18

Schaumann, W. 1958. Zusammenhange zwischen der Wirkung der Analgetica und Sympathicomimetica auf den Meerschweinchen Dunndarm. *Naunyn Schmiedebergs Arch. Exp. Pathol. Pharmakol.* 233:112–24

Schofield, G. C. 1968. Anatomy of muscular and neural tissues in the alimentary canal. See Code & Carlson 1968, pp. 1579–1627

Schultzberg, M., Dreyfus, C. F., Gershon, M. D., Hökfelt, T., Elde, R. P., Nilsson, G., Said, S., Goldstein, M. 1978. VIP-, enkephalin-, substance P-, and somatostatin-like immunoreactivity in neurons intrinsic to the intestine: Immunohistochemical evidence from organotypic tissue cultures. *Brain Res.* 155:239–48

Schulz, R., Cartwright, C. 1974a. Effect of morphine on the serotonin release from the myenteric plexus of the guinea pig. *J. Pharmacol. Exp. Ther.* 190:420–30

Schulz, R., Cartwright, C. 1974b. Supersensitivity with PCPA pretreatment on the sensitivity of normal and morphine tolerant muscle strips of the guinea pig ileum. *Fed. Proc.* 33:A502

Schulz, R., Wuster, M., Simantov, R., Snyder, S., Herz, A. 1977. Electrically stimulated release of opiate-like material from the myenteric plexus of the guinea-pig ileum. *Eur. J. Pharmacol.* 41:347–48

Simionescu, N., Simionescu, M., Palade, G. E. 1971. Permeability of intestinal capillaries. Pathways followed by dextrans and glycogens. *J. Cell Biol.* 53:365–92

Simionescu, M., Simionescu, N., Palade, G. E. 1974. Morphometric data on the endothelium of blood capillaries. *J. Cell Biol.* 60:128–52

Sosa, R. P., McKnight, A. T., Hughes, J., Kosterlitz, H. W. 1977. Incorporation of labeled amino acids into the enkephalins. *FEBS Lett.* 84:195–98

Suszkiw, J. B., Pilar, G. 1976. Selective localization of a high affinity choline uptake system and its role in ACh formation in cholinergic nerve terminals. *J. Neurochem.* 26:1133–38

Sutherland, S. D. 1967. The neurons of the gall bladder and gut. *J. Anat. London* 10:701–9

Szerb, J. C. 1975. Endogenous acetylcholine release and labeled acetylcholine formation from [³H]-choline in the myenteric plexus of the guinea-pig ileum. *Can. J. Physiol. Pharmacol.* 53:566–74

Szerb, J. C. 1976. Storage and release of labelled acetylcholine in the myenteric plexus of the guinea-pig ileum. *Can. J. Physiol. Pharmacol.* 54:12–22

Szurszewski, J. H. 1979. Dopamine. See Burnstock 1979a, p. 459

Szurszewski, J. H., Weems, W. A. 1976. A study of peripheral input to and its control by post-ganglionic neurones of the inferior mesenteric ganglion. *J. Physiol. London* 256:541–56

Takayanagi, I., Sato, T., Takagi, K. 1977. Effects of sympathetic nerve stimulation on electrical activity of Auerbach's plexus and intestinal smooth muscle tone. *J. Pharm. Pharmacol.* 29:376–77

Tamir, H., Gershon, M. D. 1979. Storage of serotonin and serotonin binding protein in synaptic vesicles. *J. Neurochem.* 33:35–44

Tamir, H., Kuhar, M. J. 1975. Association of serotonin-binding protein with projections of the midbrain raphe nuclei. *Brain Res.* 83:164–72

Toh, C. C. 1954. Release of 5-hydroxytryptamine (serotonin) from the dog's gastrointestinal tract. *J. Physiol. London* 126:248–54

Trendelenburg, P. 1917. Physiologische und pharmakologische Versuche uber die Dunndarm peristaltick. *Naunyn Schmiedebergs Arch. Exp. Pathol. Pharmakol.* 81:55–129

Uddman, R., Alumets, J., Edvinsson, L., Hakanson, R., Sundler, F. 1978. Peptidergic (VIP) innervation of the esophagus. *Gastroenterology* 75:5–8

Valenzuela, J. E. 1976. Dopamine as a possible neurotransmitter in gastric relaxation. *Gastroenterology* 71:1019–22

Vane, J. R. 1957. A sensitive method for the assay of 5-hydroxytryptamine. *Br. J. Pharmacol. Chemother.* 12:344–49

Vanhoutte, P. M. 1977. Cholinergic inhibition of adrenergic transmission. *Fed. Proc.* 36:2444–49

Vizi, V. A., Vizi, E. S. 1978. Direct evidence for acetylcholine releasing effect of serotonin in the Auerbach plexus. *J. Neural Transm.* 42:127–38

von Euler, U.S. 1951. The nature of adrenergic nerve mediators. *Pharmacol. Rev.* 3:247–77

Waterfield, A. A., Smockum, R. W. J., Hughes, J., Kosterlitz, H. W., Henderson, G. 1977. In vitro pharmacology of the opioid peptides, enkephalins and endorphins. *Eur. J. Pharmacol.* 43:107–16

Watson, A. H. D. 1979. Fluorescent histochemistry of the teleost gut: Evidence for the presence of serotonergic neurones. *Cell Tissue Res.* 197:155–64

Weber, L. J. 1970. p-Chlorophenylalanine depletion of gastrointestinal 5-hydroxytryptamine. *Biochem. Pharmacol.* 19:2169–72

Webster, W. 1974. Aganglionic megacolon in piebald-lethal mice. *Arch. Pathol.* 97:111–17

Welch, A. S., Welch, B. L. 1968. Effect of stress and parachlorophenylalanine upon brain serotonin 5-hydroxyindole acetic acid and catecholamines in grouped and isolated mice. *Biochem. Pharmacol.* 17:699–708

Welsh, J. H., Hyde, J. E. 1944. Acetylcholine content of the myenteric plexus and resistance to anoxia. *Proc. Soc. Exp. Biol. Med.* 55:256–57

Wikberg, J. 1977a. Release of ³H-acetylcholine from isolated guinea pig ileum. A radiochemical method for studying the release of the cholinergic neurotransmitter in the intestine. *Acta Physiol. Scand.* 101:302–17

Wikberg, J. 1977b. Localization of adrenergic receptors in guinea pig ileum and rabbit jejunum to cholinergic neurons and to smooth muscle cells. *Acta Physiol. Scand.* 99:190–207

Williams, J. T., North, R. A. 1979. Vasoactive intestinal polypeptide excites neurones of the myenteric plexus. *Brain Res.* 175:174–77

Wong, W. C. 1975. Degeneration of adrenergic axons in the longitudinal muscle coat of the rat duodenum following treatment with 6-hydroxydopamine. *Experentia* 31:1080–82

Wong, W. C. 1977. Ultrastructural localization of adrenergic nerve terminals in the circular muscle layer and muscularis mucosae of rat duodenum after acute treatment with 6-hydroxydopamine. *J. Anat.* 124:637–42

Wong, W. C., Helme, R. D., Smith, G. C. 1974. Degeneration of noradrenergic nerve terminals in submucous ganglia of rat duodenum following treatment with 6-hydroxydopamine. *Experentia* 30:282–84

Wood, J. D. 1970. Electrical activity from single neurons in Auerbach's plexus. *Am. J. Physiol.* 219:159–69

Wood, J. D. 1972. Excitation of intestinal muscle by atropine, tetrodotoxin and xylocaine. *Am. J. Physiol.* 222:118–25

Wood, J. D. 1973. Electrical discharge of single enteric neurons in guinea-pig small intestine. *Am. J. Physiol.* 225:1107–13

Wood, J. D. 1975a. Neurophysiology of Auerbach's plexus and control of intestinal motility. *Physiol. Rev.* 55:307–24

Wood, J. D. 1975b. Effects of elevated magnesium on discharge of myenteric neurons of cat small bowel. *Am. J. Physiol.* 229:657–62

Wood, J. D. 1979. Neurophysiology of the enteric nervous system. In *Integrative Functions of the Autonomic Nervous System*, ed. C. McBrooks, K. Koizumi, A. Sato, pp. 177–93. Amsterdam: Elsevier/North-Holland

Wood, J. D., Mayer, C. J. 1978a. Intracellular study of electrical activity of Auerbach's plexus in guinea-pig small intestine. *Pfluger's Arch.* 374:265–75

Wood, J. D., Mayer, C. J. 1978b. Slow synaptic excitation mediated by serotonin in Auerbach's plexus. *Nature* 276:836–37

Wood, J. D., Mayer, C. J. 1978c. Electrical activity of myenteric neurons: Comparison of results obtained with intracellular and extracellular methods of recording. In *Gastrointestinal Motility in Health and Disease*, ed. H. Duthie, pp. 311–20. Lancaster, Pa.: MTP Press

Wood, J. D., Mayer, C. J. 1979a. Intracellular study of tonic-type enteric neurons in guinea pig small intestine. *J. Neurophysiol.* 42:569–81

Wood, J. D., Mayer, C. J. 1979b. Serotonergic activation of tonic-type enteric neurons in guinea pig small bowel. *J. Neurophysiol.* 42:582–93

Wood, J. D., Mayer, C. J. 1979c. Adrenergic inhibition of serotonin release from neurons in guinea pig Auerbach's plexus. *J. Neurophysiol.* 42:594–603

Yamamura, H. I., Snyder, S. H. 1974. Muscarinic cholinergic receptor binding in the longitudinal muscle of the guinea-pig ileum with [³H]-quinuclidinyl benzilate. *Mol. Pharmacol.* 10:861–67

Ann. Rev. Neurosci. 1981. 4:273–99

PLASTICITY IN THE
VESTIBULO-OCULAR REFLEX:
A NEW HYPOTHESIS[+]

❖11555

F. A. Miles and S. G. Lisberger

Laboratory of Neurophysiology, National Institute of Mental Health,
Bethesda, Maryland 20205

INTRODUCTION

Eye movement recordings reveal that we do not scan our visual surround-ings like the cinecamera panning across a scene, but rather take in the view in a sequence of still-shots. Visual search is a speculative business and several times a second throughout the waking hours we generate rapid saccadic eye movements that relocate the image of the world on the retina. In this way, images that may prove to be of particular interest are positioned in the fovea—the region of most acute vision—for detailed scrutiny. If this detailed processing of the retinal images is to proceed effectively, then the eyes must be stabilized with respect to the surroundings. Indeed, visual acuity begins to deteriorate appreciably when the retinal images drift at more than a few degrees per second (Westheimer & McKee 1975). Head movements pose a serious potential threat to ocular stability and are dealt with by a reflex that generates compensatory eye movements to offset them: the vestibulo-ocular reflex (VOR).

A striking feature of the VOR is that it operates without the benefit of immediate feedback. The input to the system, head rotation, is sensed by the semicircular canals, whose primary afferent discharges supply the brain-stem with frequency-coded information effectively describing the head's angular velocity. A variety of direct and indirect neural pathways then convey these signals to the motoneurons innervating the external eye mus-cles that generate the ouput, compensatory eye rotation (see Figure 1). Because the output of the reflex does not influence the receptors sensing the input, the VOR is said to operate as an open-loop control system. A major problem shared by all such systems is calibration: How is the performance

[+]The US Government has the right to retain a nonexclusive, royalty-free license in and to any copyright covering this paper.

regulated so that the counter-rotations of the eyes do indeed accurately compensate for the rotations of the head? Although this problem had been clearly enunciated by Rønne in 1923, it was not generally recognized until very recently. The prevailing assumption—implicit if not explicit—was that the reflex was "hard-wired" and presumably genetically specified; the problem of establishing and maintaining appropriate performance levels was not generally appreciated. Gonshor & Melvill Jones (1971) first demonstrated the fallacy of this notion about a decade ago, when they found that the VOR underwent extensive changes when subjects wore left-right, reversing-prism spectacles and even showed reversal under some circumstances (Gonshor & Melvill Jones 1976b). This led to the suggestion that the VOR was subject to long-term regulation by some visually mediated adaptive process whose normal function was to keep the system appropriately calibrated. The seemingly machine-like quality of the VOR engendered the hope that it might eventually prove to be a model system for studying cellular mechanisms underlying memory and learning in the central nervous system.

An adaptive capability has now been demonstrated in the VOR using a range of optical techniques in a variety of animals. However, some species differences concerning the detailed mechanisms are beginning to emerge, and in order to avoid confusion we have structured our review around the

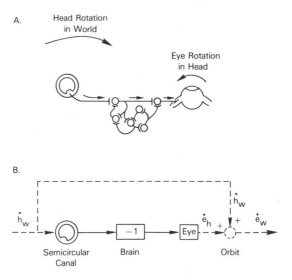

Figure 1   The VOR as an open-loop control system. *A:* Basic neuronal organization (much simplified). *B:* Block diagram; *continuous lines* denote signal flow channels within the nervous system; *discontinuous* lines represent external, physical links denoting head velocity with respect to the world ($\dot{h}_w$), eye velocity with respect to the head ($\dot{e}_h$) and eye velocity with respect to the world ($\dot{e}_w$, gaze velocity).

work on the monkey; in general, only passing reference is made to the work on other animals. As yet, no biophysical or morphological correlates of these adaptive changes have been found in any species and the major concerns remain the general nature of the adaptive mechanism and the site of the modifiable elements. At least insofar as the monkey is concerned, many of the initial ideas about the adaptive mechanism, particularly the role of the cerebellum, do not apply, and we shall propose a new hypothesis.

GENERAL NATURE OF THE VOR

Short-Term Operation

Before proceeding to consider the long-term regulation of the VOR, it is imperative to understand the general workings of the normal reflex. Insofar as the structural organization of the system is concerned, we need consider only the rudiments. Extensive treatments of the neural circuitry and signal processing are available in several excellent reviews (Carpenter 1977, Cohen 1974, Precht 1979, Wilson & Melvill Jones 1979).

MACHINE-LIKE OPERATION   Three notable features of the primate VOR that together give it a machine-like appearance are its high speed, accuracy, and consistency. The combination of a virtually instantaneous inertial receptor organ, short central pathways with few synaptic delays, and an extremely rapid plant (the extraocular muscles and eyeball) produces a total delay of only 12 msec from head movement to compensatory eye movement (F. A. Miles and S. G. Lisberger, in preparation). (Note that a car travelling at 55 mph would manage to travel only 1 ft in that time!) It is usual to assess the performance of the reflex by recording the compensatory eye movements resulting from passive sinusoidal oscillation of the whole animal about the vertical axis in the dark. Over the frequency range of 0.1 to 1.0Hz, the gain of the rhesus monkey's VOR, defined as peak eye velocity divided by peak head velocity, is usually 0.9 to 1.0; with standard deviations typically less than 5% of the response, the system can be said to deserve the description "machine-like" (Miles & Eighmy 1980). In the rhesus monkey, therefore, preservation of a stable retinal image during head turns can be accounted for almost entirely by the operation of the VOR. Other mechanisms—proprioception, preprogramming, and vision—are very minor (Bizzi et al 1972).

VISUAL FEEDBACK: A POOR SUBSTITUTE FOR THE VOR   In the light, visual feedback mechanisms in the form of the *pursuit* and *optokinetic* systems are also available to stabilize the eyes with respect to the surroundings. Operating in the manner of negative-feedback tracking systems, they respond to the slippage of retinal images by generating eye movements that tend to reduce that slip. All animals with mobile eyes appear to have an optokinetic system that responds to wide-field image motion across the whole retina. Since it is responsive to movements of the background images such as those that would occur if the VOR failed to completely stabilize the eyes during head turns, the optokinetic system operates to reduce any residual retinal image slip and therefore can be regarded as a back-up to the VOR; however, the optokinetic system is so slow that it remains a poor substitute for an appropriate VOR. The same is true, though perhaps to a lesser extent, of the pursuit system, which so far has only been found in higher vertebrates. This system is especially responsive to small moving images in the foveal region and allows the animal to track small objects even when they cross a featured background. Success here means overriding any optokinetic "resistance" that the movement of the eyes relative to the stationary background would normally elicit. However, for the present this is of less concern to us than the fact that, when not being used to track moving objects, the pursuit system will be available to act in concert with the optokinetic system to aid stabilization of the eyes with respect to the surroundings. Even when the two systems operate together in this fashion, minimal latencies from slip to onset of tracking eye movement are still in excess of 80 msec (Lanman et al 1978). Thus, extensive dependence upon visual feedback would clearly mean that the animal would have to endure considerable motion of its visual world during fast head turns, which would severely compromise its visual acuity. These limitations of the visual feedback mechanisms obviously make it all the more imperative that the VOR should be appropriately calibrated.

An obvious additional advantage of a nonvisual mechanism such as the VOR for maintaining ocular stability is that it continues to function in low-luminance conditions. Furthermore, the VOR has great utility in situations where the light levels are adequate but the environment lacks the textural features needed for visual stabilization, e.g. the open-water habitat of the pelagic fish. These situations emphasize the feedforward nature of the reflex since its purpose here is obviously not to stabilize nonexistent or featureless images but rather to insure that, if something should come into view, its image will fall upon a reasonably stable retina.

VOR FUNCTIONS TO STABILIZE RETINAL IMAGES BY PRESERVING GAZE VELOCITY   Ordinarily, when we refer to "eye position," we mean "position of the eyes with respect to the head." However, in discussing eye-head coordination it is also important to consider "position of the eyes with respect to the surrounding world," to which we refer as "gaze position." For various reasons that will gradually become apparent, the VOR is most readily dealt with in the velocity domain so that we shall mostly be referring to eye velocity, gaze velocity, etc. Algebra defines gaze velocity (abbreviated to $\dot{e}_w$, to denote eye velocity with respect to the world) as the sum of (a) the velocity of the eyes with respect to the head (eye velocity, $\dot{e}_h$) and (b) the velocity of the head with respect to the world (head velocity, $\dot{h}_w$):

$$\text{Gaze velocity } (\dot{e}_w) = \dot{e}_h + \dot{h}_w. \qquad\qquad 1.$$

The function of the VOR in simple situations where the animal is viewing the stationary surroundings can now be restated: to keep gaze velocity ($\dot{e}_w$) zero by insuring that any changes in head velocity ($\dot{h}_w$) are always offset by converse changes in eye velocity ($\dot{e}_h$). The ultimate aim, of course, is to prevent movements of the head from disturbing the stability of the retinal image of the surroundings. However, the VOR continues to be equally important when the monkey uses its pursuit system to track small moving objects, a situation in which the monkey is now more interested in stabilizing the image of the moving target than the image of the stationary surroundings. The animal's performance in such situations is measured by how well it manages to match its gaze velocity (eye velocity with respect to the world) to the target's velocity (measured with respect to the world), and this is known to be protected by the VOR against any disturbance resulting from movements of the head (Lanman et al 1978). Two systems sharing the common final pathway are at work here (see Figure 2): (a) the negative-feedback pursuit system, which attempts to match gaze velocity ($\dot{e}_w$) to target velocity ($\dot{t}_w$), and (b) the open-loop VOR, which operates to protect the gaze velocity ($\dot{e}_w$) generated by the pursuit system against changes in head velocity ($\dot{h}_w$). The extraordinarily brief delay in the vestibulo-ocular response (about 12 msec) ensures that any disturbance of the head will cause only brief interruptions—largely trivial—in the animal's pursuit of the target.

   Of course, the monkey may choose to pursue moving targets by combining smooth movements of both its eyes and head (Lanman et al 1978), although the latter will be offset by the compensatory eye movements generated by the VOR. Indeed, the VOR might seem to be counter-productive here, since in effect it cancels any contribution that the head movement might otherwise have made to the pursuit. Thus, at first glance, one might think that it would be in the system's interest to reduce the gain of the VOR in these circumstances. However, any such change would mean that unforseen disturbances of the head would be inadaequately compensated and so would severely compromise tracking performance. This is not only undesirable but also contrary to experimental findings (F. A. Miles and S. G.

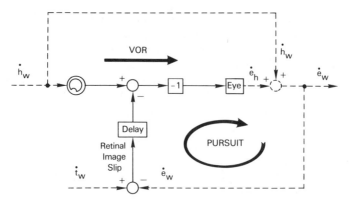

Figure 2   Block diagram of the open-loop VOR and the negative-feedback, pursuit tracking system. *Continuous lines* denote signal flow channels within the nervous system; *discontinuous* lines represent external, physical links denoting head velocity with respect to the world ($\dot{h}_w$), eye velocity with respect to the head ($\dot{e}_h$), eye velocity with respect to the world ($\dot{e}_w$, gaze velocity), and target velocity with respect to the world ($\dot{t}_w$).

Lisberger, in preparation). In the short term, the gain of the monkey's VOR seems to be fixed and immutable: No matter what the ongoing tracking behavior or the manner in which it is being achieved, the VOR is always active.

In sum, the VOR functions always to protect the preexisting gaze velocity —whether zero or otherwise—thereby stabilizing whatever retinal images are of primary interest at the time the head is disturbed. Only animals with a pursuit system have an interest in stabilizing the retinal images of moving targets. The rabbit, for example, does not seem able to track small targets moving across a featured background and is, therefore, thought to lack a pursuit system. Clearly, visual feedback in this animal always operates to stabilize the retinal image of the stationary surroundings and, hence, the preexisting gaze velocity that its VOR functions to protect is always zero.

Long-Term Regulation: Adaptive Gain Control

If for any reason the gain of the VOR deviates appreciably from unity, ocular stability will be lost during head turns and persistent retinal image slip will result. As already pointed out, because of their long latency, visual feedback mechanisms could ameliorate but never totally eradicate this ocular instability—particularly during fast head movements. Clearly, vision would be severely compromised during head movements. Even assuming that the genetic specifications for the VOR were exactly appropriate, it is surely inconceivable that the reflex could remain calibrated over a whole lifetime. If the VOR did not possess a built-in adaptive capability, it would be unacceptably vulnerable to minor diseases, traumas, and aging.

LONG-TERM SENSITIVITY OF VOR TO ALTERED VISUAL INPUT   It is important to realize that the retinal events during head turns provide the system's only truly reliable way of determining not only that a VOR calibration problem exists but also the exact nature of the problem, i.e. whether the gain is too high or too low. Persistent retinal image slip clearly signals the need for a VOR gain adjustment. If the gain were too low (compensatory eye movements fail to fully offset head rotations) then the image of the stationary surroundings would consistently slip in a direction contrary to the head movement, whereas the reverse would be experienced if the gain were too high. The simplest direct test of the system's adaptive gain control capability would be to use optical means to induce such retinal events during head turns and then measure their effects on the gain of the VOR. This has been achieved using magnifying or reducing telescopic spectacles (Miles & Fuller 1974, Gauthier & Robinson 1975, Miles & Eighmy 1980).

If an animal wearing, say, 2 X magnifying telescopic spectacles keeps its head perfectly still, no new demands are placed upon its oculomotor system. A target that appears through the spectacles to be 10° off to one side can be foveated with a 10° saccadic eye movement (although in reality the target is only 5° off to one side); likewise, a target seen moving across the field of view at 10° per sec can be tracked with a matching eye velocity of 10° per sec (although in reality the target is only moving at 5° per sec). Only if the animal moves its head will it face a new challenge: if the head rotates at, say, 10° per sec, the world seen through the spectacles will move at twice that rate. Thus, in order to keep the images on its retina stable during head turns, the animal would have to double the usual velocity of its compensatory eye movements.

With 2 X magnifying telescopic spectacles, the gain of the monkey's VOR (measured in the dark) shows a gradual increase that is roughly exponential and achieves an asymptote about 1.7 times normal in a few days. Conversely, 0.5 X reducing spectacles bring about a gradual reduction in the gain of the VOR to about 0.7 times normal (Miles & Eighmy 1980). Because these changes always operate to improve stability of the retinal image during head turns, the system is said to possess *adaptive gain control.*

It should be appreciated that in order to disturb the visual input associated with head movements in a way that will challenge the adaptive capability of the VOR, the optical device must move with the head and not with the eyes. Magnifying contact lenses (even assuming such an optical arrangement is possible) that move with the eyes would call for a change in the "gain" of the system that translates retinal image eccentricity into a rapid saccadic eye movement, but would not present a new challenge to the VOR: retinal image stability during head turns would still require a VOR gain of unity.

A variety of optical means have been employed in a variety of different animal species to demonstrate adaptability in the VOR; all have in common the characteristic of disrupting the visual input associated with head movements and all result in gradual, adaptive changes in the gain of the reflex.[1] However, it is clear that gain is not the only regulated parameter in the VOR. For example, optical-reversal of vision can elicit highly complex adaptions that affect phase as well as gain (Gonshor & Melvill Jones 1976b, Melvill Jones & Davies 1976, Miles & Eighmy 1980); our understanding of the mechanisms operating in such situations is at present rudimentary.

THE SYSTEM HAS MEMORY   Once an appropriate VOR gain has been achieved, it would clearly be advantageous for the system to retain the gain without needing continual recalibration. A system that is modifiable and has the ability to retain the modified state without reinforcement is said to be *plastic.* That the VOR has "memory" and undergoes changes of a long-term nature is evident from its ability to endure prolonged periods (days) without any "visual reinforcement," e.g. total darkness or head immobilization (Robinson 1976, Miles & Fuller 1974, Miles & Eighmy 1980). The VOR is particularly stable in its normal or low-gain state and shows no significant gain changes during head immobilization at least for periods up to one week. The high-gain state is somewhat more labile, showing some very gradual recovery back toward a more normal gain (unity) in the absence of coordinated visual-vestibular experience; however, some consolidation process seems to be involved here since the recovery becomes less severe and less rapid following longer exposures to the magnifying spectacles.

CHANGES ARE "SEMI-PERMANENT" AND NOT SIMPLY A LEARNED STRATEGY   In the absence of any biophysical or morphological evidence we can only speculate about the mechanisms subserving this adaptive capability of the reflex. However, the long-term nature of the adaptive gain changes in the VOR encourages us to believe that this system contains "modifiable synapses" and that it will eventually yield insights into some of the cellular mechanisms underlying memory and learning. In this regard, it is particularly important to try to establish that the adaptive changes are

[1]Species include humans (Gonshor & Melvill Jones 1971, 1976a, 1976b, Gauthier & Robinson 1975), monkeys (Miles & Fuller 1974, Miles & Eighmy 1980), cats (Robinson 1976, Melvill Jones & Davies 1976, Keller & Precht 1979), rabbits (Collewijn & Grootendorst 1978, 1979, Ito et al 1974, Ito et al 1979a), birds (Green & Wallman 1978) and fish (Schairer & Bennett 1977, 1978).

due to genuine "modifications" in the basic reflex and are not due to some immediate parametric adjustment resulting only from some learned strategy. Such "behavioral" influences clearly would not involve any synaptic modifications in the reflex.

In the monkey, there are several pieces of evidence to suggest that the adaptive gain changes associated with telescopic spectacle experience reflect "semi-permanent" changes in the fundamental reflex (Miles & Eighmy 1980):

1. Neither the rate of adaptation to telescopic spectacles nor the rate of recovery following their removal speeds up with repeated exposures, as might have been expected if the animals had merely been learning some strategy that could be used at any time to adjust the magnitude of their compenstory eye movements. Indeed, the changes in gain usually proceed in an almost machine-like way, the rate being very similar even in different individuals.

2. Standard caloric tests of vestibulo-ocular function in the adapted animals reveal changes in responsiveness that closely parallel the gain changes seen with the usual passive oscillation tests. Given the vastly different contextual information available to the animal in these two test situations, one might have expected very different results if strategy-specific parametric adjustments had been involved.

3. In the head immobilization experiments mentioned earlier to test for retention, several days may have elapsed since the animal last wore the adapting spectacles, yet its VOR gain still retains the adapted state. It seems doubtful that the animal would continue to employ a strategy for which there is no longer any apparent, continuing need.

4. The magnitude of the short-latency compensatory eye movements generated by the VOR in response to sudden, unexpected disturbances of the head is simply a function of the measured gain of the reflex. Even an elevated VOR gain (following adaptation to magnifying spectacles) is unaffected in situations where zero gain might seem preferable, e.g. in tracking a target that moves with the head, requiring that the animal attempt to keep its eyes stationary in its moving head (S. G. Lisberger and F. A. Miles, in preparation). Thus, in the short term, the gain of the VOR is fixed and immutable.

Unfortunately, such tests have not yet been carried out in any species other than monkey and the situation in man, in particular, is far from clear. The magnitude of the compensatory eye movements generated by human subjects in the usual VOR test situation is known to depend upon the instructions given to them (Barr et al 1976). Whatever mechanism is at work here makes it difficult to estimate the gain of the "basic" reflex in man, and renders the interpretation of human adaptation data problematical.

A STORED PATTERN? It has been suggested that in the rabbit, VOR adaptation involves the formation of memory traces representing specific motor patterns that can be emitted when the requisite stimulus conditions recur (Collewijn & Grootendorst 1979). This suggestion arose in part from the finding that adaptation elicited by passive sinusoidal oscillation of the head was relatively specific for the oscillation frequency used and that very occasionally the eyes continued to oscillate at that same frequency even when the animal was stationary in total darkness (Collewijn & Grootendorst 1979). Although the frequency-specificity of adaptation is also apparent in rhesus monkey (Miles & Lisberger 1980), we have often looked for —but never observed—spontaneous oscillations of the eyes following adaptation with passive oscillation in this species (F. A. Miles and S. G. Lisberger, unpublished observations, 1980). This is only one of several apparent differences between the adaptive gain control processes in the monkey and the rabbit, and others will be touched upon later.

ADAPTIVE GAIN CONTROL AS A NORMAL, EVERYDAY FUNCTION There is reason to believe that the adaptive gain changes are achieved through the extrapolation of normal physiological processes and do not represent pathology. In the normal monkey, gain and phase of the VOR are relatively insensitive to changes in stimulus amplitude and frequency (at least over the range of 0.1 to 1.0 Hz); because this continues to be true following adaptation to telescopic spectacles, it seems that the dynamic characteristics of the adapted reflex are relatively normal, which reinforces the idea that we are dealing with pure gain changes and minimal side effects (Miles & Eighmy 1980). In fact, it seems reasonable to assume that the challenge of telescopic spectacles is met by built-in adaptive mechanisms whose normal, everyday function is the long-term regulation of ocular stability: we are not dealing with a laboratory curiosity. It is also clear that the adaptive capability is not just a feature of the developing nervous system and may even be retained throughout life (Gonshor & Melvill Jones 1976b).

SITE OF THE MODIFIABLE ELEMENTS IN THE VOR

If the adaptive gain control mechanism in the VOR is to contribute to our understanding of the cellular mechanisms underlying motor learning in the central nervous system, there is a most urgent need to locate the site of its modifiable elements. Largely as a result of the work and the writings of Ito (1970, 1972, 1974, 1977) interest has centered initially on the vestibular part of the cerebellum.

The Cerebellar Hypothesis

Ito applied the rather general Marr-Albus cerebellar model of motor learning (Marr 1969, Albus 1971) to the specific case of the VOR. Based on his knowledge of the anatomical arrangements in this system in the rabbit, Ito developed the idea of the vestibular cerebellum—or more particularly, the flocculus—as a side-loop of the VOR that might act as the variable gain element in the system. Thus, he envisaged that the cerebellar cortical networks, carrying vestibular signals originating from mossy-fiber inputs, would contain modifiable synapses whose efficacy was under climbing-fiber control. Furthermore, he recognized the utility of the retinal image slip associated with each head turn as a potential error signal that the system might use to guide adaptation, and he suggested that the climbing fibers might, therefore, transmit visual information. The subsequent finding that, in the rabbit at least, the climbing fibers not only received visual inputs (Maekawa & Simpson 1973) but were directionally selective (Simpson & Alley 1974), was nicely consistent with the hypothesis. However, the major evidence that sustained the plausibility of this hypothesis was the effect of cerebellar lesions: in all of the species examined, lesions of the cerebellum that included the vestibular portion resulted in a loss or severe deficit in adaptive gain control in the VOR (Ito et al 1974, Robinson 1976, Schairer & Bennett 1980, Optican et al 1980). However, recent single unit studies in the monkey flocculus argue strongly against this hypothesis in this species and suggest alternative explanations for the lesion deficits.

THE PRIMATE FLOCCULUS: NORMAL PHYSIOLOGY   In order to review this recent evidence satisfactorily it is first necessary to establish the normal physiological role of the flocculus. In the monkey, the flocculus is much more than a side-loop of the VOR. Single unit studies of the activity that originates from mossy-fiber inputs (Lisberger & Fuchs 1978, Miles et al 1980b), reveal that, in addition to carrying vestibular signals encoding angular head velocity (measured with respect to the surroundings), its Purkinje output cells (P-cells) also discharge in relation to eye velocity (measured with respect to the head). In fact, these two components of P-cell discharge are, on the average, of similar strength, have the same directional preference, and sum algebraically to effectively encode gaze velocity (movement of the eyes with respect to the surroundings).

Since the monkey's VOR gain is normally close to unity, its gaze velocity during head turns is close to zero and the P-cells in the flocculus show little if any modulation of their discharge. Thus, these P-cells normally make little or no contribution to the VOR. In fact, appreciable modulation of the activity of these cells is evident only when the animal tracks moving targets; furthermore, being related to gaze velocity, this modulation is relatively independent of whatever combination of eye and head movements the monkey opts to use. Lesions of the primate flocculus result in large deficits in pursuit that are roughly equally severe whether the animal attempts to employ movements of its eyes alone or in combination with head movements (Zee et al 1978); this is consistent with the idea that these P-cells normally function to boost the performance of the smooth tracking system by contributing a gaze velocity signal.

In Figure 3 we have incorporated these gaze-velocity P-cells into the signal flow model of the VOR and pursuit system presented earlier. The head velocity component of P-cell discharge seems to be vestibular in origin and is configured in the model as part of an inhibitory side-loop of the vestibulo-ocular pathway, whereas the eye velocity component is assumed to be related to the motor command to move the eyes (though it may result from afferent input) and is configured as part of a positive-feedback loop. The existence of the latter has long been predicted by theoreticians intent upon modeling the pursuit system (Fender & Nye 1961, Fender 1962, Young et al 1968, Robinson 1971), and positive-feedback is commonly employed by engineers to boost the gain of negative-feedback control systems. However, as discussed earlier, monkeys often accomplish pursuit tasks with combined eye-head movements, substituting part of the eye movement with a head movement. It is in this situation, where the ocular component of pursuit (and hence the eye-velocity, positive-feedback boost through the flocculus) is diminished, that the vestibular (head velocity) component of P-cell discharge plays a crucial role in the process: it makes up for the drop in the eye velocity component. This ensures that the output from the flocculus continues to provide the same gaze velocity signal in support of tracking whether the head is stationary or moving. It will also be apparent from Figure 3 that the vestibular component of P-cell discharge effectively operates to counter-balance the brainstem vestibular drive generated by the head component of the tracking gesture.

In summary, the primate flocculus does not normally contribute significantly to the gain of the VOR but is concerned rather with boosting the gain of the pursuit system; since this boost is in the form of a gaze velocity signal, the pursuit system derives equal benefit no matter what combination of eye and head movement the animal chooses to deploy.

THE PRIMATE FLOCCULUS: VOR ADAPTATION   Although it has become clear that the primate flocculus does not operate in the fashion envisaged by Ito (1972) for the rabbit it might still contain the modifiable elements subserving adaptive gain control in the VOR. In the signal flow diagram in Figure 3, it is evident that the gain of the VOR is actually dependent upon the gains of several elements, designated A, B, C, and D

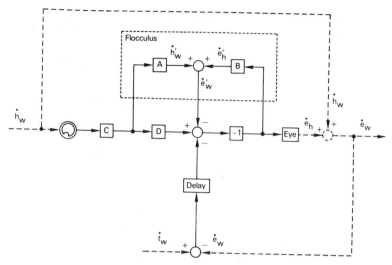

Figure 3 Block diagram of the VOR and pursuit system, incorporating the gaze velocity P-cells in the flocculus. *Continuous lines* denote signal flow channels within the nervous system; *discontinuous lines* represent external, physical links denoting head velocity with respect to the world ($\dot{h}_w$), eye velocity with respect to the head ($\dot{e}_h$), eye velocity with respect to the world ($\dot{e}_w$, gaze velocity), and target velocity with respect to the world ($\dot{t}_w$). *A, B, C,* and *D* represent gain elements; $\dot{h}'_w$ and $\dot{e}'_h$ denote the head velocity and eye velocity inputs, respectively, to the P-cells in the flocculus whose discharge modulations effectively encode gaze velocity ($\dot{e}'_w$).

in the figure, any one or more of which might be "modifiable." The following simple expression describes the gain of the VOR:

$$\text{Gain of VOR} = \frac{-C(D - A)}{1 - B}.$$

2.

An increase in the gain of the VOR might be achieved by increases in C, D, or B, and/or a decrease in A. (Note that the gain of the reflex has a negative value since its output is opposed in direction to its input.)

Single unit recordings in the primate flocculus (Miles et al 1980a) have revealed that the average strength of the head velocity (vestibular) signals carried by the P-cells varies with the gain of the VOR. Adaptation to magnifying spectacles, for instance, was associated with an increase in the average gain of this component, which was very comparable in magnitude to the increase in the gain of the VOR. In Figure 3, the strength of the vestibular signal recorded at the P-cell depends upon the gains of elements A and C. However, Eq. 2 shows that an increase in the gain of element A in the flocculus would operate to decrease the VOR gain; it would be necessary to invoke additional changes outside the flocculus—in element D

—to achieve the observed increase in VOR gain. In that case, element D would be properly regarded as the modifiable element subserving adaptive gain control in the reflex. If, on the other hand, the changes recorded at the P-cell were due to modifications in element C, then the flocculus and the brainstem relay would both receive a vestibular signal of changed gain; such modifications would, therefore, account for the observed increases in the gain of both the VOR and the vestibular component of P-cell discharge. Although the available data do not distinguish between the two possibilities, it is nonetheless clear that the modifiable elements subserving VOR gain changes lie outside the flocculus (elements C and/or D in Figure 3).

It will be recalled that these P-cells normally contribute little to the VOR, their activity being only very weakly modulated during head turns, and unit recordings reveal that this remains true in the adapted animals, at least in the high-gain state. The increase in the magnitude of the compensatory eye movements coupled to head turns in the high-gain animal means that the eye velocity input to the P-cell associated with a given head turn will be proportionately higher than normal; however, this potential threat to the usual balance between head- and eye-velocity signals at the P-cell is offset by the above-mentioned concomitant increase in the P-cells' sensitivity to head velocity. Thus, the increased eye velocity input to the flocculus, which is due directly to the increase in the gain of the VOR, is counter-balanced by a proportional increase in the head velocity input. The net result is that the P-cells continue to show little discharge modulation during head turns in the adapted animal and hence continue to make little contribution to the VOR.

If the observed changes in the P-cells' vestibular signals are not responsible for the change in the gain of the VOR, then it is pertinent to ask what purpose they accomplish. In fact, if such changes did not occur, the flocculus would be less effective in carrying out its important supporting role in combined eye-head tracking. It was pointed out above that the vestibular component of P-cell discharge is important in counter-balancing the vestibular drive generated in the brainstem by the head component of the tracking maneuver. Accordingly, any changes in the "gain" of the brainstem vestibular signal should be matched by corresponding changes in the "gain" of the flocculus vestibular signal: the one should be regulated in harmony with the other. In summary, the changes in the vestibular sensitivity of the P-cells are a secondary consequence, rather than a primary cause, of the change in VOR gain, and these changes enable the flocculus to continue to provide appropriate support for the pursuit system during head movements in the adapted animal.

FLOCCULUS RECORDINGS IN OTHER SPECIES   It has been reported that the P-cells in the rabbit's flocculus change their discharge modulation during oscillation in the dark in association with changes in the gain of the VOR (Dufossé et al 1978a, Ito 1977). Unfortunately, the signal content of the P-cells in the rabbit flocculus has not been determined, hence the etiology of the recorded changes is uncertain. In particular, it is not known if these P-cells are like those in the primate and carry an eye-velocity signal. At the present time, therefore, it is not possible to ascertain the functional significance of these findings.

WHY DO FLOCCULUS LESIONS COMPROMISE ADAPTIVE GAIN CONTROL?   It is pertinent at this point to ask why flocculus lesions, even in the primate, eliminate adaptive gain control in the VOR so effectively if, as we assert above, the single unit recording evidence is against the view that the flocculus of monkeys contains the modifiable elements. One possibility is that the known deficits in the animal's pursuit system undermine its ability to stabilize the retinal images seen through the telescopic spectacles so severely as to effectively disrupt the adaptive mechanism. In this event, the loss of adaptive capability is secondary to an oculomotor insufficiency. A second, and in our view more likely, explanation is that the output from the flocculus is important in the induction of the modifications underlying adaptive gain control in the VOR pathways, i.e. it provides all or part of the error signal guiding recalibration of the reflex (Miles et al 1980a). We shall return to this point in some detail below.

Similar explanations might be advanced to explain the elimination of adaptive gain control in the VOR by olivary lesions that destroy the climbing fiber input to the flocculus (Ito & Miyashita 1975, Haddad et al 1980). It has recently been shown that destruction of the climbing fibers in the rabbit has a dramatic effect on the target P-cells, which lose their customary inhibitory action on the vestibular relay cells in the brainstem (Dufossé et al 1978b, Ito et al 1978, 1979b). Thus, lesions of the olive may be expected to mimic lesions of the flocculus.

The Brainstem Hypothesis

In the monkey, all of the pertinent evidence suggests that the modifiable elements underlying gain changes in the VOR are not in the flocculus (Miles et al 1980a). This, and the fact that VOR gain changes are evident in even the shortest-latency vestibulo-ocular responses that follow sudden disturbances of the head (S. G. Lisberger and F. A. Miles, in preparation), focuses new attention on the brainstem and suggests that the modifiable elements might even be found in the direct VOR pathways having only two or three synapses. Furthermore, if the adaptation is specific to the VOR, then one might expect that in general the neural changes would occur at an early stage in the pathway, before the points at which other signals related to the generation of other kinds of eye movements are introduced. It would seem potentially most disruptive for the modifiable elements to be located in those parts of the pathway that are shared with other systems. We assume, after all, that these other systems did not come under any adaptive pressure. However, compelling though these arguments may seem, recent evidence suggests that some nonvestibular oculomotor control signals are routed through these modifiable gain elements (Lisberger et al 1980). Indeed, the effect of VOR gain changes on various quantitative measures of oculomotor performance has revealed some new aspects of the central organization of the oculomotor system.

CONVERGENCE OF OCULOMOTOR CONTROL SIGNALS: "BEHAV-IORAL ANATOMY"   Population studies examining the vestibular sensitivity of semicircular canal primary afferents and of cells in the medial vestibular nuclei that receive inputs from these afferents have failed to reveal any evidence for significant neural changes associated with four-fold changes in the gain of the VOR (Miles & Braitman 1980, Lisberger & Miles 1980). This implies that the modifiable elements must lie beyond the first central synapse in the VOR pathway. However, it is generally assumed that the optokinetic and vestibulo-ocular pathways converge at the level of the second-order neurons, and discharges related to optokinetic stimulation have been extensively documented in the very cells of the medial vestibular nuclei that were unaffected by VOR adaptation (Waespe & Henn 1977). It follows that the signals responsible for all or part of the optokinetic response should share the modifiable element, and that changes in VOR gain should be accompanied by parallel changes in the gain of the optokinetic responses.

A complication here is that, as already mentioned in an earlier section, the primate has two visual stabilization mechanisms: the pursuit and optokinetic systems. Both are thought to be operative in the usual optokinetic test situation, when it is customary to record the eye movements elicited by a continuously-moving visual field, achieved by surrounding the animal with a rotating, striped cylinder. The resulting optokinetic nystagmus consists of smooth, tracking eye movements in the direction of the seen movement, interrupted at intervals by fast, resetting saccades in the opposite direction. Smooth eye velocity shows an initial rapid rise that lasts for a few hundred milliseconds, followed by a much more gradual increase that may take 10 to 15 sec to reach asymptote (Cohen et al 1977, Raphan et al 1979). It is thought that the rapid changes are due mainly to the pursuit system, whereas the subsequent slow changes are due to build-up in a velocity storage mechanism that represents the optokinetic system proper (Zee et al 1976, Lisberger et al 1980). If the lights are extinguished after this build-up has commenced, then the nystagmus persists as optokinetic afternystagmus (OKAN), the initial intensity of which is assumed to be a direct index of the "optokinetic" component of the prior response (Raphan et al 1979, Lisberger et al 1980).

Changes in the gain of the VOR have been shown to result in dramatic and almost proportional changes in the intensity of the slow "optokinetic" component as indicated by the initial OKAN velocity, but not in the fast "pursuit" response (Lisberger et al 1980). That the changes in the "optokinetic" component so nearly mirrored those in the gain of the VOR is consistent with the view that they are a secondary consequence of the latter and that the modifiable elements are for the most part located in the later stages of the vestibular pathway that are shared with the optokinetic system. The failure to see changes in the "pursuit" component of the optokinetic response suggests that the signals driving it do not access the final oculomotor pathways through the modifiable elements.

These new data have been incorporated into our signal flow diagram in Figure 4. In keeping with the arguments made in the previous section, the putative modifiable elements are located at C and D in the brainstem. The optokinetic signal must converge on the vestibulo-ocular pathway above the level of these elements and the pursuit signals at some later stage. However, since flocculus lesions have a devastating effect upon the gain of pursuit (Zee et al 1978) and the "pursuit" component of the optokinetic response (Optican et al 1980), these eye movements must be mediated by inputs that converge upon the VOR pathways before the eye velocity signal is fed back to the flocculus.

OTHER SPECIES   Data from other species present a somewhat different picture. Keller & Precht (1979) have reported significant changes in the vestibular sensitivity of cells in the cat medial vestibular nuclei in association with changes in the gain of the reflex. At present it is not clear if this apparent difference between cat and monkey reflects a genuine species difference or some methodological inconsistency. Furthermore, although changes have been reported in the optokinetic responses of the rabbit in association with adaptive changes in the gain of the VOR (Collewijn & Kleinschmidt 1975, Collewijn & Grootendorst 1979), they appear to have a very different etiology from those in the monkey. Whereas the changes in the primate optokinetic responses were viewed merely as a secondary consequence of the variable gain element being in the later, shared portion of the optokinetic and vestibular pathways, the changes in the rabbit's optokinetic responses were viewed as independent of the gain changes in the

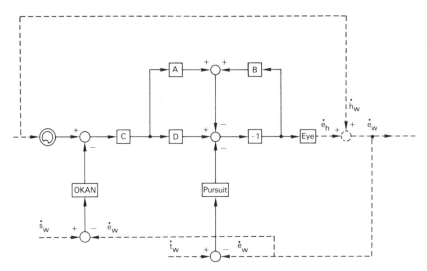

Figure 4   Block diagram of the VOR together with the "pursuit" and "optokinetic" visual feedback systems. *Continuous lines* denote signal flow channels within the nervous system; *discontinuous lines* represent external, physical links denoting head velocity with respect to the world ($\dot{h}_w$), eye velocity with respect to the head ($\dot{e}_h$), eye velocity with respect to the world ($\dot{e}_w$, gaze velocity), target velocity with respect to the world ($\dot{t}_w$), and velocity of the visual surround with respect to the world ($\dot{s}_w$).

vestibular system. Thus, in the rabbit, regardless of whether the gain of the VOR increased or decreased, the optokinetic system always showed a significant increase in gain. Since the optokinetic system is a negative-feedback mechanism, such increases in gain would always operate to improve the system's performance and hence could be regarded as "adaptive." It is difficult to generate a convincing teleological argument to justify the need for such "adaptive improvements," since it is not clear why such a system would not be calibrated optimally (i.e. maximally) at the outset of the experiment. However, before accepting these data as yet further documentation of species differences, it should be noted that, in the primate studies, optokinetic responses were assessed using prolonged unidirectional stimulation, whereas in the rabbit studies, sinusoidal stimuli were used. A further complication is that when occasionally both testing techniques (prolonged unidirectional and sinusoidal) were used in the rabbit studies, they gave contrary answers (Collewijn & Grootendorst 1979).

NATURE OF THE VOR GAIN ERROR SIGNAL

Accurate calibration of the VOR requires reliable "error" information about its performance. As already pointed out, any shortcomings in the gain of the VOR will result in persistent retinal slip during head turns. Of course, the system cannot diagnose gain errors from retinal slip alone and must relate it in some way to the ongoing head movement. The directional relationship between the slip and the head movement reliably indicates the direction in which the VOR gain needs to change to achieve better image stabilization, and most models of the adaptive mechanism derive their error information from a comparison of visual and vestibular inputs (Ito 1972, Robinson 1976, Gonshor & Melvill Jones 1976b). The ultimate dependency on visual and vestibular inputs is clearly obligate since they alone provide information to the central nervous system that relates directly to the external reference—the surrounding world. However, this need not mean that the system necessarily uses visual and vestibular inputs per se to guide calibration: any reliable central correlates of the same might serve equally well.

Substitutes for Vestibular (Head Velocity) Signals?

In an animal like the rabbit that never tracks moving objects, smooth eye velocities are used solely to compensate for head movements and operate always to stabilize the eyes with respect to the stationary surroundings. In such a system, eye velocity signals could substitute for head velocity (vestibular) signals to achieve adaptive VOR gain changes: When the retinal slip is in the same direction as the eye movement, it would indicate that the eye velocity was not large enough, a situation that could be remedied by increasing the gain of the VOR; conversely, if slip and eye velocity are in opposing directions, it would indicate the need for a decrease in the gain of the VOR. There is evidence that, indeed, this is the modus operandi in the rabbit, i.e. the regulated variable is an eye velocity signal rather than a head velocity signal. With optical situations calling merely for an increase or decrease in the gain of the VOR (e.g. magnifying or reducing telescopic spectacles), this mechanism would operate appropriately and be indistinguishable from one deriving error information from retinal slip and head velocity signals. However, optical situations producing left-right mirror reversal of the visual input have proved to be most revealing. When the rabbit was passively oscillated, Collewijn & Grootendorst (1979) observed that for the first hour or so the gain of the VOR decreased but, thereafter, showed a gradual and consistent increase. Upon examining the original eye movement records, they observed that for the first hour or so the animal's compensatory eye movements in the light were attenuated and normally directed but, thereafter, showed reversal. Thus, although the directional relationship between retinal slip and head movement remained unchanged throughout, the directional relationship between retinal slip and eye movement underwent a reversal that coincided with the reversal in the direction of the VOR gain change. When slip and eye movements were in opposing directions, the VOR gain decreased; when the two were in the same direction, it increased. Further evidence that the velocity of the eyes rather than of the head was the relevant parameter was the demonstration that the gain of the rabbit's VOR could be increased merely by prolonged exposure to sinusoidal optokinetic stimulation with the head fixed in position, i.e. head movements are not even necessary (Collewijn & Grootendorst 1979).

Prolonged optokinetic stimulation is also known to increase the gain of the VOR in fish (Schairer & Bennett 1978) but is ineffective in the monkey (Miles & Lisberger 1980). Furthermore, passive oscillation of the monkey in left-right reversing prisms results in a decrease in VOR gain even though the compensatory eye movements show consistent "reversal" in the light (Miles & Eighmy 1980). Thus, the adaptive mechanism in the monkey responds to the reversed-vision situation in the same way that it responds to reducing spectacles; that both optical situations share the same kind of relationship insofar as slip and head velocity are concerned, while differing in regard to slip and eye velocity, indicates once more that eye movements per se are irrelevant. In the monkey, therefore, all of the available evidence is consistent with the view that the system uses some direct measure of head velocity to deduce the errors in the gain.

Substitutes for Retinal Slip Signals?

Despite our contention that the flocculus is probably not the site of the modifications underlying changes in VOR gain, lesion data belie the conclusion that the primate flocculus has no role in adaptation. Indeed, lesions of the flocculus have a far more pronounced effect on the adaptive capability of the primate VOR (Optican et al 1980) than was reported in the original studies in the cat that are so widely cited in favor of the cerebellar hypothesis (Robinson 1976). This leads to the proposition that the primate flocculus might play an inductive role in producing changes in VOR gain. At this point it becomes necessary to ask whether the adaptive gain control mechanism uses retinal slip signals per se to guide recalibration or substitutes some reliable central correlate. In particular, could a gaze velocity signal such as that encoded by the P-cells in the primate flocculus provide such a correlate and so function to regulate VOR gain?

When suddenly confronted with, say, magnifying telescopic spectacles, the normal monkey must at first use its pursuit system during head turns to supplement its now inadequate VOR in the attempt to stabilize the shifting images seen through the spectacles. Thus, the challenge of telescopic spectacles is initially met by the pursuit system, an important part of which is the gaze velocity signal contributed by the flocculus P-cells (Zee et al 1978, Lisberger & Fuchs 1978, Miles et al 1980b). So long as the VOR gain is inappropriate for the new visual conditions, retinal slip will persist during head turns and with it reliance upon the pursuit system and the modulation of the flocculus P-cells' discharge. Indeed, the extent to which pursuit is consistently used during head turns is itself an index of the "error" in the VOR, and the associated discharge modulation of the gaze-velocity P-cells might be used by the adaptive mechanism as a substitute for frank visual signals.

An underlying assumption here is that so long as the animal must employ pursuit during head turns, then its VOR gain must be in need of adjustment. In some as yet undefined way, the resultant modulation of P-cell discharge, if persistent and associated with vestibular stimulation, is assumed to induce appropriate changes in the modifiable gain element. It will be recalled that the discharges of the gaze-velocity P-cells modulate consistently during head turns only when the VOR is inappropriate for the visual conditions and cease when adaptation is complete (Miles et al 1980a). Such behavior is entirely consistent with the view that these cells signal VOR gain error.

The idea that the modulation of gaze velocity P-cell discharge during head movements may be sufficient to induce VOR gain changes receives some indirect support from recent experiments involving only foveal pursuit targets (Miles & Lisberger 1980). Water-deprived monkeys were reinforced with fluids for fixating a small dim target presented against otherwise totally dark surroundings. Matters were arranged so that during the periods of fixation, the monkey and the target were moved either in-phase or 180° out of phase to mimic the gaze velocity events, and presumably also the P-cell discharge modulation, normally associated with wearing magnifying or reducing telescopic spectacles. This resulted in "adaptive" changes in the gain of the VOR equal to as much as 50% of the changes produced by an equivalent amount of experience with telescopic spectacles. These findings are consistent with—though hardly a pure test of—the hypothesis that the error signal guiding recalibration of the VOR is in part the modulation of gaze-velocity P-cell discharge; it is certainly clear, however, that appreciable gain changes can be observed in the absence of peripheral retinal slip.

Nonetheless, some influence of visual inputs clearly seems to be indicated and the "error" in the VOR would seem to be a combination of retinal slip and gaze velocity signals. In fact, it is possible that the flocculus P-cells provide both signals, especially since there is evidence that some of these cells discharge in relation to motion of the visual surround (Noda & Suzuki 1978, Warabi et al 1979); this would also help to explain the extreme severity of the deficits in adaptation following flocculus lesions. All of the available evidence concerning the role of the flocculus in the adaptive control of the primate VOR, including single unit recordings, lesion data, and even the above-mentioned long-term effects of pursuit on the reflex, are consistent with the notion that the flocculus provides the error signal that regulates the gain.

Adaptive Gain Control in the Primate VOR: A New Hypothesis

We have proposed that the modifiable elements mediating adaptive gain control in the primate VOR are (*a*) located in the brainstem pathway and (*b*) regulated by error signals generated at least in part by the flocculus P-cells. It remains to link these two ideas into a plausible single hypothesis.

In our scheme (Figure 5), the gaze-velocity P-cells have been assigned two quite separate functions with but one purpose: to improve the stability of the retinal image. As a part of the pursuit system, their discharge modulation has an immediate influence upon their target neurons through conventional (inhibitory) synaptic mechanisms. As a part of the system regulating VOR gain, their persistent discharge modulation during head turns indicates errors in the VOR and has a long-term influence on the efficacy of synaptic transmission in the vestibulo-ocular pathway.

An important feature of our hypothesis is that the P-cells exert their two different influences at two different points in the vestibular pathway. Since the gain of the pursuit system is very sensitive to lesions of the flocculus but not to VOR gain changes, we assume that the P-cells' short-term synaptic action in support of pursuit is exerted down-stream of their long-term modulatory action. In fact, in our scheme, the P-cells' long-term influence must be exerted above the point from which the P-cells derive their own vestibular input. This is assumed to be at C in Figure 5, though parallel changes at A and D, which would be operationally equivalent, albeit more complex, cannot be ruled out. It is vital for the functioning of our model that there be reciprocal connections between the P-cells and the vestibular relay cells in the brainstem, and that they be organized in essence as an internal negative-feedback loop that is concerned with long-term regulation rather than with short-term compensation. Thus, we envisage a local feedback loop operating (very slowly) to achieve zero modulation of P-cell discharge during head turns: persistent modulations of P-cell discharge ("errors") induce changes in the gain of the vestibular input that in turn operate to reduce those modulations. The single unit data show that the output of the P-cells is indeed regulated in this manner and produces minimal discharge modulation during head turns in the fully adapted state. The problem now is to explain how a control loop that is entirely internal and functions to regulate P-cell discharge can at the same time effectively achieve an appropriate VOR gain.

In the example that follows it will be seen that the crucial factor is the eye-velocity input to the P-cell: this signal is assumed to retain a reliable,

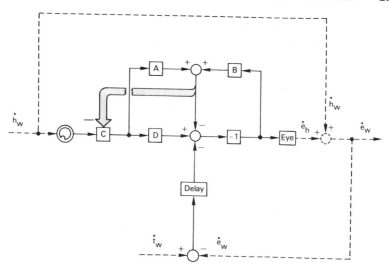

Figure 5 Block diagram of the VOR and the pursuit system, indicating the hypothesized long-term, regulatory influence of the gaze-velocity P-cells (broad stippled arrow) on the postulated modifiable gain elements in the vestibular pathway (element *C*). *Continuous lines* denote signal flow channels within the nervous system; *discontinuous lines* represent external, physical links denoting head velocity with respect to the world ($\dot{h}_w$), eye velocity with respect to the head ($\dot{e}_h$), eye velocity with respect to the world ($\dot{e}_w$, gaze velocity), and target velocity with respect to the world ($\dot{t}_w$).

constant relationship to eye velocity (something that could be achieved by an internal negative-feedback network within the cerebellar cortex) and to provide the internal standard against which everything else must be calibrated.

Consider the chain of events set in motion when the normal monkey first wears magnifying, telescopic spectacles. Initially, there is considerable retinal slip during head turns and the pursuit system takes on the immediate responsibility of augmenting the compensatory eye movements generated by the now-inadequate VOR. In the flocculus, this registers as an elevated eye velocity signal input to the P-cell, whose discharge as a consequence now modulates during head turns. In effect, this eye velocity signal is the P-cells' external reference and, together with any visual inputs that they may also receive (not indicated in Figure 5), it provides the P-cells' only link (albeit indirect, via the pursuit system) with the external visual world. If the monkey continues to wear the spectacles, the persistent modulation of P-cell discharge due to the elevated eye velocity signal input is assumed to induce a gradual increase in the gain of element C. This would result in (*a*) an increase in the gain of the VOR, which would gradually transfer the burden of generating larger compensatory eye movements from the pursuit system to the VOR, and (*b*) an increase in the head velocity signal input to the flocculus P-cells, which would operate to offset the augmented eye velocity signal and, hence, gradually reduce the P-cells' discharge modulation. We assume that this process would continue until eventually the P-cells are receiving head-velocity inputs that once more balance their eye velocity inputs during head turns in the light. Note that at all times during the adaptation, P-cell discharge modulation will be minimal during head turns in the dark. When VOR gain becomes appropriate for the visual conditions, head turns in the light will be associated with the same compensatory eye velocities as head turns in the dark, and modulation of P-cell discharge will be minimal. At this point, we have come full circle: the responsibility for generating the now-larger compensatory eye movements has shifted from the pursuit system back to the vestibular, and the error signal (P-cell discharge modulation) has returned to zero.

SUMMARY

The vestibulo-ocular reflex functions to prevent head movements from disturbing retinal images by generating compensatory eye movements to offset the head movements. In the monkey—the species mainly under consideration here—this reflex is machine-like and very effective. In the short-term, the VOR operates as an open-loop control system without the benefit of feedback and its performance is fixed and immutable: No matter what pattern of eye-head coordination the animal uses to view external objects, there is a continuing need for the VOR and it continues to operate; however, should the VOR consistently fail to stabilize the retinal images during head turns, it will gradually undergo long-term adaptive gain changes that restore that stability. This adaptive capability is ultimately dependent upon vision, and a variety of optical devices that disturb the visual input normally associated with head turns have been used to induce large changes in the reflex. Insofar as the monkey is concerned, all of the available evidence suggests to us that the modifiable elements underlying these long-term adjustments are located in the brainstem vestibular pathways and not, as previously suggested by others, in the floccular lobes of the cerebellum. However, the flocculus does appear to have an important, inductive role in the adaptive process, providing at least part of the error signal guiding the long-term adjustments in the brainstem. In our view, the VOR is a particularly well-defined example of a plastic system and promises to be a most useful model for studying the cellular mechanisms underlying memory and learning in the central nervous system.

Literature Cited

Albus, J. S. 1971. A theory of cerebellar function. *Math. Biosci.* 10:25–61

Barr, C. C., Schultheis, L. W., Robinson, D. A. 1976. Voluntary, non-visual control of the human vestibulo-ocular reflex. *Acta Oto Laryngol.* 81:365–75

Bizzi, E., Kalil, R. E., Morasso, P., Tagliasco, V. 1972. Central programming and peripheral feedback during eye-head coordination in monkeys. In *Cerebral Control of Eye Movements and Motion Perception,* ed. J. Dichgans, E. Bizzi, pp. 220–32. Basel: Karger. 403 pp.

Carpenter, R. H. S. 1977. *Movements of the Eyes.* London: Pion. 420 pp.

Cohen, B. 1974. The vestibulo-ocular reflex arc. In *Handbook of Sensory Physiology, Pt. 1: Vestibular System,* ed. H. H. Kornhuber, 6:477–540. New York: Springer-Verlag

Cohen, B., Matsuo, V., Raphan, T. 1977. Quantitative analysis of the velocity characteristics of optokinetic nystagmus and optokinetic after-nystagmus. *J. Physiol. London* 270:321–44

Collewijn, H., Grootendorst, A. F. 1978. Adaptation of the rabbit's vestibulo-ocular reflex to modified visual input: Importance of stimulus conditions. *Arch. Ital. Biol.* 116:273–80

Collewijn, H., Grootendorst, A. F. 1979. Adaptation of optokinetic and vestibulo-ocular reflexes to modified visual input in the rabbit. In *Reflex Control of Posture and Movement. Progress in Brain Research,* ed. R. Granit, O. Pompeiano, 50:772–81. Amsterdam. Elsevier

Collewijn, H., Kleinschmidt, H. J. 1975. Vestibulo-ocular and optokinetic reactions in the rabbit: Changes during 24 hours of normal and abnormal interaction. In *Basic Mechanisms of Ocular Motility and Their Clinical Implications,* ed. G. Lennerstrand, P. Bach-y-Rita, pp. 477–83. New York. Pergamon. 584 pp.

Dufossé, M., Ito, M., Jastreboff, P. J., Miyashita, Y. 1978a. A neuronal correlate in rabbit's cerebellum to adaptive modification of the vestibulo-ocular reflex. *Brain Res.* 150:611–16

Dufossé, M., Ito., M., Miyashita, Y. 1978b. Diminution and reversal of eye movements induced by local stimulation of rabbit cerebellar flocculus after partial destruction of the inferior olive. *Exp. Brain Res.* 33:139–41

Fender, D. H. 1962. The eye-movement control system: Evolution of a model. In *Neural Theory and Modeling,* ed. R. F.

Reiss, pp. 306–24. Stanford: Stanford Univ. Press. 427 pp.

Fender, D. H., Nye, P. W. 1961. An investigation of the mechanisms of eye movement control. *Kybernetic* 1:81–88

Gauthier, G. M., Robinson, D. A. 1975. Adaptation of the human vestibulo-ocular reflex to magnifying lenses. *Brain Res.* 92:331–35

Gonshor, A., Melvill Jones, G. 1971. Plasticity in the adult human vestibulo-ocular reflex arc. *Proc. Can. Fed. Biol. Soc.* 14:11

Gonshor, A., Melvill Jones, G. 1976a. Short-term adaptive changes in the human vestibulo-ocular reflex arc. *J. Physiol. London* 256:361–79

Gonshor, A., Melvill Jones, G. 1976b. Extreme vestibulo-ocular adaptation induced by prolonged optical reversal of vision. *J. Physiol. London* 256:381–414

Green, A. E., Wallman, J. 1978. Rapid change in gain of vestibulo-ocular reflex in chickens. *Neurosci. Abstr.* 4:163

Haddad, G. M., Demer, J. L., Robinson, D. A. 1980. The effect of lesions of the dorsal cap of the inferior olive on the vestibulo-ocular and optokinetic systems of the cat. *Brain Res.* 185:265–75

Ito, M. 1970. Neurophysiological aspects of the cerebellar motor control system. *Int. J. Neurol.* 7:162–76

Ito, M. 1972. Neural design of the cerebellar motor control system. *Brain Res.* 40:81–84

Ito, M. 1974. The control mechanisms of cerebellar control systems. In *The Neurosciences Third Study Program,* ed. F. O. Schmitt, F. G. Worden, pp. 293–303. Boston: MIT Press

Ito, M. 1977. Neuronal events in the cerebellar flocculus associated with an adaptive modification of the vestibulo-ocular reflex of the rabbit. In *Control of Brain Stem Neurons, Developments in Neuroscience,* ed. R. Baker, A. Berthoz, 1:391–98. Amsterdam: Elsevier. 514 pp.

Ito, M., Jastreboff, P. J., Miyashita, Y. 1979a. Adaptive modification of the rabbit's horizontal vestibulo-ocular reflex during sustained vestibular and optokinetic stimulation. *Exp. Brain Res.* 37:17–30

Ito, M., Miyashita, Y. 1975. The effects of chronic destruction of the inferior olive upon visual modification of the horizontal vestibulo-ocular reflex of rabbits. *Proc. Jpn. Acad.* 50:716–20

Ito, M., Nisimaru, N., Shibuki, K. 1979b. Destruction of inferior olive induces rapid depression in synaptic action of cerebellar Purkinje cells. *Nature* 277:568–69

Ito, M., Orlov, I., Shimoyama, I. 1978. Reduction of the cerebellar stimulus effect of rat deiters neurons after chemical destruction of the inferior olive. *Exp. Brain Res.* 33:143–45

Ito, M., Shiida, T., Yagi, N., Yamamoto, M. 1974. The cerebellar modification of rabbit's horizontal vestibulo-ocular reflex induced by sustained head rotation combined with visual stimulation. *Proc. Jpn. Acad.* 50:85–89

Keller, E. L., Precht, W. 1979. Adaptive modification of central vestibular neurons in response to visual stimulation through reversing prisms. *J. Neurophysiol.* 42:896–911

Lanman, J., Bizzi, E., Allum, J. 1978. The coordination of head and eye movement during smooth pursuit. *Brain Res.* 153:39–53

Lisberger, S. G., Fuchs, A. F. 1978. Role of primate flocculus during rapid behavioral modification of vestibuloocular reflex. I. Purkinje cell activity during visually guided horizontal smooth-pursuit eye movements and passive head rotation. *J. Neurophysiol.* 41:733–63

Lisberger, S. G., Miles, F. A. 1980. Role of primate medial vestibular nucleus in long-term adaptive plasticity of vestibuloocular reflex. *J. Neurophysiol.* 43:1725–45

Lisberger, S. G., Miles, F. A., Optican, L. M., Eighmy, B. B. 1980. Effect of long-term changes in the vestibulo-ocular reflex (VOR) on optokinetic responses (OKR) in monkey. *Neurosci. Abstr.* 6: In press

Maekawa, K., Simpson, J. I. 1973. Climbing fiber responses evoked in vestibulocerebellum of rabbit from visual system. *J. Neurophysiol.* 36:649–66

Marr, D. 1969. A theory of cerebellar cortex. *J. Physiol. London* 202:437–70

Melvill Jones, G., Davies, P. 1976. Adaptation of cat vestibulo-ocular reflex to 200 days of optically reversed vision. *Brain Res.* 103:551–54

Miles, F. A., Braitman, D. J. 1980. Long-term adaptive changes in primate vestibuloocular reflex. II. Electrophysiological observations on semicircular canal primary afferents. *J. Neurophysiol.* 43:1426–36

Miles, F. A., Braitman, D. J., Dow, B. M. 1980a. Long-term adaptive changes in primate vestibuloocular reflex. IV. Electrophysiological observations in flocculus of adapted monkeys. *J. Neurophysiol.* 43:1477–93

Miles, F. A., Eighmy, B. B. 1980. Long-term adaptive changes in primate vestibuloocular reflex. I. Behavioral observations. *J. Neurophysiol.* 43:1406–25

Miles, F. A., Fuller, J. H. 1974. Adaptive plasticity in the vestibulo-ocular responses of the rhesus monkey. *Brain Res.* 80:512–16

Miles, F. A., Fuller, J. H., Braitman, D. J., Dow, B. M. 1980b. Long-term adaptive changes in primate vestibuloocular reflex. III. Electrophysiological observations in flocculus of normal monkeys. *J. Neurophysiol.* 43:1437–76

Miles, F. A., Lisberger, S. G. 1980. The "error signal" subserving long-term adaptive gain control in the vestibulo-ocular reflex (VOR) of monkey. *Neurosci. Abstr.* 6:In press

Noda, H., Suzuki, D. A. 1978. Purkinje cell activity in the monkey flocculus during smooth pursuit eye movement. *Neurosci. Abstr.* 4:167

Optican, L. M., Zee, D. S., Miles, F. A., Lisberger, S. G. 1980. Oculomotor deficits in monkeys with floccular lesions. *Neurosci. Abstr.* 6:In press

Precht, W. 1979. Vestibular mechanisms. *Ann. Rev. Neurosci.* 2:265–89

Raphan, T., Matsuo, V., Cohen, B. 1979. Velocity storage in the vestibulo-ocular reflex arc (VOR). *Exp. Brain Res.* 35:229–48

Robinson, D. A. 1971. Models of oculomotor neural organization. In *The Control of Eye Movements,* ed. P. Bach-y-Rita, C. C. Collins. pp. 519–38. New York: Academic. 560 pp.

Robinson, D. A. 1976. Adaptive gain control of vestibuloocular reflex by the cerebellum. *J. Neurophysiol.* 39:954–69

Rønne, H. 1923. False movements appearing during vision through spectacle glasses; their significance with respect to experience in wearing spectacles and their connection with the vestibular apparatus. *Acta Ophthalmol.* 1:55–62

Schairer, J. O., Bennett, M. V. L. 1977. Adaptive gain control in vestibulo-ocular reflex of goldfish. *Neurosci. Abstr.* 3:157

Schairer, J. O., Bennett, M. V. L. 1978. VOR gain changes produced by target rotation without head movement in goldfish. *Neurosci. Abstr.* 4:167

Schairer, J. O., Bennett, M. V. L. 1980. Cerebellectomy in goldfish prevents adaptive gain control of the VOR without affecting the optokinetic system. In *Neuroscience Satellite Symposium on Vestibular Function and Morphology,* ed. T.

Gualtierotti. New York: Springer-Verlag. In press

Simpson, J. I., Alley, K. E. 1974. Visual climbing fiber input to rabbit vestibulo-cerebellum: A source of direction-specific information. *Brain Res.* 82: 302–8

Waespe, W., Henn, V. 1977. Neuronal activity in the vestibular nuclei of the alert monkey during vestibular and optokinetic stimulation. *Exp. Brain Res.* 27:523–38

Warabi, T., Noda, H., Ishii, N. 1979. Effect of retinal image motion upon flocculus Purkinje cell activity during smooth pursuit eye movements. *Neurosci. Abstr.* 5:108

Westheimer, G., McKee, S. P. 1975. Visual acuity in the presence of retinal image

motion. *J. Opt. Soc. Am.* 65:847–50

Wilson, V. J., Melvill Jones, G. 1979. *Mammalian Vestibular Physiology.* New York: Plenum. 365 pp.

Young, L. R., Forster, J. D., Van Houtte, N. 1968. A revised stochastic sampled data model for eye tracking movements. *Proc. 4th Annu. NASA-Univ. Conf. Manual Control, NASA* SP-192:489–509

Zee, D. S., Yamazaki, A., Gucer, G. 1978. Ocular motor abnormalities in trained monkeys with floccular lesions. *Neurosci. Abstr.* 4:168

Zee, D. S., Yee, R. D., Robinson, D. A. 1976. Optokinetic responses in labyrinthine-defective human beings. *Brain Res.* 113:423–28

Ann. Rev. Neurosci. 1981. 4:301–50

EVOLUTION OF THE TELENCEPHALON IN NONMAMMALS

♦11556

R. Glenn Northcutt

The University of Michigan, Division of Biological Sciences,
Ann Arbor, Michigan 48109

INTRODUCTION

Comparative studies of the vertebrate telencephalon began in the late eighteenth and early nineteenth centuries with descriptions of gross morphology (Cuvier 1809, Owen 1866); however, not until the late nineteenth and early twentieth centuries was the internal anatomy of the telencephalon described for a wide variety of vertebrates (Johnston 1906, Edinger 1908, Ramón y Cajal 1908, Papez 1929, Ariëns Kappers et al 1936). This period of intensive study yielded a number of hypotheses regarding the evolution of the vertebrate telencephalon. These hypotheses were based on the anatomy revealed by existing methods—methods that allow what is now referred to as descriptive anatomy—and this anatomy could not be confirmed experimentally because the appropriate experimental techniques did not yet exist. In addition, these hypotheses reflected anatomical assumptions grounded in *scala naturae,* which held that vertebrates form one linear series and reflect increasing complexity.

The relatively sophisticated armamentarium of neurobiological techniques available today allows us to establish more accurately the anatomy of the telencephalon; these data, data from the fossil record, and a more sophisticated view of vertebrate phylogeny allow us to propose and test new hypotheses regarding the evolution of the vertebrate telencephalon.

VERTEBRATE PHYLOGENY

More recent interpretations of the general morphology of living vertebrates and their fossil record (Romer 1966, Schaeffer 1969, Olson 1971, Hotton

301

0147-006X/81/0301-0301$01.00

1976) invalidate the assumption that vertebrates represent one linear series of ever increasing complexity. More accurately, four distinct radiations, separated from a common ancestor for at least 400 million years, have evolved at different rates, and each radiation (clade) has produced new and more complex groups (Figure 1).

One radiation, the agnathans or jawless fishes, is represented today by lampreys and hagfishes, collectively termed the cyclostomes. These fishes possess neither jaws nor paired fins; they are small in size and are restricted in their prey strategies and locomotor efficiency. The living cyclostomes are believed to be evolved from early ostracoderms, having lost external dermal armor, elongated the trunk, and reorganized head morphology for semiparasitic feeding habits. Thus, living cyclostomes are greatly modified from ancestral ostracoderms, represent a separate radiation paralleling the other vertebrates, and cannot be viewed as an ancestral stock from which other vertebrates evolved.

Gnathostomes, or jawed vertebrates, occur slightly later in the fossil record than do agnathans and are assumed to have evolved from some group of early agnathans (Romer 1966, Hotton 1976). The early gnathostomes rapidly evolved into three distinct groups: placoderms, chondrichthians, and osteichthians (Figure 1). All three groups possessed jaws, which increased feeding efficiency as well as offering a wider range of prey, and paired fins, which increased locomotive stability and maneuverability. These developments mark the evolution of new and more active vertebrate predators.

There are no living placoderms, which are thought to have been replaced by chondrichthians (chimaeras, sharks, skates, and rays). The chondrichthians are primarily marine, whereas the osteichthians (bony fishes) are both marine and freshwater. Both groups are long-separated radiations within which new species have evolved to represent distinct and more complex grades of organization. For example, the chondrichthian skates and rays arose at approximately the same time in vertebrate phylogeny as the osteichthian teleosts, birds, and mammals. Thus, each group represents the most recent grade of organization to arise within parallel radiations, and neither can be considered ancestral to another.

The bony fishes are frequently interpreted as comprising three major groups: actinopterygians (ray-finned fishes), dipnoans (lungfishes), and crossopterygians (lobe-finned fishes). The ray-finned fishes comprise most of the living fishes (approximately 30,000 species), whereas there are only three genera of lungfishes and a single extant crossopterygian (*Latimeria*).

Crossopterygians are thought to have given rise to amphibians early in the Devonian period; thus the evolutionary histories of amphibians and bony fishes are almost equally long (Figure 1). Living bony fishes cannot

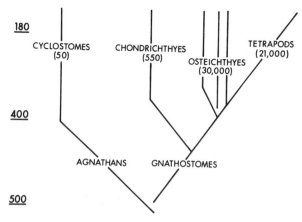

Figure 1 Phylogeny of vertebrates illustrating time of origin of major radiations in millions of years (ordinate) and approximate number of living species (in parentheses). (Reproduced from Northcutt 1980.)

be considered simpler than living amphibians, but both radiations probably retain some ancestral features inherited from their common ancestor.

One group of early amphibians (possibly anthracosaurs) gave rise to the first reptiles some 200 million years ago. The origin of reptiles marks the emergence of truly terrestrial vertebrates, an event made possible by the evolution of the amniotic egg, which eliminated an aquatic larval stage. Almost immediately after their origin, reptiles split into two major groups: sauropsid reptiles, which led to all modern reptiles and birds; and theropsid reptiles, which led through a series of now extinct intermediate forms to modern mammals. Again, the evolutionary history of modern reptiles is as long as that of the theropsid radiation leading to mammals; modern reptiles cannot be viewed as representing a more primitive grade of tetrapodal organization than mammals.

EARLIER THEORIES OF TELENCEPHALIC EVOLUTION

When vertebrate evolution was viewed as one linear series of increasing complexity, this resulted, not surprisingly, in theories that the vertebrate telencephalon evolved in unilinear complexity (Papez 1929, Ariëns Kappers et al 1936, Romer 1970). It was assumed that the telencephalon of the earliest vertebrates was a single vesicle that later formed paired hemispheres in response to the origin of paired olfactory organs (Papez 1929, Ariëns Kappers et al 1936). This theory is based on the following two lines of evidence:

1. The "brain" vesicle of amphioxus represents such a simple condition.
2. A single vesicle stage occurs early in the embryonic development of the telencephalon of all living vertebrates; thus it must represent the primitive adult condition.

From such a simple beginning, it was assumed that the telencephalon underwent an increase in size and complexity, induced first by the origin of olfactory organs, and continuing through fishes, amphibians, reptiles, and mammals by the addition of new thalamic pathways (Figure 2). Although most texts report a gradual increase in telencephalic size through succeedingly complex vertebrate classes (Papez 1929, Ariëns Kappers et al 1936, Romer 1970, Jerison 1973), there is not a single quantitative study to support this supposition. Data presented later in this review, in fact, clearly refute it.

A correlative linear increase in complexity has been said to characterize both the telencephalic roof (pallium) and striatum (Figure 2). The pallium of fishes is supposedly dominated by secondary and tertiary olfactory projections and thus believed to represent a paleopallium (Ariëns Kappers 1921, Papez 1929, Ariëns Kappers et al 1936, Romer 1970). Some fishes and amphibians are said to have evolved an additional olfactory pallium, the archipallium, which supposedly receives higher order olfactory input; however, what selective forces may have produced further differentiation of the

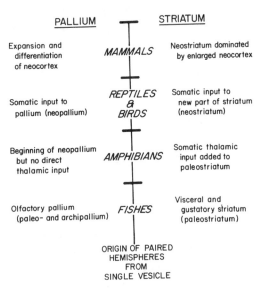

Figure 2 Linear model of telencephalic evolution summarizing hypotheses of Ariëns Kappers et al (1936).

paleopallium are never suggested. Reptiles are said to be the first vertebrates to possess a third pallial component, the neopallium, that receives direct thalamic input. The origin of visual and somatic inputs into the neopallium is said to have enlarged this structure in mammals, in which the neopallium differentiates into a laminated cortex (neocortex).

Similarly, the striatum of fishes supposedly consists of a paleostriatum receiving olfactory and gustatory inputs, with ascending somatic inputs first occurring in amphibians (Ariëns Kappers et al 1936). Reptiles are said to have evolved the first thalamo-striatal projections, which resulted in new differentiation of the striatum (neostriatum); it greatly enlarged in birds, accounting for most of the telencephalon. Mammals are said to be characterized by more conservative neostriatal development, with the neocortex enlarging instead and subjugating the neostriatum.

The last fifteen years have seen renewed interest in comparative neurobiology, and a vast number of new experimental neuroanatomical, histochemical, and physiological data have accumulated. Not only do we possess new experimental information on vertebrate telencephalic organization, but perhaps more importantly, the methods by which comparative biologists analyze data have changed. The very framework of modern comparative studies would itself force reinterpretations of the older data. The added proliferation of new data, produced by more precise and revealing techniques, force us to reject every single one of the tenets proposed in a linear theory of telencephalic evolution.

These newer methods of comparison are summarized briefly, telencephalic size differences among nonmammalian vertebrates are reviewed, and the telencephalic organization within each major vertebrate radiation is discussed. These comparisons then form the basis for a new interpretation of telencephalic evolution.

CONCEPTS OF MORPHOLOGICAL COMPARISONS

In a *scala naturae* view of phylogeny, the origin of a particular character —any definable, variable attribute of an organism that is the product of the genome and the interaction of the environment—is simply defined by its first appearance. Thus, if the neostriatum is first identified in reptiles (Figure 2), it must have originated with reptiles; however, the realization that vertebrates comprise a series of parallel radiations, each evolving at different rates, requires a very different comparative strategy. The characters must be defined and their variation (character states) documented. Characters held in common by most or all members of a major group underlie the adaptive zone of that group. ["Adaptive zone" is used here to mean a way of life (Simpson 1953), not just a place where life is led.] For example, all mammals possess a telencephalon characterized by a large laminated pallium defined as neocortex. This character, in conjunction with many other mammalian characters (endothermy, hair, differentiated teeth, a particular jaw articulation, etc), allows mammals to utilize the environment in a unique way and thus reduce competition with other major vertebrate groups.

Similar characters or character states frequently occur in more than one major vertebrate radiation. When this happens, the morphologist must discern which of several evolutionary events has occurred. The characters held in common may have been inherited from a common ancestor (homology), or they may have evolved independently (homoplasy). Homologous characters are defined as two or more characters, in two or more populations of organisms, that are believed to have arisen from a single character in the common ancestral population (Mayr 1969). Such homologous characters are inherited, with modification, from the common ancestor; the criteria for recognizing suspected homologues is the multitude and degree of similarities between these characters (Simpson 1961). The rationale is that two or more characters possessing a continuous history and traceable to a common character ancestrally must be read out of genomes that are almost identical and, thus, should show similarities that are greater than chance.

On the other hand, homoplastic characters are characters that appear to be similar but have evolved independently (Mayr 1969). Two types of homoplasy are particularly common: parallelism and convergence. Parallelism involves the independent evolution of similar characters in organisms having relatively recent common ancestry. Thus, characters not present in a common ancestor may appear independently and be based on similar parts of the genome in closely related lineages, if some members of each lineage encounter similar selective pressures. Convergence involves the independent evolution of similar characters in organisms possessing distant common ancestry; thus the similar characters are believed to be based on very different portions of the genomes.

In comparing similar characters, the major problem faced by a morphologist is distinguishing homologous characters from homoplastic characters due to parallelism, as both types of characters appear highly similar due to similarities in the genomes. There is no single criterion for recognizing homologous from homoplastic characters; however, the chances of discovering homoplastic characters increase with the number of characters examined due to the incompatability of multiple character sets in which the characters cannot be linked together without the independent origin of at least one character (Kluge 1977).

Similarly, comparison of character variation within and between major groups can reveal homoplastic characters. An examination of many character states reveals that they are oriented, in the sense that some states are more primitive than others. Thus, character states can be arranged into clines in which a determined primitive character state leads to one or more derived states. A character state is considered primitive when it is the state most widely distributed among similar groups, and among dissimilar subgroups within the group being studied. Of necessity, it represents the condition from which other homologues must have evolved (Kluge 1977). If similar derived states occur in more than one radiation, they are by definition cases of homoplasy. For example, examination of the pallium in land vertebrates and cartilaginous fishes might reveal that the primitive character state is a simple undifferentiated pallium, but some members of both radiations might exhibit an expanded, highly differentiated pallium. In this case, the derived character states must be considered homoplastic.

It is clear from these examples that comparisons of characters necessitates the examination of many species of a radiation if the polarity of clines, primitive vs derived, is to be established, and if we hope to distinguish homologous from homoplastic characters. Most comparative neuroanatomical studies possess a strong flavor of establishing the origin of a structure, leading to a history of structure x from fish to man. There is a far more important reason, however, for recognizing homologous and homoplastic characters. Once we recognize that two characters are homologous, it is possible to list how they differ; these differences are a measure of evolutionary change and adaptation. Therefore, the recognition of homologous characters, and how they differ, gives us direct insight into adaptation. Similarly, recognition of homoplastic characters reveals that similar solutions to biological problems have occurred independently. Thus, when we recognize homoplastic characters, we can ask if the organisms are coping with the same set of biological problems.

FOREBRAIN: BODY DATA

Most comparative studies of telencephalic organization suggest that the telencephalon increases in size with succeeding vertebrate classes; actually, there is little quantitative date regarding telencephalic size among vertebrates. Figure 3 illustrates a minimum convex polygon treatment (Jerison 1973) of the forebrain data reported by Ebbesson & Northcutt (1976) for 26 vertebrate species, as well as more recent data on agnathans (R. G. Northcutt, unpublished observations). These data reveal that each major group of vertebrates exhibits considerable variation (approximately two to

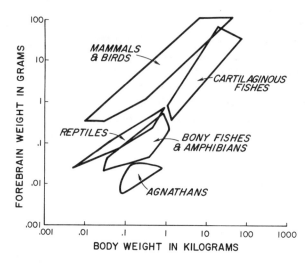

Figure 3   Forebrain and body weights for major vertebrate groups expressed as minimum convex polygons.

six-fold). Agnathans possess smaller forebrains than do other vertebrates; however, hagfishes approach bony fishes and amphibians in their forebrain development. Hagfishes possess forebrains clearly four times larger than lampreys of comparable body size and must be considered the advanced members of the agnathan radiation.

Amphibians and bony fishes possess forebrains of comparable size, and both classes exhibit comparable variation in forebrain size. Reptiles generally possess forebrains that are one to two times larger than amphibians and bony fishes of comparable body size.

Mammals and birds possess the largest forebrains for a given body size, some five to twenty-fold larger than most anamniotes; however, cartilaginous fishes appear to have independently evolved large forebrains. Telencephalic data on cartilaginous fishes (Northcutt 1978a) reveal that primitive sharks and skates possess telencephalic volumes comparable to those of amphibians and fishes, whereas advanced sharks and skates reveal a four to fifteen-fold increase in telencephalic volumes, which overlap those of birds and mammals. Figure 3 includes few avian data, and more extensive plots will probably reveal greater overlap of cartilaginous fishes with birds and mammals.

Although the present data base is very small, it does indicate that increases in forebrain size have occurred among some members of every vertebrate radiation. Agnathans possess the smallest forebrains among living vertebrates, but hagfishes appear to have independently evolved forebrain sizes comparable to those of amphibians and bony fishes. Similar forebrain increases characterize advanced sharks and skates as well as all amniotes.

TELENCEPHALIC VARIATION

Telencephalic organization and variation in the different vertebrate radiations are reviewed with particular attention to possible homologous and homoplastic characters and their distribution.

Agnathans

Reviews of telencephalic structure in lampreys (Nieuwenhuys 1977) and hagfishes (Bone 1963) indicate that the telencephalon consists of olfactory bulbs, cerebral hemispheres, and a telencephalon medium. The telencephalon of hagfishes differs from that of lampreys in that the forebrain ventricular system is reduced to a remnant of the preoptic recess, whereas that of lampreys is well developed and extends into the olfactory bulbs.

In lampreys, the olfactory bulbs are as large as the cerebral hemispheres, and together they constitute paired evaginations of the rostral half of the telencephalic vesicle. The caudal half of the vesicle is not involved in the evagination and constitutes the telencephalon medium.

The olfactory bulbs of both lampreys and hagfishes consist of several concentric layers. Superficially, the olfactory nerve fibers terminate in a layer of glomeruli located immediately beneath the entering olfactory fibers. The glomeruli receive the dendrites of more centrally located mitral cells and several cell types located in the deepest cellular layer of the bulb, the granular layer. Axons of many of the granular cells leave the bulb and form secondary olfactory efferents along with the mitral cell axons.

Four roof (pallial) cell groups and three floor (subpallial) cell groups are recognized in lampreys (Figure 4A). The pallium consists of medial (primordium hippocampi), submedial (subhippocampal lobe), dorsal, and lateral (primordium piriforme) cell groups. The subpallium consists of a preoptic nucleus, which runs the rostrocaudal length of the telencephalic floor, overlain by a rostral septal mass that is replaced more caudally by a corpus striatum extending into the telencephalon medium.

The cerebral hemispheres of hagfishes (Figure 4B) are greatly expanded, solid masses of neural tissue characterized by marked cellular migration, including distinct cellular lamination. The hemispheres form embryonically by evagination, but the ventricular system is secondarily reduced so there is no distinct boundary with the telencephalon medium or, for that matter, with the diencephalon. The hemispheres are divided into a lateral series of cellular and fibrous laminae, a mediodorsal primordium hippocampi, and a medioventral area basalis.

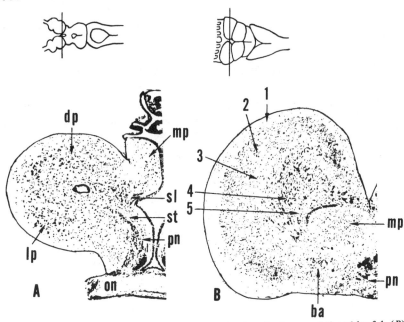

Figure 4   Transverse sections through the telencephalon of a lamprey (*A*) and hagfish (*B*) illustrating relative differentiation of the various cell groups. *Ba,* area basalis; *dp,* dorsal pallium; *lp,* lateral pallium; *mp,* medial pallium; *on,* optic nerve; *pn,* preoptic nucleus; *sl,* submedial lobe; *st,* striatum; *1–5,* pallial laminae.

At present, there are no experimental neuroanatomical studies of the telencephalon in either lampreys or hagfishes. Earlier descriptive studies (Jansen 1930, Heier 1948) claim that secondary olfactory pathways arise in the olfactory bulb and distribute to all parts of the ipsilateral telencephalon. Secondary olfactory efferents also reportedly decussate through the lamina terminalis and terminate in the contralateral hemisphere and hypothalamus.

Earlier studies suggest that all telencephalic centers receive fibers from the olfactory bulb and that these centers, in turn, primarily project to the diencephalon, particularly the epithalamus and hypothalamus; however, olfactory dominance must be viewed with scepticism until the olfactory projections have been experimentally determined, as a similar claim for other fishes is now known to be invalid. Similarly, the actual number of functionally distinct cell groups and their boundaries and connections await experimental determination.

Heier (1948) also claimed to have found ascending dorsal thalamic and hypothalamic connections with the caudal pallium of lampreys. In support of this claim, visual and somatic slow wave activity has been evoked in the telencephalon of lampreys (Bruckmoser 1973).

An understanding of agnathan telencephalic organization is critical to determining what telencephalic characters were present in the earliest vertebrates, as agnathans are the only sister radiation paralleling jawed vertebrates. Many of the characters held in common by agnathans and gnathostomes are probably characters inherited from the earliest vertebrates. An understanding of telencephalic organization in hagfishes is likely to be particularly instructive, as these fishes appear to have independently evolved pallial lamination and may possess well-developed thalamic connections.

Cartilaginous Fishes

These fishes comprise two separate groups: the holocephalons (chimaeras) and the elasmobranchs (sharks, skates, and rays). The telencephalon of chimaeras, or ratfishes, differs from those of elasmobranchs in that the olfactory bulbs arise from the rostral pole of the telencephalon, rather than far laterally as in most elasmobranchs. The chimaeras also lack pallial formations bridging the two telencephalic hemispheres (Figure 5), formations which characterize the telencephalon of all living elasmobranchs.

There is no agreement concerning the number of cellular groups comprising the telencephalon of chimaeras, or their homologues with telencephalic cell groups in other vertebrates (Northcutt 1978a). Unfortunately, there is no experimental information on the telencephalon of chimaeras, and such information is critical to our understanding of the evolution of the telencephalon of cartilaginous fishes, as chimaeras represent a sister radiation paralleling the elasmobranchs.

Northcutt (1978a) and Ebbesson (1980) have reviewed earlier literature on the telencephalon of elasmobranchs. In these fishes the telencephalon consists of olfactory bulbs, paired evaginated cerebral hemispheres, and a caudal telencephalon medium. Nieuwenhuys (1967) reviewed the histological organization of the olfactory bulbs in elasmobranchs and recognized the following four distinct bulbar layers: (a) an outer layer of olfactory nerve fibers, (b) a glomerular layer formed by the dendrites of deeper mitral and large triangular cells, (c) a mitral layer formed by the cell bodies of the mitral cells and by efferent secondary olfactory fibers, and (d) a deep granular cell layer. The granular cells, like those of agnathans, possess axons and form part of the secondary olfactory efferents.

Until recently, all parts of the elasmobranch telencephalon were believed to receive secondary olfactory fibers (Bäckström 1924, Ariëns Kappers et al 1936). New experimental studies, however, reveal the olfactory projections to be as restricted as in land vertebrates (Ebbesson & Heimer 1970, Bruckmoser & Dieringer 1973, Bodznick & Northcutt 1979).

The cerebral hemispheres of elasmobranchs consist of a pallium and

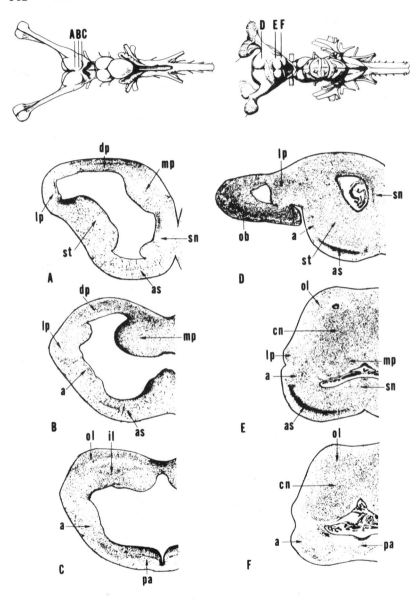

Figure 5 Transverse sections through comparable levels of the telencephalon in a squalomorph shark, *Notorynchus* (*A–C*), and a galeomorph shark, *Mustelus* (*D–F*). *A,* area a; *as,* area superficialis basalis; *cn,* central nucleus; *dp,* dorsal pallium; *il,* inner cellular lamina of dorsal pallium; *lp,* lateral pallium; *mp,* medial pallium; *ob,* olfactory bulb; *ol,* outer cellular lamina of dorsal pallium; *pa,* preoptic area; *sn,* septal nuclei; *st,* striatum.

subpallium, each of which is subdivided into major formations. The pallium consists of lateral, dorsal, and medial formations (Figure 5). The lateral pallium receives the main olfactory input via a lateral olfactory tract and is probably homologous to the lateral pallium of land vertebrates. All secondary olfactory projections experimentally determined in elasmobranchs are ipsilateral, and no crossed projections to the contralateral hemisphere appear to exist. The lateral pallium of elasmobranchs, unlike that of mammals, does not appear to project to cell groups outside the telencephalon. Projections from the lateral pallium to the ipsilateral area superficialis basalis, ipsilateral and contralateral olfactory bulbs, and contralateral dorsal and lateral pallia are known (Ebbesson 1972, Bodznick & Northcutt 1979).

The dorsal pallium is divided into inner and outer laminae continuing across the midline to form an interhemispheric bridge (Figure 5). The inner lamina undergoes extensive evolution in elasmobranchs. In many squalomorph sharks the inner lamina is poorly developed, forming a slight bulge in the roof of the lateral ventricle (Figure 5*C*). In most sharks and all batoids, the inner lamina forms a thickened mass termed the central nucleus (Figure 5*E*). In most galeomorph sharks the central nucleus is greatly enlarged, resulting in a thickened interhemispheric bridge (Figure 5*E,F*). The central nucleus receives substantial ascending sensory projections from the thalamus (Ebbesson & Schroeder 1971, Schroeder & Ebbesson 1974, Northcutt & Wathey 1979). This pallial center is known to receive visual, lateral line, trigeminal, and, possibly, auditory inputs (Cohen et al 1973, Platt et al 1974, Bullock & Corwin 1979), but the precise anatomy of these projections remains to be determined.

A medial pallial cell mass, the medial pallium, borders the dorsal pallium and also fuses across the interhemispheric bridge. Nothing is known of its connections, but its topography suggests it may be homologous to the medial pallium of other vertebrates.

A fourth pallial cell group, the pallial amygdala (area *a*, Figure 5*B,E*), may exist in elasmobranchs. This cell group arises as a ventral extension of the lateral pallium and is particularly well developed in the caudal hemisphere where it replaces a more medioventral cell group, the area superficialis basalis. Area *a* also receives ipsilateral secondary olfactory fibers via the lateral olfactory tract.

Area superficialis basalis of elasmobranchs has been homologized with the olfactory tubercle of land vertebrates; however, it does not receive secondary olfactory fibers but efferents from the lateral pallium (Ebbesson 1972). These connections and its topographical position in the ventral floor of the telencephalon suggest it may be homologous to the basal amygdala of other vertebrates (Northcutt 1978a).

Two nuclei lie above area superficialis basalis ventromedially: lateral and medial septal nuclei (*sn*, Figure 5). These nuclei receive secondary olfactory fibers via a medial olfactory tract (R. G. Northcutt and D. Bodznick, unpublished observations) and are probably homologous to the similarly named nuclei in other vertebrates.

The ventrolateral telencephalic wall contains two cell groups that may be homologous to the corpus striatum, as suggested by their topography and histochemistry. In primitive sharks such as *Notorynchus,* a ventrolateral cell group occurs rostrally as a cellular ridge ventral to the lateral pallium (Figure 5*A*). As the ridge is traced caudally it divides into dorsal and ventral components. The dorsal component is rapidly replaced by the expansion of nucleus *a*, but the ventral component continues caudally where it merges into the forebrain bundles. At present, nothing is known about the connections of these cell groups.

The telencephalon medium consists of the caudal continuation of nucleus *a,* the forebrain bundles, and the preoptic nucleus. The forebrain bundles consist of both ascending and descending pathways connecting the telencephalon with other regions of the brain. We know there are extensive ascending projections from the thalamus (Schroeder & Ebbesson 1974, Northcutt & Wathey 1979) and long descending telencephalic pathways to the thalamus, optic tectum, medulla, and even cervical spinal cord (Ebbesson 1980), but the exact origins of these pathways remain to be determined. There are sufficient experimental data to indicate that the telencephalons of some sharks and skates are characterized by large, well-differentiated cell groups with restricted olfactory input and well-developed thalamic input. Many of these telencephalic cell groups also possess long descending pathways, but too few species have been examined in too little detail to characterize telencephalic variation, and we lack many of the details of telencephalic organization needed to determine characters that are homologous or homoplastic with other vertebrate radiations.

Bony Fishes

The bony fishes comprise at least three major radiations: the ray-finned fishes, the lungfishes, and the crossopterygian fishes (Figure 6). Telencephalons of the lungfishes and the single living crossopterygian fish, *Latimeria,* consist of olfactory bulbs, paired evaginated hemispheres, and a telencephalon medium. The organization of the telencephalon of these fishes has been reviewed by Nieuwenhuys (1969). Nieuwenhuys recognized a single pallial zone in lungfishes and *Latimeria* and argued that there is no sound basis for subdividing the pallium into medial, dorsal, and lateral zones as earlier workers had done (Holmgren & van der Horst 1925); however, the verdict on the soundness of Nieuwenhuys' interpretation also awaits experimental studies. Nieuwenhuys followed earlier workers in dividing the subpallium into a lateral corpus striatum, a ventral olfactory tubercle, and a medial septal complex.

Again, experimental studies on these forms—almost certainly limited to the lungfishes because of the rarity of *Latimeria*—are badly needed if we hope to understand telencephalic variation within bony fishes and the telencephalic characters they hold in common with land vertebrates.

The telencephalon of ray-finned fishes appears strikingly different from that of other fishes and land vertebrates (Figure 6). Although ray-finned fishes possess rostrally located olfactory bulbs, the bulk of the telencephalon consists of two solid masses flanking a median ventricular space. Early interpretations of these paired masses suggested that they are homologous to the subpallium of land vertebrates and that the overlying ependyma is homologous to the pallium of other vertebrates (Goldstein 1905, Edinger 1908). Gage (1893) and Studnicka (1896) concluded that the paired masses

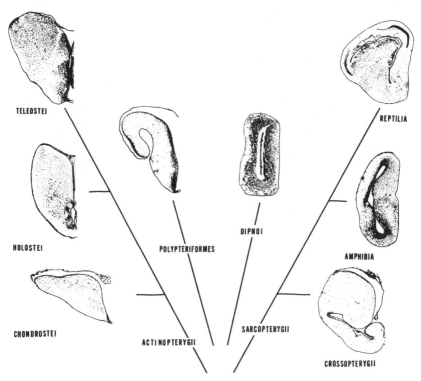

Figure 6   Dendrogram illustrating telencephalic variation in bony fishes and their derivatives. Each major category is represented by a transverse section through one-half of the cerebral hemisphere stained to reveal neuron distribution (from Northcutt & Braford 1980).

represent both the pallium and subpallium and that this unusual configuration results from an outward bending or eversion of the pallium (Figure 7). A major consequence of such an eversion is a mediolateral reversal of the pallium relative to the inverted and evaginated condition seen in other vertebrates. Most subsequent studies have supported the eversion hypothesis (Johnston 1911, Holmgren 1922, Källén 1951a, Nieuwenhuys 1962, Northcutt & Braford 1980, Schroeder 1980); however, a few workers have offered other interpretations (Crosby et al 1967, Schnitzlein 1968, Morgan 1975). Recent embryological observations (Nieuwenhuys et all 1969), as well as experimental anatomical and histochemical studies (Braford & Northcutt 1974, Finger 1975, Bass 1979, Northcutt & Braford 1980), overwhelmingly support an eversion process.

All studies for the past fifty years have recognized both pallial and subpallial fields within the telencephalon of ray-finned fishes, but there is no general agreement regarding the boundary between the pallium and subpallium, or the number of cell groups comprising these fields (see Northcutt & Braford 1980 for a recent review). This problem is further compounded by the fact that ray-finned fishes do not constitute a single grade of organization but consist of at least three or four grades. The polypteriform fishes, sometimes treated as a separate radiation (Figure 6), possess the least-differentiated telencephalon, with the number of telencephalic cell groups increasing through chondrostean, holostean, and teleos-

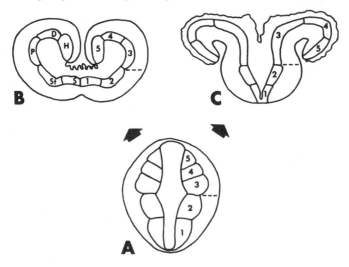

Figure 7   Schematic representation of the way in which the topology of the major subdivisions of the embryonic telencephalon (*A*) might be preserved following evagination and inversion (*B*) or eversion (*C*). *D*, dorsal pallium; *H*, hippocampus or medial pallium; *P*, piriform or lateral pallium; *S*, septal nuclei; *St*, striatum; *1–5*, major telencephalic subdivisions (from Northcutt & Braford 1980).

tean grades (Figure 6, Northcutt & Braford 1980). For this reason, the nomenclature used to describe the telencephalon of the polypteriforms, particularly the pallium, differs from that for other ray-finned fishes.

The everted pallium of these fishes is divided into three longitudinal zones: medial (P1), dorsal (P2), and lateral (P3) formations. The medial pallium is the major recipient zone of the secondary olfactory projections (Braford & Northcutt 1974), whereas the dorsal and lateral pallial formations receive ascending thalamic input (Braford & Northcutt 1978). At least a portion of these ascending thalamic inputs are now known to be visual (R. G. Northcutt and W. Saidel, unpublished observations). Based on topography and experimental studies, Northcutt & Braford (1980) proposed the following pallial homologies between polypteriforms and land vertebrates: the polypteriform medial pallium is homologous to the lateral pallium of land vertebrates; the dorsal pallia of both radiations are homologous; and the polypteriform lateral pallium is homologous to the medial pallium of land vertebrates.

The subpallium of polypteriforms is similar to that of other ray-finned fishes, and the nomenclature of Nieuwenhuys (1963) is generally used. The rostral subpallium consists of dorsal and ventral nuclei (Vd and Vv), occupying the periventricular wall, and a migrated lateral subpallial nucleus (V1). More caudally, the dorsal and ventral subpallial nuclei end at the level of the lamina terminalis. At this level, the dorsal subpallial nucleus is replaced by a supracommissural nucleus (Vs). Similarly, the supracommissural nucleus is replaced more caudally by the postcommissural nucleus (Vp), and a laterally situated entopeduncular complex is embedded in the forebrain bundle at these caudal telencephalic levels.

Although the subpallium of teleosts is slightly more complex (Figure 8), the same major subpallial nuclei are recognized as in polypteriform fishes. Teleosts possess one additional migrated subpallial nucleus rostrally, the commissural subpallial nucleus (Vc, Figure 8); and two additional nuclei caudally, an intermediate subpallial nucleus (Vi) and a nucleus taenia (NT).

Northcutt & Braford (1980) proposed the following subpallial homologies (Figure 9) between ray-finned fishes and land vertebrates, based on topological, connectional, and histochemical data:

1. The dorsal and ventral subpallial nuclei of ray-finned fishes are homologous to the lateral and medial septal nuclei, respectively, of land vertebrates.
2. The lateral subpallial nucleus of ray-finned fishes is homologous to the olfactory tubercle of land vertebrates.
3. The entopeduncular nucleus of ray-finned fishes is homologous to the same-named nucleus in land vertebrates.

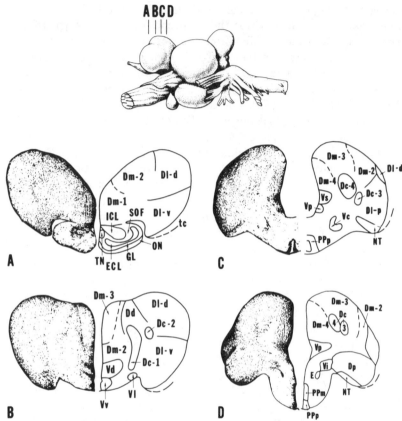

Figure 8 Transverse sections through the telencephalon of the sunfish, *Lepomis. Dc-1–4,*
parts of the central zone of the area dorsalis; *Dd,* dorsal zone of area dorsalis; *Dl-d,* dorsal
part of lateral zone of area dorsalis; *Dl-v,* ventral part of lateral zone of area dorsalis; *Dm-1–4,*
parts of medial zone of area dorsalis; *Dp,* posterior zone of area dorsalis; *E,* entopeduncular
nucleus; *ECL,* external cellular layer of olfactory bulb; *GL,* glomerular layer of olfactory bulb;
ICL, internal cellular layer of olfactory bulb; *NT,* nucleus taenia; *ON,* olfactory nerve; *PPm,*
magnocellular part of periventricular preoptic nucleus; *PPp,* Parvocellular part of periventricular preoptic nucleus; *SOF,* secondary olfactory fiber layer of olfactory bulb; *tc,* tela
chorioidea; *TN,* terminal nerve; *VC,* commissural nucleus of area ventralis; *Vd,* dorsal nucleus
of area ventralis; *Vi,* intermediate nucleus of area ventralis; *Vl,* lateral nucleus of area ventralis;
Vp, postcommissural nucleus of area ventralis; *Vs,* supracommissural nucleus of area ventralis; *Vv,* ventral nucleus of area ventralis.

4. The caudal subpallial nuclei (Vs and Vc) of ray-finned fishes are homologous to the basal amygdala of land vertebrates. Northcutt & Braford also
suggested that Vi, Vp, and NT are not subpallial nuclei, but part of the
pallium and homologous to the pallial amygdala of land vertebrates;
however, very different interpretations of these cell groups have been
offered by Nieuwenhuys (1967) and Wright (1967).

Among ray-finned fishes, pallial differentiation is most complex in teleosts, and most recent reviews (Northcutt & Braford 1980, Schroeder 1980) have used a modified nomenclature originated by Nieuwenhuys (1963). Northcutt & Braford (1980) recognized four major dorsal (D) longitudinal cell groups [medial (DM), dorsal (Dd), lateral (Dl), and posterior, (Dp)] surrounding a more central cell group (Dc). Each of these major groups (Figure 8) can be subdivided into a number of divisions based on differences in cell size and density as well as histochemistry. Northcutt & Braford (1980) proposed the following dorsal homologies (Figure 9) between ray-finned fishes and land vertebrates:

1. Dm of ray-finned fishes is not part of the pallium, but is actually homologous to the caudoputamen of land vertebrates.
2. Dd plus the dorsal subdivision of D1 is homologous to the dorsal pallium of land vertebrates.
3. The ventral and posterior subdivisions of D1 are homologous to the medial pallium of land vertebrates.
4. Dp of fishes is homologous to the lateral pallium of land vertebrates. These conclusions are based primarily on topological and histochemical considerations (Parent et al 1978, Northcutt & Braford 1980) but are compatible with the available experimental data on the dorsal areas.

Several experimental studies (Scalia & Ebbesson 1971, Finger 1975, Bass 1979) indicate that Dp is the major pallial target of the secondary olfactory input, though at least one subdivision of Dc also receives olfactory input. Ito & Kishida (1977) and Bass (1979) reported that Dc neurons constitute one of the major telencephalic efferent sources. Dc neurons are known to project to the optic tectum and olfactory bulb. Northcutt & Braford (1980) argued that Dc is not a single nucleus but a number of populations associated with several of the roof zones and representing the major efferent neurons of these zones, as well as a population that is associated with Dm and is probably homologous to part of the globus pallidus of land vertebrates. Few data are available regarding afferents to pallial areas other than Dp. Preliminary studies from several laboratories indicate that there are ascending thalamic pathways to different dorsal areas and that these pathways involve vision, lateral line, and auditory information (Ito & Kishida 1978, Finger 1979; W. Saidel, personal communication; M. R. Braford and R. G. Northcutt, unpublished observations). The details regarding exactly which thalamic cell groups project to the pallium and the exact sites of their termination are, however, obscure.

Far more extensive information is available on the organization of the olfactory bulbs and the secondary olfactory projections in teleosts (Nieuwenhuys 1967, Scalia & Ebbesson 1971, Finger 1975, Bass 1979, Northcutt & Braford 1980). The olfactory bulbs display the general pattern of concentric laminae, characteristic of all vertebrates, in the following centripetal order: (*a*) a layer of primary olfactory fibers, (*b*) a glomerular layer, (*c*) an external cellular layer including the prominent somas of the mitral cells, (*d*) a layer of secondary olfactory fibers leaving the bulb, and (*e*) an internal cellular layer. Some neurons located in the internal cellular layer also contribute axons to the secondary olfactory pathways (Bass 1979), as in agnathans and cartilaginous fishes.

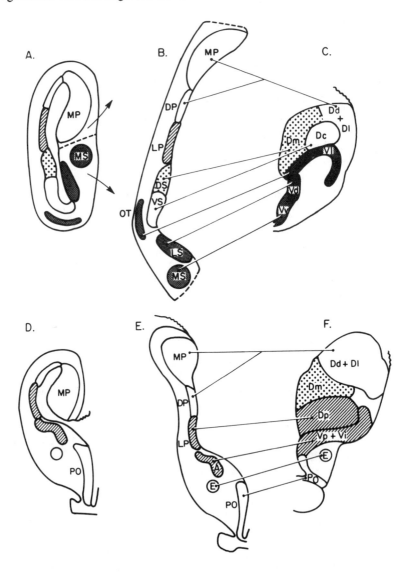

The secondary olfactory projections of teleosts form lateral and medial olfactory tracts; most projections are bilateral and pass to the contralateral hemisphere, posterior tuber of the diencephalon, and olfactory bulb via the anterior and habenular commissures (Finger 1975, Bass 1979). Bilateral secondary olfactory projections are reported to ventral telencephalic nuclei (Vc, Vd, Vi, Vp, Vs, Vv, and NT), dorsal nuclei (part of Dc, parts of Dl, and Dp), the preoptic area, and the posterior tuber. Although the secondary olfactory projections to the telencephalon are extensive, large areas (Dm, Dd, parts of Dl and Dc) do not receive secondary olfactory fibers but do receive other sensory input. Thus, there is no basis for describing the hemispheres of fishes as simple olfactory lobes divided into paleo- and archipallium, as was previously done (Ariëns Kappers et al 1936).

Little is known regarding the efferents of the telencephalon in ray-finned fishes. Vanegas & Ebbesson (1976) reported that ablation of the entire telencephalon in teleosts results in ipsilateral pathways to the thalamus, hypothalamus, inferior lobe, and optic tectum. They reported no descending pathways to the medulla or spinal cord. Karten & Finger (1976) reported that a telencephalo-thalamo-cerebellar pathway probably exists in teleosts, but the exact origin of the telencephalic cells that project to the caudal thalamus (pretectum) are unknown. Ito & Kishida (1977) and Bass (1979) reported that at least one population of Dc neurons projects to the optic tectum.

The picture of telencephalic organization in ray-finned fishes is changing very rapidly. These fishes reveal as much variation in their telencephalons as do land vertebrates. Some groups, such as the polypteriforms, possess simple hemispheres divided into a few zones, whereas many teleosts are characterized by greatly expanded and differentiated hemispheres with restricted olfactory projections. Those zones not receiving olfactory input are

---

Figure 9 Comparison of suspected homologous telencephalic cell groups in an evaginated and an everted hemisphere. Rostral (*A*) and caudal (*D*) transverse sections through an amphibian hemisphere are on the left; rostral (*C*) and caudal (*F*) sections through an holostean hemisphere are on the right. Direct comparison can be made only if the amphibian hemisphere is "unfolded" so that its cell masses approximate their relative positions in the unevaginated telencephalic vesicle. An unfolded pattern can be achieved by a transverse cut between the septal nuclei and medial pallium (*dashed lines* in *A*). The unfolded amphibian hemisphere (*B* and *E*) can then be directly compared to the holostean pattern. *A*, amygdala; *Dc*, central zone of *D; Dd*, dorsal zone of *D; Dl*, lateral zone of *D; Dm*, medial zone of *D; Dp*, dorsal pallium of amphibians or posterior zone of *D* in ray-finned fishes; *DS*, dorsal striatum; *E*, entopeduncular nucleus; *LS*, lateral septal nucleus; *LP*, lateral pallium; *MP*, medial pallium; *MS*, medial septal nucleus; *OT*, olfactory tubercle; *PO*, preoptic area; *Vd*, dorsal nucleus of *V; Vi*, intermediate nucleus of *V; Vl*, lateral nucleus of *V; Vp*, posterior nucleus of *V; VS*, ventral striatum; *Vv*, ventral nucleus of *V*.

the target of ascending projections from the thalamus and inferior lobe. Many of these hemispheric zones also appear to possess complex projections to the diencephalon and midbrain; however, the exact details of this organization must still be established. Similarities in the telencephalic organization of some ray-finned fishes, cartilaginous fishes, and amniotes, and the evolutionary bases for those similarities, must be determined by more extensive experimental study. This is one of the most exciting areas of comparative neurobiology.

Amphibians

The telencephalon of amphibians (Figure 10A–C) is similar to that of primitive sharks (Figure 5A–C) and polypteriform fishes (Figure 6) in the number of telencephalic cell groups and development of the pallium, but it is distinctly less complex than that of reptiles (Figure 10D–F) and other land vertebrates. This is probably due to the fact that amphibians are the only terrestrial anamniotes—not that their telencephalons are secondarily simple as a result of degenerative processes. Amphibian telencephalons exhibit character states that are typical of a broad range of other anamniotes; this suggests that the most striking changes characterizing the telencephalons of land vertebrates probably occurred during the amphibian-reptilian transition rather than the crossopterygian-amphibian transition.

Living amphibians comprise three different groups or orders: anurans (frogs and toads, approximately 2600 species), urodels (salamanders, approximately 300 species), and apodans (caecilians, approximately 150 species). There are recent reviews of telencephalic organization in anurans (Kicliter & Ebbesson 1976, Northcutt & Kicliter 1980) as well as urodels and apodans (Northcutt & Kicliter 1980). No significant differences in telencephalic cell groups among these three orders have been determined.

→

Figure 10 Transverse sections through comparative levels of the telencephalon of an amphibian, *Rana* (A–C), and a reptile, *Gekko* (D–F). A, anterior thalamic nucleus; *ac,* anterior commissure; *advr,* anterior dorsal ventricular ridge; *ae,* anterior entopeduncular nucleus; *b,* nucleus of Bellonci; *bn,* bed nucleus of pallial commissure; *dc,* dorsal cortex; *dm,* dorsomedial thalamic nucleus; *don,* dorsal optic nucleus; *dp,* dorsal pallium; *hc,* habenular commissure; *hy,* hypothalamus; *in,* intercalated optic nucleus; *la,* anterior lateral thalamic nucleus; *lc,* lateral cortex; *lgn,* lateral geniculate nucleus; *lp,* lateral pallium; *ls,* lateral septal nucleus; *m,* medial thalamic nucleus; *mc,* medial cortex; *mp,* medial pallium; *ms,* medial septal nucleus; *na,* nucleus accumbens; *ns,* nucleus sphericus; *on,* optic nerve; *ot,* olfactory tubercle; *pa,* preoptic area; *pc,* pallial commissure; *pdvr,* posterior dorsal ventricular ridge; *pl,* pars lateralis of the amygdala; *pm,* pars medialis of the amygdala; *r,* nucleus rotundus; *sn,* septal nuclei; *st,* striatum; *vm,* nucleus ventromedialis; *vt,* ventral thalamus.

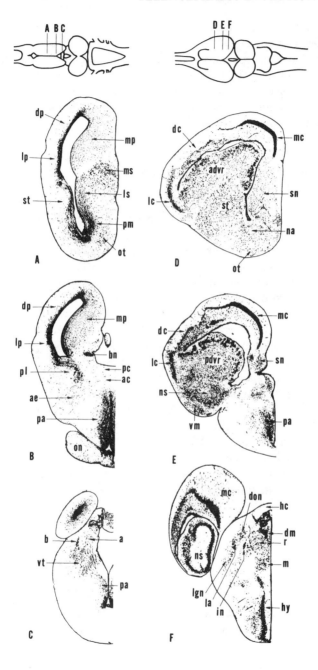

Anurans are emphasized in this review, as there are few experimental data for urodels and apodans.

Amphibian telencephalons consist of olfactory bulbs arising from the rostral poles of the paired evaginated hemispheres, and a caudal unevaginated telencephalon medium. The pallium of the hemisphere consists of four major formations: dorsal, lateral, and medial pallial sheets, and a pars lateralis of the amygdala (Figure 10A–C).

The medial pallium, unlike the other parts of the pallium, is characterized by extensive cell migrations away from the ependyma of the lateral ventricle and by a thickening of its walls. Four fiber systems are associated with the medial pallium of amphibians: (a) the dorsal association tract, (b) the fornix system, (c) the pallial commissure, and (d) the ascending anterior thalamic tract. The lateral pallium is the main target of secondary olfactory fibers. Axons of the cells of the lateral pallium project to the medial pallium by arching over the hemispheric roof and passing through the molecular layer of the dorsal pallium (Kokoros 1973, Scalia 1976a); this pathway is termed the dorsal association tract. The main efferent pathway of the medial pallium is the fornix system, which consists of pre- and postcommissural components (Ronan & Northcutt 1979). The precommissural fornix terminates in the postolfactory eminence, a small subpallial nucleus located immediately caudal to the olfactory bulbs, and in the olfactory bulb, the olfactory tubercle, and the septal nuclei. The postcommissural fornix terminates in the amygdala, preoptic area, anterior thalamic nucleus, and caudal hypothalamus. The amphibian medial pallium receives projections from all areas of the telencephalon except the striatum, pars lateralis of the amygdala, and preoptic area; it projects directly upon all telencephalic areas except the striatum (Ronan & Northcutt 1979).

There are extensive commissural connections between the two medial pallia via the pallial commissure and via neurons (bed nucleus of the pallial commissure) located dorsolaterally above the pallial commissure.

The medial pallium receives an ascending bilateral projection from the anterior thalamic nucleus and the medullar raphe system (Scalia & Colman 1975, Kicliter 1979, Ronan & Northcutt 1979). The anterior thalamic nucleus (Figures 10C and 11A receives visual (Scalia & Gregory 1970), hypothalamic (Neary & Wilczynski 1977a), dorsal column (Neary & Wilczynski 1977b), and possibly auditory (Mudry & Capranica 1978) inputs.

The dorsal pallium of amphibians (Figure 10A,B) consists of a periventricular cell plate receiving secondary olfactory fibers via the lateral olfactory tract (Northcutt & Royce 1975), as well as input via the ipsilateral lateral and medial pallia (unpublished observations) and the anterior thalamic nucleus (T. Neary, personal communication). The efferent connec-

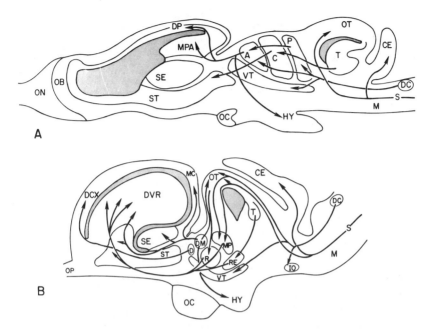

Figure 11 Summary of ascending sensory systems in amphibians (*A*) and reptiles (*B*). *A*, anterior thalamic nucleus; *C*, central thalamic nucleus; *CE*, cerebellum; *D*, dorsal optic nucleus; *DC*, dorsal column nuclei; *DCX*, dorsal cortex; *DM*, dorsomedial thalamic nucleus; *DP*, dorsal pallium; *DVR*, dorsal ventricular ridge; *HY*, hypothalamus; *IO*, inferior olivary nucleus; *M*, medulla; *MC*, medial cortex; *MP*, medial posterior thalamic nucleus; *MPA*, medial pallium; *OB*, olfactory bulb; *OC*, optic chiasm; *ON*, olfactory nerve; *OP*, olfactory peduncle; *OT*, optic tectum; *P*, posterior thalamic nucleus; *R*, nucleus rotundus; *RE*, nucleus reuniens; *S*, ascending spinal pathways; *SE*, septal nuclei; *ST*, striatum; *T*, torus semicircularis; *VT*, ventral thalamus.

tions of the dorsal pallium do not leave the telencephalon but appear to terminate in the ipsilateral lateral and medial pallia (Kokoros 1973).

The lateral pallium is the main target of the secondary olfactory efferents (Scalia et al 1968, Northcutt & Royce 1975), but it also receives input from the ipsilateral dorsal and medial pallia and, possibly, ascending input from the dorsal thalamus (Kicliter & Northcutt 1975). The lateral pallium projects to the ipsilateral olfactory bulb, striatum, olfactory tubercle, septum, and dorsal and medial pallia but does not appear to possess other efferents leaving the telencephalon (R. G. Northcutt, unpublished observations).

The amygdala of amphibians consists of a pallial pars lateralis (Figure 10 *B*) and a subpallial pars medialis (Figure 10*A*). The pars lateralis is the main, if not sole, target of the accessory olfactory bulb (Scalia 1972), which receives its input from the vomeronasal organ. The pars lateralis of the amygdala also receives bilateral medial pallial efferents (Ronan & Northcutt 1979) and projects to the ipsilateral hypothalamus (Scalia 1976a) and dorsal and ventral thalamus (Halpern 1972). Kokoros & Northcutt (1977) reported long descending projections to the medulla and cervical spinal cord following lesions of the ventrolateral caudal hemispheric wall. It is not clear, however, whether these projections to the thalamus and brainstem arise from the pars lateralis of the amygdala or from a portion of the lateral pallium immediately dorsal to the amygdala.

The pars medialis of the amygdala possesses reciprocal connections with the ipsilateral septum (Kicliter 1979) and both medial pallia (Ronan & Northcutt 1979), and an ascending input from the hypothalamus (Neary & Wilczynski 1977a); however, other efferents are presently unknown.

In addition to the pars medialis of the amygdala, the subpallium of amphibians (Figure 10A,B) consists of a laterally situated striatum, medially situated septal nuclei, and the olfactory tubercle.

The striatum of amphibians is the target of ipsilaterally ascending thalamic pathways (Gruberg & Ambros 1974, Kicliter & Northcutt 1975, Kicliter 1979, Wilczynski & Northcutt 1979) as well as projections arising in the midbrain tegmentum (Rubinson & Colman 1972, Wilczynski & Northcutt 1979) and the lateral pallium (R. G. Northcutt, unpublished observations). The striatum receives afferents from the central and lateral thalamic (pars anterior) nuclei. The central nucleus is an auditory thalamic nucleus receiving its main input from the torus semicircularis of the midbrain (Neary 1974), whereas the pars anterior of the lateral nucleus is a visual thalamic nucleus receiving its main input from the optic tectum (Rubinson 1968). The striatum (Figure 10C) projects to the anterior entopeduncular nucleus, the superficial isthmal nucleus, and the midbrain tegmentum (Halpern 1972, Wilczynski & Northcutt 1979). The anterior entopeduncular and superficial isthmal nuclei project to the optic tectum and torus semicircularis, respectively. Thus, the striatum possesses efferent pathways that project back upon the midbrain centers, from which it receives its main afference, and efferents to tegmental motor centers.

The septal nuclei of amphibians probably receive secondary olfactory input via the medial olfactory tract (Northcutt & Royce 1975), bilateral input from the medial pallium (Ronan & Northcutt 1979), and ascending input from the hypothalamus (Neary & Wilczynski 1977a). The septal nuclei are known to project to the medial pallia (Ronan & Northcutt 1979) and to the preoptic area, ventral thalamus, and hypothalamus (Halpern 1972). Thus, the septal nuclei form one of the major telencephalic efferent pathways in amphibians.

There are few data on the ipsilateral connections of the amphibian olfactory tubercle. It receives secondary olfactory fibers from the ipsilateral olfactory bulb via the medial olfactory tract, as well as input from the contralateral olfactory bulb in the form of fibers that run caudally to enter the habenular commissure and then turn rostrally in the hemisphere to enter the ipsilateral medial olfactory tract and terminate in the olfactory tubercle. The tubercle also receives bilateral input from the medial pallia and projects, in turn, to the ipsilateral medial pallium (Ronan & Northcutt 1979). Other connections of the tubercle are presently unknown.

Earlier studies (Ariëns Kappers et al 1936, Herrick 1948) assumed that the entire amphibian hemisphere received secondary or tertiary olfactory fibers and that ascending thalamic inputs were restricted to the striatum. Recent experimental studies reveal that secondary olfactory projections are primarily restricted to the lateral pallium and medial subpallial wall (Scalia et al 1968, Northcutt & Royce 1975). Higher order olfactory projections interconnect the amygdala, septum, and medial pallium; each of these telencephalic centers possesses efferents that terminate in many diencephalic areas. Although these "olfactory" pathways are extensive, the pallial formations and the striatum are both characterized by well-developed thalamic inputs carrying most sensory modalities previously believed to be restricted to amniotes. Finally, there are telencephalic efferents to midbrain, hindbrain, and even cervical spinal levels, although we still do not know the exact source(s) of these long descending pathways. There are sufficient experimental data to argue that many of the amphibian telencephalic areas are probably homologous to the telencephalic areas of the same name in amniotes (Northcutt & Kicliter 1980); however, two particularly complex areas of controversy exist. Amphibians, like many other vertebrates, possess a retino-thalamo-telencephalic pathway suggested to be homologous to the retino-lateral geniculate-visual cortical pathway of mammals (Scalia 1976b); however, the anterior thalamic nucleus of amphibians that gives rise to the visual telencephalic pathway also receives other input, which suggests similarities with the anterior and midline thalamic nuclei of mammals. In amphibians, the anterior thalamic nucleus projects not only to the telencephalon bilaterally via the medial forebrain bundle, but also to the hypothalamus (Neary 1980). In contrast, the geniculo-cortical pathway in mammals is solely ipsilateral and projects to the telencephalon via the lateral forebrain bundle. Clearly, there are a number of problems in comparing the visual telencephalic pathways of amphibians and mammals, and more data are needed before a satisfactory phyletic comparison can be completed.

Similarly, the dorsal pallium of amphibians has traditionally been homologized to mammalian isocortex, but it is the caudal ventrolateral hemispheric wall (lateral pallium and amygdala) in amphibians, not the dorsal pallium, that is characterized by long descending efferents. It is possible that the amphibian lateral pallium is a field homologue of both olfactory cortex and part of isocortex in amniotes (Northcutt 1974), or that these similarities are due to homoplasy. Again, additional experimental study should resolve this question.

Reptiles

The living reptiles comprise four different orders: Chelonia (turtles), Rhynchocephalia (tuataras), Squamata (amphisbaenians, lizards, and snakes), and Crocodilia (alligators and crocodiles). Early studies described the telencephalons of representatives of all reptilian groups, except amphisbaenians; several recent reviews (Butler 1978, 1980, Northcutt 1978b, Halpern 1980) summarize these descriptive studies as well as more recent experimental work. Experimental studies are rapidly proliferating, but there are still many telencephalic areas for which the connections and internal organization are completely unknown. Crocodilians, turtles, lizards, and snakes have all been the object of experimental studies, but we still do not possess comparable information on the connections of a single nucleus or the course of a specific pathway in all reptilian groups. There is, however, sufficient data to suggest that different telencephalic cell groups and many of their related pathways reveal considerable variation in their development.

The reptilian telencephalon (Figure 10D–F) consists of olfactory bulbs that arise rostrally from the paired evaginated hemispheres, and a caudally unevaginated telencephalon medium comprising the preoptic area and the forebrain bundles. The olfactory bulbs are frequently located some distance from the hemispheres and connected to them by elongated tracts, the olfactory peduncles. The olfactory bulbs consist of the following centripetal layers: olfactory nerve fibers, glomerular layer, external plexiform layer, mitral cell layer, internal plexiform layer (which consists of the secondary olfactory efferents), and a periventricular internal granular layer (Crosby & Humphrey 1939).

Most reptiles, except crocodilians and many turtles, also possess an accessory olfactory bulb that receives its input from the vomeronasal nerves arising as axons of bipolar cells in the vomeronasal epithelium. The accessory olfactory bulb replaces the caudomedial portion of the main olfactory bulb; in many snakes and lizards it may be as large as the main olfactory bulb. The accessory olfactory bulb appears to possess the same layers as the main olfactory bulb in reptiles (Crosby & Humphrey 1939, Halpern 1980). The secondary olfactory projections have been experimentally determined in a number of reptiles: *Lacerta* (Goldby 1937, Gamble 1952), *Testudo* (Gamble 1956), *Tupinambis* (Heimer 1969), *Chrysemys* (Northcutt 1970), *Caiman* (Scalia et al 1969), and *Thamnophis* (Halpern 1976). The main olfactory bulb gives rise to lateral and medial olfactory tracts that terminate ipsilaterally in a retrobulbar region frequently termed the anterior olfactory nucleus and in the olfactory tubercle, the lateral cortex, and perhaps the extreme rostral portion of the medial cortex and septal nuclei. Contralateral secondary olfactory projections occur in reptiles, as in amphibians; fibers pass caudally in the lateral olfactory tract, enter the stria medullaris, decussate in the habenular commissure, and then reenter the contralateral hemisphere to distribute to the lateral cortex and internal granular layer of the main olfactory bulb.

The efferents of the accessory olfactory bulb (Halpern 1976) are restricted to the ipsilateral hemisphere and terminate in an interstitial nucleus of the accessory olfactory tract and in nucleus sphericus (Figure 10*F*), a prominent nucleus in the caudal hemisphere.

The hemispheric pallium of reptiles (Figure 10*D–F*) consists of medial, dorsal, and lateral cortices and the dorsal ventricular ridge, a prominent cellular mass bulging into the lateral ventricle. The subpallium or basal region of the hemisphere consists of a medial septal complex, a ventral olfactory tubercle, and a lateral corpus striatum.

The medial cortex of reptiles appears to be homologous to at least part of the hippocampal complex of mammals and receives afferents from the dorsomedial and dorsolateral thalamic nuclei (Lohman & van Woerden-Verkley 1978, Balaban & Ulinski 1980), the superior raphe nucleus and mammillary body (Lohman & van Woerden-Verkley 1978), the ipsilateral dorsal and lateral cortices (Lohman & Mentink 1972, Ulinski 1976, Butler 1980, Halpern 1980), and the contralateral medial cortex via the pallial commissure (Northcutt 1968, Voneida & Ebbesson 1969, Lohman & Mentink 1972, Butler 1976, Ulinski 1976). Many of these workers also reported that medial cortex projects to the ipsilateral dorsal and lateral cortices and to the contralateral medial cortex. Projections to subpallial formations occur via a precommissural fornix to the septal nuclei bilaterally, and a postcommissural fornix to the ipsilateral dorsomedial thalamic nucleus and caudal hypothalamus.

The dorsal cortex of reptiles has been variably defined and subdivided in different studies (Dart 1934, Goldby & Gamble 1957, Northcutt 1967, Platel 1969, Lohman & Mentink 1972, Butler 1980). Opinions range from that of Platel, who claimed that dorsal cortex does not exist as a distinct cortical division, to that of Lohman & Mentink, who recognized three distinct dorsal cortical subdivisions. Similarly, there is no agreement on the homology of this cortical area to other vertebrate cortices. Afferents to the dorsal cortex have been reported from the ipsilateral lateral and medial cortices (Lohman & Mentink 1972, Ulinski 1976, Butler 1980), the ipsilateral septal nuclei (Ulinski 1975, Lohman & van Woerden-Verkley 1978), dorsomedial and dorsolateral thalamic nuclei (Lohman & van Woerden-Verkley 1978, Balaban & Ulinski 1980), and the contralateral dorsal cortex via the pallial commissure (Voneida & Ebbesson 1969, Northcutt 1970, Ware 1974, Butler 1975). The dorsal cortex of turtles (Hall & Ebner 1970) receives an ipsilateral ascending visual input that arises in a rostrally located retino-recipient nucleus usually termed the pars dorsalis of the lateral geniculate nucleus. The medial cortex of lizards, however, appears to be the primary cortical visual area, as it is the main region driven by visual stimulation (Gusel'nikov & Supin 1964, Andry & Northcutt 1976). Lohman & van Woerden-Verkley (1978) experimentally examined telencephalic projections of the pars dorsalis of the lateral geniculate nucleus in the lizard *Tupinambis;* they claimed that the rostral part of the pars dorsalis does not project to the telencephalon, but that the caudal part projects to a rostrolateral region of the dorsal ventricular ridge. Similar data have been reported for snakes (Wang & Halpern 1977); if correct, they suggest that the rostral retino-recipient thalamic nucleus of reptiles projects to totally different telencephalic formations in different species. The rostral thalamus of reptiles is extremely complex, and a number of different cell groups receive direct retinal input (Northcutt et al 1974, Bass 1976, Butler & Northcutt 1978, Northcutt 1978b); in addition, more than one cell group has been defined as the pars dorsalis of the lateral geniculate nucleus.

At least four distinct rostral cell groups (don, in, la, lgn; Figure 10*F*) receive retinal input in many reptiles. The highest optic terminal density (Butler & Northcutt 1978) occurs in a cell group (1a; Figure 10*F*) located between a dorsal cell group (don) and a more ventral cell group (lgn). The dorsal cell group, here termed the dorsal optic nucleus, appears to correspond to a cell group reportedly projecting to the telencephalon; there is no evidence that the more ventral cell groups (la and lgn) give rise to telencephalic projections. A similar pattern exists in amphibians and birds. The highest optic terminal density occurs in a cell group termed the nucleus and neuropil of Bellonci in amphibians (Scalia & Gregory 1970), and in the avian anterior lateral nucleus (Repérant 1973). Neither the nucleus of Bellonci (Kicliter & Northcutt 1975) nor the anterior lateral nucleus (Miceli et al 1979) projects to the telencephalon; however, a more dorsally located nucleus (anterior thalamic nucleus of amphibians and dorsolateral thalamic nucleus of birds) does project to the dorsal or medial cortices in these classes (Karten & Nauta 1968, Scalia & Colman 1975, Ronan & Northcutt 1979). Thus, although the dorsal cortex of reptiles is said to receive retino-thalamo-telencephalic input, it is not clear whether this input exists in all reptilian orders, or exactly which cell group(s) gives rise to the projection. Equally important, although a similar pathway exists in both amphibians and birds, there is insufficient data to determine whether these pathways are homologous to each other or to the retino-geniculo-cortical pathway of mammals.

Efferents of the reptilian dorsal cortex terminate in the ipsilateral lateral and medial cortices (Butler 1976, Halpern 1980), the ipsilateral dorsal ventricular ridge (Northcutt 1970, Lohman & van Woerden-Verkley 1976, Butler 1980, Halpern 1980) and septal nuclei (Lohman & Mentink 1972, Halpern 1980), and the contralateral dorsal cortex (Northcutt 1970, Ware 1974, Butler 1975). The dorsal cortex of turtles also projects to the ipsilateral rostral thalamus and to the optic tectum (Hall et al 1977). In turtles these efferents pass down the lateral and medial walls of the telencephalon, however, rather than passing laterally via the lateral forebrain bundle as in mammals, and they may represent independently evolved long dorsal cortical efferents.

The lateral cortex of reptiles is the main target of the ipsilateral and contralateral secondary olfactory projections, as well as projections from the ipsilateral dorsal and medial cortices. There are lateral cortical efferents to the ipsilateral olfactory bulb and retrobulbar area (Halpern 1980), as well as reciprocal connections with the ipsilateral dorsal and medial cortices. Projections to the dorsal ventricular ridge are also known (Butler 1976, Lohman & van Woerden-Verkley 1976).

The dorsal ventricular ridge of reptiles is divided into anterior and posterior portions (Figure 10D,E). The anterior dorsal ventricular ridge is further divided into a medio-lateral series of zones, each of which is the main target of an ascending sensory pathway (Figure 11B). The medial zone receives projections from nucleus medialis (reuniens), a thalamic nucleus that receives bilateral projections from the torus semicircularis (Pritz 1974a, 1974b, Foster & Hall 1978, Balaban & Ulinski 1980). The intermediate zone receives projections from nucleus medialis posterior, a thalamic nucleus that receives projections from the spinal cord and dorsal column nuclei (Northcutt & Pritz 1978, Balaban & Ulinski 1980). The lateral zone receives projections from nucleus rotundus, a thalamic nucleus that receives bilateral projections from the optic tectum (Hall & Ebner 1970, Pritz 1973, Lohman & van Woerden-Verkley 1978). The dorsomedial thalamic nucleus (Figure 10B) also projects diffusely to all three zones of the anterior dorsal ventricular ridge, as well as to several other brain areas (Balaban & Ulinski 1980).

Efferent projections from the anterior dorsal ventricular ridge have been reported to the ipsilateral posterior dorsal ventricular ridge, striatum, nucleus accumbens, anterior olfactory nucleus, lateral cortex, and the contralateral striatum (Hoogland 1977, Ulinski 1978, Voneida & Sligar 1979). The posterior dorsal ventricular ridge of reptiles includes a posterior division of the ridge proper, as well as the nucleus sphericus and the ventromedial nucleus (Figure 10*E,F*). Although this entire region has been homologized to the mammalian amygdalar complex (Curwen 1939), there are few experimental data regarding its organization (Halpern & Silfen 1974, Voneida & Sligar 1979). The organization and connections of the reptilian amygdala have received scant attention, but it is likely that nucleus sphericus and the ventromedial nucleus form part of the amygdala. The posterior ridge proper, however, may be more closely related to the anterior ridge. Preliminary experiments (R. G. Northcutt, unpublished observations) indicate that several regions of the anterior ridge project to the posterior ridge. This suggests that the posterior ridge may function as a higher sensory association area.

Nucleus sphericus is the main target of the accessory olfactory bulb and projects to the ipsilateral accessory olfactory bulb, the contralateral nucleus sphericus via the anterior commissure, and the ipsilateral preoptic area and caudal hypothalamus (Halpern & Silfen 1974, Halpern 1980). The connections of the ventromedial nucleus are unknown.

Birds, like reptiles, possess a well-developed dorsal ventricular ridge that has been homologized to the mammalian caudoputamen (Ariëns Kappers et al 1936) and to portions of mammalian isocortex (Källén 1951 a,b, Karten 1969, Northcutt 1969, 1970). This controversy and the embryological, histochemical, and connectional data bearing on the problem have been recently reviewed (Cohen & Karten 1974, Northcutt 1978b, Benowitz 1980). All available data indicate that the dorsal ventricular ridge in reptiles and birds is homologous to mammalian isocortex, excluding the primary occipital visual cortex, which may be homologous to part of the dorsal cortex in reptiles and birds. These homologies should be viewed with some reservation, as it is possible that the cynodont reptiles leading to mammals may not have possessed a dorsal ventricular ridge. Endocasts of these cynodonts (Hopson 1979) suggest that their cerebral hemispheres were much longer than they were broad, unlike those of modern reptiles and birds. The casts of the hemispheres look far more like those of modern amphibians, and there is a real possibility that cynodont reptiles may have possessed cerebral hemispheres more like those of modern amphibians than modern reptiles. In any case, it is clear that the dorsal ventricular ridge of reptiles and birds is not homologous to the caudoputamen of mammals. All available evidence indicates that the ridge, like isocortex, is pallial in origin. Whether the ridge is homologous or homoplastic to mammalian isocortex is a determination requiring additional data; the resolution may finally rest on our interpretation of the probable telencephalic organization in cynodonts.

The lateral floor of the reptilian cerebral hemisphere consists of a number of cell masses termed the corpus striatum. There is no agreement on the subdivisions of the striatum (Platel 1971, Northcutt 1978b, Halpern 1980, Reiner et al 1980); however, most workers have recognized a lateral striatum proper, frequently divided into dorsolateral and ventromedial subdivisions, and a more medially situated nucleus accumbens (Figure 11D). The striatum proper receives input from all dorsal thalamic nuclei that project to the dorsal ventricular ridge, as well as efferents from the ridge itself. Efferents from the dorsolateral, dorsomedial, and ventral thalamic nuclei, as well as cell groups in the rostral and caudal midbrain tegmentum, also project to the striatum (Parent 1976).

Striatal efferents project to the entopeduncular nuclei, nucleus rotundus, dorsal nucleus of the posterior commissure, midbrain tegmentum, substantia nigra, mesencephalic central grey, and lateral cerebellar nucleus (Hoogland 1977, Voneida & Sligar 1979, Reiner et al 1980). Although most of these efferents are similar to the projections of the mammalian ansa lenticularis, the striatal efferents to the dorsal nucleus of the posterior commissure (Reiner et al 1980) and the lateral cerebellar nucleus (Hoogland 1977) are unlike any pathways reported for mammals; however, similar striatal efferents to the pretectum are reported in amphibians (Kokoros & Northcutt 1977, Wilczynski & Northcutt 1979) and birds (Karten & Dubbeldam 1973, Brecha et al 1976, Brecha 1978). These data suggest that there are at least two separate efferent striatal pathways in most land vertebrates: a strio-tegmental pathway similar to the classical ansa lenticularis of mammals and a strio-pretecto-tectal pathway. A similar strio-lateral cerebellar nuclear projection has not been reported in birds, and the significance of this pathway in amphibians and reptiles is unknown.

Distel & Ebbesson (1975; H. Distel and S. O. E. Ebbesson, personal communication), reported that nucleus accumbens receives afferents from the dorsolateral thalamic nucleus, but not from rotundus or medialis. Other connections of nucleus accumbens are presently unknown.

The septal complex of reptiles is a distinct medioventral complex located beneath the medial cortex throughout most of its rostrocaudal extent (Figure 10D–F). Halpern (1980) has provided the most extensive analysis of this complex in reptiles and recognized nine distinct nuclei. The connections of each of these nuclei remain to be determined, but afferents to the septal complex are known to arise from the dorsal and medial cortices (Northcutt 1970, Ulinski 1975, Butler 1976, Lohman & van Woerden-Verkley 1976, Repérant 1976, Halpern 1980) and the dorsomedial thalamic nucleus (Balaban & Ulinski 1980). Efferents from the septal complex project to the ipsilateral dorsal cortex, nucleus of the anterior commissure, preoptic area, dorsolateral and dorsomedial thalamic nuclei, hypothalamus, and midbrain central grey (Hoogland et al 1978).

Earlier studies of reptilian telencephalic organization (Ariëns Kappers et al 1936) suggested that the cerebral hemispheres are characterized by a massive neostriatum (dorsal ventricular ridge) and poorly developed cortices. Recent experimental studies reveal that the dorsal ventricular ridge is not part of the striatum, but is homologous to parts of the pallium of other vertebrates. Furthermore, the ridge is organized into a number of separate zones receiving auditory, somatosensory, and visual information relayed by the dorsal thalamus. Much experimental work is still needed to determine the exact details of reptilian hemispheric organization, but many details of this organization are remarkably similar to those seen in birds and mammals. The most notable exception is the apparent absence of a motor pallial area giving rise to long descending efferents to the brainstem or spinal cord.

Birds

Modern birds consist of two major assemblages: palaeognathous and neognathous birds, which differ in the details of their palatal and jaw organization. The first group consists of the ratites and kiwis, whereas the second group comprises most modern birds. The palaeognathous birds have been largely ignored; most of our information regarding avian telencephalic organization comes from representatives of only two of the twenty orders of neognathous birds. Nevertheless, it is obvious that there is considerable variation in the relative development and position of the various cell groups forming the avian telencephalon (Stingelin 1958). This variation has not been explored experimentally, but our understanding of avian telencephalic organization has advanced, and a number of recent reviews exist (Karten 1969, Cohen & Karten 1974, Benowitz 1980).

The avian telencephalon consists of paired olfactory bulbs arising from evaginated cerebral hemispheres and a caudal telencephalon medium formed by the preoptic area and forebrain bundles. The olfactory bulbs are homologous to the main olfactory bulbs of reptiles—accessory olfactory bulbs do not appear to exist—and seven layers are recognized (Rieke & Wenzel 1978): olfactory nerve layer, glomerular layer, external plexiform layer, mitral cell layer, internal plexiform layer, granule cell layer, and periventricular ependymal layer. The olfactory bulb projects ipsilaterally to a small prepiriform cortex, to the parolfactory lobe, and to part of the hyperstriatum ventrale (Rieke & Wenzel 1978). Contralateral olfactory projections via the anterior commissure to the paleostriatum primitivum and caudal portion of the parolfactory lobe are also reported.

The organization of the avian cerebral hemisphere is similar to that seen in reptiles, but in birds the dorsal ventricular ridge has expanded to the point where it is difficult to recognize boundaries between the ridge and other portions of the pallium. Ariëns Kappers et al (1936) believed the

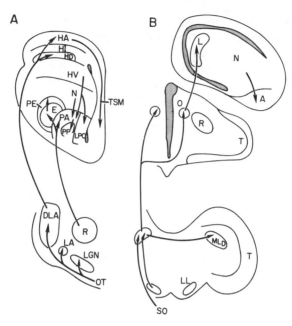

Figure 12   Avian telencephalic visual (*A*) and auditory (*B*) pathways. *A*, archistriatum; *DLA*, anterior dorsolateral thalamic nucleus; *E*, ectostriatum; *HA*, hyperstriatum accessorium; *HD*, hyperstriatum dorsale; *HI*, hyperstriatum intercalatus superior; *HV*, hyperstriatum ventrale; *L*, field L of the neostriatum; *LA*, anterior lateral thalamic nucleus; *LGN*, lateral geniculate nucleus; *LL*, nucleus of the lateral lemniscus; *LPO*, parolfactory lobe; *MLD*, pars dorsalis of the lateral mesencephalic nucleus; *N*, neostriatum; *O*, nucleus ovoidalis; *OT*, optic tract; *PA*, paleostriatum augmentatum, *PE*, periectostriatal belt; *PP*, paleostriatum primitivum; *R*, nucleus rotundus; *SO*, superior olivary nucleus; *T*, optic tectum; *TSM*, septo-mesencephalic tract.

dorsal ventricular ridge to be part of the striatum and divided it into a series of subdivisions bearing the suffix "-striatum"; even though the dorsal ventricular ridge is not, in fact, part of the striatum, this terminology remains in use. Embryological studies (Kuhlenbeck 1938, Källén 1951a, 1962) suggest that the medial pallium in birds consists of thin hippocampal and parahippocampal cortices. The dorsal pallium is divided into a rostral thick segment, termed the Wulst, and a thinner caudal dorsolateral corticoid plate. The Wulst consists of a series of dorso-ventrally oriented layers termed the hyperstriatum accessorium, hyperstriatum intercalatus superior, and hyperstriatum dorsale (Figure 12). The lateral pallium of birds accounts for a very small part of the total hemisphere and consists of laterally situated prepiriform and piriform cortices. The bulk of the pallium consists of the dorsal ventricular ridge, which is divided into a dorsal nucleus, termed the hyperstriatum ventrale, and a ventrolateral ectostriatum and ventromedial neostriatum (Figure 12). The ectostriatum is frequently divided into a core, ectostriatum proper, and a more peripheral band of cells termed the periectostriatal belt (Karten 1969). The ectostriatum is replaced more caudally by a complex series of nuclei termed the archistriatum. Recent studies suggest that part of the archistriatum is formed by amygdalar nuclei, whereas the other part is more similar to mammalian isocortex than to mammalian amygdalar nuclei. This part is closely associated with the dorsal ventricular ridge and gives rise to long descending pathways (Zeier & Karten 1971).

The avian subpallium consists of a ventrolateral striatal complex divided into a dorsal paleostriatum augmentatum and a ventral paleostriatum primitivum. The paleostriatum primitivum is, in turn, subdivided into a dorsal group of large neurons and a ventral group of smaller cells termed the intrapeduncular nucleus. The remaining portion of the subpallium consists of a more medially located parolfactory lobe and the septal nuclei. Histochemical and experimental neuroanatomical studies suggest that the paleostriatum augmentatum is homologous to the mammalian caudoputamen and that the paleostriatum primitivum is homologous to the mammalian globus pallidus (Karten 1969, Karten & Dubbeldam 1973, Brauth et al 1978). The relationship of the avian parolfactory lobe to telencephalic structures in other vertebrates is unclear.

One additional subpallial telencephalic nucleus (nucleus basalis) occurs rostral to the paleostriatum in birds. This nucleus receives a direct bilateral input from the principal sensory trigeminal nucleus via a quintofrontal tract (Zeigler & Karten 1973). Thus, in birds the immediate ascending target of the principal trigeminal nucleus is a telencephalic nucleus, rather than nucleus ventralis posterior medialis of the thalamus as in mammals. The nucleus basalis, in turn, projects to a region of the hemisphere dorsal to the archistriatum (Zeier & Karten 1971).

The hippocampal and parahippocampal cortices receive afferents from the hypothalamus, the diagonal band, and the septal nuclei (Benowitz & Karten 1976, Krayniak & Siegel 1978a) and project back upon the septal nuclei and nucleus of the diagonal band via a precommissural fornix system (Krayniak & Siegel 1978b). Birds do not appear to have a postcommissural fornix pathway to the hypothalamus as do other vertebrates; however, the septal nuclei are characterized by extensive descending projections to the hypothalamus, ventrolateral and dorsomedial thalamic nuclei, lateral habenular nucleus, and midbrain tegmentum (Krayniak & Siegel 1978a).

The avian Wulst consists of lateral and medial divisions. The lateral division constitutes a highly complex visual area, whereas the medial division seems more closely associated with the medial cortical areas, but its connections are poorly understood.

The lateral or visual Wulst (Figure 12*A*) receives a bilateral input from the pars lateralis (frequently termed the principal optic nucleus) of the anterior dorsolateral thalamic nucleus (Karten & Nauta 1968, Hunt & Webster 1972, Karten et al 1973, Meier et al 1974, Mihailović et al 1974, Miceli et al 1975, 1979). Recent studies indicate that the pars lateralis consists of at least five subdivisions receiving contralateral retinal input and that these subdivisions form ipsilateral, contralateral, and bilateral efferents to the visual Wulst. Pettigrew & Konishi (1976) reported that the visual Wulst of the barn owl is retinotopically organized; its cells are selective for orientation and direction of movement, and many of them exhibit binocular disparity similar to the cells of mammalian visual cortex. The avian Wulst also possesses long descending efferents similar to mammalian visual cortex. Wulst efferents pass down the lateral telencephalic wall and project to the neostriatum and the periectostriatal belt, as well as to the lateral geniculate nucleus, anterior dorsolateral thalamic nucleus, pretectal nuclei, and the optic tectum via a medially situated pathway, the septomesencephalic tract (Karten 1971, Karten et al 1973, Miceli et al 1979).

Birds also have a second visual telencephalic area (Figure 12*A*). Ascending tectal efferents project bilaterally to nucleus rotundus of the thalamus, which in turn projects ipsilaterally to the paleostriatum and the ectostriatum (Karten & Revzin 1966, Revzin & Karten 1966, Karten & Hodos 1970). The ectostriatum, unlike the visual Wulst, does not possess long efferents leaving the telencephalon but projects only to the periectostriatal belt, which in turn projects to parts of neostriatum, hyperstriatum ventrale, and archistriatum (Ritchie & Cohen 1977). Several of these areas project, in turn, to the paleostriatum which, along with the archistriatum, gives rise to long descending efferents to the brainstem (Zeier & Karten 1971, Karten & Dubbeldam 1973).

Avian telencephalic auditory pathways appear to be organized in a manner similar to the visual ectostriatal system. The pars dorsalis of the lateral mesencephalic nucleus, a midbrain structure, receives auditory input from a number of medullar auditory nuclei (Boord 1968) and projects bilaterally to a thalamic nucleus termed nucleus ovoidalis (Karten 1967), which in turn projects ipsilaterally to the paleostriatum and a caudal portion of the neostriatum termed field L (Karten 1968, Nottebohm et al 1976, Bonke et al 1979). The avian field L exhibits tonotopic organization (Zaretsky & Konishi 1976, Scheich et al 1979) as does mammalian primary auditory cortex; field L consists of at least three congruent tonotopic subdivisions that appear to be organized in a manner similar to higher order auditory cortices in mammals (Scheich et al 1979). Field L, in turn, projects to hyperstriatum ventrale and to part of the archistriatum (Nottebohm et al 1976, Bonke et al 1979). Hyperstriatum ventrale, in turn, projects to a portion of the parolfactory lobe (nucleus x) and to the archistriatum. Both nucleus x and the archistriatum (nucleus robustus) give rise to long descending pathways to the thalamus and midbrain, and nucleus robustus terminates directly on the hypoglossal motor nucleus, which innervates the avian syrinx, the organ of vocalization. Additional pathways interconnect field L with other regions of the telencephalon. Bonke et al (1979) reported that field L and the hyperstriatum ventrale both project to the paleostriatum as well as to a portion of the neostriatum that does not receive direct auditory input. The paleostriatum projects back upon the neostriatum, hyperstriatum ventrale, and the archistriatum, and gives rise to a descending pathway to diencephalic and midbrain centers.

There is little information on ascending somatosensory pathways to the avian telencephalon. Delius & Bennetto (1972) reported that the anterior hyperstriatum and medial neostriatum caudale are both responsive to somatosensory stimuli. Karten et al (1978) reported that the anterior Wulst of owls is characterized by a somatosensory map of the contralateral toes. This information reaches the Wulst via projections from nucleus dorsalis intermedius ventralis anterior, a thalamic relay nucleus that receives its input from the dorsal column nuclei. Further studies will probably reveal extensive areas related to somatosensory stimuli. All present sensory information reveals that there are primary sensory areas within various portions of the Wulst and ridge but that these areas are widely separated from the populations giving rise to long descending pathways (hyperstriatum accessorium, archistriatum, and paleostriatum) by complex populations (neostriatum, hyperstriatum ventrale, and dorsolateral corticoid area) best described as higher order integrative areas of the dorsal cortex and dorsal ventricular ridge.

Cohen & Karten (1974) suggested that these subdivisions of the avian dorsal ventricular ridge are homologous to neuronal populations of single laminae of mammalian neocortex, rather than entire portions of the neocortex. For example, their model suggests that neural populations homologous to individual laminae of mammalian peristriate cortex are scattered throughout various portions of the dorsal ventricular ridge, rather than a specific ridge nucleus (ectostriatum) being homologous to a major subdivision of mammalian neocortex (peristriate cortex). It is not yet clear whether the dorsal ventricular ridge of reptiles is organized in a similar manner to that of birds. *Sphenodon,* for example, possesses a dorsal ventricular ridge composed of a single cortical plate. If this condition is the primitive character state for the reptilian ridge, it is likely that ridge evolution has proceeded independently in different orders of reptiles, birds, and mammals; while the ridge of all these vertebrates may be homologous, its internal organization and many of its neuronal cell types would, therefore, be cases of homoplasy.

Studies on the efferent pathways of the avian telencephalon reveal four major systems: occipitomesencephalic tract, septomesencephalic tract, medial forebrain bundle, and the ansa lenticularis (Adamo 1967, Karten 1971, Zeier & Karten 1971, 1973, Karten & Dubbeldam 1973, Karten et al 1973, Brauth et al 1978). The occipitomesencephalic tract arises from different nuclei in the archistriatum. The anterior two-thirds of the archistriatum (archistriatum anterior and intermedium) give rise to the main portion of the occipitomesencephalic tract whose ipsilateral and contralateral components project to the following centers: (*a*) dorsal thalamus, (*b*) midbrain centers (including the optic tectum), (*c*) lateral reticular formation, (*d*) locus coeruleus, (*e*) lateral pontine nuclei, (*f*) sensory nuclei of the brainstem (including the dorsal column nuclei), and (*g*) dorsal horn of the cervical spinal cord.

The most caudal and medial parts of the archistriatum (archistriatum posterior and mediale) give rise to the pars hypothalami of the occipitomesencephalic tract, which terminates in the medial and lateral hypothalamus. Based on these projections, Zeier & Karten (1971) suggested that only the caudal and medial archistriatum should be homologized to the mammalian amygdala; the anterior and intermediate archistriatum should be homologized to parts of mammalian cortex; and the main portion of the avian occipitomesencephalic tract should be homologized to parts of the mammalian pyramidal system.

The septomesencephalic tract, as already described, arises from the hyperstriatum accessorium and projects to several retino-recipient nuclei in the thalamus and optic tectum, as well as to brain stem and cervical spinal cord nuclei. Similarities between these efferents of the hyperstriatum accessorium and the visual and somatosensory cortices of mammals have been noted, and it has been suggested that these pathways are homologous (Zeier & Karten 1971, Cohen & Karten 1974); however, the apparent absence of comparable long descending telencephalic pathways in reptiles suggests that these pathways in birds and mammals may be homoplastic characters.

The medial forebrain bundle consists of both ascending and descending pathways (Karten & Dubbeldam 1973, Benowitz & Karten 1976, Krayniak & Siegel 1978a,b). The major efferents appear to arise from the septal nuclei and the parolfactory lobe and project mainly to the hypothalamus.

The ansa lenticularis is the major efferent pathway of the paleostriatum. Many areas of the dorsal ventricular ridge project upon the paleostriatum augmentatum, which, in turn, projects upon the paleostriatum primitivum proper and nucleus intrapeduncularis. The latter areas give rise to the ansa lenticularis, which projects to thalamic, pretectal, and midbrain tegmental nuclei (Karten & Dubbeldam 1973, Brauth et al 1978). These nuclei, in turn, form extensive connections with the optic tectum as well as feed-back loops with the paleostriatum (Brecha et al 1976, Brauth et al 1978, Brecha 1978). The overall pattern of connections suggests that the avian paleostriatum augmentatum is homologous to the mammalian caudoputamen and that the avian paleostriatum primitivum and nucleus intrapeduncularis are homologous to the outer and inner segments, respectively, of the mammalian globus pallidus.

A NEW INTERPRETATION OF TELENCEPHALIC EVOLUTION

The presence of paired evaginated hemispheres and olfactory bulbs in both agnathan and gnathostome radiations suggests that such hemispheres were also present in the common ancestor. Thus, there is no evidence that the earliest vertebrates possessed a forebrain consisting of a single vesicle. Although it is possible that the hemispheres arose from a single vesicle in protochordates, recent analysis of the brain vesicle of amphioxus (Meves 1973) indicates that the rostral half of this structure consists of only a single layer of ependyma. Nor is there any trace of an unpaired olfactory organ as claimed by Ariëns Kappers et al (1936). It is likely that both paired olfactory organs and telencephalic hemispheres arose with the origin of the other sense organs (Northcutt 1980). In vertebrates, all special sense organs involve a complex interaction of neural crest and ectoderm, and it is possible that these sense organs, as well as the forebrain, arose with the orgin of vertebrates and are all due to a new ectodermo-neural crest interaction.

Similarly, there are no data to support the notion that the vertebrate telencephalon increases in size linearly through successively more complex vertebrate classes. Increases in telencephalic size, and differentiation of telencephalic cell masses, occur in some members of the agnathan, chondrichthian, osteichthian, and amniotic radiations. This trend is clearly independent within each radiation, and there are insufficient data to establish exactly which telencephalic regions account for most of the change; however, in each of the above radiations, some taxa have clearly enlarged and differentiated parts of the telencephalic pallium. At present these changes appear to be generally restricted to the dorsal or lateral pallium and may be correlated with the origin or hypertrophy of thalamo-telencephalic pathways.

Earlier theories of telencephalic evolution suggested that second and third-order olfactory pathways dominate the cerebral hemispheres of anamniotes and that thalamic pathways first reach the pallium in amniotes (Ariëns Kappers et al 1936). Experimental data are not available for any agnathans, so it is impossible to determine the extent of the olfactory projections to the telencephalon, or whether thalamotelencephalic projections exist in these forms. However, sufficient data exist for anamniotic gnathostomes to positively refute earlier claims of olfactory dominance and absence of ascending thalamo-pallial projections. If future studies reveal a similar pattern in agnathans, we must conclude that the telencephalon was never solely concerned with olfaction and that it possessed diverse sensory integrative functions from its phylogenetic origin.

Earlier theories also claimed that new cell groups gradually emerged out of preexisting, poorly organized, pallial and subpallial "primordia," so that archi- and neostructures developed from preexisting paleostructures; however, three to four cytologically distinct cell groups characterize the pallium and subpallium in every vertebrate radiation. Again, the presence of these telencephalic subdivisions in both agnathans and gnathostomes suggests that they were present in the common ancestor of both radiations and that they characterized the earliest vertebrates. Thus, certain telencephalic characters—such as the presence of a pallium divided into lateral, dorsal, and medial formations and a subpallium divided into striatum and septum—appear to characterize all vertebrates. They are primitive characters and are homologous among all vertebrates. Other characters—such as an amygdalar complex and olfactory tubercle, or character states such as specific differentiation of the pallium and the presence of certain ascending and descending pathways—are less widely distributed among vertebrate radiations and are probably derived characters or character states. Between-group similarities of these characters may be examples of homology or homoplasy.

In examining variation, there is a strong tendency for one to emphasize similarities over differences, and to construct patterns based on those similarities. Neuroanatomists are no exception; for the most part, we have viewed all experimentally revealed similarities among telencephalic cell groups and their connections as homologous characters; however, as more data become available, it is becoming clear that there is a great deal of variation in the differentiation of telencephalic cell masses and pathways. Many of these similarities will probably prove to be examples of homoplasy. Some character states that are almost certainly homoplastic include: (a) the differentiation and lamination of the hagfish pallium, (b) the hypertrophy of the inner lamina of the dorsal pallium, forming a central nucleus, in galeomorph sharks, (c) the hypertrophy of the dorsal pallium in teleost fishes, and (d) the formation of a dorsal ventricular ridge in reptiles. In each case, the pallium itself is a homologous vertebrate character, but the character state (enlargement and differentiation of various parts of the pallium) appears to have occurred independently within each radiation, as indicated by the distribution of this character state among the members of each radiation.

Further examination of descending telencephalo-medullar pathways and their distribution among non-mammalian vertebrates is likely to reveal a number of homoplastic cases. Galeomorph sharks possess pallio-medullar and cervical spinal pathways, but similar pathways do not appear to exist in bony fishes, anuran amphibians or reptiles. Thus, it is likely that these pathways have evolved independently in galeomorph sharks, urodel amphibians, birds, and mammals.

Similar cases have probably occurred during the evolution of ascending thalamo-telencephalic pathways. In amniotes and galeomorph sharks, the major thalamo-telencephalic pathways project to the pallium, whereas in anuran amphibians they project to the striatum. At present we do not know if the anuran condition is an isolated example among anamniotes or if it characterizes several different groups. If the anuran pattern characterizes lampreys, squalomorph sharks, and polypteriform fishes, it is possible that ascending thalamic pathways have invaded the pallium a number of times independently.

The probability of correctly assessing the similarities among any set of telencephalic characters will increase as our knowledge of the variation increases. This means that far more vertebrate species must be examined. It is simply not sufficient to examine one or two species of each vertebrate class and hope to extrapolate a coherent, valid picture of telencephalic evolution. The need to examine many representatives of each class extends even to mammals. Because we incorrectly believe that we know far more about telencephalic organization in mammals than in other vertebrate groups, we usually use mammals as our reference point for comparisons rather than cross-comparing radiations with each other. The limitations of using any single vertebrate group as a "standard" for comparison can be illustrated by examining the data presently available on retino-thalamo-telencephalic pathways among vertebrates. Much recent comparative research has been directed toward a search for possible homologues of the mammalian retino-geniculo-visual cortex system. Most non-mammalian vertebrates possess a retino-anterior thalamo-pallial system that has been homologized to the mammalian system. However, this pathway may be ipsilateral or bilateral; the anterior retino-recipient thalamic complex may project to the medial pallium, dorsal pallium, or, apparently, to part of the dorsal ventricular ridge; the ascending pathway may form part of the lateral or medial forebrain bundles. We have focused on the few similarities at the expense of a number of differences, and no serious attempt has been made to explain these differences. This has happened, in part, because only a single visual pathway was known to exist in mammals. More recently, Conrad & Stumpf (1975) reported a retino-anterior thalamo-retrosplenial pathway in the tree shrew. Thus, it is possible that the non-mammalian retino-anterior thalamic pathway is homologous to this newly discovered pathway in tree shrews, and not to the more classical geniculate pathway. Several other interpretations are also possible; what is important about this example is that the variation observed in the non-mammalian pathway should have suggested several equally plausible interpretations initially, and these interpretations should have led to new experiments both in mammals and in other vertebrates.

A coherent picture of telencephalic evolution still does not exist, but as more species are examined and the data analyzed within the context of modern morphological theory, we will be able to recognize those telencephalic characters that are conservative and homologous across vertebrate radiations. Equally important, we will begin to recognize homoplastic characters and character states. These cases may prove to be even more valuable by directing our attention to the limited ways in which vertebrate nervous systems respond to similar environmental factors. Thus, an examination of homoplastic characters will reveal the "rules" of neural adaptation and how the nervous system evolves.

CONCLUSIONS

The vertebrate telencephalon consists of a caudal telencephalon medium, formed by the preoptic area and the lateral and medial forebrain bundles, and a rostral pair of masses termed the cerebral hemispheres. In all vertebrates, except ray-finned fishes, these hemispheres form by evagination and inversion. In ray-finned fishes, the hemispheres form by eversion, resulting in a mediolateral reversal of the roof or pallium relative to the pallium of other vertebrates. Paired olfactory bulbs arise by evagination from the hemispheres in all vertebrates, but there is considerable variation in the laterality of this evagination. Paired olfactory bulbs and cerebral hemispheres probably arose with the first vertebrates and may be the result of new embryonic interactions between ectoderm and neural crest.

Quantitative data indicate that telencephalic size increases independently in each vertebrate radiation and that this increase occurs, in part, due to the expansion and differentiation of parts of the pallium.

The cerebral hemispheres of all vertebrates are formed by homologous roof (pallial) and floor (subpallial) regions. A comparable number of pallial and subpallial subdivisions occur in all vertebrate radiations; these subdivisions are probably homologous. Recent experimental studies do not support earlier theories suggesting olfactory dominance and absence of ascending thalamo-pallial pathways among anamniotic vertebrates. Instead, they reveal that olfactory projections to the telencephalon of non-mammalian vertebrates are as restricted as they are in mammals, and both ascending thalamo-telencephalic and descending telencephalo-medullar and spinal pathways are widely distributed among non-mammals. As increasing numbers of species within and among the separate vertebrate radiations are examined, it appears that many similarities in telencephalic cell groups and their connections have evolved independently and cannot be considered homologous. These homoplastic telencephalic characters should be more closely analyzed, for they indicate that neural systems respond to similar selective pressures in limited ways, thus revealing the rules of neural adaptation.

ACKNOWLEDGMENTS

I am grateful to Drs. M. Braford, S. Brauth, A. Butler, W. Hodos, H. Karten, T. Neary, A. Reiner, and P. Ulinski for discussing various topics covered in this review and/or permitting me to cite work in press. I also thank Springer-Verlag for allowing me to reproduce Figure 1, and Plenum Press for Figures 6 and 7. Mary Sue Northcutt assisted in many phases of the review. The preparation of the manuscript and some of the work presented was supported by NIH grants NS11006 and EYO2485.

Literature Cited

Adamo, N. J. 1967. Connections of efferent fibers from hyperstriatal areas in chicken, raven and African love-bird. *J. Comp. Neurol.* 131:337–56

Andry, M. L., Northcutt, R. G. 1976. Telencephalic visual responses in the lizard *Gekko gecko, Neurosci. Abstr.* 2:176

Ariëns Kappers, C. U. 1921. *Vergleichende Anatomie des Nervensystems,* Vol. 2. Haarlem: Bohn

Ariëns Kappers, C. U., Huber, G. C., Crosby, E. C. 1936. *The Comparative Anatomy of the Nervous System of Vertebrates, Including Man.* Reprinted 1960. New York: Hafner. 1845 pp.

Bäckström, K. 1924. Contributions to the forebrain morphology in selachians. *Acta Zool.* 5:123–240

Balaban, C. D., Ulinski, P. S. 1980. Organization of thalamic afferents to anterior dorsal ventricular ridge in turtles. I. Projections of thalamic nuclei. Submitted for publication

Bass, A. H. 1976. Retinal projections in the painted turtle, *Chrysemys picta. Neurosci. Abstr.* 2:177

Bass, A. H. 1979. *Telencephalic afferents and efferents in the channel catfish, Ictalurus punctatus.* PhD thesis. Univ. Mich., Ann Arbor. 214 pp.

Benowitz, L. I. 1980. Functional organization of the avian telencephalon. In *Comparative Neurology of the Telencephalon,* ed. S. O. E. Ebbesson, 13:389–421. New York: Plenum. 506 pp.

Benowitz, L. I., Karten, H. J. 1976. The tractus infundibuli and other afferents to the parahippocampal region of the pigeon. *Brain Res.* 102:174–80

Bodznick, D., Northcutt, R. G. 1979. Some connections of the lateral olfactory area of the horn shark. *Neurosci. Abstr.* 5:139

Bone, Q. 1963. The central nervous system. In *The Biology of Myxine,* ed. A. Brodal, R. Fänge, pp. 50–91. Oslo: Universitetsforlaget. 588 pp.

Bonke, B. A., Bonke, D., Scheich, H. 1979. Connectivity of the auditory forebrain nuclei in the guinea fowl (*Numida meleagris*). *Cell Tissue Res.* 200:101–21

Boord, R. L. 1968. Ascending projections of the primary cochlear nuclei and nucleus laminaris in the pigeon. *J. Comp. Neurol.* 133:523–41

Braford, M. R. Jr., Northcutt, R. G. 1974. Olfactory bulb projections in the bichir, *Polypterus. J. Comp. Neurol.* 156:165–78

Braford, M. R. Jr., Northcutt, R. G. 1978. Correlation of telencephalic afferents and SDH distribution in the bony fish *Polypterus. Brain Res.* 152:157–60

Brauth, S. E., Ferguson, J. L., Kitt, C. A. 1978. Prosencephalic pathways related to the paleostriatum of the pigeon (*Columba livia*). *Brain Res.* 147:205–21

Brecha, N. C. 1978. *Some observations on the organization of the avian optic tectum: Afferent nuclei and their tectal projections.* PhD thesis. State Univ. New York, Stony Brook. 155 pp.

Brecha, N. C., Hunt, S. P., Karten, H. J. 1976. Relations between the optic tectum and basal ganglia in the pigeon. *Neurosci. Abstr.* 2:1069

Bruckmoser, P. 1973. Beziehungen zwischen Struktur und Funktion in der Evolution des Telencephalon. *Verhandlungen der Deutschen Zoologischen Gesellschaft* 66:219–29

Bruckmoser, P., Dieringer, N. 1973. Evoked potentials in the primary and secondary olfactory projection areas of the forebrain in Elasmobranchia. *J. Comp. Physiol.* 87:65–74

Bullock, T. H., Corwin, J. T. 1979. Acoustic evoked activity in the brain in sharks. *J. Comp. Physiol.* 129:223–34

Butler, A. B. 1975. Telencephalic afferents in *Gekko gecko. Anat. Rec.* 181:323 (Abstr.)

Butler, A. B. 1976. Telencephalon of the lizard, *Gekko gecko* (Linnaeus): Some connections of the cortex and dorsal ventricular ridge. *Brain Behav. Evol.* 13:396–417

Butler, A. B. 1978. Forebrain connections in lizards and the evolution of sensory systems. In *Behavior and Neurology of Lizards,* ed. N. Greenberg, P. D. Maclean, pp. 65–78. Rockville, Md: Natl. Inst. Mental Health. DHEW Publ. No. (ADM) 77–491

Butler, A. B. 1980. Cytoarchitectonic and connectional organization of the lacertilian telencephalon with comments on vertebrate forebrain evolution. See Benowitz 1980, 10:297–329

Butler, A. B., Northcutt, R. G. 1978. New thalamic visual nuclei in lizards. *Brain Res.* 149:469–76

Cohen, D. H., Duff, T. A., Ebbesson, S. O. E. 1973. Electrophysiological identification of a visual area in shark telencephalon. *Science* 182:492–94

Cohen, D. H., Karten, H. J. 1974. The structural organization of avian brain: An overview. In *Birds Brain and Behavior,* ed. I. J. Goodman, M. W. Schein, pp. 29–73. New York: Academic. 273 pp.

Conrad, C. D., Stumpf, W. E. 1975. Direct visual input to the limbic system: Crossed retinal projections to the nucleus anterodorsalis thalami in the tree shrew. *Exp. Brain Res.* 23:141–49

Crosby, E. C., DeJonge, B. R., Schneider, R. C. 1967. Evidence for some of the trends in the phylogenetic development of the vertebrate telencephalon. In *Evolution of the Forebrain,* ed. R. Hassler, H. Stephan, pp. 117–35. New York: Plenum. 464 pp.

Crosby, E. C., Humphrey, T. 1939. Studies of the vertebrate telencephalon. I. The nuclear configuration of the olfactory and accessory olfactory formations and of the nucleus olfactorius anterior in certain reptiles, birds and mammals. *J. Comp. Neurol.* 71:121–213

Curwen, A. O. 1939. The telencephalon of *Tupinambis nigropunctatus.* III. Amygdala. *J. Comp. Neurol.* 71:613–36

Cuvier, G. 1809. *Lecons d'Anatomie Comparee, die Erste Auglage,* deutsch v. Leipzig: Meckel. 171 pp.

Dart, R. A. 1934. The dual structure of the neopallium: Its history and significance. *J. Anat. London* 69:1–19

Delius, J. D., Bennetto, K. 1972. Cutaneous sensory projections to the avian forebrain. *Brain Res.* 37:205–22

Distel, H., Ebbesson, S. O. E. 1975. Connections of the thalamus in the monitor lizard. *Neurosci. Abstr.* 1:559

Ebbesson, S. O. E. 1972. New insights into the organization of the shark brain. *Comp. Biochem. Physiol. A* 42:121–29

Ebbesson, S. O. E. 1980. On the organization of the telencephalon in elasmobranchs. See Benowitz 1980, 1:1–16

Ebbesson, S. O. E., Heimer, L. 1970. Projections of the olfactory tract fibers in the nurse shark (*Ginglymostoma cirratum*). *Brain Res.* 17:47–55

Ebbesson, S. O. E., Northcutt, R. G. 1976. Neurology of anamniotic vertebrates. In *Evolution of Brain and Behavior in Vertebrates,* ed. R. B. Masterton, M. E. Bitterman, C. B. G. Campbell, N. Hotton, 7:115–46. Hillsdale, NJ: Lawrence Erlbaum. 482 pp. (Distr. Halsted Press, John Wiley, NY)

Ebbesson, S. O. E., Schroeder, D. M. 1971. Connections of the nurse shark's telencephalon. *Science* 173:254–56

Edinger, L. 1908. *Vorlesungen über den Bau der nervösen Zentralorgane.* Leipzig: Vogel. 334 pp.

Finger, T. E. 1975. The distribution of the olfactory tracts in the bullhead catfish, *Ictalurus nebulosus. J. Comp. Neurol.* 161:125–42

Finger, T. E. 1979. A thalamic relay nucleus for the lateral line system in teleost fish. *Neurosci. Abstr.* 5:141

Foster, R. E., Hall, W. C. 1978. The organization of central auditory pathways in a reptile, *Iguana iguana*. *J. Comp. Neurol.* 178:783–831

Gage, S. P. 1893. *The Brain of Diemyctilus viridescens from Larval to Adult Life and Comparison with the Brain of Amia and Petromyzon*, pp. 259–314. Ithaca, NY: Wilder

Gamble, H. J. 1952. An experimental study of the secondary olfactory connexions in *Lacerta viridis*. *J. Anat. London* 86:180–96

Gamble, H. J. 1956. An experimental study of the secondary olfactory connexions in *Testudo graeca*. *J. Anat. London* 90:15–29

Goldby, F. 1937. An experimental investigation of the cerebral hemispheres of *Lacerta viridis*. *J. Anat. London* 71:332–55

Goldby, F., Gamble, H. J. 1957. The reptilian cerebral hemispheres. *Biol. Rev.* 32:383–420

Goldstein, K. 1905. Untersuchungen über das Vorderhirn und Zwischenhirn einiger Knochenfische. *Arch. Mikro. Anat.* 66:135–219

Gruberg, E. R., Ambros, V. R. 1974. A forebrain visual projection in the frog (*Rana pipiens*). *Exp. Neurol.* 44:187–97

Gusel'nikov, V. I., Supin, A. Y. 1964. Visual and auditory regions in hemispheres of lizard forebrain. *Trans. Supp. Fed. Am. Soc. Exp. Biol.* 23: (3) Pt. II, pp. 641–46

Hall, J. A., Foster, R. E., Ebner, F. F., Hall, W. C. 1977. Visual cortex in a reptile, the turtle (*Pseudemys scripta* and *Chrysemys picta*). *Brain res.* 130:197–216

Hall, W. C., Ebner, F. F. 1970. Thalamo-telencephalic projections in the turtle *Pseudemys scripta*. *J. Comp. Neurol.* 140:101–22

Halpern, M. 1972. Some connections of the telencephalon of the frog, *Rana pipiens*. An experimental study. *Brain Behav. Evol.* 6:42–68

Halpern, M. 1976. The efferent connections of the olfactory bulb and accessory olfactory bulb in the snakes, *Thamnophis sirtalis* and *Thamnophis radix*. *J. Morphol.* 150:553–78

Halpern, M. 1980. The telencephalon of snakes. See Benowitz 1980, 9:257–95

Halpern, M., Silfen, R. 1974. The efferent connections of the nucleus sphericus in the garter snake, *Thamnophis sirtalis*. *Anat. Rec.* 178:368 (Abstr.)

Heier, P. 1948. Fundamental principles in the structure of the brain: A study of the brain of *Petromyzon fluviatilis*. *Acta Anat.* 5: Suppl. 8, pp. 1–213

Heimer, L. 1969. The secondary olfactory connections in mammals, reptiles and sharks. *Ann. NY Acad. Sci.* 167:129–46

Herrick, C. J. 1948. *The Brain of the Tiger Salamander*. Chicago: Univ. Chicago Press. 409 pp.

Holmgren, N. 1922. Points of view concerning forebrain morphology in lower vertebrates, *J. Comp. Neurol.* 34:391–459

Holmgren, N., van der Horst, C. J. 1925. Contributions to the morphology of the brain of *Ceratodus*. *Acta Zool.* 6:59–165

Hoogland, P. V. 1977. Efferent connections of the striatum in *Tupinambis nigropunctatus*. *J. Morphol.* 152:229–46

Hoogland, P. V., ten Donkelaar, H. J., Cruce, J. A. F. 1978. Efferent connections of the septal area in a lizard (*Tupinambis nigropunctatus*). *Neurosci. Lett.* 7:61–65

Hopson, J. A. 1979. Paleoneurology. In *Biology of the Reptilia*, Vol. 9, *Neurology A*, ed. C. Gans, R. G. Northcutt, P. Ulinski, 2:39–146. London: Academic. 462 pp.

Hotton, N. III. 1976 Origin and radiation of the classes of poikilothermous vertebrates. See Ebbesson & Northcutt 1976, 1:1–24

Hunt, S. P., Webster, K. E. 1972. Thalamo-hyperstriate interrelations in the pigeon. *Brain Res.* 44:647–51

Ito, H., Kishida, R. 1977. Tectal afferent neurons identified by the retrograde HRP method in the carp telencephalon. *Brain Res.* 130:142–45

Ito, H., Kishida, R. 1978. Telencephalic afferent neurons identified by the retrograde HRP method in the carp diencephalon. *Brain Res.* 149:211–15

Jansen, J. 1930. The brain of *Myxine glutinosa*. *J. Comp. Neurol.* 49:359–507

Jerison, H. J. 1973. *Evolution of the Brain and Intelligence*. New York: Academic. 482 pp.

Johnston, J. B. 1906. *The Nervous System of Vertebrates*. Philadelphia: Blakiston's. 370 pp.

Johnston, J. B. 1911. The telencephalon of ganoids and teleosts. *J. Comp. Neurol.* 21:489–591

Källén, B. 1951a. Embryological studies on the nuclei and their homologization in the vertebrate forebrain. *K. Fysiogr. Saellsk. Handl.* 62:1–36

Källén, B. 1951b. On the ontogeny of the reptilian forebrain. Nuclear structures and ventricular sulci. *J. Comp. Neurol.* 95:307–47

Källén, B. 1962. Embryogensis of brain nuclei in the chick telencephalon. *Ergeb. Anat. Entwicklungsgesch.* 36:62–82

Karten, H. J. 1967. The organization of the ascending auditory pathway in the pigeon (*Columba livia*). I. Diencephalic projections of the inferior colliculus (nucleus mesencephalicus lateralis, pars dorsalis). *Brain Res.* 6:409–27

Karten, H. J. 1968. The ascending auditory pathway in the pigeon (*Columba livia*). II. Telencephalic projections of the nucleus ovoidalis thalami. *Brain Res.* 11:134–53

Karten, H. J. 1969. The organization of the avian telencephalon and some speculations on the phylogeny of the amniote telencephalon. *Ann. NY Acad. Sci.* 167:164–79

Karten, H. J. 1971. Efferent projections of the wulst of the owl. *Anat. Rec.* 169:353 (Abstr.)

Karten, H. J., Dubbeldam, J. L. 1973. The organization and projections of the paleostriatal complex in the pigeon (*Columba livia*). *J. Comp. Neurol.* 148:61–90

Karten, H. J., Finger, T. E. 1976. A direct thalamo-cerebellar pathway in pigeon and catfish. *Brain Res.* 102:335–38

Karten, H. J., Hodos, W. 1970. Telencephalic projections of the nucleus rotundus in the pigeon (*Columba livia*). *J. Comp. Neurol.* 140:35–52

Karten, H. J., Hodos, W., Nauta, W. J. H., Revzin, A. M. 1973. Neural connections of the "visual wulst" of the avian telencephalon. Experimental studies in the pigeon (*Columba livia*) and owl (*Speotyto cunicularia*). *J. Comp. Neurol.* 150:253–77

Karten, H. J., Konishi, M., Pettigrew, J. 1978. Somatosensory representation in the anterior wulst of the owl (*Speotyto cunicularia*). *Neurosci. Abstr.* 4:554

Karten, H. J., Nauta, W. J. H. 1968. Organization of retinothalamic projections in the pigeon and owl. *Anat. Rec.* 160:373 (Abstr.)

Karten, H. J., Revzin, A. M. 1966. The afferent connections of the nucleus rotundus in the pigeon. *Brain Res.* 2:368–77

Kicliter, E. 1979. Some telencephalic connections in the frog, *Rana pipiens J. Comp. Neurol.* 185:75–86

Kicliter, E., Ebbesson, S. O. E. 1976. Organization of the "non-olfactory" telencephalon. In *Frog Neurobiology*, ed. R. Llinás, W. Precht, 34:946–72. New York: Springer-Verlag. 1046 pp.

Kicliter, E., Northcutt, R. G. 1975. Ascending afferents to the telencephalon of ra-

nid frogs: An anterograde degeneration study. *J. Comp. Neurol.* 161:239–54

Kluge, A. G. 1977. Concepts and principles of morphologic and functional studies. In *Chordate Structure and Function*, ed. A. G. Kluge, 1:1–27. New York: Macmillan. 628 pp. 2nd ed.

Kokoros, J. J. 1973. *Efferent connections of the telencephalon in the toad, Bufo marinus, and the tiger salamander, Ambystoma tigrinum.* PhD thesis. Case Western Reserve Univ., Cleveland, Ohio. 141 pp.

Kokoros, J. J., Northcutt, R. G. 1977. Telencephalic efferents of the tiger salamander *Ambystoma tigrinum tigrinum* (Green). *J. Comp. Neurol.* 173:613–28

Krayniak, P. F., Siegel, A. 1978a. Efferent connections of the septal area in the pigeon. *Brain Behav. Evol.* 15:389–404

Krayniak, P. F., Siegel, A. 1978b. Efferent connections of the hippocampus and adjacent regions in the pigeon. *Brain Behav. Evol.* 15:372–88

Kuhlenbeck, H. 1938. The ontogenetic development and phylogenetic significance of the cortex telencephali in the chick. *J. Comp. Neurol.* 69:273–302

Lohman, A. H. M., Mentink, G. M. 1972. Some cortical connections of the tegu lizard (*Tupinambis teguixin*). *Brain Res.* 45:325–44

Lohman, A. H. M., van Woerden-Verkley, I. 1976. Further studies on the cortical connections of the tegu lizard. *Brain Res.* 103:9–28

Lohman, A. H. M., van Woerden-Verkley, I. 1978. Ascending connections to the forebrain in the tegu lizard. *J. Comp. Neurol.* 182:555–94

Mayr, E. 1969. *Principles of Systematic Zoology.* New York: McGraw-Hill, 428 pp.

Meier, R. E., Mihailović, J., Cuénod, M. 1974. Thalamic organization of the retino-thalamo-hyperstriatal pathway in the pigeon (*Columba livia*). *Exp. Brain Res.* 19:351–64

Meves, A. 1973. Elektronenmikroskopische Untersuchungen über die Zytoarchitektur des Gehirns von *Branchiostoma lanceolatum*. *Z. Zellforsch. Mikrosk. Anat.* 139:511–32

Miceli, D., Gioanni, H., Repérant, J., Peyrichoux, J. 1979. The avian visual wulst. I. An anatomical study of afferent and efferent pathways. II. An electrophysiological study of the functional properties of single neurons. In *Neural Mechanisms of Behavior in the Pigeon*, ed. A. M. Granda, J. H. Maxwell, 13:223–54. New York: Plenum. 436 pp.

Miceli, D., Peyrichoux, J., Repérant, J. 1975. The retino-thalamo-hyperstriatal pathway in the pigeon (*Columba livia*). *Brain Res.* 100:125–31

Mihailović, J., Perisić, M., Bergonzi, R., Meier, R. E. 1974. The dorsolateral thalamus as a relay in the retino-wulst pathway in pigeon (*Columba livia*). An electrophysiological study. *Exp. Brain Res.* 21:229–40

Morgan, G. C. Jr. 1975. The telencephalon of the sea catfish *Galeichthys felis. J. Hirnforsch.* 16:131–50

Mudry, K. M., Capranica, R. R. 1978. Electrophysiological evidence for auditory responsive areas in the diencephalon and telencephalon of the bullfrog, *Rana catesbeiana. Neurosci. Abstr.* 4:101

Neary, T. J. 1974. Diencephalic afferents of the torus semicircularis in the bullfrog, *Rana catesbeiana. Anat. Rec.* 178:425 (Abstr.)

Neary, T. J., 1980. Hypothalamic afferents in ranid frogs. *Anat. Rec.* 196:135A (Abstr.)

Neary, T. J., Wilczynski, W. 1977a. Autoradiographic demonstration of hypothalamic efferents in the bullfrog, *Rana catesbeiana. Anat. Rec.* 187:665 (Abstr.)

Neary, T. J., Wilczynski, W. 1977b. Ascending thalamic projections from the obex region in ranid frogs. *Brain Res.* 138:529–33

Nieuwenhuys, R. 1962. Trends in the evolution of the actinopterygian forebrain. *J. Morphol.* 111:69–88

Nieuwenhuys, R. 1963. The comparative anatomy of the actinopterygian forebrain. *J. Hirnforsch.* 6:171–92

Nieuwenhuys, R. 1967. Comparative anatomy of olfactory centres and tracts. *Progr. Brain Res.* 23:1–64

Nieuwenhuys, R. 1969. A survey of the structure of the forebrain in higher bony fishes (Osteichthyes). *Ann. NY Acad. Sci.* 167:31–64

Nieuwenhuys, R. 1977. The brain of the lamprey in a comparative perspective. *Ann. NY Acad. Sci.* 299:97–145

Nieuwenhuys, R., Bauchot, R., Arnoult, J. 1969. Le développement du télencéphale d'un poisson osseux primitif, *Polypterus senegalus* Cuvier. *Acta Zool.* 50:101–25

Northcutt, R. G. 1967. Architectonic studies of the telencephalon of *Iguana iguana. J. Comp. Neurol.* 130:109–47

Northcutt, R. G. 1968. Descending axon degeneration following ablation of the telencephalic cortex in the horned liz-

ard, *Phrynosma cornutum. Anat. Rec.* 160:400 (Abstr.)

Northcutt, R. G. 1969. A re-evaluation of the evolution of the tetrapod telencephalon. *Anat. Rec.* 163:318 (Abstr.)

Northcutt, R. G. 1970. The telencephalon of the western painted turtle, *Chrysemys picta belli.* Ill. Biol. Monogr. No. 43. Urbana: Univ. Ill. Press. 113 pp.

Northcutt, R. G. 1974. Some histochemical observations on the telencephalon of the bullfrog, *Rana catesbeiana* Shaw. *J. Comp. Neurol.* 157:379–90

Northcutt, R. G. 1978a. Brain organization in the cartilaginous fishes. In *Sensory Biology of Sharks, Skates, and Rays,* ed. E. S. Hodgson, R. F. Mathewson, pp. 117–93. Arlington: Off. Naval Res. 666 pp.

Northcutt, R. G. 1978b. Forebrain and midbrain organization in lizards and its phylogenetic significance. See Butler 1978, pp. 11–64

Northcutt, R. G. 1980. Central auditory pathways in anamniotic vertebrates. In *Comparative Studies of Hearing in Vertebrates,* ed. A. N. Popper, R. Fay, 3:79–118. New York: Springer-Verlag

Northcutt, R. G., Braford, M. R. Jr. 1980. New observations on the organization and evolution of the telencephalon of actinopterygian fishes. See Benowitz 1980, 3:41–98

Northcutt, R. G., Braford, M. R. Jr., Landreth, G. E. 1974. Retinal projections in the tuatara *Sphenodon punctatus:* An autoradiographic study. *Anat. Rec.* 178:428 (Abstr.)

Northcutt, R. G., Kicliter, E. 1980. Organization of the amphibian telencephalon. See Benowitz 1980, 8:203–55

Northcutt, R. G., Pritz, M. B. 1978. A spinothalamic pathway to the dorsal ventricular ridge in the spectacled caiman, *Caiman crocodilus. Anat. Rec.* 190:618–19 (Abstr.)

Northcutt, R. G., Royce, G. J. 1975. Olfactory bulb projections in the bullfrog *Rana catesbeiana. J. Morphol.* 145:251–68

Northcutt, R. G., Wathey, J. C. 1979. Some connections of the skate dorsal and medial pallia. *Neurosci. Abstr.* 5:145 (Abstr.)

Nottebohm, F., Stokes, T. M., Leonard, C. M. 1976. Central control of song in the canary, *Serinus canarius. J. Comp. Neurol.* 165:457–86

Olson, E. C. 1971. *Vertebrate Paleozoology.* New York: Wiley, 839 pp.

Owen, R. 1866. *On the Anatomy of Verte-*

brates, Vol. 1, *Fishes and Reptiles.* London: Longmans, Green. 650 pp.

Papez, J. 1929. *Comparative Neurology.* New York: Crowell. 518 pp.

Parent, A. 1976. Striatal afferent connections in the turtle (*Chrysemys picta*) as revealed by retrograde axonal transport of horseradish peroxidase. *Brain Res.* 108:25–36

Parent, A., Dube, L., Braford, M. R. Jr., Northcutt, R. G. 1978. The organization of monoamine-containing neurons in the brain of the sunfish (*Lepomis gibbosus*) as revealed by fluorescence microscopy. *J. Comp. Neurol.* 182:495–516

Pettigrew, J. D., Konishi, M. 1976. Neurons selective for orientation and binocular disparity in the visual wulst of the barn owl (*Tyto alba*). *Science* 193:675–78

Platel, R. 1969. Étude cytoarchitectonique qualitative et quantitative des aires corticales d'un Saurien: *Scincus scincus* (L.) Scincidés. *J. Hirnforsch.* 11:31–66

Platel, R. 1971. Étude cytoarchitecturale qualitative et quantitative des aires basales d'un Saurien Scincide: *Scincus scincus* (L.). I. Étude cytoarchitecturale qualitative. *J. Hirnforsch.* 13:65–87

Platt, C. J., Bullock, T. H., Czéh, G., Kovačević, N., Konjević, D., Gojković, M. 1974. Comparison of electroreceptor, mechanoreceptor, and optic evoked potentials in the brain of some rays and sharks. *J. Comp. Physiol.* 95:323–55

Pritz, M. B. 1973. Connections of the alligator visual system: Telencephalic projections of nucleus rotundus. *Anat. Rec.* 175:416 (Abstr.)

Pritz, M. B. 1974a. Ascending connections of a midbrain auditory area in a crocodile, *Caiman crocodilus, J. Comp. Neurol.* 153:179–98

Pritz, M. B. 1974b. Ascending connections of a thalamic auditory area in a crocodile, *Caiman crocodilus. J. Comp. Neurol.* 153:199–214

Ramón y Cajal, S. 1908–11. *Histologie du Système Nerveux de l'Homme et des Vértébrés.* Paris: Maloine. Repr. Consejo Superior de Investigaciones Cientificas, Madrid, 1972. 1979 pp.

Reiner, A., Brauth, S. E., Kitt, C. A., Karten, H. J. 1980. Basal ganglionic pathways to the tectum: Studies in reptiles. *J. Comp. Neurol.* 193:565–89

Repérant, J. 1973. Nouvelles données sur les projections visuelles chez le pigeon (*Columba livia*). *J. Hirnforsch.* 14:151–87

Repérant, J. 1976. Afférences et efférences télencéphaliques du cortex dorsal de la vipère (*Vipera aspis* L.), données préliminaires. *C. R. Acad. Sci. Ser. D.* 283:809–11

Revzin, A. M., Karten, H. J. 1966. Rostral projections of the optic tectum and the nucleus rotundus in the pigeon. *Brain Res.* 3:264–76

Rieke, G. K., Wenzel, B. M. 1978. Forebrain projections of the pigeon olfactory bulb. *J. Morphol.* 158:41–56

Ritchie, T. C., Cohen, D. H. 1977. The avian tectofugal visual pathway: Projections of its telencephalic target, the ectostriatal complex. *Neurosci. Abstr.* 3:94

Romer, A. S. 1966. *Vertebrate Paleontology.* Chicago: Univ. Chicago Press. 468 pp.

Romer, A. S. 1970. *The Vertebrate Body.* Philadelphia: Saunders. 601 pp. 4th ed.

Ronan, M. C., Northcutt, R. G. 1979. Afferent and efferent connections of the bullfrog medial pallium. *Neurosci. Abstr.* 5:146

Rubinson, K. 1968. Projections of the tectum opticum of the frog. *Brain Behav. Evol.* 1:529–61

Rubinson, K., Colman, D. R. 1972. Designated discussion: A preliminary report on ascending thalamic afferents in *Rana pipiens. Brain Behav. Evol.* 6:69–74

Scalia, F. 1972. The projection of the accessory olfactory bulb in the frog. *Brain Res.* 36:409–11

Scalia, F. 1976a. Structure of the olfactory and accessory olfactory systems. See Kicliter & Ebbesson 1976, 6:213–33

Scalia, F. 1976b. The optic pathway of the frog: Nuclear organization and connections. See Kicliter & Ebbesson 1976, 11:386–406

Scalia, F., Colman, D. R. 1975. Identification of telencephalic-afferent thalamic nuclei associated with the visual system of the frog. *Neurosci. Abstr.* 1:46

Scalia, F., Ebbesson, S. O. E. 1971. The central projections of the olfactory bulb in a teleost (*Gymnothorax funebris*). *Brain Behav. Evol.* 4:376–99

Scalia, F., Gregory, K. 1970. Retinofugal projections in the frog: Location of the postsynaptic neurons. *Brain Behav. Evol.* 3:16–29

Scalia, F., Halpern, M., Knapp, H., Riss, W. 1968. The efferent connexions of the olfactory bulb in the frog: A study of degenerating unmyelinated fibers. *J. Anat.* 103:245–62

Scalia, F., Halpern, M., Riss, W. 1969. Olfactory bulb projections in the South American caiman. *Brain Behav. Evol.* 2:238–62

Schaeffer, B. 1969. Adaptive radiation of the

fishes and the fish-amphibian transition. *Ann. NY Acad. Sci.* 167:5–17

Scheich, H., Bonke, B. A., Bonke, D., Langner, G. 1979. Functional organization of some auditory nuclei in the guinea fowl demonstrated by the 2-deoxyglucose technique. *Cell Tissue Res.* 204:17–27

Schnitzlein, H. N. 1968. Introductory remarks on the telencephalon of fish. In *The Central Nervous System and Fish Behavior*, ed. D. Ingle, pp. 97–100. Chicago: Univ. Chicago Press. 292 pp.

Schroeder, D. M. 1980. The telencephalon of teleosts. See Benowitz 1980, 4:99–115

Schroeder, D. M., Ebbesson, S. O. E. 1974. Nonolfactory telencephalic afferents in the nurse shark (*Ginglymostoma cirratum*). *Brain Behav. Evol.* 9:121–55

Simpson, G. G. 1953. *The Major Features of Evolution.* New York: Columbia Univ. Press. 434 pp.

Simpson, G. G. 1961. *Principles of Animal Taxonomy.* New York: Columbia Univ. Press. 247 pp.

Stingelin, W. 1958. *Vergleichend Morphologische Untersuchungen am Vorderhirn der Vögel auf Cytologischer und Cytoarchitektonischer Grundlage.* Basel: Helbing & Lichtenhahn. 123 pp.

Studnicka, F. K. 1896. Beiträge zur Anatomie und Entwicklungsgeschichte des Vorderhirns der Cranioten. Sitzungsber. *Akad. Boehm. Wiss. Math. Nat. Kl.*, Abt. 2, pp. 1–32

Ulinski, P. S. 1975. Corticoseptal projections in the snakes *Natrix sipedon* and *Thamnophis sirtalis.* *J. Comp. Neurol.* 164:375–88

Ulinski, P. S. 1976. Structure of anterior dorsal ventricular ridge in snakes. *J. Morphol.* 148:1–22

Ulinski, P. S. 1978. A working concept of the organization of the anterior dorsal ven-tricular ridge. See Butler 1978, pp. 121–32

Vanegas, H., Ebbesson, S. O. E. 1976. Telencephalic projections in two teleost species. *J. Comp. Neurol.* 165:181–96

Voneida, T. J., Ebbesson, S. O. E. 1969. On the origin and distribution of axons in the pallial commissures in the tegu lizard (*Tupinambis nigropunctatus*). *Brain Behav. Evol.* 2:467–81

Voneida, T. J., Sligar, C. A. 1979. Efferent projections of the dorsal ventricular ridge and the striatum in the tegu lizard, *Tupinambis nigropunctatus.* *J. Comp. Neurol.* 186:43–64

Wang, R. T., Halpern, M. 1977. Afferent and efferent connections of thalamic nuclei of the visual system of garter snakes. *Anat. Rec.* 187:741–42 (Abstr.)

Ware, C. B. 1974. Projections of dorsal cortex in the side necked turtle (*Podocnemis unifilis*). *Soc. Neurosci. 4th Ann. Meet., St. Louis, 1974,* p. 466 (Abstr.)

Wilczynski, W., Northcutt, R. G. 1979. Striatal efferents in the bullfrog, *Rana catesbeiana.* *Neurosci. Abstr.* 5:147

Wright, J. D. 1967. The telencephalon of the bichir, *Polypterus.* *Ala. J. Med. Sci.* 4:252–73

Zaretsky, M. D., Konishi, M. 1976. Tonotopic organization in the avian telencephalon. *Brain Res.* 111:167–71

Zeier, H., Karten, H. J. 1971. The archistriatum of the pigeon: Organization of afferent and efferent connections. *Brain Res.* 31:313–26

Zeier, H. J., Karten, H. J. 1973. Connections of the anterior commissure in the pigeon (*Columba livia*). *J. Comp. Neurol.* 150:201–16

Zeigler, H. P., Karten, H. J. 1973. Brain mechanisms and feeding behavior in the pigeon (*Columba livia*). Quinto-frontal structures. *J. Comp. Neurol.* 152:59–81

Ann. Rev. Neurosci. 1981. 4:351–79
Copyright © 1981 by Annual Reviews Inc. All rights reserved

SYNAPTIC PLASTICITY IN THE MAMMALIAN CENTRAL NERVOUS SYSTEM

❖11557

Nakaakira Tsukahara

Department of Biophysical Engineering, Faculty of Engineering Science, Osaka University, Toyonaka, Osaka, and National Institute for Physiological Sciences, Okazaki, Japan

INTRODUCTION

More than a decade has passed since Kandel & Spencer (1968) in their excellent review discussed cellular neurophysiological approaches to the study of learning. The understanding of plasticity, defined as any persistent change in the functional properties of single neurons or neuronal aggregates, is a prerequisite for the neurophysiological study of learning. Since 1968, we have increased greatly our understanding of the nature of these plastic changes in the central nervous system. This paper reviews this recent literature. In particular I focus on four areas.

1. Plasticity of synaptic transmission: I review in detail the remarkable phenomenon of potentiation.
2. Axonal sprouting and formation of new synapses: Although this phenomenon was first demonstrated in the peripheral nervous system in the 1950s, it has been difficult until recently to demonstrate unequivocally the sprouting of central axonal connections in the central nervous system.
3. Plasticity of neuronal networks: It has long been postulated that properties of reverberating circuits form a basis for short-term memory; our present knowledge of such circuits is examined in detail.
4. I review the studies that attempt to link behavioral and cellular plasticity. Because a major goal of the study of neuronal plasticity is to explain behavioral changes (such as "learning and memory") in terms

351

0147-006X/81/0301-0351$01.00

of cellular or network plasticity in the central nervous system, these studies provide a framework for analysis of the biological significance of plastic changes.

Because of space limitation, no attempt is made to provide an extensive coverage of the literature; discussion is confined to studies of synaptic and network plasticity in the mammalian central nervous system. Recent advances in the study of plasticity of the visual system after neonatal modification of retinal inputs, adaptive modification of vestibular-ocular reflex, and effects of functional deprivation or stimulation on the development of dendritic morphology are not reviewed; however, many reviews covering portions of this general problem have appeared, to which the reader may refer (Raisman 1977, Cotman & Lynch 1976, Teyler 1978, Björklund & Stenevi 1979, Eccles 1976, Kandel 1977, Thompson et al 1972, Lund 1978).

PLASTICITY OF SYNAPTIC TRANSMISSION

Well-known examples of synaptic transmission in the mammalian nervous system are frequency potentiation (Eccles et al 1961, Langren et al 1962) and posttetanic potentiation (Lloyd 1949, Curtis & Eccles 1960, Hughes et al 1956). These phenomena have been reported at a variety of central synapses (for review see Kandel & Spencer 1968). Attempts have been made to demonstrate longer duration posttetanic potentiation. In the spinal cord, Lloyd (1949) found potentiation lasting for several minutes. With very long trains of tetanus, Spencer & Wigdor (1965) produced posttetanic potentiation (PTP) that lasted for hours. Attempts to prolong the duration of the PTP in the spinal cord still further have not been successful, partly because longer trains of stimuli produce depression of nerve conduction at the terminals.

Long-Term Potentiation

In 1973, Bliss & Lømo found a long-lasting PTP in the granule cells of dentate gyrus following stimulation of the perforant path in anesthetized rabbits. The potentiation lasted for many hours and hence has been termed *long-term potentiation* (LTP). Because of its time course and the small number of tetanic stimuli required, LTP is an extraordinarily interesting example of synaptic plasticity. The perforant path constitutes a monosynaptic pathway from entorhinal cortex, terminating the outer two-thirds of the dendrites of the granule cells. After stimulating this pathway, Bliss & Lømo found potentiation lasting hours of both extracellularly recorded monosynaptic excitatory postsynaptic potentials (EPSPs) and the postsynaptic population spikes of granule cells. Subsequently, Bliss & Gardner-Medwin (1971, 1973) reported long-term potentiation in unanesthetized rabbits that lasted up to three days. Further work by Douglas & Goddard

(1975) demonstrated even longer-lasting potentiation, lasting up to 12 days, in unanesthetized rats.

Long-term potentiation differs from posttetanic potentiation in several ways: LTP starts later than PTP; it lasts longer than PTP: and it is produced by relatively low frequency stimulus trains (10 to 20 Hz). Possible mechanisms for LTP are (a) an increase in the size of afferent volley, (b) an increase in the excitability of the cell population, (c) increased synchrony of cell firing, and (d) increased efficacy of synaptic transmission due to increased postsynaptic sensitivity to transmitter, increased transmitter release, or increased postsynaptic membrane resistance.

Schwartzkroin & Wester (1975) succeeded in producing long-term potentiation in hippocampal slices; they and others have used this preparation as a simplified system for the investigation of synaptic mechanisms underlying LTP. Schwartzkroin & Wester found that LTP is not limited to the granule cells of dentate gyrus, but also occurs in the pyramidal cells of CA_1 or CA_3 (Schwartzkroin & Wester 1975, Alger & Teyler 1976). Because antidromic field potentials remained unchanged, the authors concluded that LTP is specific to the synapses tetanized.

Andersen and co-workers (1977) subsequently utilized the synaptic organization of CA_1 to examine the synaptic specificity of LTP in more detail. Because fibers in the stratum oriens make synaptic contacts with the basal dendrites of pyramidal cells, whereas those in the stratum radiatum terminate on the apical dendrites, the contribution of these inputs to LTP can be assessed independently. Andersen et al (1977) found that stimulation of each of these inputs produced monosynaptic EPSPs preceded by respective presynaptic volleys. The EPSP amplitudes were linearly related to the presynaptic volleys, followed high frequency stimulation, and were unchanged by removal of Ca^{2+} from the bathing fluid. Synaptic transmission in the radiatum pathway was greatly facilitated after tetanic stimulation of the radiatum fibers, and a greatly enhanced population spike with reduced latency was observed for more than one hour. By contrast, no consistent change in amplitude of the presynaptic volley or extracellular EPSPs was found in response to stimulation of oriens fibers. With this and other evidence, Andersen and co-workers concluded that the long-term potentiation of CA_1 pyramidal cells is input-specific and not due to the change in the afferent volleys.

By contrast, Lynch and co-workers (1977) reported that in a similar hippocampal slice experiment responses to stimulation in the nontetanized pathway showed a depression. The depression was more evident in the population spike than in the slow field potentials. Persistent depression of the postsynaptic cells after tetanic stimulation was also reported by Lynch et al (1976), who showed that long-term potentiation is accompanied by a persistent decrease in discharges of CA_1 pyramidal cells in response to electrophoretically applied glutamic acid. Increased accumulation of tritiated glutamate in the region of the potentiated synapses was reported (Wieraszko et al 1979). Furthermore, LTP was observed more frequently in calcium-rich media (2.5 mM calcium) than in low calcium (1.0 mM) media, which suggests an involvement of calcium in the development of long-term potentiation (Dunwiddie & Lynch 1978).

Yamamoto & Chujo (1978) also raised the possibility that the potentiation is not specific to the tetanized afferent but is partly nonspecific to the input. They reported that tetanization of a group of mossy fibers in the CA$_3$ region in the hippocampal sections potentiated responses induced by another group of presynaptic fibers, thus confirming the observation of Andersen and co-workers (1977) that there is a group of CA$_3$ neurons in which EPSPs can be augmented. However, Yamamoto & Chujo also found that IPSPs were depressed in another group of CA$_3$ neurons. This depression of IPSPs may account for part of the nonspecific potentiation.

Enlargement of dendritic spines of dentate granule cells has been proposed as a morphological correlate of LTP. Van Harreveld & Fifkova (1975) reported that stimulation of the entorhinal area induced swelling of dendritic spines in the dentate granule cells of hippocampus. The mean area of the spines in the proximal one-third of the granule cell dendrites, where no endings of perforant fibers from entorhinal cortex are present, was compared with that of the spines in the distal one-third, where endings of perforant fibers exist. In preparations in which the perforant fibers were tetanized, the mean area of the spines on the distal one-third of dendrites was significantly larger than in the nontetanized control. Spine enlargement was evident 2 min after the end of tetani and did not significantly decrease after 60 min. In a second study, Fifkova & Van Harreveld (1977) reported that following a single tetanus of 30 sec duration, spines increased by 15% and 38%, with poststimulation intervals of 2 to 6 min and 10 to 60 min, respectively. After 4 to 8 hr and 23 hr, the area of spines was still larger (35% and 23%) than control. Fifkova & Van Harrevold suggested that the initial enlargement of spines was due to a glutamate-induced increase in the sodium permeability of the spine membrane and that the long-lasting enlargement of dendritic spines resulted from an increase in protein synthesis. The increased size of spines and spine stems may provide an increased spread of synaptic currents from spine to dendritic shaft (Rall & Rinzel 1973) and result in an increased amplitude of the postsynaptic potentials.

AXONAL SPROUTING AND FORMATION OF NEW SYNAPSES

It is well established that in the peripheral nervous system collateral sprouts of intact motor fibers reinnervate deafferented muscle fibers after partial transection of motor nerve (Edds 1953). This phenomenon was also found by Murray & Thompson (1957) in sympathetic ganglion cells. By contrast, evidence of sprouting in the central nervous system has been obtained only recently. In 1958, Liu & Chambers demonstrated for the first time that dorsal root fibers can sprout in cat spinal cord in which all but one dorsal root or the descending pathways are cut; however, their light microscopic observations could not establish whether the sprouted fibers make synaptic contacts with their targets. This failure has been remedied by more recent electron microscopic work of Raisman (1969) and Raisman & Field (1973) in the rat septum which provided dramatic evidence of synaptic formation on septal neurons after interruption of either the fimbrial or medial forebrain bundle fibers. Subsequently, Moore and co-workers (1971) traced cathecholaminergic fibers by using fluorescence histochemistry and provided further support for new synapse formation in the septum. Lynch and co-workers (1974, 1975) subsequently also found prominent sprouting of synapses on granule cells of dentate gyrus of hippocampus after removal of ipsilateral entorhinal cortex. There is now a growing list of investigations that have confirmed the existence of sprouting in the central nervous system.

Morphological Studies of Sprouting After Partial Denervation

SPROUTING DURING EARLY DEVELOPMENTAL STAGES   It is generally agreed that the degree and extent of sprouting is more marked after denervation at the neonatal stage. In some cases, sprouting occurs only after neonatal deafferentiation (Lund & Lund 1971, Guillery 1972). In the lateral geniculate body, sprouting of the remaining optic fibers after unilateral enucleation occurs only in kittens (partially deafferented before 9 days after birth; Guillery 1972, Robson et al 1978, Hickey 1975). In the superior colliculus, most of the retinotectal fibers terminate contralaterally and a small portion of the uncrossed retinotectal fibers end in a small area of the ipsilateral superior colliculus. Unilateral lesion of the retinotectal fibers results in the expansion of the projection of the uncrossed, intact retinotectal fibers (Lund 1972, Lund et al 1973, Lund & Lund 1973, 1976, Lund & Miller 1975). This expansion of the projective area of uncrossed retinotectal fibers is not observed following lesions of retinal fibers made later than the twelfth postnatal day. Similarly, neonatal lesions of corticotectal fibers before the twentieth postnatal day result in sprouting of the intact corticotectal fibers from the remaining cortex (Mustari & Lund 1976).

Other sprouting phenomena also appear to be limited to neonatal life. Hicks & D'Amato (1970) found that unilateral hemispherectomy in neonatal rats resulted in the appearance of an uncrossed corticospinal tract, which is never observed in normal rats. This uncrossed corticospinal tract was not found after ablation of a cerebral hemisphere in adult rats. Similarly, after unilateral cerebral hemispherectomy in neonatal rats, corticopontine and corticotectal fibers crossed the midline and terminated in the contralateral pontine nucleus and superior colliculus (Leong & Lund 1973, Leong 1976a), making synaptic contacts on dendrites of the target neurons (Nah & Leong 1976b, Leong 1976b). Finally, Lim & Leong (1975) found that unilateral lesions of the cerebellar interpositus and dentate nuclei in neonatal rats induced aberrant projections of uncrossed cerebello-rubral and cerebello-thalamic fibers from the remaining ipsilateral cerebellar nuclei. These projections terminated in the red nucleus and ventrolateral thalamic nucleus. Similar findings were reported in kittens (Kawaguchi et al 1979, Kawaguchi et al 1979, Fujito et al 1980). Furthermore, unilateral ablation of the sensorimotor cortex in newborn rats resulted in the appearance of a crossed corticorubral projection from the intact contralateral cortex (Nah & Leong 1976a, b), and the aberrant corticorubral fibers were found to make synaptic contact with the distal dendrites of the red nucleus cells (Nah & Leong 1976b, Fujito et al 1980).

Destruction of target neurons during a neonatal stage also results in aberrant projections. For example, unilateral lesions of the superior colliculus in neonatal hamsters or neonatal fetal rats result in the aberrant retinothalamic (Baisinger et al 1977, Kalil & Schneider 1975, Schneider 1970, 1973) and retinotectal (Miller & Lund 1975) projections.

SPROUTING IN THE ADULT CENTRAL NERVOUS SYSTEM   In contrast to sprouting after neonatal lesions, which takes place over a considerable physical distance, sprouting in the adult CNS is restricted to the territory of the dendritic field of the deafferented neurons. Since sprouting is also less extensive in adult than in young animals, it is more difficult to detect experimentally.

As I discuss above, Raisman (1969) and Raisman & Field (1973) have shown that sprouting occurs in the septal nucleus of adult rats by transecting one of two convergent inputs, either fimbrial afferents to dendrites of septal cells or medial forebrain bundle afferents to dendrites and somata. Following transection of the medial forebrain bundle, axonal sprouts from fimbrial fibers make synaptic contacts on somatic synaptic sites formerly occupied by medial forebrain bundle afferents. Removal of the fimbrial fibers results in multiple synapse formation on dendritic sites formerly occupied by fimbrial synapses. The sprouting of medial forebrain bundle axons causes an increased intensity of staining of adrenergic nerve fibers in the septum following unilateral transection of the fimbria (Moore et al 1971).

Cotman & Lynch (Cotman & Lynch 1976, Lynch & Cotman 1975) have used the laminar afferent organization of the dentate gyrus to investigate axonal sprouting in adult rats. Because the perforant pathway from the ipsilateral entorhinal cortex terminates on the distal two-thirds of dendritic fields of granule cells and both associational and commissural fibers end on the proximal one-third, sprouting into restricted synaptic territories can be investigated following interruption of different afferent pathways. After ipsilateral entorhinal lesions in adult rats, the synaptic territory of the associational and commissural fibers expands into the inner one-fourth of the synaptic territory occupied by the perforant fibers (Lynch et al 1973). The dentate granule cells also receive a small projection from contralateral cortex (crossed perforant fibers). After unilateral lesions of entorhinal cortex, the projection expands to form synaptic contacts on parts of dendritic field formerly occupied by the ipsilateral entorhinal fibers (Steward et al 1974). After bilateral lesions of entorhinal cortex, the synaptic territory of associational fibers expands up to the inner 35 to 38% of proximal dendrites, contrasted with a distribution restricted to the inner 25% of granule cell dendrites in normal rats (Lynch et al 1972, 1974, 1976.) Thus, removal of major synaptic input to the dentate gyrus results in expansion of intact afferent systems onto deafferented parts of dendritic fields of granule cells.

Sprouting in the dentate gyrus has also been confirmed electron microscopically. After removal of ipsilateral entorhinal cortex in adult rats, 86% of the synapses on the distal two-thirds of granule cell dendrites disappear. However, by 240 days after lesion, synaptic density returns to up to 80% of normal values (Matthews et al 1976a). This recovery appears to be primarily due to synapse formation of dendritic spines (Matthews et al 1976b). Thus, the new synapses appear to occupy deafferented synaptic sites.

Extrinsic afferents to the dentate gyrus from the septum and brain stem consist of relatively few fibers whose zones of termination are not well delimited. Acetylcholine histochemistry has revealed that after entorhinal lesions cholinergic septohippocampal fibers proliferate in the dentate gyrus (Storm-Mathisen 1974). The response of brain stem afferents has not been described.

There also is evidence of plasticity of intrinsic hippocampal circuitry following entorhinal lesions. Granule cells in the dentate gyrus receive inhibitory inputs from basket cell axons. The inhibitory action is mediated by gamma-aminobutyric acid (GABA). After ipsilateral entorhinal lesions, GABAergic interneurons show an increase in both the level of the synthetic enzyme (Nadler et al 1974) and the amount of GABA they release (Nadler et al 1977). This may be attributed either to an increase in the effectiveness of synaptic transmission or sprouting of inhibitory synapses.

The extent of axonal sprouting in the dentate gyrus after entorhinal lesions is far less prominent in adults than in neonates. In neonates, the commissural fibers grow out nearly to the superficial margin of the molecular layer, whereas associational fibers expand to the margin of the layer. Septohippocampal fibers also expand to form a dense plexus along the outer margin of the molecular layer. Thus, neonatal lesions of entorhinal cortex induce more marked sprouting than do adult lesions.

Although the mechanisms underlying these differences between neonates and adults are not clear, it appears that both the extent and the rate of sprouting varies with age. The rate and extent of the outward expansion of the commissural-associational fibers of dentate gyrus in 3-month-old rats was compared to that of 25-month-old rats (Cotman & Scheff 1979). Twelve days after the lesion, the extent of commissural-associational projection in younger animals (3 months old) showed an increase of approximately 22% compared to normal controls; by contrast, older (25 months old) animals showed only a 10 to 11% increase. Septo-hippocampal fibers in the younger animals showed significantly less intense staining. The reasons for these differences are not clear.

Since the pioneering investigation of Liu & Chambers (1958), sprouting of intact axons in the spinal cord has been confirmed by electron microscopy (Bernstein & Bernstein 1971, 1973a, b). New synapses are formed on partially deafferented spinal neurons after spinal hemisections in adult rats, monkeys, and humans. Similarly, after spinal hemisection at the mid-thoracic level in newborn and weanling rats, axons from descending nerve tracts show increased collateral growth rostral to the hemisection (Prendergast & Stelzner 1976). Pullen & Sears (1978) reported that "C" synapses, which are one of the seven categories of presynaptic terminals found in spinal motoneurons, on the thoracic motoneurons are modified after spinal hemisection. Furthermore, "C" synapses, which originate from propriospinal neurons, increase in numbers following partial central deafferentation and are found postoperatively in the neuropile on dendrites not normally exhibiting this particular synaptic type.

Synaptic reorganization in the red nucleus was reported in adult cats (Nakamura et al 1974, Hanaway & Smith 1978). After a unilateral lesion of nucleus interpositus of the cerebellum, a second lesion was placed in the frontal cortex. Degeneration of axosomatic and axodendritic terminals was much more dense in the red nucleus contralateral to the first cerebellar lesion. Corticorubral endings normally found both on distal dendrites and in the rostral portion of the red nucleus were greatly increased on dendritic shafts and somata and their concentration was greater in the caudal quadrants of the nucleus than in the rostral quadrants (Hanaway & Smith 1978). The slight increase in number of somatic terminals was observed (Nakamura et al 1978).

TIME COURSE OF SPROUTING   The time course of sprouting is characterized by an initial rapid phase followed by a much slower one lasting for months. The initial signs of sprouting are observed within two weeks in the dentate gyrus (West et al 1975, Steward et al 1974), the red nucleus (Tsukahara et al 1974, 1975a, b), the septum (Moore et al 1971), and the lateral geniculate body (Stenevi et al 1972). This rapid phase of sprouting is followed by a much slower one that continues well over several months. The time course may also be modified by "priming" lesions. This was demonstrated by placing a small priming lesion in the entorhinal cortex, followed by removal of remaining entorhinal cortex. The reactive sprouting in the dentate gyrus was greatly accelerated after the second entorhinal lesion (Scheff et al 1978). Thus, whether or not denervation is sufficient to trigger significant growth, it can prime the system for sprouting after additional damage.

SELECTIVITY OF SPROUTING   Sprouting is highly selective both in terms of where it takes place and the types of fibers that display the phenomenon. Even in the same neurons in the same sites, some fiber systems have potentiality to sprout and others do not. For example, in the dentate granule cells of hippocampus, commissural, associational, and septo-hippocampal fibers sprout after ipsilateral entorhinal cortical lesions, whereas the entorhinal input does not sprout after lesions of associational and commissural fibers (Cotman & Nadler 1978).

Potentiality for sprouting in sensory relay nuclei has been reported to be very weak or entirely absent after interruption of the major synaptic input. For example, there is no evidence of sprouting in the spinal trigeminal (Kerr 1972, Beckerman & Kerr 1976), lateral cuneate (O'Neal & Westrum 1973), and dorsal column nuclei (Rustioni & Molenaar 1975). Similarly, there is no evidence of sprouting in the third-order avian auditory nucleus, nucleus laminaris (Benes et al 1977). In the lateral geniculate nucleus, sprouting occurs only in kittens partially deafferented before the ninth postnatal day (Guillery 1972, Robson et al 1978). However, other studies reveal no evidence of sprouting (Stelzner & Keating 1977, Hickey 1975). There is some evidence that suggests that sprouting occurs in the ventrobasal nucleus of thalamus after cortical ablations (Donoghue & Wells 1977).

In the motor system, sprouting of bulbospinal noradrenergic (Björklund et al 1971) and indolaminergic (Nobin et al 1973) fibers has been reported, and there is some evidence of sprouting of corticospinal fibers (Kalil & Reh 1978) but rubrospinal fibers do not appear to sprout (Castro et al 1977, Castro 1978).

In addition to the hippocampus, evidence of sprouting and synaptic reorganization has been reported in prepyriform cortex (Westrum 1969) and the cerebral cortex (Rose et al 1960, Purpura 1961).

Physiological Studies of Sprouting After Partial Denervation

PHYSIOLOGICAL EFFECTIVENESS   Wall & Egger (1971) reported that the ventral posterior lateral nucleus (VPL) in rats is organized somatotopically, with the forelimb represented in the medial two-thirds of the nucleus and the hindlimb in the lateral third. After the nucleus gracilis was destroyed, extracellular unit recordings indicated that the forelimb representation had expanded to occupy two-thirds of the area previously responding to hindlimb stimulation. The expansion was first observed three days after the lesion and was complete by seven days. If the nucleus gracilis was left intact and was deafferented by transection of the thoracic dorsal column, there was no expansion of the forelimb representation. Wall & Egger suggested that one possible mechanism for this phenomenon is sprouting of the remaining cuneate fibers to the deafferented VPL area.

Lynch et al (1973) reported that the laminar profile of the negative field potential of hippocampal granule cells to commissural stimulation expanded to span the depth of the molecular layer in rats with a neonatal entorhinal cortical lesion. In normal rats the maximum negativity to commissural stimulation is found only in the inner molecular layer. This finding suggested that commissural axons that sprout into the outer molecular layer after neonatal entorhinal lesions form functional synaptic connections in that region. However, this may not be a feature of all systems that display sprouting following partial deafferentation. Chow et al (1973) recorded activity of neurons in the superior colliculus of neonatally enucleated rabbits. Anatomical investigations showed that axonal spreading of the ipsilateral retinal projection in the rabbit superior colliculus occurs following neonatal enucleation. Chow and co-workers demonstrated that unit responses of neurons in the area occupied by new axonal growth do not respond to either stimulation of the intact eye or electric shock stimulation of the optic nerve, a result, they concluded, that brings into question the functional significance of demonstrations of axonal "sprouting" in mammalian systems.

In order to establish the functional effectiveness and the properties of synaptic transmission of newly-formed synapses unequivocally, it is necessary to record the synaptic potentials mediated by newly formed synapses. If new synapses are formed at sites different from normal, one could approach the problem by examining the time course of synaptic potentials, which depend on the distance from the synapses to the soma. The red nucleus is a good system for examining this problem. Red nucleus (RN) neurons receive two kinds of synaptic inputs, one from the nucleus interpositus (IP) of the cerebellum on their somata and the other from the sensorimotor cortex (SM) on the distal dendrites. The cortico-rubral dendritic EPSPs are characterized by a slow-rising time course, while the somatic IP-EPSPs are characterized with a fast-rising time course (Tsukahara & Kosaka 1968). A change of synaptic location relative to the soma should produce corresponding changes in the electrotonic distortion of the waveform of the EPSPs. Tsukahara and co-workers (1974, 1975a, b) have shown that a new fast-rising component is superimposed on the slow-rising corticorubral dendritic EPSPs after a lesion of the IP nucleus in adult cats. After determining that the slight change of the cable properties (electrotonic length) of dendrites of RN neurons following IP lesions accounts for only a minor portion (less than 5%) of the observed change in the time to peak of the corticorubral EPSPs, Tsukahara et al concluded that new and active synapses are formed at the proximal portion of the soma-dendritic membrane of RN cells.

An analysis of unitary corticorubral EPSPs before and after lesions of the nucleus interpositus (IP) were also consistent with this interpretation (Murakami et al 1977a, Tsukahara 1978, 1980). Corticorubral unitary EPSPs in normal cats are characterized by a slow rising time course and small amplitude (Figure 1C). Two groups of the unitary EPSPs were found in cats with chronic IP lesions, one consisting of corticorubral EPSPs with a shorter time-to-peak than in normal cats (Figure 1B), and the other consisting of unitary EPSPs within the normal range. The former group was found to be more sensitive to membrane potential displacement than the latter (Figure 2), which suggests a proximal dendritic or somatic locus of origin. The relationship of the time-to-peak and amplitude of the corticorubral unitary EPSPs before and after chronic IP lesions can be fitted to a theoretical relation derived from Rall's compartment model (Rall 1964) (Figure 1D). There is a tendency of the larger unitary EPSP to display the shorter time-to-peak. This tendency is predicted theoretically by Rall's compartment model calculating EPSPs produced at each of five compartments, as illustrated in the inset of Figure 1D. A good concordance between the theoretical curve and the experimental points support the view that new synapses were formed at the proximal portion of the soma-dendritic membrane of red nucleus cells close to the soma after IP lesion.

PROPERTIES OF SYNAPTIC TRANSMISSION   Properties of synaptic transmission of newly formed corticorubral synapses have been investigated in adult cats. Murakami et al (1977b) recorded intracellularly from RN neurons after IP lesions and found that the degree and the time course of facilitation at newly-formed corticorubral synapses displayed no major differences from those at the normal corticorubral synapses; however, the mean facilitation decayed more slowly in the newly-formed synapses. Posttetanic potentiation with a similar time course was also observed both

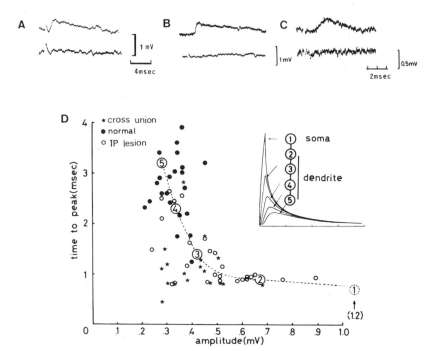

Figure 1 Corticorubral unitary EPSPs. *A:* Intracellular EPSP evoked by stimulation of cerebral peduncle in a red nucleus cell of a cross-innervated cat. *B:* Same as *A* but evoked by stimulation of sensorimotor cortex in a cat with lesion of the nucleus interpositus 27 days before acute experiment. *C:* Same as in *B* but in a normal cat. *Upper traces,* intracellular potentials. *Lower traces,* extracellular field corresponding to the *upper traces. D:* Relation between time-to-peak and amplitude of the unitary EPSPs. *Open circles* represent unitary EPSPs of IP-lesioned cats; *stars* represent those of cross-innervated cats more than two months before recording; *filled circles* represent those of normal cats. *Large open circles* represent time-to-peak and amplitude of theoretical EPSPs derived by Rall's compartment model initiated at each compartment of a five-compartment chain. The time course of the theoretical EPSPs generated in these compartments is shown in the inset of the figure (from Tsukahara 1980).

at newly-formed and normal corticorubral synapses (Murakami 1979, Murakami et al 1976).

Steward and co-workers (1977) demonstrated that field potentials of dentate granule cells in rats with lesions of ipsilateral entorhinal cortex exhibit facilitation following paired stimuli to the contralateral entorhinal cortex. This paired pulse potentiation (facilitation) had a comparable time course to that produced by stimulating the normal ipsilateral entorhinal afferent system, but the magnitude of the potentiation was somewhat less marked in the reinnervated dentate gyrus. This result should be interpreted with caution, however, because subsequent studies have shown that there

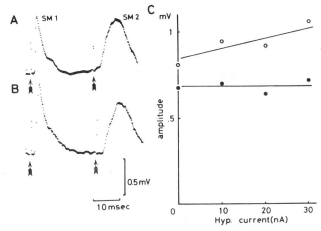

Figure 2   Sensitivity of amplitude of the EPSPs to membrane polarization. *A:* Fast rising *(SM1)* and slow rising *(SM2)* corticorubral EPSP recorded in the same cell by stimulating different points of sensorimotor cortex in a cat with lesion of the nucleus interpositus. *B:* Same as *A* but the EPSPs were recorded during hyperpolarization produced by injecting current through a microelectrode. *Upward arrows* indicate the onset of stimulation. *C:* Relation between amplitude of EPSP and injected current. *Open circles* correspond to the fast-rising EPSPs and *closed circles* to slowly rising ones (from Murakami et al 1977a).

are previously undetected connections from contralateral entorhinal cortex to dentate granule cells in normal rats (Cotman & Nadler 1978).

SPROUTING AND RECOVERY OF FUNCTION AFTER LESIONS   A relationship between collateral sprouting and functional recovery has had some experimental support (McCouch et al 1958, McCouch 1961), but the evidence is circumstantial and fails to identify the primary sites responsible for recovery of function. Murray & Goldberger (1974) reported that in adult cats intrinsic reflexes elicited by ipsilateral dorsal root input increased after partial hemisection of the spinal cord between T12 and L1. In some cases, reflexes that were depressed initially became hyperactive two weeks after the operation. Anatomical examination revealed signs of collateral sprouting from dorsal roots in response to degeneration of descending tracts on the same side. Both degeneration and radioautographic methods have shown that the site of the sprouting is both in Rexed's lamina VI and laterally in lamina VII, where interneurons mediating cutaneous reflexes and stretch reflex facilitation are located. Goldberger & Murray (1974) found that unilateral hindlimb deafferentation induced collateral sprouting of descending fibers, especially onto the Clark's nucleus, and they attributed this sprouting to the observed recovery of some descending reflexes. Furthermore, Goldberger (1977) reported that recovery of locomotor patterns after hindlimb deafferentation can be explained by sprouting of ipsilateral descending fibers.

Dieringer & Precht (1977, 1979a, b) reported a shortening of the time-to-peak of EPSPs in frog vestibular neurons following unilateral chronic labyrinthectomy. In these preparations, the commissurally evoked EPSPs had shorter times-to-peak and larger amplitudes than those of nonoperated frogs. The compensation of postural and locomotor disturbances after unilateral vestibular nerve transection is well known, and the authors suggested that these changes in vestibular EPSPs may be an underlying cellular mechanism of the behavioral compensation.

Research in rhesus monkeys indicated that changes in unitary EPSPs in motoneurons produced by group Ia stimulation may also be associated with functional recovery after spinal hemisection. The rise time of unitary EPSPs in motoneurons produced by Ia stimulation three to six months postoperatively was slightly shorter on the hemisected side than on the control side (Aoki & Mori 1978).

Bromberg & Gilman (1978) reported that multiunit activities in the red nucleus show an initial decrease of response and a subsequently slow recovery after lesions of the contralateral interpositus nucleus. To evaluate whether the restoration of activity depends upon intact corticofugal fibers, ipsilateral pericruciate cortex was ablated six weeks after the contralateral nucleus interpositus lesion. After the ablation, there was an immediate drop in amplitude of the activity. Slow restoration of multiunit activities was also observed after lesions of the nucleus interpositus and the cerebral cortex.

Loesche & Steward (1977) reported that recovery of alternation performance in a T-maze (for a food reward) occurs following unilateral entorhinal cortex lesions. They suggested that this recovery may depend upon the reinnervation of the denervated granule cells by contralateral entorhinal fibers. Bilateral lesions of the entorhinal cortex resulted in a persistent performance deficit. A secondary contralateral entorhinal lesion resulted in a deficit in alternation performance similar to that following one stage bilateral lesions.

A well-known procedure for testing the recovery and readjustment of behavior is cross-innervation or cross-connection of muscles. Sperry (1947) reported that motor readjustment occurred after cross-innervation of antagonistic muscles in monkeys during certain tasks practised for a long period of time. This readjustment had never been observed in rats (Sperry 1942). In cats, cross-innervation of forelimb nerves results in some modification of activity when tested in quadripedal locomotion (Tsukahara 1978). Yumiya and co-workers (1979) reported that motor readjustment occurred after cross-connection of forearm muscles when tested electromyographically during walking. Readjustment of the tactile placing reaction was also reported.

The neuronal basis of this behavioral readjustment after cross-innervation is not yet clear. Eccles and co-workers (1962) and Mendell & Scott (1975) investigated the reorganization of monosynaptic excitatory input to motoneurons from muscle afferents, but failed to detect a major change. Tsukahara & Fujito (1976) examined the possible synaptic reorganization in the red nucleus after cross-innervation of antagonistic muscles in cats. They found a change in the rise time of corticorubral EPSPs after cross-innervation and suggested that new synapses were formed on proximal parts of red nucleus neurons. Yumiya and co-workers (1979) examined the input-output relationship in cortical efferent zones after cross-connection of forearm muscles in cats and failed to detect any significant changes.

Axonal Sprouting Without Degeneration

Most studies of sprouting have been concerned with those induced after a lesion of a major synaptic input of central neurons. These studies even led to a simplified hypothesis that degeneration is essential for sprouting to occur and that vacated synaptic sites are reoccupied by new synaptic connections (Raisman & Field 1973); however, recent reports indicate that the number of vacated postsynaptic sites also changes after functional denervation without degeneration (Cotman & Nadler 1978). Aguilar et al (1973) reported that in the peripheral nervous system of salamanders interruption of axoplasmic transport in the peripheral nerve by colchicine application induced sprouting of adjacent peripheral nerves. This sprouting occurred in the absence of nerve degeneration. Thus, the question arises of whether the presence of nerve degeneration is necessary for sprouting and, consequently, to what extent sprouting occurs in circumstances other than denervation.

Tsukahara & Fujito (1976) found that a new, fast-rising component in the time-to-peak of corticorubral EPSPs appeared in cats whose flexor and extensor forelimb nerves were cross-innervated (Figure 1). Since the electrotonic length as well as the membrane time constant did not change after cross-innervation, appearance of the new fast rising component was not attributable to changes in cable properties of dendrites. The new component appeared after postoperative periods varying from three to ten months, and was found in cells innervating upper spinal segments. Red nucleus cells innervating the unoperated, lower spinal segments showed a less prominent change in the shape of corticorubral EPSPs. The possibility that neurons in nucleus interpositus degenerated after cross innervation was excluded because there was no appreciable change in the physiological estimates of the number of cerebellorubral fibers converging on red nucleus cells. Tsukahara & Fujito concluded that sprouting occurs without the presence of nerve degeneration at central synapses.

Other studies also suggest that degeneration is not necessary for induction of sprouting. Brown & Ironton (1977) found that a few days of muscle inactivity gives rise to a sprouting stimulus. In an experiment reminiscent of Aguilar et al (1973), Goldowitz & Cotman (1980) applied colchicine to the fimbria, which includes fibers of the hippocampal commissural system. The region of commissural termination in the molecular layer of the dentate gyrus was monitored electron microscopically for changes in number of synapses per unit area. At both 11 days and 60 to 70 days after colchicine treatment, the number of synapses per unit area increased. These data suggest that sprouting is not exclusively a reaction to neuronal deafferentation, but may be a mechanism that permits neurons to react to normal environmental changes.

The nature of signals that initiate sprouting and their method of transmission to neuronal somata are not known; however, two possible mechanisms have been proposed.

1. Some signals may be produced by target neurons and transmitted retrogradely to the presynaptic fibers. The inducing signals might be released from degenerating tissue (Raisman & Field 1973), or from inactive muscle (Brown & Ironton 1977), or from nerve growth factor in the sympathetic ganglion (Purves 1975). Furthermore, there is some evidence to suggest that the postulated signals that induce sprouting at neuromuscular junctions are related to acetylcholine receptors at the extrajunctional sites (Pestronk & Drachman 1978).
2. Nerve terminals may normally release a factor that prevents sprouting, so that when these nerves are injured or the concentration of this factor decreases, inhibition of sprouting no longer takes place (Diamond et al 1976).

Pickel et al (1974) found, 30 days after partial lesioning of the superior cerebellar peduncle, enhanced terminal flourescence indicative of sprouting of noradrenergic fibers from nucleus locus coeruleus, was found not only in the cerebellum, but also in the hippocampus. The increased terminal density in the hippocampus suggests that sprouting of the hippocampal terminals may be initiated, not from the denervation of postsynaptic cells, but by more remote signals such as neurons in locus coeruleus or collaterals of the cut axons in the cerebellum. This is consistent with observations of Andén and co-workers (1966) that lesions of the norepinephrine pathway in the mesencephalon results in increased terminal flourescence in the cerebellum and other brain stem areas after 2 to 4 weeks. Studies by Rotshenker & McMahan (1976) and Rotshenker (1979) suggest that, in frogs, induction of sprouting of a nerve innervating an intact pectoris muscle after the contralateral nerve innervating the same muscle is crushed may be communicated in the spinal cord from injured motoneurons to intact motoneurons.

In the hypoglossal nucleus, there is evidence suggesting that signals transmitted retrogradely from the target organ to the cell soma influence the number of synaptic connections the target cell receives. Neurons lose synaptic connections when their axons are severed; the number of synapses returns to normal when the axons regenerate and make synaptic contacts on their targets (Sumner 1977). Interruption of neuromuscular contact due to blockage of transmitter release by botulinum toxin (Sumner 1977) also produces presynaptic bouton loss (Sumner 1977, Cull 1975).

There appear to be regulatory mechanisms for the growth rate, magnitude of sprouting, and loss of sprouted synapses after formation (Roper & Ko 1978). In both the dentate molecular layer after entorhinal lesions and the septum after fimbrial lesions, the average density of synapses is restored to normal by sprouting (Raisman & Field 1973, Matthews et al 1976a). Similar strict control of the number of synapses has also been reported in the cardiac ganglion of frog (Roper & Ko 1978).

In summary, there has been rapid progress in our understanding of synaptic plasticity in the central nervous system. Sprouting, it is now acknowledged, occurs widely in the central nervous system and may provide an important neuronal basis, not only for the recovery of function after brain damage, but also for physiological phenomena such as learning and memory.

PLASTICITY OF NEURONAL NETWORKS

Not all plasticity can be attributed to changes at the cellular level. There may be circumstances where a certain form of plasticity appears at the network level. It has been frequently postulated that short-term memory is attributable to properties of neuronal networks such as reverberating circuits. However, there is little experimental evidence relevant to this hypothesis.

Properties of Reverberating Circuits

The concept of reverberating circuits is not new. In 1922, Forbes postulated the presence of a reverberating circuit in the spinal cord and suggested that reflex after-discharges are produced by the circulation of impulses along this closed loop. Subsequently, Lorente de Nó (1936) extended this concept in his analysis of interneuronal chains of oculomotor neurons. Furthermore, Bishop (1936) specifically suggested thalamo-cortical reverberating circuits as the neuronal basis of brain waves. Later, Burns (1950) analyzed the electrical activities of isolated cortical slabs and reported that a single shock applied to the slab set up a series of prolonged discharges of cortical neurons.

The reverberating circuit is used frequently as a hypothetical basis for short-term memory. Although this idea seems attractive, it is probably unrealistic, or at least very uncommon. The brain contains a wealth of inhibitory neurons, within closed loops, that are likely to prevent reverberatory activity. It is therefore unreasonable to assume impulse reverberation in any network unless it can be experimentally demonstrated.

There are, however, some experimental results that suggest the possibility of impulse reverberation in cerebellar circuits. Mutually excitatory connections exist between cerebellar nuclei and precerebellar nuclei, such as the paramedian reticular nucleus, pontine tegmental reticular nucleus (nucleus reticularis tegmenti pontis), lateral reticular nucleus, and the inferior olive (see Allen & Tsukahara 1974 for review). Furthermore, neurons in the cerebellar nuclei are subject to the inhibitory action of Purkinje cells. Thus, the presence of subcortical reverberatory activities may provide an excitatory background for the inhibitory sculpturing of the cerebellar cortical output. In view of the absence of clear demonstrations of reverberatory activities in mammalian brains, it is of interest to know whether reverberatory activities of cerebellar nuclear neurons can be demonstrated when Purkinje cell inhibition is removed experimentally. If reverberation is observed, this system may provide a useful model to characterize the dynamic properties of reverberating circuits.

Tsukahara and associates (1971) have shown that depolarization can be produced in neurons in the red nucleus after removal of Purkinje cell inhibition by injecting picrotoxin intravenously. Similar prolonged depolarization of neurons in the red nucleus and nucleus interpositus was induced in cats by surgical ablation of the intermediate cerebellar cortex (Tsukahara et al 1978). Because transection of the spinal cord or ablation of the cerebral sensorimotor cortex did not abolish this prolonged depolarization, it is probably not mediated by loops via the spinal cord or cerebral sensorimotor cortex. The prolonged depolarization exhibited regenerative properties with a threshold, an important property of positive feedback systems. Furthermore, the rise of the prolonged depolarization became faster as intensity of stimulation of one of the constituent nuclei was increased. These features were reproduced by a mathematical model simulating a reverberating circuit (Tsukahara et al 1973). It was further reported that the prolonged depolarization and associated repetitive discharges could be reversibly abolished by local cooling of the inferior and middle cerebellar peduncles (Tsukahara et al 1978). This indicates that the prolonged depolarization is not due to pacemaker properties of neurons of the cerebellar nuclei, but to a loop passing through the inferior and middle cerebellar peduncles.

Systematic investigations suggest that pontine tegmental reticular nucleus and paramedian reticular nucleus are probably candidates for sources of the other arm of this reverberating circuit. Both of these nuclei receive extensive input from nucleus interpositus and nucleus dentatus via the brachium conjunctivum descendens; however, the pontine tegmental reticular nucleus receives a heavier projection than the paramedian reticular nucleus. Stimulation of the pontine tegmental reticular nucleus and the paramedian reticular nucleus produces monosynaptic EPSPs in neurons of nucleus interpositus and nucleus dentatus. However, excitation is less prominent by stimulating the reticular neucleus than the paramedian nucleus (N. Tsukahara, T. Bando, F. Murakami, and N. Ozawa, 1980, personal communication). Therefore, pontine tegmental reticular and paramedian reticular nuclei may be candidates for the other arm of the reverberating circuit.

BEHAVIORAL CORRELATES OF NEURONAL PLASTICITY

A major goal of the study of neuronal plasticity is to provide a neuronal basis for behavioral plasticity, specifically as regards learning and memory. This section presents a brief overview of studies of simple behavioral phenomena, classical conditioning phenomena, and attempts to correlate synaptic and network plasticity with behavioral plasticity.

One strategy for investigating the correlation between neuronal and behavioral plasticity is to study the simplest forms of behavioral plasticity, such as "habituation." By recording intracellularly, Spencer and his associates (1966a, b) determined that habituation of the spinal flexion reflex in cats is produced by a change in amplitude of polysynaptic EPSPs following stimulation of afferent nerves. There was no change in the threshold of motoneurons, which suggested that the decreased amplitude of polysynaptic EPSPs was due to a depression of excitatory transmission in the flexion reflex pathway. Unfortunately, the flexion reflex pathway in cats contains many interneurons, and this hypothesis still remains untested. Further advances in our understanding of the neuronal mechanisms of habituation have been made in simpler invertebrate systems, such as that of *Aplysia* (see Kandel 1976 for a review).

A large body of studies deals with the neuronal correlates of classical conditioning (for reviews, see Kandel & Spencer 1968, Kandel 1977, Thompson et al 1972, 1978). In the mammalian central nervous system it is not possible to distinguish local neuronal changes at the site of the recording from those that occur at a distant site and are reflected at the recorded neuron as a result of changed synaptic input. Attention in this review, therefore, focuses on the following four studies based on strategies specifically designed to isolate the primary sites of conditioning:

1. Simplifying afferent pathways for the conditioned responses: Woody and his associates (Woody & Black-Cleworth 1973, Woody & Engel 1972, Woody et al 1970, Black-Cleworth et al 1975) investigated changes in neuronal activity associated with a conditioned eye blink in cats. After pairing a click with the glabella tap, they found that conditioning clicks gave rise to (a) responses in the facial nucleus with a latency of 17 msec and (b) activity in orbicularis oculi muscles at a 20 msec latency. By recording unit activities in the motor cortex, they obtained responses to clicks with a latency of 13 msec. Stimulation of motor cortex produced an eye blink with a latency of 7 msec. Thus, the latency of the conditioned responses, 20 msec, appears to consist of (a) latency of cortical neurons fired by conditioning clicks, 13 msec, (b) conduction from cortex to facial nucleus, 4 msec, and (c) time spent from facial nucleus to orbicularis oculi muscles, 3 msec. In a second group of studies responses of eye muscles to microstimulation of motor cortex after click-conditioned eye blinks was examined electromyographically. Cortical responses to the conditioned stimulus (CS) were also recorded from the same microelectrode used for stimulation. Neurons with the lowest microstimulation threshold for evoking EMG exhibited the largest responses to the CS (Woody et al 1970). Woody & Engel (1972) trained cats to induce either eye blink or nose twitch, and the threshold to evoke a response by microstimulation of the motor cortex was measured. Thresholds were lower for areas projecting to eyelid muscles in cats trained to blink their eyes, and cats trained to twitch their noses had lower thresholds for areas controlling nose movements. Woody & Black-Cleworth (1973) used intracellular stimulation to demonstrate a lower threshold for initiating spikes in neurons projecting to eyelid muscles when an eye blink was the conditioned response. In view of the inevitable difficulty in measuring threshold currents in unanesthetized cortical cells with the large amount of synaptic bombardment, the results should be interpreted cautiously. However, Woody and co-workers concluded that the postsynaptic changes in cortical neurons account for the increase of excitability.

2. Examining the role of an unconditioned stimulus for the acquisition of conditioned responses by simplifying the afferent pathway for unconditioned responses: Black-Cleworth and co-workers (1975) conditioned eye blinks by using direct electrical stimulation of the facial nerve as an unconditioned stimulus instead of glabellar tap and using click as a conditioned stimulus. Since the stimulation of the facial nerve is equally effective for conditioning the conditioned click-induced eye blink, they concluded that (a) sensory excitation produced by the unconditioned stimulus is not neces-

sary for processing the CR and (*b*) the firing of the postsynaptic cells is essential as the unconditioned stimulus. O'Brien and co-workers (1977) used antidromic activation of cortical neurons as the unconditioned stimulus and used hind paw stimuli as conditioned stimuli. They found that activation of neurons produced by antidromic stimulation was important for conditioning. This again suggests that the role of unconditioned stimuli may be simply to activate a neuron at an appropriate interval following conditioning stimuli. Furthermore, their results suggest that the local sites are present in neocortex. This does not, of course, exclude the possibility that neurons in other areas of the brain have similar properties, such that local learning sites may be distributed widely throughout the brain. Olds and associates (1972) attempted to localize "learning centers" by measuring the latencies of learned unit responses that either arose de novo or increased during conditioning by auditory signals. If response latencies were equal to or shorter than those of sensory responses in the inferior colliculus, the de novo responses were considered to be in "learning centers". The authors identified "learning centers" in both the pontine reticular formation and the central tegmentum: the largest proportions were in the posterior nucleus of the thalamus. In the telencephalon, "learning centers" were found in CA3 of the hippocampus and in several parts of neocortex.

3. Simplifying afferent pathways of both conditioned and unconditioned responses: Voronin (1971) attempted to analyze single-unit activity in the sensorimotor cortex of unanesthetized rabbits with a classical conditioning paradigm. Direct electrical stimulation of the cortical surface near the location of the microelectrode was used as the unconditioned stimulus (UCS), with the stimulation of a remote (3 to 12 mm) cortical point as the conditioned stimulus (CS). A majority of units tested showed significant modification of responses to the CS after one to several trials. The increase in amplitude and duration of the EPSPs produced by the CS, lasting up to 30 sec, was the most prominent feature of this modification. Since a few units showed significantly greater facilitation after a CS-UCS combination than after unpaired repetition of the UCS alone, the observed modification was considered to be analogous to behavioral pseudoconditioning or the dominant focus effect. Baranyi & Fehér (1978) recorded intracellularly from pyramidal tract cells and nonpyramidal tract cells during paired stimuli to the pyramidal tract and the ventrolateral thalamic nucleus (VL). Pyramidal stimuli paired to thalamic stimulation facilitated VL-evoked EPSPs in 13.7% of the cortical cells examined. This facilitation improved the efficacy of VL-EPSPs, shortened their latency, and enhanced the firing of cortical neurons. Similar results were observed by conditioning intracellularly evoked spikes on VL-EPSPs. However, there are several problems in interpreting these results. Because the conditioning pyramidal stimuli preceded

the testing VL-EPSPs, the experiments did not use a classical conditioning paradigm. Furthermore, the time course of the change was much faster than that of usual conditioning phenomena. Therefore, the relationship of this phenomenon to the classical conditioning paradigm is uncertain.

4. Correlating known examples of synaptic or network plasticity with behavioral plasticity: This comparatively little explored approach seeks to use known examples of synaptic or network plasticity as elements in reconstruction of behavioral phenomena. Tsukahara et al (1979) took this approach, using classically conditioned response. A conditioning stimulus (electrical shock) was applied to the cerebral peduncle in cats with lesions of corticofugal fibers below the red nucleus. The lesion eliminated the pyramidal tract, cortico-ponto-cerebellar, cortico-olivo-cerebellar fibers, and descending fibers to the pontine and medullary reticular formation. However, cortico-rubrospinal fibers were spared. After pairing of cerebral peduncle stimulation (CS) with forearm electrical shocks (UCS), an initially ineffective stimulus to the cerebral peduncle gave rise to flexion of the elbow. Extinction of the conditioned response is produced by reversing the order of stimuli or by giving only the CS. Because the threshold and the strength of the elbow flexion response induced by stimulation of nucleus interpositus was identical with the control animal, the interposito-rubrospinal system could not be the site of the neuronal change. Since the conditioned response is most probably mediated by the cortico-rubrospinal system, it is possible that modification at corticorubral synapses underlies this behavioral change. Although none of the above four approaches for studying neuronal mechanisms of learning and memory have specified the neuronal basis for behavioral plasticity associated with learning and memory, it is hoped that further research will clarify these important phenomena in the near future.

CONCLUSION

The 1970s witnessed a rapid progress in our understanding of the neuronal plasticity in the mammalian central nervous system. Examples of cellular plasticity, such as long-term potentiation, provide insights into mechanisms of functional plasticity and are a useful model for future studies using both in vivo and in vitro preparations. Axonal sprouting and formation of new synapses, including changes with much longer duration, are now accepted as widespread phenomena in the central nervous system after partial denervation. In some cases, sprouting has also been documented in the absence of denervation. These phenomena may provide a morphological basis for recovery of function after brain damage, and, more generally, of learning and memory in normal animals. Network plasticity, such as reverberatory activities of populations of neurons, may constitute a neuronal basis for short-term memory storage.

During this decade, several forms of neuronal plasticity in the mammalian central nervous system have been described in detail at the phenomenological level. In the next few years, research should probably proceed in two main directions: studies should (*a*) be directed at subcellular and molecular mechanisms underlying neuronal plasticity and (*b*) continue to attempt to link cellular and network plasticity with behavioral plasticity, using combined techniques of experimental psychology, anatomy, and physiology. These two approaches should yield a clearer view of the relationship of neuronal plasticity to normal and pathological functions of the mammalian central nervous system.

Literature Cited

Aguilar, C. E., Bisby, M. A., Cooper, E., Diamond, J. 1973. Evidence that axoplasmic transport of trophic factors is involved in the regulation of peripheral nerve fields in salamanders. *J. Physiol. London* 234:449–64

Alger, B. E., Teyler, T. J. 1976. Long-term and short-term plasticity in the CA1, CA3 and dentate regions of the rat hippocampal slice. *Brain Res.* 110:463–80

Allen, G. I., Tsukahara, N. 1974. Cerebrocerebellar communication systems. *Physiol. Rev.* 54:957–1006

Andén, N.-E., Fuxe, K., Larsson, K. 1966. Effect of large mesencephalic-diencephalic lesions on the noradrenalin, dopamine and 5-hydroxytryptamine neurons of the central nervous system. *Experientia* 22:842–43

Andersen, P., Sundberg, S. H., Sveen, O., Wigstrom, H. 1977. Specific long-lasting potentiation of synaptic transmission in hippocampal slices. *Nature* 266:736–37

Aoki, M., Mori, S. 1978. Changes in monosynaptic EPSPs of quadriceps motoneurons in monkeys with spinal cord chronically hemisected at the thoracic level. In *Integrative Control Functions of the Brain*, ed. M. Ito, N. Tsukahara, K. Kubota, K. Yagi, 1:170–71. Tokyo/Amsterdam:Kodansha/Elsevier. 457 pp.

Baisinger, J., Lund, R. D., Miller, B. 1977. Aberrant retinothalamic projections resulting from unilateral tectal lesions made in fetal and neonatal rats. *Exp. Neurol.* 54:369–82

Baranyi, A., Fehér, O. 1978. Conditioned changes of synaptic transmission in the motor cortex of the cat. *Exp. Brain Res.* 33:283–98

Beckerman, S. B., Kerr, F. W. L. 1976. Electrophysiologic evidence that neither

sprouting nor neuronal hyperactivity occur following long-term trigeminal or cervical primary deafferentation. *Exp. Neurol.* 50:427–38

Benes, F. M., Parks, T. N., Rubel, E. W. 1977. Rapid dendritic atrophy following deafferentation: An EM morphometric analysis. *Brain Res.* 122:1–13

Bernstein, J. J., Bernstein, M. E. 1971. Axonal regeneration and formation of synapses proximal to the site of lesion following hemisection of the rat spinal cord. *Exp. Neurol.* 30:336–51

Bernstein, J. J., Bernstein, M. E. 1973a. Neuronal alteration and reinnervation following axonal regeneration and sprouting in mammalian spinal cord. *Brain Behav. Evol.* 8:135–61

Bernstein, M. E., Bernstein, J. J. 1973b. Regeneration of axons and synaptic complex formation rostral to the site of hemisection in the spinal cord of the monkey. *Int. J. Neurosci.* 5:15–26

Bishop, G. H. 1936. The interpretation of cortical potentials. *Cold Spring Harbor Symp. Quant. Biol.* 4:305–19

Björklund, A., Katzman, R., Stenevi, U., West, K. A. 1971. Development and growth of axonal sprouts from noradrenaline and 5-hydroxytryptamine neurons in the rat spinal cord. *Brain Res.* 31:21–33

Björklund, A., Stenevi, U. 1979. Regeneration of monoaminergic and cholinergic neurons in the mammalian central nervous system. *Physiol. Rev.* 59:62–100

Black-Clearworth, P., Woody, C. D., Nieman, J. 1975. A conditioned eyeblink obtained by using electrical stimulation of the facial nerve as the unconditioned stimulus. *Brain Res.* 90:45–56

Bliss, T. V. P., Gardner-Medwin, A. R. 1971. Long-lasting increases of synaptic influence in the unanesthetized hippocampus. *J. Physiol. London* 216:P32–33

Bliss, T. V. P., Gardner-Medwin, A. R. 1973. Long-lasting potentiation of synaptic transmission in the dentate area of the unanesthetized rabbit following stimulation of the perforant path. *J. Physiol. London* 232:357–74

Bliss, T. V. P., Lømo, T. 1973. Long-lasting potentiation of synaptic transmission in the dentate area of the anesthetized rabbit following stimulation of the perforant path. *J. Physiol. London* 232: 331–56

Bromberg, M. B., Gilman, S. 1978. Changes in rubral multiunit activity after lesions in the interpositus nucleus of the cat. *Brain Res.* 152:353–57

Brown, M. C., Ironton, R. 1977. Motor neurone sprouting induced by prolonged tetrodotoxin block of nerve action potentials. *Nature* 265:459–61

Burns, B. D. 1950. Some properties of the cat's isolated cerebral cortex. *J. Physiol. London* 111:50–68

Castro, A. J. 1978. Analysis of corticospinal and rubrospinal projections after neonatal pyramidotomy in rats. *Brain Res.* 144:155–58

Castro, A. J., Clegg, D. A., McClung, J. R. 1977. The effect of large unilateral cortical lesions on rubrospinal tract sprouting in newborn rats. *Am. J. Anat.* 149:39–46

Chow, K. L., Mathers, L. H., Spear, P. D. 1973. Spreading of uncrossed retinal projection in superior colliculus of neonatally enucleated rabbits. *J. Comp. Neurol.* 151:307–22

Cotman, C. W., Lynch, G. S. 1976. Reactive synaptogenesis in the adult nervous system: The effects of partial deafferentation on new synapses formation. In *Neuronal Recognition,* ed. S. Barondes, pp. 69–108. New York: Plenum

Cotman, C. W., Nadler, J. V. 1978. Reactive synaptogenesis in the hippocampus. In *Neuronal Plasticity,* ed. C. W. Cotman, pp. 227–72. New York: Raven. 335 pp.

Cotman, C. W., Scheff, S. W. 1979. Compensatory synapse growth in aged animals after neuronal death. *Mech. Ageing Dev.* 9:103–17

Cull, R. E. 1975. Role of axonal transport in maintaining central synaptic connections. *Exp. Brain Res.* 24:97–101

Curtis, D. R., Eccles, J. C. 1960. Synaptic action during and after repetitive stimulation. *J. Physiol. London* 150:374–98

Diamond, J., Cooper, E., Turner, C., MacIntyre, L. 1976. Trophic regulation of nerve sprouting—Neuron-target interactions and spatial relations control sensory nerve fields in salamander skin. *Science* 193:371–77

Dieringer, N., Precht, W. 1977. Modification of synaptic input following unilateral labyrinthectomy. *Nature* 269:431–33

Dieringer, N., Precht, W. 1979a. Mechanisms of compensation for vestibular deficits in the frog. I. Modification of the excitatory commissural system. *Exp. Brain Res.* 36:311–28

Dieringer, N., Precht, W. 1979b. Mechanisms of compensation for vestibular deficits in the frog. II. Modification of the inhibitory pathways. *Exp. Brain Res.* 36:329–41

Donoghue, J. P., Wells, J. 1977. Synaptic rearrangement in the ventrobasal complex of the mouse following partial cortical deafferentation. *Brain Res.* 125:351–55

Douglas, R. M., Goddard, G. V. 1975. Long-term potentiation of the perforant path-granule cell synapse in the rat hippocampus. *Brain Res.* 86:205–15

Dunwiddie, T. V., Lynch, G. S. 1978. Calcium involvement in long term potentiation in the hippocampal slice preparation. *Soc. Neurosci. Abstr.* 4:424

Eccles, J. C. 1976. The plasticity of the mammalian central nervous system with special reference to new growths in response to lesions. *Naturwissenschaften* 63:8–15

Eccles, J. C., Eccles, R. M., Shealy, C. N., Willis, W. D. 1962. Experiments utilizing monosynaptic excitatory action on motoneurons for testing hypotheses relating to specificity of neuronal connections. *J. Neurophysiol.* 25:559–80

Eccles, J. C., Hubbard, J. I., Oscarsson, O. 1961. Intracellular recording from the cells of the ventral spinocerebellar tract. *J. Physiol. London* 158:486–516

Edds, M. V. Jr. 1953. Collateral nerve regeneration. *Q. Rev. Biol.* 28:260–76

Fifkova, E., Van Harreveld, A. V. 1977. Long-lasting morphological changes in dendritic spines of dentate granular cells following stimulation of the entorhinal area. *J. Neurocytol.* 6:211–30

Forbes, A. 1922. The interpretation of spinal reflexes in terms of present knowledge of nerve conduction. *Physiol. Rev.* 2:361–414

Fujito, Y., Tsukahara, N., Yoshida, M. 1980. Synaptic plasticity in chronically hemicerebellectomized or hemispherectomized kittens. *Neurosci. Lett.* 4: Suppl., p. 42

Goldberger, M. E. 1977. Locomotor recovery

after unilateral hindlimb deafferentation in cats. *Brain Res.* 123:59–74

Goldberger, M. E., Murray, M. 1974. Restitution of function and collateral sprouting in the cat spinal cord: The deafferented animal. *J. Comp. Neurol.* 158:37–54

Goldowitz, D., Cotman, C. W. 1980. Do neurotrophic interactions control synapse formation in the adult rat brain? *Brain Res.* 181:325–44

Guillery, R. W. 1972. Experiments to determine whether retinogeniculate axons can form translaminar collateral sprouts in the dorsal lateral geniculate nucleus of the cat. *J. Comp. Neurol.* 146:407–20

Hanaway, J., Smith, J. 1978. Sprouting of corticorubral terminals in the cerebellar deafferented cat red nucleus. *Soc. Neurosci. Abstr.* 4:1507

Hickey, T. L. 1975. Translaminar growth of axons in the kitten dorsal lateral geniculate nucleus following removal of one eye. *J. Comp. Neurol.* 161:359–82

Hicks, S. P., D'Amato, C. J. 1970. Motor-sensory and visual behavior after hemispherectomy in newborn and mature rats. *Exp. Neurol.* 29:416–38

Hughes, J. R., Evarts, E. V., Marshall, W. H. 1956. Posttetanic potentiation in the visual system of cats. *Am. J. Physiol.* 186:483–87

Kalil, K., Reh, T. 1978. Plasticity in the corticospinal tract after early lesions of the medullary pyramid. *Soci. Neurosci. Abstr.* 4:476

Kalil, R. E., Schneider, G. E. 1975. Abnormal synaptic connections of the optic tract in the thalamus after midbrain lesions in newborn hamsters. *Brain Res.* 100:690–98

Kandel, E. R. 1976. *Cellular Basis of Behavior, an Introduction to Behavioral Neurobiology.* San Francisco: Freeman. 727 pp.

Kandel, E. R. 1977. Neuronal plasticity and the modification of behavior. In *Handbook of Physiology,* ed. E. R. Kandel, 1:1137–82. Baltimore: Williams & Williams. 1182 pp.

Kandel, E. R., Spencer, W. A. 1968. Cellular neurophysiological approaches in the study of learning. *Physiol. Rev.* 48:65–134

Kawaguchi, S., Yamamoto, T., Samejima, A. 1979. Electrophysiological evidence for axonal sprouting of cerebellothalamic neurons in kittens after neonatal hemicerebellectomy. *Exp. Brain Res.* 36:21–39

Kawaguchi, S., Yamamoto, T., Samejima, A., Itoh, K., Mizuno, N. 1979. Morphological evidence for axonal sprouting of cerebellothalamic neurons in kittens after neonatal hemicerebellectomy. *Exp. Brain Res.* 35:511–18

Kerr, F. W. L. 1972. The potential of cervical primary afferents to sprout in the spinal nucleus of V following long term trigeminal denervation. *Brain Res.* 43:547–60

Langren, S., Phillips, C. G., Porter, R. 1962. Minimal synaptic action of pyramidal impulses on some alpha motoneurones of the baboon's hand and forearm. *J. Physiol. London* 161:91–111

Leong, S. K. 1976a. An experimental study of the corticofugal system following cerebral lesions in the albino rats. *Exp. Brain Res.* 26:235–47

Leong, S. K. 1976b. A qualitative electron microscopic investigation of the anomalous corticofugal projections following neonatal lesions in the albino rats. *Brain Res.* 107:1–8

Leong, S. K., Lund, R. D. 1973. Anomalous bilateral corticofugal pathways in albino rats after neonatal lesions. *Brain Res.* 62:218–21

Lim, K. H., Leong, S. K. 1975. Aberrant bilateral projections from dentate and interposed nuclei in albino rats after neonatal lesions. *Brain Res.* 96:306–9

Liu, C. N., Chambers, W. W. 1958. Intraspinal sprouting of dorsal root axons. *Arch. Neurol. Psychiatry* 79:46–61

Lloyd, D. P. C. 1949. Post-tetanic potentiation of response in monosynaptic reflex pathways of the spinal cord. *J. Gen. Physiol.* 33:147–70

Loesche, J., Steward, O. 1977. Behavioral correlates of denervation and reinnervation of the hippocampal formation of the rat: Recovery of alternation performance following unilateral entorhinal cortex lesions. *Brain Res. Bull.* 2:21–39

Lorente de Nó, R. 1936. Discussion remark to H. H. Jasper: Cortical excitatory state and synchronism in the control of bioelectric autonomous rhythms. *Cold Spring Harbor Symp. Quant. Biol.* 4:336–37

Lund, R. D. 1972. Anatomic studies of the superior colliculus. *Invest. Ophthal.* 11:434–41

Lund, R. D. 1978. *Development and Plasticity of the Brain.* New York: Oxford

Lund, R. D., Cunningham, T. J., Lund, J. S. 1973. Modified optic projections after unilateral eye removal in young rat. *Brain Behav. Evol.* 8:51–72

Lund, R. D., Lund, J. S. 1971. Synaptic adjustment after deafferentation of the superior colliculus of the rat. *Science* 171:804–7

Lund, R. D., Lund, J. S. 1973. Reorganization of the retinotectal pathway in rats after neonatal and retinal lesions. *Exp. Neurol.* 40:377–90

Lund, R. D., Lund, J. S. 1976. Plasticity in the developing visual system: The effects of retinal lesions made in young rats. *J. Comp. Neurol.* 169:133–54

Lund, R. D., Miller, B. F. 1975. Secondary effects of fetal eye damage in rats on intact central optic projections. *Brain Res.* 92:279–89

Lynch, G., Cotman, C. W. 1975. The hippocampus as a model for studying anatomical plasticity in the adult brain. In *Hippocampus,* ed. R. L. Isacson, K. H. Pribram, 1:123–55 New York: Plenum

Lynch, G., Deadwyler, S., Cotman, C. W. 1973. Postlesion axonal growth produces permanent functional connections. *Science* 180:1364–66

Lynch, G., Dunwiddie, T., Gribkoff, V. 1977. Heterosynaptic depression: A postsynaptic correlate of long-term potentiation. *Nature* 266:737–39

Lynch, G., Gall, C., Mensah, P., Cotman, C. W. 1974. Horseradish peroxidase histochemistry: A new method for tracing efferent projections in the central nervous system. *Brain Res.* 65:373–80

Lynch, G., Gall, C., Rose, G., Cotman, C. W. 1976. Changes in the distribution of the dentate gyrus associational system following unilateral or bilateral entorhinal lesions in the adult rat. *Brain Res.* 110:57–71

Lynch, G., Gribkoff, V. K., Deadwyler, S. A. 1976. Long term potentiation is accompanied by a reduction in dendritic responsiveness to glutamic acid. *Nature* 263:151–53

Lynch, G., Matthews, D. A., Mosko, S., Parks, T., Cotman, C. W. 1972. Induced acetylcholine-esterase-rich layer in rat dentate gyrus following entorhinal lesions. *Brain Res.* 42:311–18

Lynch, G., Stanfield, B., Cotman, C. W. 1973. Developmental differences in post-lesion axonal growth in the hippocampus. *Brain Res.* 59:155–68

Matthews, D. A., Cotman, C. W., Lynch, G. 1976a. An electron microscopic study of lesion-induced synaptogenesis in the dentate gyrus of the adult rat. I. Magnitude and time course of degeneration. *Brain Res.* 115:1–21

Matthews, D. A., Cotman, C. W., Lynch, G. 1976b. An electron microscopic study of lesion-induced synaptogenesis in the dentate gyrus of the adult rat. II. Reappearance of morphologically normal synaptic contacts. *Brain Res.* 115:23–41

McCouch, G. P. 1961. Factors in the transition to spasticity. In *The Spinal Cord,* ed. G. Austin, pp. 256–261. Springfield, Ill.: Thomas

McCouch, G. P., Austin, G. M., Liu, C. N., Liu, C. Y. 1958. Sprouting as a cause of spasticity. *J. Neurophysiol.* 21:205–16

Mendell, L. M., Scott, J. G. 1975. The effect of peripheral nerve cross-union on connections of single Ia fibers to motoneurons. *Exp. Brain Res.* 22:221–34

Miller, B., Lund, R. D. 1975. The pattern of retinotectal connections in albino rats can be modified by fetal surgery. *Brain Res.* 91:119–25

Moore, R. Y., Björklund, A., Stenevi, U. 1971. Plastic changes in the adrenergic innervation of the rat septal area in response to denervation. *Brain Res.* 33:13–35

Murakami, F. 1979. Formation of new corticorubral synapses in red nucleus neurons after denervation and the properties of their synaptic transmission. In *Neurobiology,* ed. M. Otsuka, pp. 295–305. New York: Wiley

Murakami, F., Fujito, Y., Tsukahara, N. 1976. Physiological properties of the newly formed cortico-rubral synapses of red nucleus neurons due to collateral sprouting. *Brain Res.* 103:147–51

Murakami, F., Tsukahara, N., Fujito, Y. 1977a. Analysis of unitary EPSPs mediated by the newly-formed corticorubral synapses after lesion of the interpositus nucleus. *Exp. Brain Res.* 30:233–43

Murakami, F., Tsukahara, N., Fujito, Y. 1977b. Properties of synaptic transmission of the newly formed corticorubral synapses after lesion of the nucleus interpositus of the cerebellum. *Exp. Brain Res.* 30:245–58

Murray, J. C., Thompson, J. W. 1957. The occurrence and function of collateral sprouting in the sympathetic nervous system of the cat. *J. Physiol. London* 135:133–62

Murray, M., Goldberger, M. E. 1974. Restitution of function and collateral sprouting in the cat spinal cord: The partially hemisected animal. *J. Comp. Neurol.* 158:19–36

Mustari, M. J., Lund, R. D. 1976. An aberrant crossed visual corticotectal pathway in albino rats. *Brain Res.* 112: 37–44

Nadler, J. V., Cotman, C. W., Lynch, G. S. 1974. Biochemical plasticity of short-axon interneurons: Increased glutamate decarboxylase activity in the denervated area of rat dentate gyrus following entorhinal lesion. *Exp. Neurol.* 45:403–13

Nadler, J. V., White, W. F., Vaca, K. W., Cotman, C. W. 1977. Calcium-dependent-aminobutyrate release by interneurons of rat hippocampal lesions: Lesion-induced plasticity. *Brain Res.* 131: 241–58

Nah, S. H., Leong, S. K. 1976a. Bilateral corticofugal projection to the red nucleus after neonatal lesions in the albino rat. *Brain Res.* 107:433–36

Nah, S. H., Leong, S. K. 1976b. An ultra-structural study of the anomalous corticorubral projection following neonatal lesions in the albino rat. *Brain Res.* 111:162–66

Nakamura, Y., Mizuno, N., Konishi, A. 1978. A quantitative electron microscope study of cerebellar axon terminals on the magnocellular red nucleus neurons in the cat. *Brain Res.* 147:17–27

Nakamura, Y., Mizuno, N., Konishi, A., Sato, M. 1974. Synaptic reorganization of the red nucleus after chronic deafferentation from cerebellorubral fibers: An electron microscope study in the cat. *Brain Res.* 82:298–301

Nobin, A., Baumgarten, H. G., Björklund, A., Lachenmayer, L., Stenevi, U. 1973. Axonal degeneration and regeneration of the bulbospinal indolamine neurons after 5,6-dihydroxytryptamine treatment. *Brain Res.* 56:1–24

O'Brien, J. H., Wilder, M. B., Stevens, C. D. 1977. Conditioning of cortical neurons in cats with antidromic activation as the unconditioned stimulus. *J. Comp. Physiol. Psychol.* 91:918–29

Olds, J., Disterhoft, J. F., Segal, M., Kornblith, C. L., Hirsh, R. 1972. Learning centers of rat brain mapped by measuring latencies of conditioned unit responses. *J. Neurophysiol.* 35:202–19

O'Neal, J. T., Westrum, L. E. 1973. The fine structural synaptic organization of the cat lateral cuneate nucleus. A study sequential alterations in degeneration. *Brain Res.* 51:97–124

Pestronk, A., Drachman, D. B. 1978. Motor nerve sprouting and acetylcholine receptors. *Science* 199:1223–25

Pickel, V., Segal, M., Bloom, F. E. 1974. Axonal proliferation following lesions of cerebellar peduncles. A combined fluorescence microscopy and radioauto-graphic study. *J. Comp. Neurol.* 155: 43–60

Prendergast, J., Stelzner, D. J. 1976. Increases in collateral axonal growth rostral to a thoracic hemisection in neonatal and weanling rat. *J. Comp. Neurol.* 166:145–62

Pullen, A. H., Sears, T. A. 1978. Modification of 'C' synapses following partial central deafferentation of thoracic motoneurons. *Brain Res.* 145:141–46

Purpura, D. P. 1961. Analysis of axodendritic synaptic organizations in immature cerebral cortex. *Ann. NY Acad. Sci.* 94:604–54

Purves, D. 1975. Functional and structural changes in mammalian sympathetic neurones following interrruption of their axons. *J. Physiol. London* 252: 429–63

Raisman, G. 1969. Neuronal plasticity in the septal nuclei of the adult rat. *Brain Res.* 14:25–48

Raisman, G. 1977. Formation of synapses in the adult rat after injury: Similarities and diffferences between a peripheral and a central nervous site. *Philos. Trans. R. Soc. London* 278:349–59

Raisman, G., Field, P. M. 1973. A quantitative investigation of the development of collateral reinnervation after partial deafferentation of septal nuclei. *Brain Res.* 50:241–64

Rall, W. 1964. Theoretical significance of dendritic trees for neuronal input-output relations. In *Neural Theory and Modeling*, ed. R. F. Reiss, pp. 73–87. Stanford: Stanford Univ. Press

Rall, W., Rinzel, J. 1973. Branch input resistance and steady attenuation for input to one branch of a dendritic neuron model. *Biophys. J.* 13:648–88

Robson, J. A., Mason, C. A., Guillery, R. W. 1978. Terminal arbors of axons that have formed abnormal connections. *Science* 201:635–37

Roper, S., Ko, C.-P. 1978. Synaptic remodeling in the partially denervated parasympathetic ganglion in the heart of the frog. In *Neuronal Plasticity*, ed. C. W. Cotman, pp. 1–25. New York: Raven. 335 pp.

Rose, J. E., Malis, L. I., Kruger, L., Baker, C. P. 1960. Effects of heavy, ionizing, monoenergetic particles on the cerebral cortex. *J. Comp. Neurol.* 115:243–55

Rotshenker, S. 1979. Synapse formation in intact innervated cutaneous-pectoris muscles of the frog following denervation of the opposite muscle. *J. Physiol. London* 292:535–47

Rotshenker, S., McMahan, U. J. 1976. Altered pattern of innervation in frog muscle after denervation. *J. Neurocytol.* 5:719–30

Rustioni, A., Molenaar, I. 1975. Dorsal column nuclei afferents in the lateral funiculus of the cat: Distribution pattern and absence of sprouting after chronic deafferentation. *Exp. Brain Res.* 23:1–12

Scheff, S. W., Bernado, L. S., Cotman, C. W. 1978. Effect of serial lesions on sprouting in the dentate gyrus: Onset and decline of the catalytic effect. *Brain Res.* 150:45–53

Schneider, G. E. 1970. Mechanisms of functional recovery following lesions of visual cortex or superior colliculus in neonatal and adult hamsters. *Brain Behav. Evol.* 3:295–323

Schneider, G. E. 1973. Early lesions of superior colliculus: Factors affecting the formation of abnormal retinal projections. *Brain Behav. Evol.* 8:73–109

Schwartzkroin, P. A., Wester, K. 1975. Long-lasting facilitation of a synaptic potential following tetanization in the in vitro hippocampal slice. *Brain Res.* 89:107–19

Spencer, W. A., Thompson, R. F., Neilson, D. R. Jr. 1966a. Response decrement of the flexion reflex in the acute spinal cat and transient restoration by strong stimuli. *J. Neurophysiol.* 29:221–39

Spencer, W. A., Thompson, R. F., Neilson, D. R. Jr. 1966b. Decrement of ventral root electrotonus and intracellularly recorded PSPs produced by iterated cutaneous afferent volleys. *J. Neurophysiol.* 29:253–74

Spencer, W. A., Wigdor, R. 1965. Ultra-late PTP of monosynaptic reflex responses in cat. *Physiologist* 8:278

Sperry, R. W. 1942. Transplantation of motor nerves and muscles in the forelimb of the rat. *J. Comp. Neurol.* 76:283–321

Sperry, R. W. 1947. Effect of crossing nerves to antagonistic limb muscles in the monkey. *Arch. Neurol. Psychiatry* 58:452–73

Stelzner, D. J., Keating, E. G. 1977. Lack of intralaminar sprouting of retinal axons in monkey LGN. *Brain Res.* 126:201–10

Stenevi, U., Björklund, A., Moore, R. Y. 1972. Growth of intact central adrenergic axons in the denervated lateral geniculate body. *Exp. Neurol.* 35:290–99

Steward, O., Cotman, C. W., Lynch, G. S. 1974. Growth of a new fiber projection in the brain of adult rats: Reinnervation of the dentate gyrus by the contralateral entorhinal cortex following ipsilateral entorhinal lesions. *Exp. Brain Res.* 20:45–66

Steward, O., White, W. F., Cotman, C. W. 1977. Potentiation of the excitatory synaptic action of commissural, associational and entorhinal afferents to dentate granule cells. *Brain Res.* 134:551–60

Storm-Mathisen, J. 1974. Choline acetyltransferase and acetylcholinesterase in fascia dentata following lesion of the entorhinal afferents. *Brain Res.* 80:181–97

Sumner, B. E. H. 1977. Ultrastructural responses of the hypoglossal nucleus to the presence in the tongue of botulinum toxin, a quantitative study. *Exp. Brain Res.* 30:313–21

Teyler, T. J., ed. 1978. *Brain and Learning.* Dordrecht: Reidel. 163 pp.

Thompson, R. F., Patterson, M. M., Berger, T. W. 1978. Associative learning in the mammalian nervous system. In *Brain and Learning,* ed. T. J. Teyler, pp. 51–90. Dordrecht: Reidel. 163 pp.

Thompson, R. F., Patterson, M. M., Teyler, T. J. 1972. The neurophysiology of learning. *Ann. Rev. Psychol.* 23:73–104

Tsukahara, N. 1978. Synaptic plasticity in the red nucleus. In *Neuronal Plasticity,* ed. C. W. Cotman, pp. 113–30. New York: Raven. 335 pp.

Tsukahara, N. 1980. Synaptic plasticity in the red nucleus neurons. *Proc. 28th Int. Congr. Physiol. Sci. Budapest, 1980.* In press

Tsukahara, N., Bando, T., Kitai, S. T., Kiyohara, T. 1971. Cerebello-pontine reverberating circuit. *Brain Res.* 33:233–37

Tsukahara, N., Bando, T., Kiyohara, T. 1973. Dynamics of cerebello-pontine reverberating circuit. In *Regulation and Control in Physiological Systems,* ed. A. S. Iberall, A. C. Guyton, pp. 535–37. Pittsburgh, Pa.: ISA. 607 pp.

Tsukahara, N., Bando, T., Murakami, F., Ozawa, N. 1978. Control of the cerebellar reverberatory activities by local cooling of the cerebellar peduncles. In *Integrative Control Functions of the Brain,* ed. M. Ito, N. Tsukahara, K. Kubota, K. Yagi, 1:439–40. Tokyo/Amsterdam: Kodansha/Elsevier. 457 pp.

Tsukahara, N., Fujito, Y. 1976. Physiological evidence of formation of new synapses from cerebrum in the red nucleus neurons following cross-union of forelimb nerves. *Brain Res.* 106:184–88

Tsukahara, N., Hultborn, H., Murakami, F. 1974. Sprouting of cortico-rubral synapses in red nucleus neurons after destruction of the nucleus interpositus of the cerebellum. *Experientia* 30: 57–58

Tsukahara, N., Hultborn, H., Murakami, F., Fujito, Y. 1975a. Electrophysiological study of formation of new synapses and collateral sprouting in red nucleus neurons after partial denervation. *J. Neurophysiol.* 38:1359–72

Tsukahara, N., Hultborn, H., Murakami, F., Fujito, Y. 1975b. Physiological evidences of collateral sprouting and formation of new synapses in the red nucleus following partial denervation. In *Golgi Centennial Symposium Proceedings,* ed. M. Santini, pp. 299–303. New York: Raven. 668 pp.

Tsukahara, N., Kosaka, K. 1968. The mode of cerebral excitation of red nucleus neurons. *Exp. Brain Res.* 5:102–17

Tsukahara, N., Oda, Y., Notsu, T. 1979. Associative conditioning mediated by the red nucleus in the cat. *Proc. Jpn. Acad. Ser. B* 55:537–41

Van Harreveld, A., Fifkova, E. 1975. Swelling of dendritic spines in the fascia dentata after stimulation of the perforant fibers as a mechanism of post-tetanic potentiation. *Exp. Neurol.* 49:736–49

Voronin, L. L. 1971. Microelectrode study of cellular analogs of conditioning. *Proc. 25th Int. Congr. Physiol. Sci. Munich, 1971* 8:199–200

Wall, P. D., Egger, M. D. 1971. Formation of new connexions in adult rat brains after partial deafferentation. *Nature* 232: 542–45

West, J. R., Deadwyler, S. A., Cotman, C. W., Lynch, G. S. 1975. Time-dependent changes in commissural field potentials in the dentate gyrus following lesions of the entorhinal cortex in adult rats. *Brain Res.* 97:215–33

Westrum, L. E. 1969. Electron microscopy of degeneration in the lateral olfactory tract and plexiform layer of the rat. *Z. Zellforsch.* 98:157–87

Wieraszko, A., Baudry, M., Creager, R., Finn, R., Lynch, G. 1979. Increase in ^{3}H-glutamate accumulation following induction of long-term synaptic potentiation in hippocampal slices. *Soc. Neurosci. Abstr.* 5:637

Woody, C. D., Black-Cleworth, P. 1973. Differences in excitability of cortical neurons as a function of motor projection in conditioned cats. *J. Neurophysiol.* 36:1104–16

Woody, C. D., Engel, J. Jr. 1972. Changes in unit activity and thresholds to electrical microstimulation at coronal-pericruciate cortex of cat with classical conditioning of different facial movements. *J. Neurophysiol.* 35:230–41

Woody, C. D., Vassilevsky, N. N., Engel, J. Jr. 1970. Conditioned eye blink: Unit activity at coronal-pericruciate cortex of the cat. *J. Neurophysiol.* 33:851–64

Yamamoto, C., Chujo, T. 1978. Long-term potentiation in thin hippocampal sections studied by intracellular and extracellular recordings. *Exp. Neurol.* 58:242–50

Yumiya, H., Larsen, K. D., Asanuma, H. 1979. Motor readjustment and input-output relationship of motor cortex following cross-connection of forearm muscles in cats. *Brain Res.* 177:566–70

Ann. Rev. Neurosci. 1981. 4:381–417

SLEEP AND ITS DISORDERS ❖11558

Elliot D. Weitzman

Laboratory of Human Chronophysiology, Sleep-Wake Disorders Center, Department of Neurology, Montefiore Hospital and Medical Center, The Albert Einstein College of Medicine, Bronx, New York 10467

INTRODUCTION

The application of scientific information regarding sleep physiology and biological rhythm functions to the diagnosis and treatment of sleep disorders in man has been remarkable. The often deplored gap between the accumulation of knowledge from basic science research and its application to patients is being bridged rapidly in this new field (Dement 1973, Weitzman 1974, Guilleminault et al 1976c, Hartmann 1974, Guilleminault & Dement 1978b, Lugaresi et al 1978, Mendelson et al 1977, Williams & Karacan 1975, 1978, Kleitman 1963).

This review of the sleep disorders emphasizes the scientific base of knowledge supporting the diagnostic classification and, when appropriate, the treatment modality available. It would clearly be impossible to include all relevant publications in such a vast literature; emphasis therefore is given to reports that are recent and seminal.

A special committee of the Association of Sleep Disorders Centers under Howard Roffwarg has recently compiled a nosology of the sleep disorders (Association of Sleep Disorders Centers, 1979). This nosology, because it has structured the classification primarily on the basis of clinical symptomatology, is the framework of this review. Since, with very few exceptions, there is an absence of tissue pathology to support this classification of clinical diseases, it is necessary to rely on the natural clustering of clinical symptoms and signs from which the diagnostic syndromes and disease entities derive. An additional important advance has been the development of clinical *polysomnography* for the objective measurement of physiological and autonomic changes in patients during prolonged and/or recurrent daily (nightly) sleep sequences. This important new clinical tool derives from the

381

extensive data of the last 50 years, beginning with Hans Berger (1929), using electroencephalographic and polygraphic sleep recordings. Extensive reference, therefore, is made to both normal and abnormal polygraphic patterns where appropriate for the diagnostic entity under consideration.

Sleep disorders encompass the entire life span of man, from the prematurely born infant to the senescent person. Each age group has its special diseases that derive both from maturational aspects of the sleep-waking process (e.g. nocturnal enuresis, night terrors, and somnambulism) and as idiosyncratic and cryptogenic age and sex-related processes (e.g. hypersomnia, sleep apnea syndrome). Nevertheless, certain conditions such as the narcolepsy-cataplexy syndrome and nocturnal seizures may be life-long, beginning in childhood or adolescence, never to disappear, even in the aged, but usually adequately controlled by appropriate medication.

EPIDEMIOLOGICAL CONSIDERATIONS

Before embarking on a detailed description of selected individual sleep disorders, it is valuable to outline briefly information regarding the scope of the sleep disorders problem. Although the data still remains fragmentary, certain recent statistics are striking.

In a survey of one million people in the United States over the age of 30, 13% of men and 26% of women complained of insomnia (Hammond 1964, Kripke et al 1979). A British survey of 2400 people revealed that 8% reported sleeping less than 5 hr per night, 35% reported frequent nocturnal arousals, and 25% had morning tiredness (McGhie & Russell 1962). A survey of an urban Florida county indicated that 13% of the adults "often" had problems getting to sleep or staying asleep (Karacan et al 1976). It therefore appears that 8 to 15% of the adult population have frequent and chronic complaints about the quality and amount of sleep. This high prevalence is supported by data regarding the use of hypnotic drugs in the United States. A national survey in 1971 estimated that about 3% of the adult population (ages 18 to 74) used a hypnotic in the preceding year (Balter & Bauer 1975). Another survey in 1972 estimated that 11% of the population over the age of 18 used a sedative during one year (National Institute on Drug Abuse, 1977); a repeat survey in 1977 estimated 9%. All studies indicate that the use of hypnotic-sedative drugs increases with age. In 1977 (Association of Sleep Disorders Centers, 1979), 39% of all prescriptions for hypnotics were for people over the age of 60 (National Institute on Drug Abuse, 1977). There were 26 million prescriptions written for total hypnotics (including barbiturates and nonbarbiturates) in 1977, representing sales of 82.38 million dollars (National Institute on Drug Abuse, 1977).

A thorough analysis of prospective epidemiological data of one million "healthy" American adults (over age 30) carried out by the American Cancer Society (1959–1960 and 1965–1966) revealed that the mortality ratio of observed to expected deaths in men at a six year follow-up was 2.8 for those who reported usually sleeping less than 4 hr per night and 1.8 for those sleeping 10 hr or more. The norms were 7.0 to 7.9 hr per night (Hammond 1964). For women the comparable values were 1.5 (less than 4 hr) and 1.8 (more than 10 hr). Those who reported "often" using sleeping pills had 1.5 times the mortality of those who "never" used sedative drugs (Kripke et al 1979). Therefore, not only is the prevalence of sleep disorders a major problem in regard to frequency of complaint, excessive drug usage, and financial cost, but more important, there is evidence for increased mortality risk in the population who complain of a chronic sleep disorder and who often use a hypnotic. Studies are clearly needed to determine the causes of these mortality risk factors.

The recent classification of sleep disorders (Association of Sleep Disorders Centers, 1979) serves as a useful framework for the specific clinical disorders of sleep that we consider in this review.

In this review we describe only selected disorders, primarily considering common syndromes and ones in which important understanding or treatment regimens have recently emerged. For a more complete review of all the sleep disorders outlined, the reader is referred to the special issue of *Sleep* (Association of Sleep Disorders Centers, 1979).

DISORDERS OF INITIATING AND MAINTAINING SLEEP (INSOMNIAS)

There has been a significant advance in the understanding of five syndromes associated with insomnia: (*a*) affective disorders, (*b*) drugs and alcohol, (*c*) sleep apnea and hypoventilation syndromes, (*d*) nocturnal myoclonus (periodic movements during sleep) and "restless legs" syndromes, and (*e*) the delayed sleep phase syndrome. The last entity is described in more detail under the section on disorders of the sleep-wake schedule.

Insomnia Associated with the Psychiatric Disorders of Affective States

Affective disorders associated with insomnia have been divided into two categories: *major depressive and bipolar disorders* (recurrent depression, manic-hypomanic episodes, or both) and *secondary depression* (*see Diagnostic and Statistics Manual of Mental Disorders,* 1980). The major sleep abnormalities found in these disorders consist of a chronic and recurrent inability to maintain sleep through the expected sleep period, "premature"

Table 1 Diagnostic classification of sleep and arousal disorders

DIMS: disorders of initiating & maintaining sleep (insomnias)	DOES: disorders of excessive somnolence *(continued)*
Psychophysiological	Associated with sleep-related (nocturnal) myoclonus and "restless legs"
Transient and situational	Sleep-related (nocturnal) myoclonus DOES syndrome
Persistent	"Restless legs" DOES syndrome
Associated with psychiatric disorders	Narcolepsy
Symptoms and personality disorders	Idiopathic CNS hypersomnolence
Affective disorders	Associated with other medical, toxic, and environmental conditions
Other functional psychoses	Associated with other DOES conditions
Associated with use of drugs and alcohol	Intermittent DOES (periodic) syndromes
Tolerance to or withdrawal from CNS depressants	Kleine-Levin syndrome
Sustained use of CNS stimulants	Menstrual-associated syndrome
Sustained use of or withdrawal from other drugs	Insufficient sleep
Chronic alcoholism	Sleep drunkenness
Associated with sleep-induced respiratory impairment	Not otherwise specified
Sleep apnea DIMS syndrome	No DOES abnormality
Alveolar hypoventilation DIMS syndrome	Longer sleeper
Associated with sleep-related (nocturnal) myoclonus and "restless legs"	Subjective DOES complaint without objective findings
Sleep-related (nocturnal) myoclonus DIMS syndrome	Not otherwise specified

Disorders of the sleep-wake schedule

DIMS	
"Restless legs" DIMS syndrome	Transient
Associated with other medical, toxic and environmental conditions	Rapid time zone change ("jet lag") syndrome
Childhood-onset DIMS	"Work shift" change in conventional sleep-wake schedule
Associated with other DIMS conditions	Persistent
Repeated REM sleep interruptions	Frequently changing sleep-wake schedule
Atypical polysomnographic features	Delayed sleep phase syndrome
Not otherwise specified	Advanced sleep phase syndrome

DOES: disorders of excessive somnolence

Non-24-hour sleep-wake syndrome

DOES	
Psychophysiological	Irregular sleep-wake pattern
Transient and situational	Not otherwise specified

Persistent

Dysfunctions associated with sleep, sleep stages, or partial arousals (parasomnias)

Associated with psychiatric disorders	Sleepwalking (somnambulism)
Affective disorders	Sleep terror (pavor nocturnus, incubus)
Other functional disorders	Sleep-related enuresis (bed-wetting)
Associated with use of drugs and alcohol	Other dysfunctions
Tolerance to or withdrawal from CNS stimulants	Dream anxiety attacks (nightmares)
Sustained use of CNS depressants	Sleep-related epileptic seizures
Associated with sleep-induced respiratory impairment	Sleep-related bruxism
Sleep apnea DOES syndrome	Sleep-related headbanging (jactatio capitis nocturnus)
Alveolar hypoventilation DOES syndrome	

Table 1 *(Continued)*

Dysfunctions associated with sleep, sleep stages, or partial arousals (parasomnias) *(continued)*	Dysfunctions associated with sleep, sleep stages, or partial arousals (parasomnias) *(continued)*

Other dysfunctions *(continued)*
  Familial sleep paralysis
  Impaired sleep-related penile
    tumescence
  Sleep-related painful erections
  Sleep-related cluster headaches and
    chronic paroxysmal hemicrania
  Sleep-related abnormal swallowing
    syndrome
  Sleep-related asthma
	Other dysfunctions *(continued)*
	  Sleep-related cardiovascular
	    symptoms
	  Sleep-related gastroesophageal reflux
	  Sleep-related hemolysis (paroxysmal
	    nocturnal hemoglobinuria)
	  Asymptomatic polysomnographic
	    finding
	  Not otherwise specified

arousal early in the morning, and, to a lesser degree, difficulty initiating sleep at the expected time at night.

The sleep disturbances associated with the affective disorders characteristically involve different degrees of intrusive wakefulness occuring during the course of the daily sleep period (Hauri 1977, Kales & Kales 1974, Reynolds et al 1979, Coble et al 1979, Gillin et al 1977, Gresham et al 1965, Hawkins & Mendels 1966, Kupfer 1971, Lange et al 1976, Spiker et al 1978, Hauri et al 1974). The patients typically complain of severe and persistent restlessness at night and a tired, disinterested, "washed out" feeling during the day. It is very unusual for most depressed patients to complain of true sleepiness or to actually sleep recurrently (i.e. take naps) during the day. Indeed there is not only a complaint of nocturnal insomnia, but often of daytime insomnia as well. Older individuals with affective disorders characteristically have an abbreviated sleep with a premature early morning arousal pattern.

It has been recently recognized that monopolar and bipolar depressed patients have different amounts of sleep and organization of sleep stages (Gillin et al 1977, 1979). The monopolar depressions characteristically have repeated awakenings with a premature nocturnal arousal, a shortened REM latency, and reduced stages 3 and 4 non-REM sleep. On some nights there is an associative high stage REM percentage. It has been well known for many years that REM sleep deprivation can lead to a REM rebound on recovery nights. Although this has been suggested as a possible explanation for the increased REM amounts, others have considered the changes in REM amounts and timing as a primary feature of the sleep disturbance in depression (Kupfer 1976, Lange et al 1976, Vogel et al 1975, Schulz et al 1979).

By contrast, in the bipolar depression, excessive daytime naps and extended sleep periods at night are often present (Sitaram et al 1978, Papousek 1976, Kripke et al 1978). Although the occurrence of sleep has been confirmed with polygraphs, these patients nevertheless awaken feeling unrefreshed. During times of mania or hypomania, total or partial sleeplessness will be present in association with inappropriate and excessive agitated behavior.

Measurements of physiological parameters during sleep (polysomnography) demonstrate that in the bipolar depressed patient there is also a shortened REM sleep latency and reduced sleep stages 3 and 4 (Kupfer 1971, 1976, Lange et al 1976, Schulz et al 1979, Sitaram et al 1978, Vogel et al 1975).

Patients with complaints of moderate and severe insomnia who have secondary depression or personality disorders (generalized anxiety, neurotic phobias, obsessive-compulsive disorders, and "neurotic" personality disorders) do not generally demonstrate this shortened REM latency. They do, however, have fragmented sleep with a reduction in stages 3 and 4 of sleep and often an associative reduction in stage REM sleep (Beutler et al 1978, Billiard 1978, de la Pena 1978, Kales & Kales 1974, Kales et al 1976, Roth et al 1976, Gillin et al 1979). These patients may also have a concomitant circadian rhythm disturbance especially associated with sleep onset insomnia (see the section on disorders of the sleep-wake schedule, below).

Several associated pattern abnormalities of polygraph sleep recording have been shown to accompany the disturbed sleep in patients with the chronic complaint of insomnia. These include "alpha-delta activity," i.e. the presence of coherent alpha activity (9 to 11 hz) superimposed on the sleep EEG patterns; and REM-spindle activity (Hauri 1977, Hauri & Hawkins 1973, Phillips et al 1974), i.e. the simultaneous mixture of rapid eye movements and 12 to 14 hz sleep spindles in the EEG (McGregor et al 1975). These polygraphic patterns are almost never present during normal sleep and indicate the internal dissociative sleep-wake features of chronic insomnia syndromes. These indicators of a temporally disorganized electroencephalographic pattern may be related to the common observation that upon awakening in the morning after unequivocal sustained periods of 6 to 7 hrs of polygraphically defined sleep, chronic insomniac patients insist that they either have not slept at all or only for several hours ("pseudo-insomnia syndrome") (Roth et al 1977, Hoddes et al 1972, Frankel et al 1976, Bixler et al 1973, Billiard et al 1978, Carskadon et al 1976). The perception of not having slept is at variance with the objective polygraphic evidence of sleep.

Several treatments for depression specifically use, not drugs, but manipulation of the timing of sleep stages or sleep itself within the circadian sleep-wake cycle. Thus, Vogel et al (1975) report reversing depressive symptoms following a period of several weeks of REM deprivation, and Wehr et al (1979) report reversing the depressive phase of patients with bipolar depressions by producing a progressive *phase advance* of the sleep time within the patients' circadian biological day (see section on disorders of the sleep-wake schedule, below). In addition, there have been a number of reports that paradoxically claim that sleep deprivation itself will temporarily reverse symptoms of depression (Rudolf et al 1977). All of these recent clinical studies emphasize the importance of studying the sleep-wake biological rhythm in order to understand the nature of the sleep disturbances in disorders of initiating and maintaining sleep (insomnias).

Insomnias Associated with The Use of Drugs and Alcohol

The extensive use of CNS depressants (hypnotic and sedative drugs, tranquilizers, bedtime use of alcohol) without a rational pharmacological basis has been emphasized by the excellent report entitled *Sleeping Pills, Insomnia and Medical Practice,* published by the Institute of Medicine of the National Academy of Sciences (1979). During the past ten years, there has been a marked decrease (approximately 80%) in the use of barbiturates as hypnotics with a concomitant dramatic increase in the use of flurazepam (Dalmane®) (approximately 370%). The major effect of this change has been a decrease in suicides from barbiturates (75% of all drug suicides in 1963; 30% in 1976), paralleled by a drop of barbiturate prescriptions from 41 million in 1970 to 20 million in 1976. Unfortunately this has not been accompanied by a decrease in the number of suicides caused by drugs (2,635 in 1968; 3,000 in 1976). Actually, there has been a progressive increase in the total number of suicides in the United States (21,476 in 1968; 26,832 in 1976); the most common method is by firearms (55% in 1976) (National Academy of Sciences, 1979).

Although psychopathology is present in most insomniac patients, these psychiatric diseases fail to respond to hypnotics (Oswald 1979, Wheatley 1979); indeed, the effectiveness of long-term use (greater than 30 d) of hypnotic drugs has not been scientifically established (Hartmann 1976, Kay et al 1976, Kales et al 1974). Flurazepam, which is now the most frequently used "sleeping pill" in the United States, will temporarily improve sleep patterns but it has not been studied beyond one month of continuous use (National Academy of Sciences 1979, Dement et al 1973a, 1978, Kales et al 1970a,b). In addition, although approximately one-third of all sedative drugs are used by people over the age of 60, there have been very few objective studies of the effectiveness of hypnotic drugs on the aged (Frost & DeLucchi 1979, Viukari et al 1978, Amin 1976). Alcohol is commonly used as a sedative but its effectiveness over the long term has also not been demonstrated (Williams & Salamy 1972, Baekeland et al 1974). The major sleep disturbance that results from chronic alcoholism must be differentiated from the nightly sedative use of alcoholic beverages. With heavy and sustained ingestion of alcohol, the sleep stage organization is disrupted, with abbreviated REM sleep periods and associated intrusive awakenings (Johnson et al 1970, Pokorny 1978, Rundell et al 1977). Acute withdrawal from alcohol in the chronic alcoholic leads to a lengthening of the sleep onset latency, decrease in non-REM sleep and an increase in REM sleep. The REM latency shortens and changes qualitatively as well (Gross et al 1971, 1973, Gross & Goodenough 1967). If severe, the acute toxic withdrawal syndrome (delirium tremors) will develop (Gross et al 1966). In the chronic "dried out" alcoholic, an abnormal sleep pattern may persist for many weeks, although most show a return to normal sleep within 2 weeks (Adamson & Burdick 1973, Zarcone et al 1975).

A major advance in understanding sleep disorders has been the recognition that tolerance to or withdrawal from these CNS depressants may produce an insomnia syndrome (Drug Dependency Insomnia) (Kales et al 1969a,b). The essential features of this syndrome are the development of tolerance and withdrawal to the chronically used drug. The short-term sleep-inducing and maintenance effects are lost and the patient and the physician often increase the dosage or combine it with other drugs. Sleep-disturbing symptoms can develop due to partial withdrawal even when use of the drugs is continued, and these symptoms are often misinterpreted as evidence of persistence of the underlying insomniac syndrome. When these drugs are acutely withdrawn, a severe sleep disturbance occurs with marked sleeplessness, which is again misinterpreted as the return of the underlying "insomnia" (Kay et al 1976, Kales et al 1974).

The use of l-tryptophan, an amino acid present in normal dietary intake, has been evaluated recently as a hypnotic agent (Hartmann et al 1971, Hartmann 1973, Hartmann 1977, Hartmann & Spinweber 1979). Although side effects and withdrawal symptoms are not present, its effectiveness as a clinically useful sedative drug has not been demonstrated (Griffiths et al 1972, Brown et al 1979, Wyatt et al 1970, Volk et al 1978).

During chronic use of a hypnotic agent, sleep is interrupted by frequent awakenings lasting more than 5 min, especially during the second half of the night (Hauri 1977). The decrease in the effectiveness of the drugs and the consequent partial withdrawal each night contribute to the early morning arousal in tolerant individuals. Recordings during sleep demonstrate that stages 3 and 4 and REM sleep are decreased; stage demarcations are less clear, with frequent stage transitions and sleep spindles; and K-complexes, delta waves, and REM sleep eye movements are decreased. There are increased "pseudo spindles" (14 to 18 Hz) and beta frequencies (20 to 30 Hz) present during all sleep stages (Hauri 1977).

Following a rapid or acute withdrawal of the chronic use of high dose daily hypnotic agents, (especially barbiturates), the sleep pattern is very disrupted with a high percentage of REM sleep and intense phasic activity.

This is presumably due to a compensatory REM rebound. Periodic movements (nocturnal myoclonus) related to sleep may also appear temporarily during the withdrawal period (see section on sleep-related nocturnal myoclonus). In addition, daytime symptoms appear. These include restlessness, nervousness, generalized muscle aches, and, in severe cases, drug withdrawal and grand mal convulsions. This is especially true for withdrawal from barbiturates and glutethimide (doriden) (Kay et al 1976). Therefore, in patients that have used high doses of several hypnotic drugs chronically, it is important that withdrawal be done gradually and under clinical supervision. Many patients will then show remarkable improvement in both objective and subjective aspects of their sleep abnormalities, although they may still not return fully to a normal sleep pattern.

Insomnia Associated with Sleep-Induced Respiratory Impairment

Although the great majority of patients with sleep apnea complain primarily of excessive daytime somnolence, in some cases sleep apnea or hypoventilation may cause frequent awakenings, restless, and unrefreshing sleep (Guilleminault et al 1978b, 1978c, Okada et al 1979; see also the section below on excessive daytime somnolence). In patients with primarily an insomniac complaint, central apnea, i.e. cessation of breathing (diaphragmatic cessation), during both REM and non-REM sleep may be the predominant abnormal respiratory pattern, although in some cases a mixed upper airway obstructive and central apneic pattern may be present (Guilleminault et al 1976c, Webb 1974). The largest apneic episodes are present during REM sleep, especially during bursts of rapid eye movements. During the course of the night, recurring arousals will occur, some associated with a gasping for air or a feeling of choking and some at the termination of an apneic episode. There does not appear to be a correlation between the degree of hypoxia or hypercapnia and an awakening. Cardiac arrhythmias may be present during sleep in some patients, although most have normal respiration and cardiac rhythm function during the waking day (Guilleminault et al 1977a).

The central alveolar hypoventilation syndrome, in which a specific etiology is more apparent, is also associated with a major change in respiratory function during sleep (Severinghaus & Mitchell 1962, Schwartz 1976, Lugliani et al 1979). Sleep-related central apnea and hypopnea is clearly present and associated with recurrent hypoxemia, hypercapnia, and a decreased tidal and minute volume. REM sleep is also the time of greatest abnormality, with a deterioration of arterial blood gas values. Certain specific conditions have been associated with this entity. These include massive obesity, chronic obstructive pulmonary disease, chronic residual poliomyelitis, muscle diseases (myotonic dystrophy), and involvement of thoracic cage bellows action and/or diaphragmatic muscle weakness, high spinal cord cordotomy, and brainstem neurological lesions of structures that control ventilation (Guilleminault et al 1976c, Rochester & Enson 1974, Sanger 1977, Krieger & Rosomoff 1974a,b, Hudgel & Shucard 1979, Guilleminault et al 1978a, Guilleminault & Dement 1978b, Carroll 1974, Weitzman et al 1978a, Cummiskey et al 1978, Strohl et al 1978, Koo et al 1975). This important interface between pulmonary medicine, neurology, and sleep physiology has only developed recently and we are likely to see more investigations in this area in the future (Weitzman 1979).

Sleep-Related (Nocturnal) Myoclonus (Periodic Movements During Sleep, PMS)

Many patients with a primary complaint of chronically disturbed sleep show periodic movement of one or both legs and feet during sleep (Lugaresi et al 1967, Guilleminault et al 1975, Coleman et al 1978, 1980). These periodic movements in sleep (PMS) are stereotyped, repetitive, and nonepileptiform, and occur primarily in non-REM sleep. The movement lasts 1.5 to 2.5 sec and consists of a dorsiflexion of the foot, extension of the big toe, and often flexion of the lower leg at the knee and hip. Although historically called "myoclonic" by Sir Charles Symonds (1953), PMS do not occur in an isolated muscle and are not as brief as true myoclonic contractions. They are remarkably periodic (approximately 20 to 30 sec) and persist for many minutes and at times for hours. Unlike most movement disorders (e.g. Parkinsonism, Chorea, Hemiballism, Dystonia, etc), which are inhibited by sleep, PMS are initiated by sleep (Freemon 1978). PMS are to be differentiated from "sleep starts" (hypnagogic jerks), which are nonperiodic isolated myoclonic movements, generally of the trunk and all extremities simultaneously, occurring at sleep onset and frequently associated with the perception of falling. These "sleep starts" can be considered to be normal since most normal children and adults have experienced them, usually on multiple occasions.

Lugaresi et al (1967, 1970) were the first to record PMS in sleep in patients complaining of the restless legs syndrome and chronic insomnia; they suggested that periodic movement during sleep might be a specific cause of insomnia. Restless legs syndrome, originally described by Ekbom (1960), is a disorder in which the afflicted individual feels an irresistible urge to move the legs, generally when sitting or lying down, especially in bed at night, just prior to falling asleep. The patient senses a discomfort deep inside the calfs that, although not truly painful, forces the patient to move, exercising the legs, or even to walk about vigorously (Coleman et al 1980, Frankel et al 1974). The restless leg symptom interferes with the onset of sleep, but may also recur during the night. Usually by morning the restless legs symptoms have either disappeared or attenuated considerably and the patient is able to fall asleep more easily. In all cases of restless legs syndrome studied polygraphically, PMS are also present; however, it is now clear that PMS occur in association with a wide variety of sleep disorders but without clinical symptomatology of the restless legs syndrome. In a recent extensive study using all-night polysomnographic recording of 409 sleep disorder patients, 53 (13%) had PMS. Only 8 of the 53 had restless legs syndrome (Coleman et al 1980). PMS were present in almost all sleep disorder diagnostic categories, including narcolepsy-cataplexy, sleep apnea, drug dependency insomnia, etc. These findings suggest that, rather than PMS causing insomnia, a chronic sleep-wake disturbance may be associated with PMS and may lead to the development of these movements. PMS may develop in patients being treated with tricyclic antidepressants and during withdrawal from a variety of drugs such as anticonvulsants, benzodiazepines, barbiturates, and other hypnotic agents as well as in patients with chronic uremia. Movements similar to PMS during sleep have been induced in Parkinsonian patients receiving high dosages of L-DOPA and in narcoleptic patients on high doses of chlorimipramine for the treatment of cataplexy (Coleman et al 1980). Although a wide variety of drugs have been tried in clinical and experimental treatment trials, none have been clearly effective in eliminating PMS (Coleman et al 1980, Matthews 1979). Although the neurophysiological mechanisms responsible for PMS are unknown, the remarkable periodicity found in all studies indicates that there is an underlying CNS pacemaker and that a chronic disturbance of the temporal organization of the daily sleep-wake schedules and specific sleep stages might "disinhibit" an underlying periodic CNS process (Coleman et al 1980).

DISORDERS OF EXCESSIVE SOMNOLENCE

The chief symptoms of disorders of excessive somnolence include undesirable and inappropriate sleepiness and actual sleep during a time of day when the patient wishes to be awake. A regular (i.e. entrained) 24 hr sleep-wake cycle is generally presupposed and, although the disorder may occur on a transient basis, it should be present for weeks to years in order to be considered pathological. Characteristic complaints include an increased amount of unavoidable napping, increase in total sleep during the 24 hr, and difficulty in achieving full alertness upon awakening at the end of the daily sleep period. These complaints should be differentiated from abnormalities of the sleep-wake schedule (see below) and not confused with complaints of "tiredness," lack of "energy," decreased alertness, or lack of motivation, interest, or "drive." Although these latter complaints may accompany sleepiness and recurrent sleep events, they are actually alterations in mood that affect the patient during the waking state. An important recent advance has been the development of the *Multiple Sleep Latency Test* at the Stanford Sleep-Disorders Center (Richardson et al 1978). This test, performed during an 8 hr "waking" period, demonstrates that normal individuals rarely fall asleep when conducive conditions are present; however, patients with a complaint of excessive daytime sleepiness, who fall into a clear diagnostic category on the basis of a clinical history and polysomnography, demonstrate recurrent sleep episodes in this test (Mitler et al 1979).

For completeness, the classification list of disorders of excessive daytime somnolence includes a large number of possible conditions; however, in actuality, from a clinical point of view, two major syndromes represent approximately 80% of the patients suffering from excessive somnolence who are seen in sleep disorders centers (Dement 1973, Williams & Karacan 1978, Weitzman et al 1979a). These are (*a*) sleep-induced respiratory impairment and (*b*) narcolepsy.

Sleep-Induced Respiratory Impairment (The Hypersomnia Sleep-Apnea Syndrome)

One of the most important advances in our understanding of normal and pathological sleep patterns has been the recognition that abnormal respiratory function may lead to serious clinical disorders that, in some cases, are life threatening (Guilleminault et al 1976c, Guilleminault & Dement 1978a). Important new studies of normal respiratory patterns and the control of breathing during sleep (Phillipson 1978) have proliferated along with clinical investigations into the mechanism of sleep-related air-flow abnormalities.

Although excessive daytime somnolence is the primary complaint, there may be considerable variation in its intensity. Almost all patients have loud intermittent snoring upon inspiration, which increases in severity with each breath during a period of approximately 20 to 50 sec, culminating in a series of loud snoring gasps. The episode is terminated by a brief arousal, with body and head movements, followed typically by a variable period of apnea with respiratory silence, only to recur in a cyclic pattern. Several sleep disorders centers have made careful radiographic and direct fiberoptic endoscopic observations of the upper airway during the apnea process and all have concluded that the site of functional airway obstruction during sleep is in the oral pharynx at the level of the velopharyngeal sphincter (Weitzman et al 1978b). The mechanism of obstruction involves a recurrent apposition of the lateral pharyngeal walls and posterior movement of the base of the tongue.

In addition to severe daytime sleepiness there are also night-time symptoms that can be very disturbing to the patient as well as to the bed partner. The patient is very restless, with frequent brief arousals, assuming unusual sleeping postures and positions in the bed, and loud intermittent snoring and snorting alternates with periods of silence (apnea). The patient will often talk during sleep (somniloquy). Nocturnal enuresis (bed-wetting) and falling out of bed during sleep also occur. The patient awakens in the morning unrefreshed, very sleepy, and often with a generalized severe headache (Guilleminault et al 1978c).

The disease tends to occur between the fourth and sixth decades and the ratio of men to women is 20 to 1. When HSA occurs in women, it does so predominantly after menopause. Approximately two-thirds of the patients are obese (Weitzman et al 1978b) and further rapid gain in weight will increase the severity of the symptoms (Guilleminault et al 1976c, Guilleminault & Dement 1978a, Lugaresi et al 1978). In moderate to severe cases, there is a very high association of systemic hypertension (Lugaresi et al 1978).

An all-night polygraphic recording is used to both diagnose the condition and quantify the frequency, severity, and type of sleep apnea (Lugaresi et al 1978, Guilleminault & Dement 1978b, Weitzman 1979). *Nonobstructive sleep apnea* is the cessation of all respiratory movements (diaphragmatic arrest); *obstructive apnea* is the absence of airflow (from nose or mouth) in association with respiratory effort; *mixed apnea* consists of an initial nonobstructive followed by obstructive apnea; a *hypopnea* is a partial decrease of respiratory effort and airflow by greater than 50%; and *subobstructive apnea* is a partial decrease of airflow in the presence of increased respiratory effort (indicating a partial functional obstruction in the pharyngeal airway). All of the above types of apnea patterns can be present in an individual patient; however, the obstructive, mixed, and subobstructive types are most often present in patients with the HSA syndrome. The recognition that subobstructive apnea is associated with HSA and recurrent oxygenation desaturation is especially important. Recent studies indicate that 30 to 40% of adults over the age of 40 are chronic habitual snorers and that there is an increased incidence of hypertension in that large group (Lugaresi 1978).

During a polygraphic sleep recording of patients with HSA, a characteristic abnormal pattern is seen. This consists of recurrent apnea (ranging from 20 to 120 sec with an average of approximately 38 sec), oxygen desaturation (with decreases often below 50%), and a bradycardia ($<$ 50 bpm) alternating with a tachycardia ($>$ 120 bpm). The patient with severe sleep apnea has little if any stage 3 or 4 sleep, but does have stages 2 and REM sleep (Guilleminault et al 1976c, Weitzman et al 1981). The duration of apneic episodes increases and oxygen desaturation is greater in REM sleep. A wide range of cardiac arrhythmias are often present in these patients, often only during sleep, and on occasion they are quite severe (e.g. cardiac arrest lasting 5 to 7 sec). The severe bradycardia and transient episodes of asystole typically occur at the end of an apnea just prior to the resumption of breathing (loud snores). Since the pattern can be blocked by atropine (Guilleminault et al 1978b), these cardiac rate changes presumably represent excessive parasympathetic vagal activity.

The cause of this respiratory abnormality during sleep is not known, but it occurs in a wide range of conditions in which there is a narrowing of the upper airway. These include (a) nasal passage deformities, (b) enlarged tonsils and adenoids, (c) mandibular shortening (Conway et al 1977, Kuo et al 1979, Sanger 1977) and tempero-mandibular joint abnormalities (Weitzman et al 1978a), (d) pharyngeal soft tissue abnormalities such as occurs in hypothyroid myxedema (Duron et al 1972, Yamamoto et al 1977) or acromegaly (Romanczuk et al 1978), (e) congenital palato-pharyngeal abnormalities, and (f) following velo-palatal reconstructive surgery for the cleft palate syndrome. In addition, conditions with respiratory muscle weakness have been associated with sleep apnea, especially the nonobstructive apnea and hypopnea types (Guilleminault & Dement 1978a, Guilleminault et al 1976c). These include chronic poliomyelitis, myotonic dystrophy, cervical spinal cord disease, etc. Studies of high altitude (hypoxic) respiratory patterns during sleep have also demonstrated the sleep apnea syndrome to be present (Guilleminault & Dement 1978a). A link between sleep-related apnea and the sudden infant death syndrome (SIDS) has been postulated (Weitzman & Graziani 1974) with recent supporting evidence from studies of "near miss" for SIDS infants (Kelly & Shannon 1979). Patients with chronic obstructive pulmonary disease also demonstrate nightly frequent sleep apnea episodes with significant oxygen desaturation (Strohl et al 1978, Koo et al 1975).

Because of the clinical importance on the control of breathing during sleep of sleep apnea, there has been a concomitant research in animal models (Phillipson 1978). It has been shown, in the dog, that the ventilatory response to CO_2 is much decreased during REM sleep—specifically during phasic REM events compared to tonic REM, non-REM sleep, and waking (Sullivan et al 1979)—whereas it remained intact to hypoxia during REM sleep despite a delayed waking response (Phillipson 1978). Elimination of vagal respiratory control in dogs leads to a profound slowing of respiration during non-REM sleep but not during wakefulness or REM sleep (Phillipson 1978). There is a 29% higher upper airway resistance in REM compared to non-REM sleep (in cats) (Orem & Dement 1975) and a marked reduction of both tonic and inspiratory related intercostal muscle activity during REM sleep similar to the REM-related inhibition of other skeletal muscles (Hagan et al 1976). These changes are associated with decreased expansion (in adults) (Tusiewicz et al 1977) and even a paradoxical motion of the rib cage (in infants) (Henderson-Smart & Read 1976). The diaphragm muscle, driven by alpha motor neurons, is apparently not normally affected in REM sleep; however, with clinical disease of anterior horn cells, phrenic nerves, neuro-muscular junctions, or the diaphragm muscle itself, the patient can be at special risk during REM sleep. In addition, evidence that the ventilatory response to a respiratory "load" (airway occlusion or inspiration from a rigid container) is diminished in REM sleep (Phillipson 1978) adds to the risk factor of patients with low hypoxic sensitivity. The use of microelectrodes to record discharges from respiratory neurons in the medulla of intact cats has demonstrated that certain cells reduce their discharge frequency during non-REM and become silent in REM sleep; others were only active during waking and still others were no longer synchronized to respiration during sleep (Orem et al 1974, Orem & Dement 1975).

The most effective treatment with severe HSA is to provide the patient with an open tracheal airway (tracheostomy) during the night's sleep (Weitzman et al 1981, Hill et al 1978). There is marked improvement in daytime sleepiness and mood, a decrease in systemic hypertension, a fall in the hematocrit, elimination of the bradytachycardia pattern, a return to maintained high values of oxygen saturation during sleep, and in most cases a marked decrease in the frequency of sleep-related cardiac arrhythmias. A recent study (Weitzman et al 1981) of ten patients with severe HSA (> 350 apneas/night) found that within the first week after performing a tracheostomy, the patients improved from a mean of 3% (pre) to 19% (post) stages 3 and 4 of total sleep, although total sleep time did not change. This was accompanied by a marked decrease (approximately four-fold) in the number of sleep stage changes per hour. It was the reversal of the fragmentary, unsustained sleep stage organization in association with a major increase in stages 3 and 4 sleep that was reversed by tracheostomy. There was also a marked decrease in the number of obstructive and mixed apneas with little change in the number or duration of central apneas and hypopneas. Oxygen saturation was also markedly improved through the night. Other treatments have been reported for mild to moderate forms of HSA using tri-cyclic drugs (Guilleminault et al 1974, Clark et al 1979), progesterone (Orr et al 1979a), and weight loss, all with only limited success (Fisher et al 1978, Hoffmeister et al 1978).

Narcolepsy-Cataplexy Syndrome

The *narcolepsy-cataplexy syndrome* is a distinct clinical entity with well-defined symptoms, age of onset, and natural history (Dement 1973, Guilleminault et al 1976a, York & Daly 1960, Zarcone 1973). The most characteristic symptom is the recurrent episodes of excessive daytime sleepiness and sleep attacks during the waking portion of the patient's day (Dement et al 1966, Hishikawa et al 1968). Although it is more characteristic for the sleep periods to occur with an episodic and irresistible character, usually the patient also complains of sleepiness and decreased vigilance between sleep attacks as well (Roth 1978). Cataplexy is also a very characteristic feature and develops at some point in the history of the illness in approximately 70 to 90% of all cases (Guilleminault 1976). It is characterized by brief sudden episodes of muscle weakness, without loss of consciousness. These episodes may be total, with flaccid paralysis of all somatic musculature (except diaphragmatic and extra-ocular muscle movements) leading to collapse when the individual is standing, or the episodes may be partial, with weakness of isolated muscle groups. If a thorough history is taken, patients with cataplexy frequently describe dropping of the head or jaw, dysarthric speech, weakness of facial muscle, weakening about the knees, and the dropping objects from their hands, all lasting only seconds or minutes. Cataplectic episodes are almost always triggered by a sudden emotional feeling such as laughter, anger, excitement, surprise, or even sadness and crying. In most instances sleepiness and sleep episodes and cataplectic attacks occur independently, but they may occur in combination, or in rapid succession so that a conscious cataplectic episode can immediately develop into a sleep episode (REM sleep) with characteristic hypnogogic hallucinations. In addition, an episode of sudden overwhelming sleepiness may be associated with or followed by muscle paralysis. These episodes will often occur more than once a day, although in some patients they may be less frequent, occurring once a week or even at longer intervals. The other characteristic symptoms are hypnogogic hallucinations, sleep paralysis, brief lapses of memory, and automatic behavior episodes. These symptoms, as well as the sleepiness sleep attacks and cataplexy, are often neither understood nor appreciated by physicians; patients typically go for many years before a correct diagnosis is made.

The condition typically begins toward the end of puberty and during early adulthood and once present is a life-long illness. The nature of the symptoms produces serious social, familial, educational, and economic consequences to the patient (Broughton & Ghanem 1976). Studies have shown that these patients do not achieve their intellectual potential and suffer frequent marital, occupational, and educational failures because of the symptoms of sleepiness and cataplexy. The family members, friends, and even the patient himself often interpret the symptoms before a diagnosis is established as indicating nonmotivation, "laziness," and/or depression.

It is estimated that the narcolepsy-cataplexy syndrome occurs in 40 per 100,000 population, with an essentially equal rate in men and women (Dement et al 1972, 1973b, Dement 1979). In the city of New York, therefore, with a population of approximately 8 million, there are approximately 3000 victims of this syndrome. Although firm statistics are not available, it is probable that less than half the cases are identified.

There have been two major advances in our understanding of narcolepsy during the past 20 years. The first is the recognition that the daytime sleep attacks, cataplexy, and hypnogogic hallucinations are manifestations and fragments of the REM sleep process (Dement et al 1966, Hishikawa et al 1968). Sleep onset REM periods during an afternoon nap or at night are characteristic and are used to establish the clinical diagnosis (Guilleminault et al 1976a, Weitzman et al 1979a). The second is the demonstration that narcolepsy-cataplexy occurs in dogs (Mitler 1976, Mitler & Dement 1977) and other animals, and can be genetically transmitted as an autosomal recessive trait (Kessler 1976, Kessler et al 1974, Mamelak et al 1979). Studies of the probability of a familial-genetic relationship have estimated that relatives of index cases of narcolepsy have a sixty-fold greater risk of having the condition than the general population. These concepts clearly establish that the narcolepsy syndrome is a genetically transmittable disease.

The present treatment for narcolepsy is the use of "stimulant" drugs, (methylphenidate, amphetamines, pemoline, etc) for the sleepiness and sleep attacks and tricyclic compounds (imipramine, chlorimipramine, etc) for cataplexy (Guilleminault et al 1976a). Gamma-hydroxy-butyrate given nightly (Broughton & Mamelak 1976) and propranalol given during the day (Kales et al 1979) have also been recently reported to be effective in the treatment of sleep episodes. Monoamine oxidase inhibitors such as phenelzine sulfate (Nardil) have been shown to totally suppress REM sleep in patients with narcolepsy (Fisher et al 1972) but, because of serious side effects, they are not used clinically.

The etiology and localization of the narcolepsy syndrome to specific areas of the brain have not been delineated. There have been several reports of structural CNS diseases that have had classic narcoleptic symptoms. These symptoms have been localized to the brain stem, especially in the area of the ponto-mesencephalic reticular formation (Hobson 1975, Legkonogov 1971). Evidence that there is (*a*) a loss of monosynaptic reflex activity (patellar reflex and H-reflex) in REM sleep and in cataplexy and (*b*) sleep paralysis in man and in dogs points to a similar neurophysiological mechanism for both conditions. In addition, when carefully recorded, the narcoleptic has a periodic (120 to 200 min) pattern of REM sleep episodes during 24-hr periods (Montplasir et al 1978, Passouant 1974).

The narcolepsy-cataplexy syndrome appears to be genetically transmitted and is thought to be the result of abnormal control of the recurrent REM sleep episodes. It is not clear, however, whether the syndrome is due to an increased "drive" for REM sleep or whether the abnormal timing of REM sleep and its characteristics during the waking day are a manifestation of a disinhibition of the control of REM sleep.

Other Conditions Associated with Excessive Daytime Somnolence

Although most (70 to 80%) patients with the complaint of excessive daytime sleepiness fall into the two categories of narcolepsy and the sleep apnea syndrome, there are other less common conditions that cause daytime sleepiness. One of these is the Kleine-Levin syndrome (periodic hypersomnia). This begins between the ages of 10 to 21, usually in males, and is characterized by recurrent periods of very prolonged sleep alternating with wakefulness (Critchley 1962, Frank et al 1974, Levin 1936, Lavie et al 1979). During these "bouts" of periodic hypersomnolence the patient develops a major change in appetite, usually excessive (voracious), and bizarre food ingestion (megaphagia). There are also often serious disturbances in social interaction, sexual hyperactivity and exhibitionism, behavioral excitation, depression, and even frank delusions, hallucinations, disorientation, and memory deficits. During the intervening (non-hypersomniac) sleep-wake periods, the patient is usually "normal," although some cases show a persistent personality disorder. The frequency of hypersomnia attacks is quite variable but usually is self-limited, rarely occurring beyond the fourth or fifth decade. The course is unknown and no specific neuropathology has been reported, although it has been postulated to represent a dysfunction of limbic and hypothalamic regions.

Many other medical and neurological conditions are associated with excessive daytime sleepiness (Guilleminault & Dement 1974, Mendelson et al 1977, Kales & Tan 1969, Freemon 1978). Endocrine and metabolic disorders will often have excessive somnolence as a prominent feature. These include uremia, liver failure, hypothyroidism (severe with myxedema), chronic pulmonary disease (with hypercapnia), and diabetes mellitus with incipient coma or with severe hypoglycemia. In addition, CNS disorders such as a neoplasm in the area of the third ventricle (glioma, pinealoma, dysgerminoma, craniopharyngioma, pituitary adenomas), obstructive hydrocephalus, viral encephalitis and other infections of the brain and its surrounding membranes (fungal, viral, and bacterial meningitis, intracerebral abscess, etc) can cause increased daytime somnalence. Finally, although uncommon, various degenerative diseases of the brain may be associated with excessive daytime somnolence. These include dementia of the Alzheimer type, multiple sclerosis, and chronic vascular disease. The post concussion syndrome has been associated with increased sleepiness (Walker et al 1969). Complaints of tiredness, fatigue, depression, lack of interest, malaise, and lethargy associated with metabolic, endocrine, neurologic, and psychiatric disease must be carefully distinguished from sleepiness per se; however, it is often difficult to distinguish excessive sleepiness from an abnormality of disordered consciousness such as occurs in stupor, obtundation, and degrees of coma. Indeed, the use of the many different descriptive words by the clinician emphasizes the very real, unresolved problem of definition of these overlapping clinical behavioral states.

DISORDERS OF THE SLEEP-WAKE SCHEDULE

An important recent development in understanding the physiology of human sleep and its disorders has been the realization that biological rhythms have important functions in man. The now well-established biological discipline of chronobiology provides the sleep researcher with a rich source of biological facts and theories upon which to build a new conceptual understanding of the human circadian time-keeping systems (Wever 1979).

To appreciate the symptoms and syndromes of circadian rhythm dyssomnias in man, it is important to understand the principles of circadian rhythms as they apply to humans. As is the case with other organisms, when humans are not entrained by 24-hr "zeitgebers," they develop daily cycles that have periods greater or less than 24 hr (Wever 1975, 1979, Aschoff & Wever 1976, Weitzman et al 1979b). Studies carried out for weeks to months in both cave and controlled laboratory conditions have consistently demonstrated a preferred "free-running" period length of approximately 25 hr. Measurements of a variety of indices, including body temperature, urinary electrolytes, plasma and urine hormones, psychological performance measures, sleep duration, and sleep stage organization, have been carried out within the structure of time-isolated and nonscheduled human laboratory environments. These studies have shown that different variables can develop independent cycle lengths, a finding that has given rise to the concept that there are multiple oscillators normally synchronized with each other that can become desynchronized under free-running conditions (Aschoff 1969).

The recent development of methods to obtain very frequent plasma samples (from weeks to months), coupled with detailed and continuous polygraphic sleep stage measurements and the application of computer data processing and analysis capability, has led to important additional information and new concepts (Weitzman et al 1979b, Czeisler 1978). The length of sleep was found to be correlated with the phase of the circadian core temperature rhythm and not with the duration of prior wakefulness (Czeisler et al 1980). During free-running conditions, sleep is frequently prolonged to 18 hr, the exact length depending on the self-selected bedtime in relation to the phase of the temperature cycle; long sleep lengths (12 to 16 hr) occur when the patient chooses to sleep at the peak and shorter sleep lengths (6 to 8 hr) occur at the nadir (Weitzman et al 1979b, Czeisler et al 1980). There is an associated phase advance of REM sleep relative to sleep onset and the temperature cycle, such that there is a significant shortening of the REM latency, an increase in REM amounts, and often a sleep onset REM period during the first third of the circadian sleep episode while the subject is in the free-running condition (Weitzman et al 1980a). A phase shift of stages 3 and 4 sleep, however, does not occur during free-running; the timing and amount of stages 3 and 4 are linked to the initiation and course of the sleep process and do not appear to be dependent on the total duration of sleep. There are no differences in the REM–non-REM cycle length comparing entrained and free-running conditions. In a study performed, not under conditions of temporal isolation, but rather under conditions allowing unrestricted sleep in normal subjects awake for different lengths prior to bedtime, there was a circadian rhythm of sleep length and phase-advanced REM sleep and body temperature similar to those obtained when subjects have a free-running sleep-wake rhythm under nonscheduled conditions (Äkerstedt & Gillberg 1981).

Measurements of plasma cortisol throughout entrained and free-running conditions demonstrated that one component of the circadian rhythm had a phase advance (6 to 8 hr) relative to sleep onset, whereas a second component clearly followed sleep onset. The temporal pattern of growth hormone secretion, on the other hand, was directly related to the first 2 hr of sleep onset. A major episode of hormonal secretion occurred just after sleep onset for almost all sleep periods during both the entrained and free-running conditions (Weitzman et al 1979b, 1980b).

Further evidence for the importance of biological rhythm functions in the understanding of sleep disturbances comes from studies of phase shifts of the sleep-wakefulness cycles either in laboratory conditions or after transmeridian rapid flights (Weitzman et al 1970, Klein et al 1977, Hume 1980). These studies also demonstrate that the daily sleep episode is disturbed after east-west or west-east shifts as well as after 180° acute inversion. Under these conditions REM sleep also phase advances relative to sleep onset, but the adaptation occurs more rapidly with a phase delay than with a phase advance. These results are considered in the discussion below of the delayed sleep phase syndrome and its treatment by a chronotherapy regimen.

The importance of these recent findings and new concepts is illustrated by their direct application to the diagnosis and treatment of affective disorders in man. Wehr et al (1979) recently reported that by phase-advancing sleep time in several patients with bipolar manic-depressive illness, they repeatedly effected an immediate switch out of depression. Similar results have been obtained by the process of REM deprivation (Vogel et al 1975) and even total sleep deprivation (Rudolf et al 1977, Papousek et al 1974). On the basis of the REM advance present during sleep in depressive illness, Vogel has postulated a chronobiological explanation for the therapeutic effect of REM deprivation (Vogel et al 1980).

The disorders of the circadian sleep-wake cycle can be divided into two major categories: *transient* and *persistent.* The *transient* disorders include the dyssomnia of a rapid time zone change ("jet-lag"), a dyssomania similar to that found following an acute "work-shift" change. The sleep disturbance that results is due to both sleep deprivation and the circadian phase-shift change. The symptoms vary considerably from subject to subject but generally consist of (*a*) an inability to have a sustained sleep period, (*b*) frequent arousals occurring primarily at the end of the sleep episode associated with excessive sleepiness, and (*c*) falling asleep at inappropriate times in relation to social or economic requirements. For varying periods of time (usually several days, up to 2 wk), the person affected is sleepy, fatigued, intermittently inattentive during the waking period, and has partial insomnia during the sleep period. Although this may not very seriously disturb vacationing tourists, it can be a major problem to travelers on important business. These individuals need to function immediately at high levels of performance during what is normally their habitual sleep phase. The wide range of critical occupations involved in these acute phase shift disturbances emphasizes the importance of the syndrome in our modern society. These occupations include doctors, nurses, airline pilots, air-traffic controllers, police, firemen, military radar operators, long distance truck drivers, etc.

The *persistent* sleep-wake cycle disorders can be divided into the following five subcategories: (*a*) frequently changing sleep-wake schedules, (*b*) delayed sleep phase syndrome, (*c*) advanced sleep phase syndrome, (*d*) non-24 hr sleep-wake syndrome and (*e*) irregular sleep-wake pattern. The subject with a frequently changing sleep-wake schedule characteristically has a mixed pattern of altered excessive sleepiness alternating with periods of arousal, often at inopportune and inappropriate times during the day. Sleep is usually shortened and disrupted and waking is associated with a decrease in performance and vigilance. The syndrome characteristically disrupts social and family life and often becomes intolerable to the chronic shift worker; however, probably through the process of self-selection, some shift workers prefer or at least adapt readily to night work and rotating shift schedules (Rentos & Shephard 1976).

The *delayed sleep phase syndrome* (DSPS) has been recently recognized as a discrete chronobiologic sleep disorder (Weitzman et al 1979c, Weitzman & Pollak 1979). Patients with DSPS report a chronic inability to fall asleep at a desired clock time to meet their required work or study schedules. They typically cannot fall asleep until 2 to 6 A.M. When not required to maintain a strict schedule (e.g. weekends, holidays, and vacations), however, the patients will sleep without difficulty and after a sleep period of normal length they will awaken spontaneously, feeling refreshed. These patients have a normal sleep length and internal organization of sleep when their clock time of sleep onset coincides with the time the patient usually falls alseep (e.g. 4 to 5 A.M.). If sleep onset is attempted at earlier times (e.g. midnight), there will usually be a long sleep onset latency. These patients have a long history of many unsuccessful attempts to fall asleep at an earlier time, i.e. phase advance their sleep onset times. They score high as a "night person" on a standard questionnaire and have been previously considered to have "sleep-onset insomnia." Attempts to phase advance the time of going to sleep by the use of hypnotic drugs, alcohol, behavior modification techniques, sleep hypnosis, psychotherapy, and a variety of home remedies have repeatedly failed in these patients.

In one series of patients, the daily sleep episode was successively phase shifted by a progressive phase delay of the sleep time (chronotherapy) (Czeisler et al 1979). By delaying the time of going to sleep by 3 hr each day (i.e. a 27 hr sleep-wake cycle), the patient's sleep timing can be "reset" to occur at the clock time described by the patient. Since chronotherapy can clearly be effected by a phase delay and not by a phase advance, we have postulated that differences in the shape of the phase-response-curve (PRC) may underly these patients' chronobiological problems (see Bünning 1973 for a description of PRC). The phase advance portion of the curve may be much less prominent than in normal subjects, whereas the phase delay portion is presumably quite intact. This concept implies that normal individuals phase advance the sleep time each day by 1 hr in order to hold a 24 hr day. Because the non-entrained biological day length in man is approximately 25 hr (Wever 1979, Weitzman et al 1979c, Czeisler 1978), the daily entrainment process requires a well-functioning advance portion of the phase response curve with a range of entrainment sufficiently broad on both sides of 24 hr to enable normal individuals to adjust to variations in bedtime and time of arising in order to maintain an appropriate phase sleep time with a period of 24 hr.

There is no clear evidence as yet that there is an advanced sleep phase syndrome; however, it has been suggested that individuals might fall into this category who have chronic sleep onset and wake times that are undesirably early but without disturbance of the sleep process itself. Since the "early to bed, early to rise" rhythm conforms very well with our social-economic and solar day timing system, these individuals probably do not seek medical or psychological help; however, it has been well documented that aging leads to a characteristic change in the timing of sleep such that older individuals find themselves awakening spontaneously earlier in the morning and going to sleep earlier in the evening (Weitzman 1980). It is possible that maturational alterations in the shape of the phase-response curve may underlie these changes and that many older individuals do, in fact, have an advanced sleep phase syndrome; however, there are other biological rhythm changes of the endogenous circadian and ultradian oscillators that occur as a function of aging, including intrusive nocturnal awakenings, fragmentation of the sleep pattern, and repeated daytime brief sleep periods (i.e. naps) (Lewis 1969).

The *non-24-hour sleep wake syndrome* (hypernychthermal) has been described recently in several patients and is of considerable interest because of its important relationship to phase shift disorders and to the free-running sleep-wake pattern of normal man living in temporal isolation (Kokkoris et al 1978, Miles et al 1977). These patients are unable to be entrained to society's 24 hr day and therefore develop a 25 to 27 hr biological day in spite of all attempts to do otherwise (hypernychthermal syndrome). Blindness or a personality disorder may predispose to this condition. No individual living in society has been described thus far who developed a consistent sleep-wake rhythm with a consistent period of less than 24 hr. The subject's repeated attempts to hold to a normal 24 hr sleep-wake schedule can lead to cyclic periods of nonsynchronizing with society's rhythm as well as to cyclic (3 to 4 wk) disrupted and delayed sleep with associative daytime sleepiness. Since patients with the delayed sleep phase syndrome also demonstrate a consistent tendency toward a progressive phase delay, it is postulated that the two syndromes may only differ in degree, the difference being that the patient with the non-24 hr sleep-wake syndrome has an even further altered phase response curve with essentially no phase advance capability.

Support for the idea that there may be large individual differences in regard to a phase advance capability derives from recent studies of phase shift after transmeridian air flight (Klein et al 1977) and adaptation after phase shifts in the laboratory (Hume 1980). It was found that an eastward flight (phase advance of 6 hr) and a laboratory advance of 8 hr took much longer in some subjects than others, whereas a phase delay (westward flight) occurred more rapidly.

Finally, a syndrome of *irregular sleep-wake pattern* is described. The pattern is one of considerable irregularity without an identifiable persistent circadian sleep-wake rhythm. The condition is presumed to be associated with frequent daytime naps at irregular times associated with a disturbed nocturnal sleep pattern. A similar pattern can be produced in animals (rats, hamsters, monkeys, etc) by bilateral destruction of the suprachiasmatic nucleus (SCN) (Rusak & Zucker 1975). The sleep-wake (activity-rest) cycle of these lesioned animals no longer demonstrates a circadian rhythm under free-running conditions. The retinohypothalamic tract to the suprachiasmatic nucleus is the presumed pathway mediating light-dark entrainment in animals (Moore 1979). Many studies have demonstrated the importance of the suprachiasmatic nucleus in the control of timing of mammalian circadian rhythms and it has been proposed that the suprachiasmatic nucleus is an endogenous CNS pacemaker. Strong support has come from the demonstration that suprachiasmatic nucleus neurons in neurally isolated hypothalamic island will discharge with a sustained free-running circadian rhythm, whereas neurons from other brain areas, outside the island, revealed no circadian rhythm (Inouye & Kawamura 1979, Kawamura & Inouye 1979). Recent data from monkeys suggests, however, that although the activity-rest cycle was lost, the circadian body temperature rhythm persists in squirrel monkeys (Moore-Ede et al 1980), as does the CSF melatonin rhythm in Macacca Mulatta (Reppert et al 1979) after bilateral suprachiasmatic nucleus lesions. It will, therefore, be of considerable interest to define the range and pattern of rhythmic abnormalities in patients with an "irregular sleep-wake pattern."

DYSFUNCTIONS ASSOCIATED WITH SLEEP, SLEEP STAGES, OR PARTIAL AROUSALS (PARASOMNIAS)

There is a group of clinical conditions in which the sleep and waking process per se is not abnormal, but an undesirable behavioral and physiological event occurs in association with or during sleep. The disturbing symptom specifically affects either motor behavior, autonomic dysfunction, psychological state, or a combination of the three. The range of abnormalities is very broad and occurs in all age groups. There may be considerable overlap with previous nosological entities (nocturnal myoclonus, sleep apnea, sleep paralysis-narcolepsy) or there may be quite independent clinical entities (somnambulism, night terrors, cluster headaches, gastro-esophageal reflex). In the present state of our knowledge, they are primarily listed as individual entities, although an attempt has been made to organize several about the concept of "disorders of the state of partial arousal" (sleep terror, sleep-related enuresis, and sleep walking) (Broughton 1968). A sleep stage relationship has been defined for several entities: somnambulism, enuresis, and night terrors (stages 3 to 4) (Fisher et al 1973); sleep-related painful penile erections (REM sleep) (Karacan 1971); and cluster headaches (REM sleep) (Dexter & Weitzman 1970). Other entities occur in several sleep stages (bruxism, somniloquy). The present list of parasomnias will certainly be augmented during the next few years as cases of disturbing sleep dysfunctions are increasingly referred to sleep disorders centers.

Sleep Walking (Somnambulism) .

Sleep walking is a sudden behavioral sequence interrupting sleep that consists of either sitting up in bed or leaving the bed and walking about, but without achieving full consciousness (Kales et al 1966). It characteristically occurs during non-REM sleep, usually in stages 3 or 4. The complex behavior includes repetitive automatic and semipurposeful motor acts such as walking, opening and closing doors, climbing stairs, dressing, etc. On occasion the patient will carry out an act that can be harmful to himself, such as climbing out a window. The patient is not easily aroused during the episode and will typically strongly resist being aroused. The episode generally lasts no more than 15 min and is terminated by either a spontaneous arousal, or by returning to the bed or going to sleep in another place. The patient awakens in the morning, surprised to find himself or herself in another place, and has no or only a dim, fragmented recollection of the episode. Dreaming is not prominent and, if aroused during the episode, the patient will usually not report hallucinatory imagery and strong emotions or will report only brief and nonsustained thoughts.

Somnambulism episodes occur normally in children and adolescents; 15% of all children have one or more episodes (Anders & Weinstein 1972). In a small percentage of children (1 to 6%), however, these will occur frequently (at times nightly). If the episodes persist into or begin in adulthood, there is a high association with serious psychopathology.

Polysomnographic studies have demonstrated that a somnambulism episode is preceded by high amplitude delta waves during stage 3 or 4 sleep and that there is no evidence of any change in heart rate or respiration just prior to the onset of motor behavior (Kales et al 1966, Broughton 1968). There is no evidence of seizure activity either preceding or during the episode of somnambulism, although clinically such activity must be distinguished from a nocturnal seizure disorder such as psychomotor epilepsy (Guilleminault et al 1977b). There is an association of somnambulism in the same patient and in families with other episodic disorders during sleep, including nocturnal enuresis and night terror attacks (Anders & Weinstein 1972).

Sleep Terror (Pavor Nocturnus, Incubus)

Similar to somnambulism, a sleep terror episode is a sudden arousal during stages 3 and 4 sleep (first third of the night), initiated by a sudden loud high pitched scream, sitting up in bed and behaving in an agitated, frightened, and panicked manner (Fisher et al 1974). Although not present during sleep just before the attack, after the onset there are major autonomic changes, including sweating, rapid pulse and respiration, and pupillary dilatation (the pupils are normally meiotic during sleep). Sleep terror, like somnambulism, also occurs primarily in children, although the condition may persist or begin in adults. During the 5 to 15 min of an attack, the individual is usually inconsolable and may repeat the name of a family member such as "mama," "grandma," etc. Complex dream imagery is not obtainable at the end of the event and therefore it is different from a nightmare. The patient typically will return to sleep very quickly when the episode is over and will have amnesia for the event the next morning. Like somnambulism, sleep terrors can occasionally occur normally in children (Anders & Weinstein 1972), but if frequent and especially if present in adults, their occurrence may indicate an underlying psychological disturbance. It is important to differentiate sleep terrors from recurrent nightmares (dream anxiety attacks). Nightmares occur during REM sleep, during the middle and latter third of the nocturnal sleep episode, and can usually be recounted in a vivid, emotionally charged, detailed story with personal fear and threats a prominent feature. A terrified scream is uncommon in a nightmare.

Sleep-Related Enuresis (Bed-wetting)

The clinical entity of nocturnal sleep-related bed-wetting has characteristics similar to those found for somnambulism and night terrors. It occurs during stages 3 and 4 non-REM sleep in the first third of the night and is associated with a brief arousal. The individual is usually confused and disoriented but does not describe a characteristic dream sequence, or a feeling of terror or anxiety. The condition is divided into "primary" enuresis, the persistence of sleep-related bed-wetting from infancy to childhood, and "secondary" enuresis, the reappearance of bed-wetting after a time of successful bladder toilet training. Both "primary" and "secondary" (idiopathic) are distinguished from symptomatic enuresis, which is caused by a known organic condition (Broughton & Gastaut 1975). Most children with idiopathic enuresis stop having the episodes by puberty, although emotional stress can reactivate the problem (Anders & Weinstein 1972). There is often a family history of nocturnal enuresis and, as indicated above, bed-wetting is interrelated with somnambulism and night terrors. There is no evidence that there is an epileptic discharge associated with idiopathic enuresis, but it must be distinguished clinically from sleep-related seizures that may be associated with urinary incontinence. Polysomnographic and electroencephalographic recording is useful in identifying this and many other parasomnias.

Other Parasomnia Dysfunctions

There is a growing list of other patho-physiological events that can occur within and disturb sleep. In accord with the Association of Sleep Disorders Centers (ASDC) classification, these include (a) dream anxiety attacks (nightmares) (Fisher et al 1970, Kramer 1979), (b) sleep-related epileptic seizures (Passouant et al 1975), (c) bruxism (Reding et al 1968), (d) sleep-related head banging and rocking (Jactatio Capitis Nocturnus) (Baldy-Moulinier et al 1970), (e) familial sleep paralysis (Hishikawa 1976), (f) impaired sleep-related penile tumescence (Karacan et al 1978), (g) sleep-related painful erections (Karacan 1971), (h) sleep-related cluster headaches (Dexter & Weitzman 1970), (i) chronic paroxysmal hemicrania (Kayed et al 1978), (j) abnormal swallowing syndrome, (k) sleep-related asthma (Kales et al 1968), (l) sleep-related cardiovascular symptoms (Smith et al 1972), (m) sleep-related gastroesophogeal reflex (Orr et al 1979b), and (n) sleep-related hemolysis (paroxysmal nocturnal hemoglobinuria) (Hansen 1968). The reader is referred to the *Diagnostic Classification of Sleep and Arousal Disorders* (Dement & Guilleminault 1979) and other selected references for more complete information regarding the above disorders. In this review three of these parasomnias are briefly described because of their special pathophysiological relevance.

Impaired Sleep-Related Penile Tumescence

Penile erections occur normally during 80 to 90% of all REM periods in males in every age group, from premature infants to the aged (eighth and ninth decades). Tumescence occurs in non-REM sleep as well, but much less frequently and, when present, it is not well sustained. The amount of nocturnal penile tumescence (NPT) per night decreases with age. In the 20 to 30-year-old age group it is present for an average of 200 min of sleep and in the 70 to 80-year-old age group, for approximately 100 min. The presence of a highly predictable recurrent REM-related cycle of NPT is used clinically to distinguish organic from psychogenic impotence. In organic diseases that affect waking erectile potency, NPT is either totally absent or markedly diminished. These organic diseases include diabetes mellitus, cardiovascular syndromes (e.g. hypertension), neurological diseases (e.g. multiple sclerosis, spinal cord injury), endocrine, and urogenital diseases (e.g. Peyronie's disease). Many pharmacological substances (tricyclics, adrenergic blockers, etc) cause organic impotence as well. In patients with psychogenic impotence, measurements show little or no reduction in NPT compared to normal age-related standards. Sleep disorder centers are therefore using the measurement of NPT to assist in the differential diagnosis between organic and psychogenic causes of impotence (Karacan 1971, Karacan et al 1978, Wasserman et al 1980).

Sleep-Related Cluster Headaches and Chronic Paroxysmal Hemicrania

It is characteristic for many patients with cluster headaches, migraine, and chronic paroxysmal hemicrania to have an attack during sleep. The cluster headache attack and the paroxysmal hemicrania awakens the patient with severe pain, either directly out of REM sleep or in the immediate post-REM period. Phase shifts of the daily sleep time demonstrate that the headache follows REM sleep itself. A cluster headache is characterized by the rapid development of severe unilateral eye and upper facial pain in association with tearing, rhinorrhea, and conjunctival redness. A Horner's syndrome may be present. Nausea and anorexia may also occur. The pain awakens the patient and he then finds himself in an already established episode. The attack will last from 30 min to several hours. Many patients have several episodes during the night, each associated with a REM sleep period (Dexter & Weitzman 1970, Kayed et al 1978). Chronic paroxysmal hemicrania attacks occur more frequently, at times up to 24 episodes per day, are short-lasting (5 to 15 min), and there is no especial nocturnal predominance; however, during sleep they occur in a "time locked" occurrence to REM sleep (Kayed et al 1978). Because it has been well established that in REM sleep, marked autonomic nervous system changes occur ("autonomic storm") (Hobson 1969), the triggering of these headache attacks in relation to altered autonomic functions is being investigated.

Sleep-Related Gastro-Esophageal Reflux

The clinical symptoms of sleep-related gastro-esophageal reflux are recurrent awakening from sleep with a burning substernal pain (heartburn), associated with a sour taste, coughing, choking, and respiratory discomfort. These are often associated during the waking state with postprandial regurgitation, dysphagia, esophagitis, laryngopharyngitis, and heartburn. There is decreased acid clearance in the lower esophagus during sleep in association with recurrent reflux of gastric juice up into the esophagus (Orr et al 1979b). This leads to a progressive esophagitis and, in serious cases, to esophageal stricture, aspiration pneumonia, bronchiectasis, and laryngopharyngitis. Recent studies during sleep have demonstrated that gastric acid will reflux up the lower esophagus and that the normal clearing mechanism is decreased, although a specific sleep stage relationship has not been established as yet.

SUMMARY

The advances in research on sleep and biological rhythms have recently been applied to the diagnosis and treatment of sleep disorders. A new clinical specialty has developed with the establishment of sleep disorder centers and a diagnostic classification of sleep and arousal disorders. This new nosological approach has evolved from an extensive base of new scientific information concerning descriptive polygraphic and analysis of clinical case series. Four major categories have been defined: (*a*) disorders of initiating and maintaining sleep (insomnias), (*b*) disorders of excessive somnolence, (*c*) disorders of the sleep-wake schedule, and (*d*) dysfunctions associated with sleep. Within this comprehensive classification certain major pathophysiological advances are described for the "insomnias." These include polysomnographic identification of altered sleep stage patterns in the major affective illnesses, insomnias related to hypnotic drugs and alcohol, sleep disturbances associated with sleep-induced respiratory impairment, and sleep-related periodic movements during sleep (nocturnal myoclonus).

Excessive daytime somnolence is primarily associated with the hypersomnia sleep-apnea syndrome and with narcolepsy. The relationship between biological rhythms (chronobiology) and disorders of the human sleep-wake schedules is very actively investigated. The recognition that sleep length, internal organization, and timing within neurophysiological circadian time-keeping systems has lead to better diagnosis of these sleep-wake disorders and new chronotherapeutic regimens. Finally, increasing identification and description of "parasomnias," i.e. dysfunctions associated with sleep, has led sleep research into important new areas that are of general physiological interest.

It is now clear that sleep disorders medicine has become a new scientific and clinical discipline in its own right.

Literature Cited

Adamson, J., Burdick, J. A. 1973. Sleep of dry alcoholics. *Arch. Gen. Psychiatr.* 28:(1)146–49

Åkerstedt, T., Gillberg, M. 1981. Proceedings of the 1979 ONR/NIOSH symposium on variation in work-sleep schedules: Effects on health and performance. In *Advances in Sleep Research*, Vol. 6, ed. D. I. Tepas, L. C. Johnson. New York: Spectrum. In press

Amin, M. M. 1976. Drug treatment of insomnia in old age. *Psychopharmacol. Bull.* 12(2):52–55

Anders, T. F., Weinstein, P. 1972. Sleep and its disorders in infants and children—A review. *Pediatrics* 50:312–24

Aschoff, J. 1969. Desynchronization and resynchronization of human circadian rhythms. *Aerospace Med.* 40:844–49

Aschoff, J., Wever, R. 1976. Human circadian rhythms: A multioscillatory system. *Fed. Proc.* 35:2326–32

Association of Sleep Disorders Centers. 1979. Diagnostic classification of sleep and arousal disorders, 1st ed., Sleep Disorders Classification Comm., H. P. Roffwarg, Chmn. *Sleep* 2:1–137

Baekeland, F., Lundwall, L., Shanahan, T. J., Kissin, B. 1974. Clinical correlates of reported sleep disturbance in alcoholics. *Q. J. Stud. Alcohol.* 35(4):1230–41

Baldy-Moulinier, M., Levy, M., Passouant, P. 1970. A study of jactatio capitis during night sleep. *Electroencephalogr. Clin. Neurophysiol.* 28:87

Balter, M. B., Bauer, M. L. 1975. Patterns of prescribing and use of hypnotic drugs in the United States. In *Sleep Disturbance and Hypnotic Drug Dependence,* ed. A. D. Clift. New York: Excerpta Medica. 352 pp.

Berger, H. 1929. Uber das Elektrenkephalogramm des Menschen. *Arch. Gen. Psychiatr.* 87:527

Beutler, L. E., Thornby, J. I., Karacan, I. 1978. Psychological variables in the diagnosis of insomnia. In *Sleep Disorders: Diagnosis and Treatment,* ed. R. L. Williams, I. Karacan, pp. 61–100. New York: Wiley. 417 pp.

Billiard, M. 1978. Insomnia: Evaluation, therapeutic strategies and results. *Adv. Biosci.* 21:95–103

Billiard, M., Besset, A., Passouant, P. 1978. What can be expected from all night polygraphic recordings in the treatment of chronic insomnia. *Sleep Res.* 7:211

Bixler, E. D., Kales, A., Leo, L. A., Slye, T. 1973. A comparison of subjective estimates and objective sleep laboratory findings in insomniac patients. *Sleep Res.* 2:143

Broughton, R. 1968. Sleep disorders: Disorders of arousal? *Science* 159:1070–78

Broughton, R., Gastaut, H. 1975. Recent sleep research on enuresis nocturna, sleep walking, sleep terrors and confusional arousals. A review of dissociative awakening disorders in slow wave sleep. In *Sleep, 1974. 2nd European Congress on Sleep Research,* ed. P. Levin, W. P. Koella, pp. 82–91. Rome/Basel: Kargel. 525 pp.

Broughton, R., Ghanem, Q. 1976. The impact of compound narcolepsy on the life of the patient. In *Advances in Sleep Research,* Vol. 3, *Narcolepsy,* ed. C. Guilleminault, W. C. Dement, P. Passouant, pp. 210–19. New York: Spectrum. 689 pp.

Broughton, R., Mamelak, M. 1976. Gamma-hydroxy-butyrate in the treatment of narcolepsy. A preliminary report. See Broughton & Ghanem 1976, pp. 659–66

Brown, C. C., Horrom, N. J., Wagman, A. M. 1979. Effects of L-tryptophan on sleep onset insomniacs. *Waking Sleeping* 3(2):101–8

Bünning, E. 1973. *The Physiological Clock.* New York: Springer-Verlag

Carroll, D. 1974. Sleep, periodic breathing and snoring in the aged: Control of ventilation in the aging and diseased respiratory system. *J. Am. Geriatr. Soc.* 22(7):307–15

Carskadon, M. A., Dement, W. C., Mitler, M. M., Guilleminault, C., Zarcone, V. P., Spiegel, R. 1976. Self reports versus sleep laboratory findings in 122 drug free subjects with the complaints of chronic insomnia. *Am. J. Psychiatr.* 133:1382–88

Clark, R. W., Schmidt, H. S., Schaal, S. F., Boudoulas, H., Schuller, D. E. 1979. Sleep apnea: Treatment with protriptyline. *Neurology* 29(9):1287–92

Coble, P. A., Kupfer, D. J., Spiker, D. G., Neil, J. F., McPartland, R. J. 1979. EEG sleep in primary depression. A longitudinal placebo study. *J. Affect. Dis.* 1(2):131–38

Coleman, R. M., Pollak, C. P., Weitzman, E. D. 1978. Periodic nocturnal myoclonus in a wide variety of sleep-wake disorders. *Trans. Am. Neurol. Assoc.* 103:230–33

Coleman, R., Pollak, C., Weitzman, E. D. 1980. Periodic movements in sleep (nocturnal myoclonus): Relation to sleep disorders. *Ann. Neurol.* In press

Conway, W. A., Bower, G. C., Barnes, M. E. 1977. Hypersomnolence and intermittent upper airway obstruction, occurrence caused by micrognathia. *J. Am. Med. Assoc.* 237(25):2740–42

Critchley, M. 1962. Periodic hypersomnia and megaphagia in adolescent males. *Brain* 85:627–56

Cummiskey, J., Lynne-Davies, P., Guilleminault, C. 1978. Sleep study and respiratory function in myotonic dystrophy. In *Sleep Apnea Syndromes,* ed. C. Guilleminault, W. C. Dement, pp. 295–308. New York: Liss. 372 pp.

Czeisler, C. A. 1978. Human circadian physiology: Internal organization of temperature, sleep-wake and neuroendocrine function in an environment free of time cues. PhD Thesis. Stanford Univ., 346 pp.

Czeisler, C. A., Richardson, G., Coleman, R., Dement, W., Weitzman, E. 1979. Successful non-drug treatment of delayed sleep phase syndrome with chronotherapy: Resetting a biological clock in man. *Sleep Res.* 8:179 (Abstr.)

Czeisler, C. A., Weitzman, E. D., Moore-Ede, M. C., Kronauer, R. E., Zimmerman, J. C., Campbell, C. 1980. Human sleep: Its duration and structure depend on the interaction of two separate circadian oscillators. *Sleep Res.* 9: In press (Abstr.)

de la Pena, A. 1978. See Beutler 1978, pp. 101–44

Dement, W. C. 1973. *Some Must Watch While Some Must Sleep.* Stanford, Ca: Stanford Alumni Assoc. 148 pp.

Dement, W. C. 1979. Narcolepsy—Not as rare as we believed. 1979. *Med. Times* 107(6):51–55

Dement, W., Rechtschaffen, A., Gulevich, G. 1966. The nature of the narcoleptic sleep attack. *Neurology* 16:18–33

Dement, W. C., Zarcone, V., Varner, V., Hoddes, E., Nassau, S., Jacobs, B., Brown, J., McDonald, A., Horan, K., Glass, R., Gonzales, P., Friedman, E., Phillips, R. 1972. The prevalence of narcolepsy. *Sleep Res.* 1:148 (Abstr.)

Dement, W. C., Zarcone, V. P., Hoddes, E., Smythe, H., Carskadon, M. 1973a. Sleep laboratory and clinical studies with flurazepam. In *The Benzodiazepines,* ed. S. Garattini, E. Mussini, L. O. Randall, pp. 599–611. New York: Raven. 685 pp.

Dement, W. C., Carskadon, M., Ley, R. 1973b. The prevalence of narcolepsy II. *Sleep Res.* 2:147

Dement, W. C., Carskadon, M. A., Mitler, M. M., Phillips, R., Zarcone, V. P. 1978. Prolonged use of flurazepam: A sleep laboratory study. *Behav. Med.* 5:25–31

Dement, W. C., Guilleminault, C., eds. 1979. *Diagnostic Classification of Sleep and Arousal Disorders. Sleep* 2(1). New York: Raven. 154 pp.

Dexter, J. D., Weitzman, E. D. 1970. The relationship of nocturnal headaches to sleep stage patterns. *Neurology* 20(5):513–18

Diagnostic & Statistics Manual of Mental Disorders 1980. Washington DC: Am. Psychiatr. Assoc. 494 pp. 3rd ed.

Duron, B., Quichaud, J., Fullana, N. 1972. The mechanism of apneic periods in the pickwickian syndrome. *Bull. Physio-pathol. Respir.* 8(5):1277–88

Ekbom, R. A. 1960. Restless legs syndrome. *Neurology* 10:868–73

Fisher, C., Byrne, J., Edwards, A., Kahn, E. 1970. A psychophysiological study of nightmares. *J. Am. Psychoanal. Assoc.* 18(4):747–82

Fisher, C., Kahn, E., Edwards, A., Davis, D. 1972. Total suppression of REM sleep with Nardil in a patient with intractable narcolepsy. *Sleep Res.* 1:153 (Abstr.)

Fisher, C., Kahn, E., Edwards, A., Davis, D. M. 1973. A psychophysiological study

of nightmares and night terrors. *J. Nerv. Ment. Dis.* 157:75–98

Fisher, C., Kahn, E., Edwards, A., Davis, D. M. Fine, J. 1974. A psychophysiological study of nightmares and night terrors. *J. Nerv. Ment. Dis.* 158:174–88

Fisher, J. G., de la Pena, A., Mayfield, D., Flickinger, R. 1978. Starvation and behavior modification as a treatment in obese patients with sleep apnea: A follow-up. *Sleep Res.* 7:222

Frank, Y., Braham, J., Cohen, B. E. 1974. The Kleine-Levin syndrome: Case report and review of the literature. *Am. J. Dis. Child* 127:412–13

Frankel, B. L., Coursey, R. D., Buchbinder, R., Snyder, F. 1976. Recorded and reported sleep in chronic primary insomnia. *Arch. Gen. Psychiatr.* 33(5):615–23

Frankel, B. L., Patten, B. N., Gillin, J. C. 1974. Restless legs syndrome. Sleep electroencephalographic and neurologic findings. *J. Am. Med. Assoc.* 230:1302–3

Freemon, F. R. 1978. Sleep in patients with organic diseases of the nervous system. See Beutler 1978, pp. 261–83

Frost, J. D. Jr., DeLucchi, M. R. 1979. Insomnia in the elderly: Treatment with flurazepam hydrochloride. *J. Am. Geriatr. Soc.* 27(12):541–46

Gillin, J. C., Duncan, W., Pettigrew, K. D., Frankel, B. L., Snyder, F. 1979. Successful separation of depressed, normal, and insomniac subjects by EEG sleep data. *Arch. Gen. Psychiatr.* 36:85–90

Gillin, J. C., Mazure, C., Post, R. M., Jimerson, D., Bunney, W. E. Jr. 1977. An EEG sleep study of bipolar (manic-depressive) patients with a nocturnal switch process. *Biol. Psychiatr.* 12:711–18

Gresham, S. C., Agnew, H. W., Williams, R. L. 1965. The sleep of depressed patients. An EEG and eye movement study. *Arch. Gen. Psychiatr.* 13:503–7

Griffiths, W. J., Lester, B. K., Coulter, J. D., Williams, H. L. 1972. Tryptophan and sleep in young adults. *Psychophysiology* 9:345–56

Gross, M. M., Goodenough, D. R. 1967. Observations and formulations regarding REM and other disturbances of sleep in the acute alcoholic psychoses and related states. Simposia Int. Psicofisiologia del Sonno E del Sogno, Sept. 11–12, Rome

Gross, M. M., Goodenough, D. R., Hastey, J., Lewis, E. 1973. Experimental study of sleep in chronic alcoholics before, during and after four days of heavy drinking, with a non-drinking comparison. *Ann. NY Acad. Sci.* 215:254–75

Gross, M. M., Goodenough, D. R., Hasty, J. M., Rosenblatt, S. M., Lewis, E. 1971. Sleep disturbances in alcohol intoxication and withdrawal. In *Recent Advances in the Study of Alcoholism,* ed. N. K. Mello, J. A. Mendelson, pp. 317–97. Washington DC: GPO. 920 pp.

Gross, M. M., Goodenough, D. R., Tobin, M., Halpert, E., LePore, D., Perlstein, A., Sirota, M., DiBianco, J., Fuller, R., Kishner, I. 1966. Sleep disturbances and hallucinations in the acute alcoholic psychosis. *J. Nerv. Ment. Dis.* 142:493–514

Guilleminault, C. 1976. Cataplexy. See Broughton & Ghanem 1976, pp. 125–43

Guilleminault, C., Dement, W. C. 1974. Pathologies of excessive sleep. In *Advances in Sleep Research,* ed. E. D. Weitzman, 1:345–90. New York: Spectrum. 424 pp.

Guilleminault, C., Dement, W. C. 1978a. Sleep apnea syndromes and related sleep disorders. See Broughton & Ghanem 1976, pp. 9–28

Guilleminault, C., Dement, W. C. 1978b. *Sleep Apnea Syndromes.* New York: Liss. 372 pp.

Guilleminault, C., Dement, W. C., Holland, J. V. 1974. Action of imipraminic medications on sleep induced apneas. *Sleep Res.* 3:136

Guilleminault, C., Raynal, D., Weitzman, E. D., Dement, W. C. 1975. Sleep-related periodic myoclonus in patients complaining of insomnia. *Trans. Am. Neurol. Assoc.* 100:19–21

Guilleminault, C., Dement, W. C., Passouant, P., eds. 1976a. *Advances in Sleep Research,* Vol. 3, *Narcolepsy.* New York: Spectrum. 689 pp.

Guilleminault, C., Eldridge, F. L., Phillips, J. R., Dement, W. C. 1976b. Two occult causes of insomnia and their therapeutic problem. *Arch. Gen. Psychiatr.* 33:1241–45

Guilleminault, C., Tilkian, A., Dement, W. C. 1976c. The sleep apnea syndromes. *Ann. Rev. Med.* 27:465–84

Guilleminault, C., Tilkian, A., Lehrman, K., Forno, L., Dement, W. C. 1977a. Sleep apnea syndrome: States of sleep and autonomic dysfunction. *J. Neurol. Neurosurg. Psychiatr.* 40(7):718–25

Guilleminault, C., Pedley, T., Dement, W. C. 1977b. Sleepwalking and epilepsy. *Sleep Res.* 6:170 (Abstr.)

Guilleminault, C., Motta, J., Flagg, W., Coburn, S., Dement, W. C. 1978a. Poli-

omyelitis and sleep apnea. *Sleep Res.* 7:230

Guilleminault, C., Hill, N. W., Simmons, F. B., Dement, W. C. 1978b. Obstructive sleep apnea: Electromyographic and fiberoptic studies. *Exp. Neurol.* 62:48–67

Guilleminault, C., van den Hoed, J., Mitler, M. 1978c. Clinical overview of the sleep apnea syndromes. See Cummiskey et al 1978, pp. 1–12

Hagan, R., Bryan, H. M., Gulston, G. 1976. The effect of sleep state on intercostal muscle activity and rib cage motion. *Physiologist* 19:214 (Abstr.)

Hammond, E. C. 1964. Some preliminary findings on physical complaints from a prospective study of 1,064,004 men and women. *Am. J. Public Health* 54:11–23

Hansen, N. E. 1968. Sleep related plasma hemoglobin levels in paroxysmal nocturnal hemoglobinuria. *Acta. Med. Scand.* 184:547–49

Hartmann, E. 1973. Effects of L-tryptophane on sleep human and animal studies. In *The Nature of Sleep,* ed. U. J. Jovanovic, pp. 290–95. Stuttgart: Fischer. 308 pp.

Hartmann, E. 1974. *The Functions of Sleep.* New Haven: Yale Univ. Press. 198 pp.

Hartmann, E. 1976. Long-term administration of psychotropic drugs: Effects on human sleep. In *Pharmacology of Sleep,* ed. R. L. Williams, I. Karacan, pp. 211–23. New York: Wiley, 354 pp.

Hartmann, E. 1977. L-tryptophan: A rational hypnotic with clinical potential. *Am. J. Psychiatr.* 134:366–70

Hartmann, E., Chung, R., Chien, C. P. 1971. L-tryptophane and sleep. *Psychopharmacologia* 19(2):114–27

Hartmann, E., Spinweber, C. L. 1979. Sleep induced by L-tryptophan. Effect of dosages within the normal dietary intake. *J. Nerv. Ment. Dis.* 167(8):497–99

Hauri, P. 1977. *The Sleep Disorders.* Kalamazoo: Upjohn. 76 pp.

Hauri, P., Chernik, D., Hawkins, D., Mendels, J. 1974. Sleep of depressed patients in remission. *Arch. Gen. Psychiatr.* 31:386–91

Hauri, P., Hawkins, D. R. 1973. Alpha-delta sleep. *Electroencephalogr. Clin. Neurophysiol.* 34:233–37

Hawkins, D. R., Mendels, J. 1966. Sleep disturbances in depressive syndromes. *Am. J. Psychiatr.* 123:682–90

Henderson-Smart, D. J., Read, D. J. C. 1976. Depression of respiratory muscles and defective response to nasal obstruction during active sleep in the newborn. *Aust. Paediatr. J.* 12:261–66

Hill, M. W., Simmons, F. B., Guilleminault, C. 1978. Tracheostomy and sleep apnea. See Cummiskey et al 1978, pp. 347–52

Hishikawa, Y. 1976. Sleep paralysis. See Broughton & Ghanem 1976. pp. 97–124

Hishikawa, Y., Nan'no, H., Tachibana, M., Furuya, E., Koida, H., Kaneko, Z. 1968. The nature of the sleep and other symptoms of narcolepsy. *Electroencephalogr. Clin. Neurophysiol.* 24:1–10

Hobson, J. A. 1969. Sleep: Physiologic aspects. *N. Engl. J. Med.* 281:1343–45

Hobson, J. A. 1975. Dreaming sleep attacks and desynchronized sleep enhancement. Report of a case of brain stem signs. *Arch. Gen. Psychiatr.* 32(11): 1421–24

Hoddes, E., Carskadon, M., Phillips, R., Zarcone, V., Dement, W. 1972. Total sleep time in insomniacs. *Sleep Res.* 1:152

Hoffmeister, J. A., Cabatingan, O., McKee, A. 1978. Sleep apnea treated by intestinal bypass. *J. Maine Med. Assoc.* 69(3): 72–74

Hudgel, D. W., Shucard, D. W. 1979. Coexistence of sleep apnea and asthma resulting in severe sleep hypoxemia. *J. Am. Med. Assoc.* 242(25):2789–90

Hume, K. I. 1980. Sleep adaptation after phase shifts of the sleep-wakefulness rhythm in man. *Sleep* 2:417–35

Inouye, S. T., Kawamura, H. 1979. Persistence of circadian rhythmicity in a mammalian hypothalamic "island" containing the suprachiasmatic nucleus. *Proc. Natl. Acad. Sci. USA* 76(11):5962–66

Johnson, L. C., Burdick, J. A., Smith, J. 1970. Sleep during alcohol intake and withdrawal in the chronic alcoholic. *Arch. Gen. Psychiatr.* 22:406–18

Kales, A., Jacobson, A., Paulson, M. J., Kales, J. D., Walter, R. D. 1966. Somnambulism: Psychophysiological correlates. *Arch. Gen. Psychiatr.* 14:586–94

Kales, A., Beall, G. N., Bajor, G. F., Jacobson, A., Kales, J. D. 1968. Sleep studies in asthmatic adults: Relationships of attacks to sleep stage and time of night. *J. Allergy* 41:164–73

Kales, A., Malinstrom, E. J., Scharf, M. B., Rubin, R. T. 1969a. Psychophysiological and biochemical changes following use and withdrawal of hypnotics. In *Sleep: Physiology and Pathology,* ed. A. Kales, pp. 33–43. Philadelphia: Lippincott. 360 pp.

Kales, A., Heuser, G., Kales, J. D., Rickles, W. H. Jr., Rubin, R. T., Scharf, M. B., Ungerleider, J. T., Winters, W. D. 1969b. Drug dependency. Investigations of stimulants and depressants. *Ann. Intern. Med.* 70(3):591–614

Kales, A., Allen, C., Scharf, M. B., Kales, J. D. 1970a. Hypnotic drugs and their effectiveness. All-night EEG studies of insomniac subjects. *Arch. Gen. Psychiatr.* 23(3):226–32

Kales, A., Kales, J. D., Scharf, M. B., Tan, T. L. 1970b. Hypnotics and altered sleep-dream patterns. 2. All night EEG studies of chloral hydrate, flurazepam, and methaqualone. *Arch. Gen. Psychiatr.* 23(3):219–25

Kales, A., Bixley, E. O., Tan, T. L., Scharf, M. B., Kales, J. D. 1974. Chronic hypnotic drug use: Ineffectiveness, drug withdrawal insomnia and dependence. *J. Am. Med. Assoc.* 227:513–17

Kales, A., Kales, J. D., 1974. Sleep disorders: Recent findings in the diagnosis and treatment of disturbed sleep. *N. Engl. J. Med.* 290:487–98

Kales, A., Caldwell, A. B., Preston, T. A., Healey, S., Kales, J. D. 1976. Personality patterns in insomnia: Theoretical implications. *Arch. Gen. Psychiatr.* 33: 1128–34

Kales, A., Cadieux, R., Soldatos, C. R., Tan, T. L. 1979. Successful treatment of narcolepsy with propranolol: A case report. *Arch. Neurol.* 36(1):650–51

Kales, A., Tan, T. L. 1969. See Kales et al 1969a, pp. 148–57.

Karacan, I. 1971. Painful nocturnal penile erections. *J. Am. Med. Assoc.* 215:1831

Karacan, I., Salis, P. J., Williams, R. L. 1978. The role of the sleep laboratory in diagnosis and treatment of impotence. See Beutler 1978, pp. 353–82

Karacan, I., Thornby, J. I., Anch, M., Holzer, C. P., Wacheit, G. J., Schwab, J. J., Williams, R. L. 1976. Prevalence of sleep disturbance in the general urban Florida county. *Soc. Sci. Med.* 10:239–44

Kawamura, H., Inouye, S.-I. T. 1979. Circadian rhythm within the hypothalamic island containing the suprachiasmatic nucleus. In *Biological Rhythms and Their Central Mechanism,* ed. M. Suda, O. Hayaishi, H. Nakagawa, pp. 335–41. Amsterdam: Elsevier/North-Holland Biomed. Press. 453 pp.

Kay, D. C., Blackburn, A. B., Buckingham, J. A., Karacan, I. 1976. See Hartmann 1976, pp. 83–210

Kayed, K., Godtlibsen, O. B., Sjaastad, O. 1978. Chronic paroxysmal hemicrania. IV. "REM Sleep Locked" nocturnal headache attacks. *Sleep* 1:91–95

Kelly, D. H., Shannon, D. C. 1979. Periodic breathing in infants with near-miss sudden infant death syndrome. *Pediatrics* 63(3):355–60

Kessler, S. 1976. See Broughton & Ghanem 1976, pp. 285–300

Kessler, S., Guilleminault, C., Dement, W. C. 1974. A family study of 50 REM narcoleptics. *Acta. Neurol. Scand.* 50:503–12

Klein, K. E., Hermann, H., Kuklinski, P., Wegmann, H.-M. 1977. Circadian performance rhythms: Experimental studies in air operation. In *Vigilance Theory, Operational Performance and Physiological Correlates,* ed. R. R. Mackie, pp. 117–32. New York/London: Plenum. 876 pp.

Kleitman, N. 1963. *Sleep and Wakefulness.* Chicago: Univ. Chicago Press. 552 pp.

Kokkoris, C. P., Weitzman, E. D., Pollak, C. P., Spielman, A. J., Czeisler, C. A., Bradlow, H. 1978. Long-term ambulatory temperature monitoring in a subject with a hypernychthermal sleep-wake cycle disturbance. *Sleep* 1(2):177–90

Koo, K. W., Sax, D. S., Snider, G. L. 1975. Arterial blood gases and PH during sleep in chronic obstructive pulmonary disease. *Am. J. Med.* 58(5):663–70

Kramer, M. 1979. Dream disturbances. *Psychiatr. Ann.* 9(7):366–76

Krieger, A. J., Rosomoff, H. L. 1974a. Sleep induced apnea. Part 1: A respiratory and autonomic dysfunction syndrome following bilateral percutaneous cervical cordotomy. *J. Neurosurg.* 40(2): 168–80

Krieger, A. J., Rosomoff, H. L. 1974b. Sleep induced apnea. Part 2: Respiratory failure after anterior spinal surgery. *J. Neurosurg.* 40(2):181–85

Kripke, D. F., Mullaney, D. J., Atkinson, M., Wolf, S. 1978. Circadian rhythm disorders in manic-depressives. *Biol. Psychiatry* 13(3):335–51

Kripke, D. F., Simons, R. N., Garfinkel, L., Hammond, E. C. 1979. Short and long sleep and sleeping pills: Is increased mortality associated? *Arch. Gen. Psychiatr.* 36:103–16

Kuo, P. C., West, R. A., Bloomquist, D. S., McNeil, R. W. 1979. The effect of mandibular osteotomy in three patients with hypersomnia sleep apnea. *Oral Surg.* 48(5):385–92

Kupfer, D. J. 1971. Sleep and depressive syndromes. *N. Engl. J. Med.* 285:1490

Kupfer, D. J. 1976. REM latency: A psychobiologic marker for primary depression disease. *Biol. Psychiatry* 11(2): 159–74

Lange, H., Burr, W., von Aswege, J. 1976.

Sleep stage shifts in depressive illness. *Biol. Psychiatry* 11(2):239–43

Lavie, P., Gadoth, N., Gordon, C. R., Goldhammer, G., Bechar, M. 1979. Sleep patterns in Kleine-Levin syndrome. *Electroencephalogr. Clin. Neurophysiol.* 47(3):369–71

Legkonogov, V. A. 1971. Cataplectoid and narcoleptic seizures in primary tumors of the 3rd brain ventricle. *Z.h. Nevropatol Psikhiatyr. im. S. S. Korsakova* 71:1299–1304

Levin, M. 1936. Periodic somnolence and morbid hunger. A new syndrome. *Brain* 59:494–515

Lewis, S. A. 1969. Sleep patterns during afternoon naps in the young and elderly. *Br. J. Psychiatr.* 115(518):107–8

Lugaresi, E. 1978. Snoring and its clinical implications. See Cummiskey et al 1978, pp. 13–21

Lugaresi, E., Coccagna, G., Gambi, D., Berti Ceroni, G., Poppi, M. 1967. Symond's nocturnal myoclonus. *Electroencephalogr. Clin. Neurophysiol.* 23:289

Lugaresi, E., Coccagna, G., Mantovani, M. 1978. Hypersomnia with periodic apneas. In *Advances in Sleep Research,* Vol. 4. New York: Spectrum. 151 pp.

Lugaresi, E., Coccagna, G., Mantovani, M., Berti Ceroni, G., Pazzaglia, P., Tassinari, C. A. 1970. The evolution of different types of myoclonus during sleep. A polygraphic study. *Eur. Neurol.* 4(6):321–31

Lugliani, R., Whipp, B. J., Wasserman, K. 1979. Doxapram hydrochloride: A respiratory stimulant for patients with primary alveolar hypoventilation. *Chest* 76(4):414–19

Mamelak, M., Caruso, V. J., Stewart, K. 1979. Narcolepsy: A family study. *Biol. Psychiatry* 14(5):821–34

Matthews, W. B. 1979. Treatment of the restless legs syndrome with clonazepam. *Br. Med. J.* 1(6165):751

McGhie, A., Russell, S. M. 1962. The subjective assessment of normal sleep patterns. *J. Ment. Sci.* 108:642–54

McGregor, P., Pollak, C. P., Fusco, R., Weitzman, E. D. 1975. Spindles and vertex waves during REM periods in sleep deprived subjects. *Sleep Res.* 4:239 (Abstr.)

Mendelson, W. B., Gillin, J. C., Wyatt, R. J. 1977. *Human Sleep and Its Disorders.* New York: Plenum. 260 pp.

Miles, L. E. M., Raynal, D. M., Wilson, M. A. 1977. Blind man living in normal society has circadian rhythms of 24.9 hours. *Science* 198:421–23

Mitler, M. M. 1976. Toward an animal model of narcolepsy-cataplexy. See Broughton & Ghanem 1976, pp. 387–409

Mitler, M. M., Dement, W. C. 1977. Sleep studies on canine narcolepsy: Pattern and cycle comparisons between affected and normal dogs. *Electroencephalogr. Clin. Neurophysiol.* 43:691–99

Mitler, M. M., van den Hoed, J., Carskadon, M. A., Richardson, G., Park, R., Guilleminault, C., Dement, W. C. 1979. REM sleep episodes during the multiple sleep latency test in narcoleptic patients. *Electroencephalogr. Clin. Neurophysiol.* 46:479–81

Montplaisir, J., Billiard, M., Takahashi, S., Bell, I. R., Guilleminault, C., Dement, W. C. 1978. Twenty-four-hour recording in REM-narcoleptics with special reference to nocturnal sleep disruption. *Biol. Psychiatry* 13(1):73–89

Moore, R. Y. 1979. The retinohypothalamic tract, suprachiasmatic hypothalamic nucleus and central neural mechanisms of circadian rhythm circulation. See Kawamura & Inouye 1979, pp. 343–54

Moore-Ede, M. C., Lydic, R., Czeisler, C. A., Tepper, B., Buller, C. A. 1980. Characterization of separate circadian oscillators driving rest-activity and body temperature rhythms. *Sleep Res.* In press

National Academy of Sciences. 1979. *Sleeping Pills, Insomnia, and Medical Practice,* Publ. IOM–79–04. Washington DC: Inst. Med., 198 pp.

National Institute on Drug Abuse. 1977. *National Survey on Drug Abuse* Vol. 1, *Main Findings.* DHEW Publication No. (ADM) 78–618. Washington DC: GPO

Okada, T., Haga, Y., Terashima, M., Fujita, K., Kayukawa, Y., Ohta, T. 1979. Pathophysiological manifestation of periodic breathing during sleep, with special reference to "insomnia with sleep apnea syndrome." *Clin. Electroencephalogr. (Osaka)* 21(5):303–14

Orem, J., Dement, W. C. 1975. Neurophysiological substrates of the changes in respiration during sleep. In *Advances in Sleep Research,* ed. E. D. Weitzman, 2:1–42. New York: Spectrum. 236 pp.

Orem, J., Montplaisir, J., Dement, W. C. 1974. Changes in the activity of respiratory neurons during sleep. *Brain Res.* 92:309–15

Orr, W. C., Imes, N. K., Martin, R. J. 1979a. Progesterone therapy in obese patients with sleep apnea. *Arch. Intern. Med.* 139:109–11

Orr, W. C., Robinson, M. G., Johnson, L. F. 1979b. Acid clearing during sleep in pa-

tients with esophagitis and controls. *Gastroenterology* 76:1213

Oswald, I. 1979. The why and how of hypnotic drugs. *Br. Med. J.* 1(6172):1167–68

Papousek, M., Frank, H. P., Stöhr, H. 1974. Sleep deprivation therapy in endogenous depression: Effects on circadian rhythms. See Broughton & Gastaut 1975, pp. 474–77

Papousek, M. 1976. Temporal structure of sleep in endogenous depression. *Arzneim. Forsch.* 26(6):1062–64

Passouant, P. 1974. REM's ultradian rhythm during 24 hours in narcolepsy. In *Chronobiology,* ed. L. E. Scheving, F. Halbers, J. E. Pauly, pp. 495–98. Tokyo: Igaku Shoin. 784 pp.

Passouant, P., Besset, A., Carriere, A., Billard, M. 1975. Night sleep and generalized epilepsies. See Broughton & Gastaut 1975, pp. 185–96

Phillips, R. L., Mitler, M. M., Dement, W. C. 1974. Alpha sleep in chronic insomniacs. *Sleep Res.* 3:143 (Abstr.)

Phillipson, E. A. 1978. Respiratory adaptations in sleep. *Ann. Rev. Physiol.* 40:133–56

Pokorny, A. D. 1978. Sleep disturbances, alcohol, and alcoholism. See Beutler et al 1978, pp. 233–60

Reding, G., Sepelin, H., Robinson, J. E. Jr., Zimmerman, S. O., Smith, V. H. 1968. Nocturnal teeth-grinding: All night psychophysiology studies. *J. Dent. Res.* 47:786–97

Rentos, P. G., Shephard, R. D., eds. 1976. *Shift Work and Health: A Symposium.* Washington DC: Natl. Inst. Occup. Safety Health

Reppert, S. M., Perlow, M. J., Mishkin, M., Tamarkin, L., Klein, D.C. 1979. Effects of damage to the suprachiasmatic area of the anterior hypothalamus on the daily melatonin rhythm in the rhesus monkey. *Endocrine Soc. Proc., 61st meet., Anaheim, Calif.* pp. 76 (Abstr.)

Reynolds, C. F. 3rd, Coble, P., Holzer, B., Carroll, R., Kupfer, D. J. 1979. Sleep and its disorders. *Primary Care* 6(2):417–38

Richardson, G., Carskadon, M. A., Flagg, W., van den Hoed, J., Dement, W. C., Mitler, M. M. 1978. Excessive daytime sleepiness in man: Multiple sleep latency measurement in narcoleptic and control subjects. *Electroencephalogr. Clin. Neurophysiol.* 45:621–27

Rochester, D. F., Enson, Y. 1974. Current concepts in the pathogenesis of the obesity-hypoventilation syndrome. *Am. J. Med.* 47:402–20

Romanczuk, B. J., Potsic, W. P., Atkins, J. P. Jr. 1978. Hypersomnia with periodic breathing (an acromegalic Pickwickian). *Otolaryngology* 86 (6 Pt.) ORL–897–903

Roth, B., 1978. Narcolepsy and hypersomnia. See Beutler 1978, pp. 29–60

Roth, T., Kramer, M., Lutz, T. 1976. The nature of insomnia: A descriptive summary of a sleep clinic population. *Compr. Psychiatry* 17(1):217–20

Roth, T., Lutz, T., Kramer, M., Tietz, E. 1977. The relationship between objective and subjective evaluations of sleep in insomniacs. *Sleep Res.* 6:178

Rudolf, G. A. E., Schilgen, B., Tolle, R. 1977. Anti-depressive behandlung mittels schlafentzug. *Nervenarzt* 48:1–11

Rundell, O. H., Williams, H. L., Lester, B. K. 1977. Sleep in alcoholic patients: Longitudinal findings. In *Alcohol Intoxication and Withdrawal,* ed. M. M. Gross, 3B:382–402. New York: Plenum. 422 pp.

Rusak, B., Zucker, I. 1975. Biological rhythms and animal behavior. *Ann. Rev. Psychol.* 26:137–71

Sanger, B. A. 1977. Retrognathia and sleep apnea. *J. Am. Med. Assoc.* 238(14): 1496–97

Schulz, H., Lund, R., Cording, C., Dirlich, G. 1979. Bimodal distribution of sleep latencies in depression. *Biol. Psychiatry* 14(4)595–600

Schwartz, B. A. 1976. Ondine's curse. *Lancet* 2(7987):695

Severinghaus, J. W., Mitchell, R. A. 1962. Ondine's curve. Failure of respiratory center automaticity while awake. *Clin. Res.* 19:122

Sitaram, N., Gillin, J. C., Bunney, W. E. Jr. 1978. The switch process in manic-depressive illness: Circadian variation in time of switch and sleep and manic ratings before and after switch. *Acta Psychiatr. Neurol. Scand.* 58:267–78

Smith, R., Johnson, L., Rothfeld, D., Zir, L., Tharp, B. 1972. Sleep and cardiac arrhythmias. *Arch. Intern. Med.* 130: 751–53

Spiker, D. G., Coble, P., Cofsky, J., Gordon-Foster, F., Kupfer, D. J. 1978. EEG sleep and severity of depression. *Biol. Psychiatry* 13(4):485–88

Strohl, K. P., Saunders, N. D., Feldman, N. T., Hallett, M. 1978. Obstructive sleep apnea in family members. *N. Engl. J. Med.* 299:969–73

Sullivan, C. E., Murphy, E., Kozar, L. F., Phillipson, E. A. 1979. Ventilatory responses to CO_2 and lung inflation in tonic versus phasic REM sleep. *J. Appl. Physiol.* 47:1304–10

Symonds, C. 1953. Nocturnal myoclonus. *J. Neurol. Neurosurg. Psychiatr.* 16: 116–71

Tusiewicz, K., Moldofsky, H., Bryan, A. C., Bryan, M. H. 1977. Mechanics of the rib cage and diaphragm during sleep. *J. Appl. Physiol.* 43:600–11

Viukari, M., Linnoila, M., Aalto, U. 1978. Efficacy and side effects of flurazepam, fosazepam, and nitrazepam as sleeping aids in psychogeriatric patients. *Acta Psychiatr. Scand.* 57(1):27–35

Vogel, G. W., Thurmond, A., Gibbons, P., Sloan, K., Boyd, M., Walker M. 1975. REM sleep reduction effects on depression syndromes. *Arch. Gen. Psychiatr.* 32(6):765–77

Vogel, G. W., Vogel, F., McAbee, R. S., Thurmond, A. J. 1980. Improvement of depression by REM sleep deprivation. *Arch. Gen. Psychiatr.* 37(3):247–53

Volk, W., Dietsch, P., Spiegelberg, U. 1978. Therapy of sleeping disorders with L-tryptophane plus oxprenolol/a double-blind study (author's transl). *Arzneim. Forsch.* 28(1):1798–1800

Walker, A. E., Caveness, W. E., Gutchley, M., eds. 1969. *The Late Effects of Head Injury.* Springfield Il: Thomas. 560 pp.

Wasserman, M., Pollak, C. P., Spielman, A. J., Weitzman, E. D. 1980. The differential diagnosis of impotence. Measurement of nocturnal penile tumescence. *J. Am. Med. Assoc.* 243:2038–42

Webb, P. 1974. Periodic breathing during sleep. *J. Appl. Physiol.* 37(6):899–903

Wehr, T., Wirz-Justice, A., Goodwin, F. K., Duncan, W., Gillin, J. C. 1979. Phase advance of the circadian sleep-wake cycle as an anti-depressant. *Science* 206:710–11

Weitzman, E. D., ed. 1974. *Advances in Sleep Research,* Vol. 1. New York: Spectrum. 424 pp.

Weitzman, E. D. 1979. The syndrome of hypersomnia and sleep-induced apnea. *Chest* 75:414–15

Weitzman, E. D. 1980. Sleep and aging. In *Neurologic Diseases of the Aged,* ed. R. Katzman, R. Terry. New York: Raven. In press

Weitzman, E. D., Graziani, L. 1974. Sleep and the sudden infant death syndrome. See Guilleminault & Dement 1974, pp. 327–44

Weitzman, E. D., Kripke, D. F., Goldmacher, D., McGregor, P., Nogeire, C. 1970. Acute reversal of the sleep-waking cycle in man. *Arch. Neurol.* 22: 483–89

Weitzman, E. D., Pollak, C. P. 1979. Disorders of the circadian sleep-wake cycle. *Med. Times* 107(6):83–94

Weitzman, E. D., Pollak, C. P., Borowiecki, B. 1978a. Hypersomnia-sleep apnea due to micrognathia: Reversal by tracheoplasty. *Arch. Neurol.* 35:392

Weitzman, E. D., Pollak, C. P., Borowiecki, B., Burack, B., Shprintzen, R., Rakoff, S. 1978b. The hypersomnia-sleep apnea syndrome: Site and mechanism of upper airways obstruction. See Cummiskey et al 1978, pp. 235–48

Weitzman, E. D., Pollak, C., McGregor, P. 1979a. The polysomnographic evaluation of sleep disorder in man. In *Electrophysiological Approaches to Neurological Diagnosis,* ed. M. J. Aminoff, pp. 496–524. New York: Churchill Livingstone. 600 pp.

Weitzman, E. D., Czeisler, C. A., Moore-Ede, M. 1979b. Sleep-wake, neuroendocrine and body temperature circadian rhythms under entrained and nonentrained (free running) conditions in man. See Kawamura & Inouye 1979, pp. 199–227

Weitzman, E., Czeisler, C., Coleman, R., Dement, W., Richardson, G., Pollak, C. 1979c. Delayed sleep phase syndrome: A biological rhythm disorder. In *Sleep Res.* 8:208

Weitzman, E. D., Czeisler, C. A., Zimmerman, J. C., Ronda, J. M. 1980a. Timing of REM and stages 3 + 4 sleep during temporal isolation in man. *Sleep* 2:391–407

Weitzman, E. D., Czeisler, C. A., Zimmerman, J. C., Moore-Ede, M. 1980b. Biological rhythms and sleep wake relationships in man of cortisol, growth hormone and temperature during temporal isolation. In *Advances in Neurology,* ed. J. B. Martin, S. Reichlin, K. Bick. New York: Raven. In press

Weitzman, E. D., Kahn, E., Pollak, C. P. 1981. A quantitative analysis of sleep and sleep apnea before and after tracheostomy in patients with the hypersomnia sleep apnea syndrome. *Sleep.* In press

Wever, R. 1975. The circadian multi-oscillator system of man. *Int. J. Chronobiol.* 3:19–55

Wever, R. 1979. *The Circadian System of Man.* New York: Springer-Verlag. 256 pp.

Wheatley, D. 1979. Clinical significance of prescribing hypnotics in general practice. *Br. J. Clin. Pharmacol.* 8(1):79–80

Williams, H. L., Salamy, A. 1972. Alcohol and sleep. In *The Biology of Alcoholism,* ed. B. Kissin, H. Bergleiter, pp. 435–83. New York: Plenum. 552 pp.

Williams, R. L., Karacan, I. 1975. Sleep disorders and disordered sleep. In *American Handbook of Psychiatry,* Vol. 4, *Organic Disorders and Psychosomatic Medicine,* ed. M. F. Reiser, pp. 854–904. New York: Basic. 980 pp.

Williams, R. L., Karacan, I., eds. 1978. *Sleep Disorders: Diagnosis and Treatment.* New York: Wiley. 417 pp.

Wyatt, R. J., Engelman, K., Kupfer, D. J., Fram, D. H., Sjoerdsma, A., Snyder, F. 1970. Effects of L-tryptophan (a natural sedative) on human sleep. *Lancet* 2:842–46

Yamamoto, T., Hirose, N., Miyoshi, K. 1977. Polygraphic study of periodic breathing and hypersomnolence in a patient with severe hypothyroidism. *Eur. Neurol.* 15(4):188–93

York, R., Daly, D. 1960. Narcolepsy. *Med. Clin. North Am.* 44:953–68

Zarcone, V. P. Jr. 1973. Narcolepsy. *N. Engl. J. Med.* 288:1156–66

Zarcone, V. P. Jr., Cohen, M., Hoddes, E. 1975. WAIS, MMPI, and sleep variables in abstinent alcoholics. *Sleep Res.* 4:123

Ann. Rev. Neurosci. 1981. 4:419–61
Copyright © 1981 by Annual Reviews Inc. All rights reserved

β-ADRENERGIC RECEPTOR SUBTYPES: PROPERTIES, DISTRIBUTION, AND REGULATION

❖11559

Kenneth P. Minneman, Randall N. Pittman, and Perry B. Molinoff

Department of Pharmacology, University of Colorado Health Sciences Center, Denver, Colorado 80262

INTRODUCTION

The multiple effects of the endogenous catecholamines epinephrine (EPI) and norepinephrine (NE) occur following the interaction of these compounds with identifiable high affinity receptors on the external surface of specific target cells. These interactions initiate a series of biochemical events leading to alterations in cell metabolism or physiology. The effects of catecholamines are often mediated through changes in the intracellular concentration of 3',5'-adenosine monophosphate (cyclic AMP). This molecule acts as a second messenger and modifies aspects of cellular metabolism, usually via selective effects on protein phosphorylation.

Ahlquist (1948) divided responses to catecholamines into those resulting from activation of α-adrenergic or β-adrenergic receptors (Table 1). This division was based on the potencies of a series of natural and synthetic catecholamines for these receptors. He concluded that α- and β-adrenergic receptors are found in a variety of peripheral tissues and that they usually mediate different physiological responses. There was no clear correlation between the excitatory or inhibitory actions of catecholamines and a particular receptor type. Stimulation of either α- or β-adrenergic receptors produced excitatory effects in some tissues and inhibitory effects in others. Ahlquist (1948) showed that the two receptor types could coexist within a single organ, such as the uterus, where α-adrenergic receptors mediate excitatory effects and β-adrenergic receptors mediate inhibitory effects.

419

0147-006X/81/0301-0419$01.00

Table 1 Classification of adrenergic receptors

Type	Potency	Characteristics	References
α-Adrenergic	NE > EPI >> ISO	Vasoconstriction, excitation of uterine contractions, contraction of nictitating membrane, pupillary dilation, inhibition of intestinal peristalsis	Ahlquist 1948
α_1-Adrenergic	Postsynaptic	Vasoconstriction	Langer 1974, Berthelsen & Pettinger 1977
α_2-Adrenergic	Presynaptic	Inhibit NE release, inhibit renin release, lower blood pressure via central action	Langer 1974, Berthelsen & Pettinger 1977
β-Adrenergic	ISO > EPI > NE	Vasodilation, inhibition of uterine contraction, myocardial stimulation	Ahlquist 1948
β_1-Adrenergic	ISO > EPI = NE	Fatty acid mobilization from adipose tissue, cardiac stimulation	Lands et al 1967a,b
β_2-Adrenergic	ISO > EPI >> NE	Bronchodilation, vasodepression	Lands et al 1967a,b

The division of adrenergic receptors into α- and β-subtypes is now supported by a large body of experimental evidence and has provided a rationale for the development of drugs with increased affinity and selectivity for each of the two major types of catecholamine receptor. It has become clear, however, that the apparent pharmacological specificity of α- and β-adrenergic receptors varies from tissue to tissue. This suggests that each of the principal types of adrenergic receptor should be subdivided into at least two subtypes, i.e. α_1/α_2 (Langer 1974, Berthelsen & Pettinger 1977) and β_1/β_2 (Lands et al 1967a,b) (see Table 1). There do not appear to be, however, an infinite number of adrenergic receptor subtypes. As discussed below, it is likely that there are only two types of β-adrenergic receptor in mammalian tissues. The properties of these two receptor subtypes are remarkably consistent in a variety of tissues (see Minneman et al 1979c).

This article is designed to provide an overview of the available evidence relating to the existence and properties of β-adrenergic receptor subtypes. Results obtained in studies with intact animals and isolated organs are summarized, and the important contributions of, as well as the problems inherent in, this approach are discussed. Because this aspect of the subject has been previously reviewed (Furchgott 1972, Ariens & Simonis 1976) our focus is on recently obtained biochemical data. The development of techniques for identifying and quantitating β-adrenergic receptors using radioligand binding assays has resulted in a great deal of new information with regard to the number, properties, cellular localization, and regulation of β-adrenergic receptor subtypes. We attempt to provide a critical review of the current status of the field as well as an indication of likely directions for future research.

PHYSIOLOGICAL EVIDENCE FOR β-ADRENERGIC RECEPTOR SUBTYPES

The first suggestion that two distinct types of β-adrenergic receptor exist came from studies of the effects of α-methyl congeners of methoxamine on β-adrenergic receptor mediated responses in various tissues. These methoxamine analogs, including isopropylmethoxamine (Burns et al 1964, Levy 1964, Salvador et al 1964), butoxamine (Levy 1966a, Burns et al 1967), dimethylisopropylmethoxamine (Levy 1966b), and 1-(4'-methylphenyl)-2-isopropylamino-propranol HCl (H 35/25, Levy 1967), blocked vascular β-receptors in doses that had no effect on β-receptor mediated responses in the myocardium.

In 1967, Lands and co-workers (Lands et al 1967a,b) showed that the rank order of potency of a series of catecholamines for a variety of β-adrenergic receptor mediated responses could be subdivided into two major groups. They postulated that these two groups reflected the existence of two types of β-adrenergic receptor, which they named β_1 and β_2. The β-adrenergic receptors controlling cardiac stimulation, fatty acid mobilization from adipose tissue, and inhibition of contraction of rabbit small intestine were stimulated by catecholamines with the rank order of potency ISO $>$ EPI = NE. These were called β_1-receptors. The β-receptors controlling bronchodilation, vasodilation, inhibition of uterine contraction, and contraction of the KCl-relaxed diaphragm were stimulated by catecholamines with the rank order of potency ISO $>$ EPI $>>$ NE. These were called β_2-receptors. Further experiments by Lands, Arnold, and McAuliff and co-workers (Lands et al 1969, Arnold et al 1968, Arnold & McAuliff 1968, 1969, Arnold & Selberis 1968) confirmed the usefulness of this classification and extended it to β-receptors in other tissues and species. The additional responses studied included calorigenesis in the rat (β_1), stimulation of chicken heart (β_1), glycogenolysis in skeletal muscle (β_2), elevation of blood glucose in the dog (β_2), stimulation of chicken rectal caecum (β_2), and stimulation of frog heart (β_2).

Physiological Function of β_1- and β_2-Adrenergic Receptors

The major physiological difference between β_1- and β_2-receptors is their differential sensitivity to NE. β_1-adrenergic receptors have approximately the same affinity for EPI as for NE, whereas β_2-receptors have a much higher affinity for EPI than for NE. Thus, β_1-receptors can be stimulated by circulating EPI or neurally released NE. On the other hand, β_2-receptors probably respond to EPI as their natural agonist. It has been suggested that the β_2-receptor is a hormonal receptor, whereas the β_1-receptor is a neuronal receptor (Ariens & Simonis 1976, Carlsson & Hedberg 1977; Figure 1); however, β_1-receptors respond to both neuronal and hormonal stimulation and β_2-adrenergic receptors are activated by high concentrations of NE. Furthermore, the actual concentration of NE in the synaptic cleft is not known and it is possible that NE is the natural agonist for at least some β_2-adrenergic receptors.

Development of Cardiac and Bronchio-Selective Drugs

The postulated existence of two types of β-adrenergic receptor suggested that drugs might be developed that would be selective for each of the two subtypes. In most species the β-receptors mediating increased heart rate and contractility seemed to be predominantly of the β_1-subtype, while those mediating bronchial relaxation seemed to be predominantly of the β_2-subtype. Therefore, selective drugs were thought to be potentially useful in the treatment of cardiac arrythmias and angina (β_1-antagonists) or bronchial asthma (β_2-agonists). It was hypothesized that increased selectivity would decrease the incidence of adverse side effects.

Over the last 15 years, a number of agonists and antagonists have been developed that show physiological selectivity for β-adrenergic receptors in different organs. Several of these agents are now used clinically or as research tools. Commonly used drugs are listed in Table 2, along with the code numbers by which they are identified in the literature. The structures of these drugs are shown in Figures 2 and 3.

β_1-SELECTIVE AGONISTS   The endogenous catecholamine NE is a β_1-selective agonist (Lands et al 1967a), but it also stimulates α-adrenergic receptors and is rapidly inactivated by degradation and by reuptake into

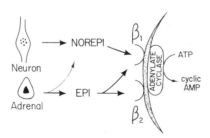

Figure 1   Schematic representation of endogenous catecholamines and the β-adrenergic receptor subtypes with which they primarily interact.

Table 2   Widely used drugs showing selectivity for β_1- or β_2-adrenergic receptors

Drug name	Code number	Source	Reference
β_1-Agonists			
Dobutamine		Eli Lilly	Tuttle & Mills 1975
Tazolol	ITP	Syntex Corp.	Strosberg 1976
Prenalterol	(–) H 80/62	Hassle	Carlsson et al 1977
	H 133/22		
β_1-Antagonists			
Practolol	ICI 50172	ICI	Dunlop & Shanks 1968
	AY 21011		
Metoprolol	H 93/26	Hassle/CIBA-GEIGY	Ablad et al 1973
Atenolol	ICI 66082	ICI	Barret et al 1973,
			Barrett 1977
para-Oxprenolol	CIBA 47920	CIBA-GEIGY	Vaughan Williams et al
			1973
β_2-Agonists			
Salbutamol	AH 3365	Allen and Hanbury's	Brittain et al 1968,
(Albuterol)			Cullum et al 1969
Soterenol	MJ 1992	Mead Johnson	Dugan et al 1968
Terbutaline	Me 501	AB Draco, Lund	Bergman et al 1969
			Persson & Olsson 1970,
Fenoterol	Th 1165a	Boehringer-Ingeheim	O'Donnell 1970
Hexoprenaline	St 1512	Chemie-Linz	Turnheim & Kraupp 1971
Rimiterol	WG 253	3M-Riker	Griffin & Turner 1971,
	R 798		Bowman & Rodger 1972
Zinterol	MJ 9184-1	Mead Johnson	Gwee et al 1972
Carbuterol	SKF 40383-A	Smith, Kline and French	Wardell et al 1974
Salmefamol		Allen and Hanbury's	Apperley et al 1976
Procaterol	OPC 2009	Otsuka	Yoshizaki et al 1976,
			Yabuuchi et al 1977
β_2-Antagonists			
—	H 35/25	Hassle	Levy 1967
Butoxamine		Wellcome	Burns & Lemberger 1965
			Levy 1966a
—	IPS 339	—	Imbs et al 1977

presynaptic terminals. Three synthetic drugs have been reported to be selective myocardial stimulants (Figure 2): dobutamine, a hydroxylated aryl alkyl derivative of dopamine (Tuttle & Mills 1975); tazolol, a thiazol compound (Strosberg 1976); and prenalterol (H 133/22), a drug containing an aminopropanolic side chain linked to a phenolic oxygen (Carlsson et al 1977). These compounds have been shown to increase myocardial contractility and heart rate at concentration causing little or no vasodilation.

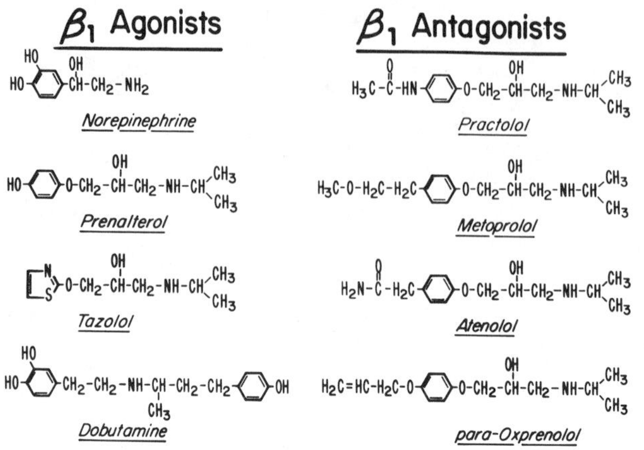

Figure 2   Structures of drugs selective for β_1-adrenergic receptors.

β_1-SELECTIVE ANTAGONISTS β_1-selective antagonists all contain an aminopropanolic side chain attached to a phenolic oxygen with a para substituent (Figure 2). Practolol was the first of these drugs to be described (Dunlop & Shanks 1968). It was found to be much more effective in blocking the positive chronotropic and inotropic effects of catecholamines than in blocking effects on blood flow to the hindlimb of the dog or effects on tracheal or bronchial smooth muscle (Dunlop & Shanks 1968). Three other β_1-selective antagonists with similar structures have been developed and are widely used (Table 2). These agents have been shown to be cardioselective and they are potentially useful for the treatment of angina and cardiac arrhythmias.

β_2-SELECTIVE AGONISTS   These drugs all contain a β-hydroxyphenethylamine moiety with two oxygen or nitrogen substituents on the meta and/or para positions of the benzene ring. Examination of the structure-activity relationships of β_2-agonists (Figure 3) has led to the conclusion that the presence of a t-butyl substituent on the amino group favors selectivity for β_2-adrenergic receptors (O'Donnell 1972, Letts et al 1977). Most of the drugs (except hexoprenaline and rimiterol) do not contain a catechol moiety and are thus resistant to degradation by catechol-O-methyltransferase (COMT).

   The first selective β_2-agonists to be discovered were salbutamol and soterenol (Brittain et al 1968, Dungan et al 1968, Cullum et al 1969). These drugs were shown to be effective bronchodilators that caused only minimal cardiac stimulation. Their selectivity has been confirmed in many tissues, and a number of other drugs showing β_2-agonist selectivity have been described (Table 2; Figure 3).

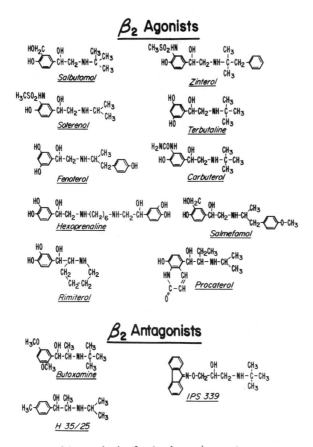

Figure 3   Structures of drugs selective for β_2-adrenergic receptors.

β_2-SELECTIVE ANTAGONISTS   As discussed above, a number of α-methyl congeners of methoxamine, particularly butoxamine and H35/25 (Burns et al 1967, Levy & Wilkenfeld 1969, 1970) selectively block vascular β-receptors at concentrations that do not affect myocardial β-receptors. Butoxamine and H35/25 do not inhibit, or only weakly block, a number of β_1-receptor mediated responses, including inhibition of intestinal contraction, positive chronotropism and inotropism, and catecholamine-mediated increases in plasma free fatty acids. These drugs do antagonize a number of β_2-receptor mediated responses, including vasodilation, uterine relaxation, increases in blood glucose and lactic acid, increase in rat skeletal muscle phosphorylase, and inhibition of the contractions of the guinea pig trachea (Levy 1966b, Levy & Wilkenfeld 1969, 1970, Burns et al 1967, Wasserman & Levy 1972, Loakpradit & Lockwood 1977).

Another drug that selectively blocks β_2-adrenergic receptors has recently been described. IPS 339 (Figure 3) contains the aminopropanolic side chain commonly seen in β-adrenergic receptor antagonists. However, in this case the side chain is fixed to fluorene through an oximino linkage (Imbs et al 1977). IPS 339 has been reported to be over one hundred times more potent as an inhibitor of the affects of catecholamines on guinea pig trachea (β_2) than on the inhibition of inotropic or chronotropic responses to catecholamines in the heart (β_1) (Imbs et al 1977).

Practical Considerations in Determining the Affinity and Selectivity of Drugs

The potency and selectivity of a drug on a receptor-mediated response in an isolated organ or tissue preparation can be assessed accurately only if certain experimental conditions, related to the pharmacokinetics of intact tissues, are satisfied. The major considerations are summarized by Furchgott (1970, 1972). In view of the modest differences in the affinities of selective drugs for β-adrenergic receptor subtypes it is particularly important to control experimental conditions rigorously.

A few examples serve to illustrate the importance of carefully controlling experimental conditions in studies with intact preparations. The order of potency of catecholamines in stimulating the isolated guinea pig atria is ISO > EPI > NE; however, when neuronal uptake is inhibited by cocaine, the order becomes ISO > NE > EPI (Furchgott 1967). Similarly, the apparent difference between the pharmacological specificity of the β_2-receptors mediating vasodilation and those mediating relaxation of tracheal muscle disappears in the presence of inhibitors of extraneuronal uptake (O'Donnell & Wanstall 1976). Enzymatic inactivation of catecholamines can also lead to serious errors in the estimation of drug potency. For example, Kaumann (1968) showed that inhibition of COMT resulted in a six-fold increase in the sensitivity of cat papillary muscle to isoproterenol.

The apparent potencies of antagonists can also be influenced by uptake processes (Furchgott 1972, Lumley & Broadley 1975). Furthermore, high concentrations of some antagonists can cause problems due to the membrane stabilizing or local anesthetic actions of these drugs. Blinks (1967) reported that some β-adrenergic receptor antagonists directly depress myocardial contractility and the spontaneous rate of isolated perfused hearts.

When distinguishing between receptors with very similar properties, small errors in estimation of drug potency become important. Such errors are likely to result where studies are performed with tissues containing a significant receptor reserve (spare receptors). In tissues containing more receptors than are necessary to elicit a maximal effect, agonist affinities calculated from EC_{50} values will be overestimated. That is to say, it will take a higher concentration of drug to half-saturate the receptors than to cause a half-maximal response. Since the degree of β-adrenergic receptor reserve may be different in different tissues (see Buckner & Saini 1975), the existence of spare receptors can lead to misleading results in assessing the pharmacological specificity of different responses. The existence of a receptor reserve also makes it difficult to evaluate the efficacy of agonists. If occupation of only a small proportion of receptors is required to induce a maximal response, then partial agonists will also cause a maximal response.

Two methods have been described to resolve problems caused by spare receptors. The first method utilizes an irreversible antagonist to progressively inactivate receptors (Furchgott 1976). This method obviously depends on the availability of a specific, irreversible β-adrenergic receptor antagonist. Several putative irreversible ligands have been synthesized (Erez et al 1975, Atlas et al 1976), but none of these compounds has yet come into widespread use. A second approach involves what is called functional antagonism (Van den Brink 1973, Buckner & Saini 1975). In this approach, the preparation is exposed to an agonist specific for a receptor that has effects opposite to those of the receptor one wishes to study. For example, in tracheal smooth muscle muscarinic cholinergic agonists cause contraction, whereas β-adrenergic receptors mediate relaxation. The overall effect of a functional antagonist is to alter the relationship between receptor occupancy and tissue response, such that more receptors must be occupied to produce the same response (Van den Brink 1973, Buckner & Saini 1975).

O'Donnell & Wanstall (1978) have used a functional antagonist to determine the efficacy of two β_2-selective agonists, fenoterol and salbutamol, on guinea pig trachea. These two drugs decreased the potency of carbachol in causing contraction of the trachea. Although both fenoterol and salbutamol were equally efficacious in causing tracheal relaxation, fenoterol caused a larger functional antagonism to carbachol than did salbutamol, which suggests that fenoterol has a higher efficacy than does salbutamol. Support for this conclusion comes from the fact that salbutamol, but not fenoterol, acted as a competitive β-receptor antagonist at high concentrations in this preparation (O'Donnell & Wanstall 1978). This behavior would be expected of a partial agonist (Furchgott 1972). Partial agonists can also be defined biochemically by measuring the ability of a compound to stimulate adenylate cyclase activity relative to stimulation caused by a reference compound such as isoproterenol (see below). In studies of adenylate cyclase activity in homogenates of rat heart and lung, salbutamol was shown to be a partial agonist at β_2-receptors, whereas fenoterol was a full agonist at these receptors (Minneman et al 1979a).

Studies of the pharmacological specificity of a response are further complicated by evidence suggesting that both β_1- and β_2-receptors are sometimes present in the same tissue and that both receptors can mediate the same physiological response (Carlsson et al 1972, Furchgott et al 1975). O'Donnell & Wanstall (1979) have performed an extensive series of experiments examining this issue: pA_2 values (defined as the logarithm of the equilibrium association constant, see Schild 1949) for propranolol, butoxamine, H35/25, and atenolol on guinea pig trachea were calculated using NE, isoproterenol, and fenoterol as agonists. The authors concluded that both β_1- and β_2-receptors are present in this tissue.

Another approach to the study of receptor subtypes relies on the fact that adrenergic receptors are highly stereoselective. Patil et al (1971) suggested that the use of optical isomers can minimize differences in physicochemical properties that may affect diffusion or access of drugs to the receptor and thus influence the apparent affinity of the receptor for the drug. Examination of the stereoselectivity of various drugs in different tissues may be useful for determining differences between β-adrenergic receptors in different tissues (Buckner & Patil 1971, Harms 1976). However, lack of differences in stereoselectivity ratios is not adequate evidence that receptors are of the same type, since two receptors might have different affinities for a drug while maintaining the same degree of stereoselectivity. On the other hand, when differences in stereoselectivity ratios are found in different tissues (Harms 1976, Harms et al 1977), it may indicate a difference in the properties or relative populations of the different receptor subtypes.

Evidence for the Existence of Multiple Subtypes of β-Adrenergic Receptor

Although there is general agreement that two types of β-adrenergic receptor exist, small differences in the observed pharmacological specificity of responses in various tissues have led some investigators to propose that there may be more than two subtypes (Bristow et al 1970, Farmer et al 1970a,b, Boissier et al 1971, Wasserman & Levy 1972, Weinstock et al 1974). Discrepancies in the literature with regard to pA_2 values for drugs in various tissues and the various pharmacological specificities observed led Ahlquist (1976) to propose that each tissue might be associated with a unique subtype of β-adrenergic receptor.

On the other hand, there are methodological problems inherent in determining the affinity of a receptor for a drug in an intact tissue. As noted above, O'Donnell & Wanstall (1976) showed that tissue-specific processes, such as extra-neuronal uptake, can lead to an inaccurate classification of receptors into multiple subtypes. Examination of the literature that supports the existence of more than two subtypes of β-receptor suggests that many of these experiments did not conform to the criteria outlined by Furchgott (1970) (see above). At present there is no convincing evidence for the existence of more than two subtypes of β-adrenergic receptor in mammalian tissues. The biochemical results discussed below also support the view that there are only two subtypes of β-adrenergic receptor in mammalian tissues.

Evidence for the Coexistence of β₁- and β₂-Adrenergic Receptors in a Single Organ

The first evidence that the β-adrenergic receptors mediating heart rate and contractility were different came from Farmer et al (1970a,b) and Brittain et al (1970), who showed that the selective β_2-agonists salbutamol and soterenol had greater effects on heart rate than on contractility in rat and guinea pig heart. Carlsson et al (1972) reported that the β_1 selective antagonist, practolol, was more effective in inhibiting the chronotropic response to the β_1 selective agonist NE than it was to the nonselective agonist EPI. Conversely, the β_2 selective antagonist H 35/25 had a greater effect on the chronotropic response to EPI than to NE. The authors interpreted their results to suggest that both β_1- and β_2-receptors mediate chronotropic responses in the cat heart (Carlsson et al 1972). Further studies by the same group (Ablad et al 1975, Carlsson et al 1977) and others (Vlietstra & Blinks 1976, Yabuuchi et al 1977, Bonnelli 1978) have produced similar evidence in studies of cat hearts as well as canine and human atria. The most plausible explanation of these results is that only β_1-receptors mediate the positive inotropic effects of catecholamines, while both β_1- and β_2-receptors mediate the positive chronotropic effects.

Furchgott (1975) has obtained evidence that β_1- and β_2-receptors coexist in, and mediate relaxation of, guinea pig tracheal muscle. These conclusions were based on the observation that the pA_2 values for β_1- and β_2-selective antagonists varied with the use of different selective and nonselective agonists. These observations were subsequently confirmed and extended by O'Donnell & Wanstall (1979).

Although there is general agreement that β_1- and β_2-receptors coexist in guinea pig trachea and in the right atria of cat, dog, and human heart (see above), there is controversy as to whether these receptors coexist in guinea pig atria. Farmer et al (1970a,b) showed that the β_2-agonists soterenol and salbutamol were more potent on chronotropic than inotropic activity in guinea pig heart. Dreyer & Offermeier (1975) examined the relative activities of drugs on heart rate and contractility in isolated guinea pig atria and concluded that the β-adrenergic receptors mediating these two effects of catecholamines were different. On the other hand, Lumley & Broadley (1977) and O'Donnell & Wanstall (1979) have performed extensive experiments examining the pharmacological specificity of the β-adrenergic receptors present in guinea pig heart. Neither of these studies disclosed evidence for the coexistence of β_1- and β_2-receptors mediating positive chronotropic effects in guinea pig atria. O'Donnell & Wanstall (1979) showed that the pA_2 values of antagonists did not depend on the specificity of the agonist used in studies of guinea pig atria, whereas such a dependence was observed in guinea pig trachea. These results are consistent with the interpretation that both β_1- and β_2-receptors coexist in guinea pig trachea, but that only β_1-receptors mediate the positive chronotropic effects of catecholamines in guinea pig atria.

BIOCHEMICAL STUDIES OF β-ADRENERGIC RECEPTOR SUBTYPES

Advantages of Broken Cell Preparations

The development of direct biochemical assays for β-adrenergic receptors has circumvented many of the pharmacokinetic problems inherent in determining the affinity of a receptor for a drug in an intact tissue. The major advantage of the use of a broken cell preparation is that problems of selective uptake or unequal access of drugs to the receptor are minimized. Since β-adrenergic receptors are generally linked to adenylate cyclase, the effects of various compounds on the activity of this enzyme can be measured. Receptor-mediated stimulation of adenylate cyclase activity by agonists, and inhibition of this stimulation by antagonists, can frequently be determined in tissue homogenates. Although measurement of adenylate cyclase activity in broken cell preparations has many advantages over measuring the responses in intact tissues, several potential problems remain. Effects of the drug distal to the initial hormone-receptor interaction, such as effects on the catalytic moiety of adenylate cyclase or on the coupling of the enzyme to the receptor, may contribute to the observed response. To circumvent these difficulties, techniques have been developed to directly monitor the initial hormone-receptor interaction by examining the binding of radiolabeled drugs to receptors on tissue fragments. Since in many cases the initial interaction between the ligand and the receptor can be shown to be a simple bimolecular interaction, it is likely to display quantitative characteristics (rates of association and dissociation, equilibrium affinity constants) that can be experimentally determined. In studies with radioligands, quantitative comparisons of the affinities of various drugs for receptors in different tissues are relatively straightforward because most pharmacokinetic problems are eliminated.

Measurement of Adenylate Cyclase Activity

A number of workers have examined the pharmacological specificity of β-adrenergic receptors linked to adenylate cyclase in broken cell preparations of a variety of tissues. These studies support the concept that β-adrenergic receptors in different tissues have differing pharmacological specificities (Burges & Blackburn 1972, Mayer 1972). For example, the β_2-selective agonists salbutamol and soterenol have been shown to activate adenylate cyclase in membranes prepared from lung but not from cardiac tissue (Burges & Blackburn 1972) or fat cells (Lefkowitz 1975). The β_1-selective antagonists practolol, metoprolol, and para-oxprenolol blocked β-adrenergic-receptor-stimulated adenylate cyclase activity with a greater potency in heart (Lefkowitz 1975, Murad 1973, Petrack & Czernik 1976, Minneman et al 1979a) and adipose tissue (Lefkowitz 1975, Murad 1973) than in liver (Lefkowitz 1975, Murad 1973), trachea (Murad 1973), or lung (Lefkowitz 1975, Minneman et al 1979a). These studies also demonstrate that both β_1- and β_2-receptors are coupled to adenylate cyclase.

Several studies have recently appeared in which the subtype of β-adrenergic receptor existing in a particular tissue or cell type has been classified on the basis of the potency and selectivity of drugs in activating adenylate cyclase. A preliminary approximation of the β-adrenergic receptor subtype present in a given type of cell can be obtained by examining the relative potencies of NE and EPI. Experiments of this type suggest, for example, that S-49 lymphoma cells (Maguire et al 1977) and C-6 glioma cells (Lucas & Bockaert, 1977) contain mainly β_1-receptors, while EH-118 astrocytoma cells (Clark et al 1975) and L6 muscle cells (R. Pittman, unpublished observations) contain a preponderance of β_2-receptors. Adenylate cyclase activity in rat peritoneal macrophages (Ikegami 1977), Ehrlich ascites tumour cells (Onaya et al 1978), and mouse epidermis (Duell 1980) is stimulated by selective β_2-agonists (salbutamol or hexoprenaline), which suggests that these tissues contain β_2-adrenergic receptors.

Radioligand Binding Assays

Several requirements must be satisfied if one is to use radioligand binding assays to study receptor subtypes. The basic technique involves monitoring the displacement of a radolabeled compound, usually an antagonist with nonradioactive agonists or antagonists. The interactions of the labeled and competing drugs with the receptor must be reversible and competitive. It is also important that the amount of radioligand bound be small relative to the total amount of radioligand in the assay so that the gradual displacement of the ligand from its binding sites does not change the concentration of free radioligand. If these requirements are satisfied, the actual equilibrium dissociation constant (K_D) of the displacing drug can be calculated by correcting for the concentration of the radioligand by the method of Cheng & Prusoff (1973). The K_D value of the radioligand should be equal to the ratio of the first order dissociation constant to the second order association constant for the drug/receptor complex. The Cheng & Prusoff (1973) correction has been shown to be experimentally valid for the inhibition of IHYP binding to rat lung membranes by various drugs (Weiland et al 1980). In this tissue it has been shown that the apparent K_D value of a drug determined by inhibition of ^{125}I-IHYP binding as calculated by the method of Cheng & Prusoff is independent of the concentration of the radioligand (Weiland et al 1980).

The presence or absence of guanosine 5'-triphosphate (GTP) is an important consideration in evaluating radioligand binding experiments (see Minneman & Molinoff 1980). GTP is an essential cofactor in the stimulation of adenylate cyclase activity by many hormone receptors (Rodbell 1980), including the β-adrenergic receptor. In addition, GTP affects the affinity of agonists, but not antagonists, for the β-adrenergic receptor (Maguire et al 1976; Lefkowitz et al 1976). Although the interactions of antagonists with the β-adrenergic receptor follow mass-action principles, in some tissues agonists display shallow displacement curves consistent with apparent negative cooperativity (Maguire et al 1976, Lefkowitz et al 1976, Hegstrand et al 1979). This agonist-specific apparent negative cooperativity disappears in the presence of GTP. Inclusion of GTP in the binding assay also leads to a four- to ten-fold reduction in the apparent affinities of agonists. Since stimulation of adenylate cyclase by β-adrenergic receptors requires the presence of GTP, the affinities of agonists measured in the presence of GTP are probably the most meaningful. In addition, the absence of GTP leads to apparently negatively cooperative displacement curves for agonists that are not related to, but can easily be confused with multiple receptor subtypes (see below; Minneman & Molinoff 1980).

The use of radioligand binding assays has provided strong evidence for the existence of two distinct classes of β-adrenergic receptor. These results have generally been consistent with results obtained in studies of receptor-mediated responses in intact organs or of adenylate cyclase activity. In several studies (U'Prichard et al 1978, Rugg et al 1978, Minneman et al 1979a, Coleman et al 1979a, Gibson et al 1979), the pharmacological specificity of β-adrenergic receptors as measured by radioligand binding has been directly compared in tissues thought to contain mostly β_1-receptors (rat heart, calf cerebral cortex, rabbit heart, rabbit lung) with that in tissues thought to contain mostly β_2-receptors (rat lung, calf cerebellum, rat corpus luteum). GTP was not included in any of the binding assays in the studies quoted above. Complexities caused by the absence of GTP in studies with agonists are discussed below (see Minneman & Molinoff 1980). In some studies the results obtained in binding assays have been directly compared to results of studies of adenylate cyclase activation in the same tissue

(Minneman et al 1979a, Coleman et al 1979a). The results confirm the existence of differences in the pharmacological specificity of the β-adrenergic receptors in tissues thought to contain mostly β_1-receptors from that observed in tissues thought to contain mostly β_2-receptors.

U'Prichard et al (1978) have performed an extensive comparison of the β_2-adrenergic receptors in rat lung and calf cerebellum with the β_1-adrenergic receptors in rat heart and calf cerebral cortex. The results suggest that a number of drugs have different affinities for the two subtypes of receptor. However, these experiments were performed in the absence of GTP and sodium. The assay conditions would increase the apparent affinity of the receptor for agonists, but not its affinity for antagonists. In addition, the studies with agonists and antagonists were performed at different temperatures, which has been shown to have selective effects on the affinities of agonists (Weiland et al 1979, 1980). Therefore, the apparent selectivity of drugs that are agonists at one receptor but antagonists at the other (see below) was overestimated. Under the conditions used, the observed selectivity of antagonists would be indicative of their physiological selectivity.

Rugg et al (1978) measured the affinities of the β_1-selective drugs practolol, atenolol, and NE and the β_2-selective drug procaterol (OPC 2009) in rat (β_2) and rabbit (β_1) lung. These authors reported that the β_1-selective drugs were more potent in rabbit lung than in rat lung, whereas OPC 2009 was more potent in rat lung than in rabbit lung.

Minneman et al (1979a) compared the ability of a large series of selective and nonselective drugs to activate or inhibit adenylate cyclase with their ability to inhibit the specific binding of ^{125}I-IHYP to β-adrenergic receptors in rat heart (β_1) and lung (β_2). A similar study was performed by Coleman et al (1979a,b) using ^{3}H-DHA in rabbit heart (β_1) and superovulated rat corpus luteum (β_2). A number of drugs showed selectivity as assessed by their potency in inhibiting IHYP binding. These drugs included β_2-selective agonists (zinterol and salmefamol), which were more potent in the lung than in the heart, and β_1-selective antagonists (practolol, metoprolol and atenolol), which were more potent in the heart than in the lung (Minneman et al 1979a). The β_1-selective antagonists practolol, atenolol, and ICI 89,406 were more potent in inhibiting ^{3}H-DHA binding in heart than in corpus luteum (Coleman et al 1979a) and the β_2-selective antagonist H 35/25 was slightly more potent in the corpus luteum than in the heart. Neither Minneman et al (1979a) nor Coleman et al (1979a,b) included GTP in their binding assays. In the studies of rat heart and lung and rabbit heart, relatively crude tissue preparations were used. These preparations apparently contained a sufficient amount of endogenous GTP to avoid the problems that can arise when the affinities of agonists are determined in the absence of GTP. The absence of endogenous GTP in the partially purified corpus luteal membranes used by Coleman et al (1979a,b) was probably responsible for a consistent ten-fold discrepancy between the affinities of agonists as measured by activation of adenylate cyclase activity and inhibition of ^{3}H-DHA binding.

Gibson et al (1979) compared dissociation constants for drugs in inhibiting the binding of ^{3}H-DHA in rat and rabbit heart and lung. They found the same pharmacological specificity in rat and rabbit heart and rabbit lung (β_1) but a different specificity in rat lung (β_2). β_2-Adrenergic receptors predominate in the lungs of most species. Rabbit lung is unusual in this regard in that it appears to contain mostly β_1-adrenergic receptors.

Comparison of Selectivity of Drugs in Biochemical and Physiological Experiments

All drugs that have been reported to show selectivity for β_1- or β_2-receptors in in vitro biochemical studies of adenylate cyclase activity or radioligand binding assays also show selectivity in physiological experiments with intact tissues or isolated organs. The converse is not true, however, since some drugs that show selectivity in physiological experiments do not show selectivity in studies of adenylate cyclase activity or radioligand binding.

β_1-SELECTIVE AGONISTS   Several agonists including dobutamine, tazolol, and H 133/22 selectively activate inotropic and chronotropic responses in the heart (β_1) but have little or no effect on lung or tracheal smooth muscle (β_2) (Tuttle & Mills 1975, Strosberg 1976, Carlsson et al 1977). However, tazolol and H 133/22 do not activate adenylate cyclase in any tissue thus far examined, including rat heart (Vauquelin et al 1976; A. Hedberg, unpublished results). Tazolol and H 133/22 are relatively potent antagonists of isoproterenol-stimulated adenylate cyclase activity; however, they are equally effective in tissues containing predominately β_1- or β_2-adrenergic receptors (Vauquelin et al 1976; A. Hedberg, unpublished results). Dobutamine does increase adenylate cyclase activity to 30 to 50% of maximal levels; however, this compound is equally potent and efficacious in rat heart and lung (Minneman et al 1979a). Dobutamine and H 133/22 have been shown to be equipotent inhibitors of radioligand binding in tissues containing a preponderance of β_1- or β_2-receptors (Minneman et al 1979a,c). NE, on the other hand, selectively activates adenylate cyclase and inhibits radioligand binding in tissues containing a preponderance of β_1-receptors (Burges & Blackburn 1972, Lefkowitz 1975, Minneman et al 1979a, Gibson et al 1979, Coleman et al 1979a).

β_1-SELECTIVE ANTAGONISTS   There is excellent agreement between the observed selectivity of β_1-antagonists in physiological experiments and the selectivity observed in studies of adenylate cyclase activity and radioligand binding. Practolol, metoprolol, atenolol, and para-oxprenolol are more potent in inhibiting adenylate cyclase or radioligand binding in tissues that contain mostly β_1-receptors than in tissues that contain primarily β_2-receptors. The observed affinities of these drugs for β_1-receptors in radioligand binding assays is generally twenty- to a hundred-fold higher than for β_2-receptors.

β_2-SELECTIVE AGONISTS   The effects of these drugs on adenylate cyclase is generally similar to their effects in intact tissues and isolated organs. However, a number of drugs, such as salbutamol and soterenol, which selectively increase adenylate cyclase activity in tissues that contain predominantly β_2-receptors, have similar affinities for β_1- and β_2-adrenergic receptors in radioligand binding assays (Minneman et al 1979a). Many of these drugs are antagonists at β_1-receptors (Minneman et al 1979a). Since binding assays do not readily distinguish between agonists and antagonists, these agents do not show selectivity in binding assays. Only drugs (such as zinterol, salmefamol, and procaterol) that are relatively weak antagonists at β_1-receptors show an apparent selectivity for β_2-receptors in in vitro binding assays.

β_2-SELECTIVE ANTAGONISTS   The correlation between physiological and biochemical experiments is not nearly as good for β_2-selective antagonists as it is for β_1-selective antagonists. A large number of studies have documented the selectivity of butoxamine and H 35/25 as β_2-selective antagonists in studies of intact organs and isolated tissues. Selectivity ratios of fourteen- to eighteen-fold have been reported for these compounds in carefully controlled physiological experiments (O'Donnell & Wanstall 1979); however, these drugs show little or no selectivity in studies of adenylate cyclase activity or radioligand binding (U'Prichard et al 1978, Nahorski 1978, Minneman et al 1979a,c, Coleman et al 1979a). On the other hand, IPS 339 (Imbs et al 1977) has been shown to be a β_2-selective antagonist in experiments on guinea pig trachea, atria, and vasculature (Imbs et al 1977) and it shows a marked selectivity for β_2-receptors in in vitro binding assays (Minneman et al 1979c). A good correlation was observed for the pA_2 of IPS 339 in guinea pig atria and trachea with its K_D values for β_1- and β_2-receptors, respectively.

Use of Radioligand Binding Assays to Measure Receptor Subtypes

Several methods are now available that use in vitro binding assays to distinguish between and quantify β-adrenergic receptor subtypes. The experimental strategy involves the use of selective drugs with different affinities for β_1- and β_2-receptors in in vitro binding assays (see above). Inhibition of specific radioligand binding by these drugs makes it possible to determine whether all the receptors in a tissue have the same affinity for the selective drug. If more than one population of receptors with different affinities for the drug exists in the tissue, the relative proportion of each binding site and its affinity for the drug can be calculated.

Several basic assumptions are involved in the use of binding assays to study β-adrenergic receptor subtypes. The major assumption is that there are only two subtypes of the receptor. Given the intrinsic limitation in the quality of the data that can be obtained, none of the techniques for data analysis can distinguish more than two or, at most, three classes of sites. The second assumption is that the interactions of drugs (both agonists and antagonists) with each receptor subtype must follow simple mass action principles and display no evidence of positive or negative cooperativity. Available techniques cannot distinguish effects due to complex negatively cooperative behavior from results due to the presence of multiple classes of receptor sites. The calculations are greatly facilitated if the radioligand has the same affinity for both putative receptor subtypes. To address this issue, Scatchard analyses of saturation isotherms of radioligand binding are performed. If the Scatchard plots are linear and the apparent K_D values do not differ significantly in tissues with apparently different populations of receptors, then it is reasonable to assume that the radioligand has the same affinity for each of the receptor subtypes.

Analysis of Heterogeneous Binding Sites

If, under appropriate conditions, the inhibition of radioligand binding by selective drugs shows evidence for more than one binding site with different affinities for the drug, then the relative proportion of the two classes of binding site and their affinities for the drug can be calculated by a mathematical analysis of the shape of the curve. This analysis, which assumes that the interactions of each of the drugs with the receptor follows mass action principles, has been discussed in detail by Minneman & Molinoff (1980).

Evaluation of heterogeneous binding sites can be accomplished either by direct analysis of the log dose response curve for inhibition of radioligand binding by the drug, or by analysis after transformation into a form that would yield a straight line if the drug interacted with a homogeneous population of receptor sites. This modified Scatchard (or Hofstee) analysis shows an apparent curvature if more than one type of receptor with different affinities for the drug is present in the preparation.

Several methods are now used to evaluate experimental data for the possible existence of multiple classes of binding sites. The direct graphic analysis of nonlinear modified Scatchard (Hofstee) plots (plot of bound vs bound/drug concentration; see Hofstee, 1952), introduced by Rugg et al (1978), is attractive in its simplicity. If drugs were available that were entirely selective for β_1- or β_2-receptors, this method would yield accurate results. Unfortunately, the selective drugs currently available have only twenty- to a hundred-fold selectivity. Use of this method therefore results in a significant overestimate in the number of high affinity sites and an underestimate of the number of low affinity sites (Minneman & Molinoff 1980). Since these errors are in opposite directions, the error in the ratio of the concentration of high affinity sites to low affinity sites is magnified.

A similar modified Scatchard (Hofstee) analysis was developed by Minneman et al (1979b). These authors subjected transformed data to a computerized iterative analysis. This approach corrects each component for the contribution of the other component, which results in an accurate analysis of theoretical data regardless of the degree of selectivity of the competing ligand (Minneman et al 1979b). An advantage of this method is that it combines the intuitive simplicity of the Hofstee plot (where the characteristics of the two components can be easily visualized as the asymptotes of the biphasic curve) with a mathematically accurate analysis. The major limitation of the method derives from its use of unweighted linear regression. Furthermore, the transformation results in the propagation of errors onto both the x and the y coordinates.

Biphasic Hofstee plots could reflect negative cooperativity rather than two distinct receptor subtypes, but several lines of evidence suggest that this is not the case. First, only drugs known to have different affinities for β_1- and β_2-receptors show curvilinear Hofstee plots (Minneman et al 1979b,c). Other drugs that have similar affinities for the two classes of receptor result in linear Hofstee plots. Furthermore, all drugs that had been shown to selectively inhibit ligand binding to membranes rich in β_1- (rat heart) or β_2-(rat lung) receptors, showed this behavior (Minneman et al 1979b, Hedberg et al 1980). Third, in tissues containing only one receptor subtype the interaction of both selective and nonselective drugs resulted in linear Hofstee plots with Hill coefficients of 1.0 (Minneman et al 1979c).

Hancock et al (1979) have analyzed displacement curves directly using a computer-aided nonlinear least squares curve-fitting technique. Since this method uses untransformed data, errors are not expressed on the abscissa. For statistical reasons, this curve-fitting technique is more accurate than the iterative graphic analysis described by Minneman et al (1979b,c). However, this method loses the intuitive simplicity inherent in the Hofstee analysis, and the computer analysis must be taken at face value.

All three published methods require data that is essentially error free. In practice, it is usually necessary to perform duplicate or triplicate determinations at each of a large number of concentrations of each competing drug. Furthermore, it is usually necessary for experiments to be repeated a significant number of times if reliable data is to be obtained. For further discussion of these problems see Minneman & Molinoff (1980).

Effects of GTP on Agonist Interactions with β-Adrenergic Receptors

Complexities inherent in the interactions of agonists with β-adrenergic receptors require that special precautions be taken to obtain meaningful results when agonists are used for the determination of receptor subtypes. In many tissues agonist inhibition of antagonist binding is associated with low Hill coefficients (0.6 to 0.7) consistent with apparent negative cooperativity (Maguire et al 1976, Lefkowitz et al 1976, Hegstrand et al 1979). This apparent negative cooperativity is not related to and should not be confused with the coexistence of β_1- and β_2-receptors in the same tissue. This agonist-specific apparent negative cooperativity disappears in the presence of GTP. Thus, in the presence of GTP, the interaction of nonselective agonists with the β-adrenergic receptor follows the principles of mass action. On the other hand, in the absence of GTP the interaction of nonselective drugs can be erroneously interpreted as reflecting binding to multiple subtypes of β-adrenergic receptor.

Tissue preparations used for studies of radioligand binding are usually washed extensively to remove endogenous hormone. This also removes most of the endogenous GTP (Lefkowitz et al 1976, Maguire et al 1976, Hegstrand et al 1979). It is therefore important to include GTP in binding assays when selective agonists are used to distinguish receptor subtypes. Since GTP affects both β_1- and β_2-receptors (Hegstrand et al 1979), failure to include GTP (Barnett et al 1978, Hancock et al 1979, U'Prichard et al 1980) can lead to forms of either or both receptor subtypes that have high and low affinities for agonists. Thus, tissues that contain only β_1- or only β_2-adrenergic receptors will erroneously be thought to contain both subtypes of receptor. Tissues that contain both β_1- and β_2-adrenergic receptors could appear to contain up to four forms of the receptor. At the present time such data cannot be analyzed in a meaningful way.

Criteria for Classification of β-Adrenergic Receptor Subtypes

The fundamental criterion in the delineation of receptor subtypes is that the properties of the receptor must be conserved. This means that the pharmacological specificity of a particular receptor subtype must be identical in every tissue examined. For example, the properties of a β_1-receptor in the heart must be the same as the properties of a β_1-receptor in the brain, in adipose tissue, or even in a tissue containing a preponderance of β_2-adrenergic receptors, such as the rat lung or cerebellum. If the pharmacological properties of a particular receptor subtype were to depend on the tissue in which the receptor is found, then receptors should not be classified by examining their pharmacological properties. Fortunately, the ready availability of a variety of drugs makes it possible to rigorously define the pharmacological properties of β-adrenergic receptor subtypes.

Co-existence of β_1- and β_2-Receptors in a Single Tissue

As discussed above, there is excellent evidence from pharmacological studies performed with isolated organs and intact tissues (Carlsson et al 1972, 1977, Furchgott 1975, O'Donnell & Wanstall 1979) and from radioligand binding studies (Barnett et al 1978, Minneman et al 1979b) to support the concept that β_1- and β_2-receptors can coexist in a single organ. In a variety of tissues the inhibition of radioligand binding by nonselective drugs follows simple mass-action principles (Rugg et al 1978, Minneman et al 1979b); however, in these same tissues inhibition of radioligand binding by β_1- and β_2-selective drugs is more complex. In rat heart, lung, spleen, cerebral cortex, cerebellum, caudate, hippocampus and diencephalon, rabbit lung, and right atria from cat and guinea pig heart, the inhibition of specific IHYP and/or DHA binding by β_1 or β_2 selective drugs results in biphasic Hofstee plots, which is indicative of two binding sites with different affinities for the selective drug (Barnett et al 1978, Minneman et al 1979b, Nahorski et al 1979a, Hedberg et al 1980, Pittman et al 1980a). Analysis of these data can provide an estimate of the relative proportion of β_1- and β_2-receptors in the tissue and the affinity of each receptor for each selective drug. The relative proportions of the two receptor subtypes range from 85% β_1 (rat heart) to 98% β_2 (cerebellum from mature rats).

No data presently available directly address the question of whether the two subtypes of β-adrenergic receptor can coexist in a single cell, or whether the coexistence of these receptors within an organ is due simply to the heterogeneity of cell types within the organ. In recent experiments carried out in a number of laboratories, several presumably homogeneous populations of cells have been studied with respect to the presence of β_1- and/or β_2-receptors. These cells were usually shown to contain only β_1- or β_2-receptors. Thus, astroglia cultured from neonatal rat cerebral cortex contain only β_1-receptors (T. K. Harden and K. D. McCarthy, unpublished observations), whereas human and rat lymphocytes (Aarons et al 1980) and L-6 myoblasts (R. N. Pittman and P. B. Molinoff, unpublished observations) contain homogeneous populations of β_2-adrenergic receptors. On the other hand, C-6 glioma cells have been reported to have both β_1- and β_2-receptors (J. Bockaert, unpublished observations).

Evidence for the Existence of Only Two Subtypes of β-Adrenergic Receptor in Mammalian Tissues

Although it is impossible to prove that there are only two subtypes of β-adrenergic receptor, the available evidence strongly supports this conclusion. The fact that the pharmacological specificity of each receptor subtype is highly conserved in the large number of mammalian tissues that have been examined is good evidence for the existence of only two subtypes of β-adrenergic receptor in mammalian tissues (Minneman et al 1979c). Furthermore, the relative proportion of the two receptor subtypes in a single tissue is the same regardless of whether a β_1- or β_2-selective agonist or antagonist is used for the determination (Minneman et al 1979b). The results are thus internally consistent. The existence of a variable amount of a third subtype would be expected to yield discrepancies in K_D values or in the calculated proportion of each subtype. It would be difficult, however, to identify a third subtype if its properties were very similar to those of either β_1- or β_2-adrenergic receptors.

Tissues Containing Only One Receptor Subtype

The strongest evidence that there are only two subtypes of β-adrenergic receptor comes from studies of tissues containing only one receptor subtype (Minneman et al 1979c). In these tissues, the inhibition of specific IHYP binding by β_1- and β_2-selective drugs results in linear Hofstee plots, which indicates that there is only a single class of binding sites (Minneman et al 1979c). In the left ventricle of cat and guinea pig heart there appears to be a homogeneous population of β_1-receptors, whereas rat liver and cat soleus muscle contain only β_2-receptors (Minneman et al 1979c, Hedberg et al 1980). In these tissues it is possible to demonstrate directly that the interaction of each drug (including the selective drugs) with a single receptor subtype follows the principles of mass action, yielding linear Hofstee plots and Hill coefficients of 1.0. In addition, it was possible to show that the affinities of each receptor for each drug in tissues that contain only one receptor subtype are the same as the affinities calculated from the computer-aided graphic analysis of biphasic Hofstee plots (Minneman et al 1979c). This evidence strongly suggests that there are only two subtypes of β-adrenergic receptor in mammalian tissues.

β-Adrenergic Receptors in Non-mammalian Tissues

In some non-mammalian tissues the kinetic properties and pharmacological specificity of β-adrenergic receptors are distinct from those of mammalian β_1- or β_2-adrenergic receptors. The β-adrenergic receptor of the turkey erythrocyte has been compared to mammalian β_1- and β_2-receptors (Gibson et al 1979, Minneman et al 1980a). Although turkey erythrocytes contain an apparently homogeneous population of β-adrenergic receptors, these receptors have major kinetic and pharmacological differences that distinguish them from mammalian β_1- or β_2-receptors. Characteristics of the β-adrenergic receptors in frog erythrocytes (Williams & Lefkowitz 1978) and hearts (Hancock et al 1979) also appear to be different from those observed for mammalian β_1- or β_2-receptors (Minneman et al 1979c). In any case, extrapolation of studies of β-adrenergic receptors in non-mammalian tissues to those in mammals should be approached with caution.

LOCALIZATION OF β-ADRENERGIC RECEPTORS

Distribution of β-Adrenergic Receptor Subtypes in Cat and Guinea Pig Heart

As discussed above, physiological evidence suggests that β_1- and β_2-receptors may coexist in the right atria of cat heart and that both subtypes are involved in the positive chronotropic effects of catecholamines. However, only β_1-receptors seem to be involved in the positive inotropic effect of catecholamines. Hedberg et al (1980) studied the distribution of β_1- and β_2-receptors in right atrium and left ventricle of cat and guinea pig hearts. Although the left ventricles of both species contained an apparently homogeneous population of β_1-receptors, the right atria contained both β_1- and β_2-receptors in a ratio of approximately 3 to 1 (Hedberg et al 1980). In the cat heart, these results are in aggreement with physiological data, which suggests that only β_1-receptors mediate the inotropic response in the left ventricle, but that both β_1- and β_2-receptors mediate the chronotropic response in the right atria (Carlsson et al 1972, 1977). In the guinea pig heart the results are less clear. The biochemical results are similar to those obtained in the cat (Hedberg et al 1980). On the other hand, physiological evidence with regard to a possible role for β_2-receptors in mediating the chronotropic response to catecholamines in the guinea pig (see above) remains controversial. It is possible that β_2-receptors, present in the right atrium of guinea pig hearts, are not functionally coupled to the chronotropic response in this species.

Distribution of β-Adrenergic Receptor Subtypes in Rat Brain

The distribution of β-adrenergic receptors in regions of the mammalian CNS has been extensively examined. Several groups of investigators (Alexander et al 1975, Sporn & Molinoff 1976, Bylund & Snyder 1976) have reported that β-adrenergic receptors are distributed relatively homogeneously in various regions of the CNS. This is surprising in that if β-adrenergic receptors are associated with noradrenergic neuronal function, one would expect them to be assymetrically distributed. Furthermore, the distribution of β-adrenergic receptors in the CNS bears no apparent relationship to the catecholamine content of particular brain regions. Some regions containing a high content of NE, such as cerebral cortex, also have a high density of β-receptors. Other regions such as the caudate nucleus, where there is little or no detectable NE, also contain a high density of β-receptors. Based on these findings, Sporn & Molinoff (1976) suggested that some of the β-adrenergic receptors in the CNS might be associated with glial rather than neuronal elements. It is also possible that some of the β-adrenergic receptors in the CNS are associated with cerebral blood vessels. It was therefore of interest to determine the distribution of the two subtypes of β-adrenergic receptor in regions of the mammalian CNS (Minneman et al 1979b, Minneman & Molinoff 1979). It seemed possible that the density of one or the other subtype might more closely parallel the noradrenergic innervation, the density of glial elements, or the degree of vascularization of various brain regions. Although no clear correlations were observed, the results disclosed an interesting pattern. The density of β_2-adrenergic receptors in different brain regions varied by only two- to three-fold. The highest density was observed in the cerebellum and the lowest in the hippocampus and diencephalon. However, depending on the age of the animals, the ratio of the densities of β_1-receptors in cerebral cortex to β_1-receptors in cerebellum ranged up to 100 to 1 (Minneman et al 1979b, Pittman et al 1980a). These results suggested that β_1-receptors might be located on heterogeneously distributed populations of neurons, while β_2-receptors might be associated with more homogeneously distributed cellular elements such as glia or blood vessels. There was no correlation, however, between the density of β_1-receptors and the tissue content of NE. For example, the density of β_1-receptors was relatively high in the caudate, which contains little or no NE. The possible significance of this finding is discussed below.

Strategies for Examining Cellular Localization of Receptors

The cellular complexity of many organs makes it difficult to interpret the results obtained when tissue homogenates are assayed for their content of receptors. This problem is particularly acute in studies of the central nervous system where catecholamine receptors may be associated with a variety of cellular elements. Differences in the regional localization of β_1- and β_2-adrenergic receptors in the mammalian brain suggest that different types of receptors may be associated with different cell types.

The complexities that may arise from the association of adrenergic receptors with both neuronal and non-neuronal elements can be circumvented through the use of histological techniques. Melamed and collaborators (1976) have described an apparently well-defined yellow band in the region of the Purkinje cell layer in the cerebellar cortex following the intravenous administration of a fluorescent compound, 9-amino-acridyl-propranolol; however, Hess (1979) has suggested that this finding is due to autofluorescence and is unrelated to the specific binding of the compound to β-adrenergic receptors. More recently an autoradiographic approach for the localization of β-adrenergic receptors has been developed (Palacios & Kuhar 1980). ^{3}H-DHA binding to β-adrenergic receptors on tissue sections mounted on glass slides was studied. High densities of receptors were found in the superficial layers of the cerebral cortex and in the molecular layer of the cerebellum. The use of selective drugs to inhibit the binding of ^{3}H-DHA should make it possible to localize $β_1$- and $β_2$-adrenergic receptors (M. Kuhar, unpublished observations).

A number of techniques are available for examining the cellular localization of β-adrenergic receptor subtypes in the brain. Tissue fractions enriched with respect to neurons or glia or particular neuronal types can be obtained using biochemical fractionation techniques (Johnson & Sellinger 1971, Sellinger et al 1974). Capillary-enriched fractions can be isolated by methods described by Joó & Karnushina (1973) or Goldstein and his collaborators (1975). Alternative techniques involve the use of selective neurotoxins, e.g., the intraventricular administration of 6-hydroxydopamine (Thoenen & Tranzer 1968), which results in the destruction of noradrenergic terminals. The administration of kainic acid (Coyle & Schwarcz 1976, McGeer & McGeer 1976) appears to result in the loss of cell bodies with relatively little effect on nerve terminals. If neurotoxin administration results in the loss of a particular class of receptor sites, then it is reasonable to conclude that the receptors are associated with those components selectively destroyed by the toxin.

It has also been documented in a variety of systems that removal of the normal neuronal input to a tissue results in a variety of compensatory changes which frequently include an increase in the density of receptors. Thus, the intraventricular administration of 6-hydroxydopamine to adult rats results in an increase in the density of β-adrenergic receptors in the cerebral cortex (Sporn et al 1976), which suggests that these receptors are associated with postsynaptic neurons that have lost their normal endogenous input. Administration of β-adrenergic receptor antagonists like propranolol can also lead to an increase in the density of β-adrenergic receptors in animals (Glaubiger & Lefkowitz 1977, Wolfe et al 1978), as well as in man (Aarons et al 1980). If the administration of an antagonist or toxin results in an increase in the density of receptors, then it is likely that these receptors are normally tonically activated and that this tonic activation has been inhibited by the administration of the agent.

Several other research paradigms result in preparations in which one or more cell types have been eliminated. Altman & Anderson (1973) described experiments in which neonatal rat pups were subjected to X-irradiation, destroying cerebellar interneurons but sparing early-developing Purkinje cells. The availability of mutant mice deficient in one or another type of cerebellar neuron may provide another useful model. Mice that lack either cerebellar Purkinje ("nervous") or granule ("weaver," "staggerer") cells can be studied (Sidman et al 1965). Many of these techniques have proven to be useful in preliminary examination of the cellular localization of β-adrenergic receptor subtypes in rat brain (see below).

Cerebral Cortex

SUPERSENSITIVITY   Increases in the sensitivity of tissues to stimulation by agonists caused by decreases in the endogenous input (denervation supersensitivity) have been shown to occur in the central nervous system. This phenomenon involves both pre- and postsynaptic components (Kalisker et al 1973, Sporn et al 1976). The presynaptic component is apparently due to the loss of reuptake sites for NE, while the postsynaptic component is at least partially due to an increase in the density of receptors. The intraventricular administration of 6-hydroxydopamine results in a decrease in the EC_{50} value for NE-stimulated cyclic AMP accumulation in slices of rat brain. This decrease occurs concomitantly with the degeneration of the presynaptic terminals (1 to 2 days). A slowly developing increase in the density of β-adrenergic receptors (postsynaptic component) was shown to occur concomitantly with an increase in the maximum stimulation of cyclic AMP accumulation by catecholamines (Sporn et al 1976). With the development of methods for the quantitative determination of β_1- and β_2-adrenergic receptor subtypes it was of interest to reinvestigate the effects of 6-hydroxydopamine on the density of receptors in rat cerebral cortex. The ability of the β_2-selective agonist zinterol to inhibit IHYP binding was assessed in control animals and in rats that had been treated with 6-hydroxydopamine on postnatal days one to four. The administration of 6-hydroxydopamine led to a marked increase in the density of β_1-adrenergic receptors but had no effect on the density of β_2-receptors (Minneman et al 1979d). These findings suggest that the β_1-adrenergic receptors in the cerebral cortex normally receive a noradrenergic input, whereas β_2-adrenergic receptors may not be normally innervated.

The selective increase in the density of β_1-receptors observed in rat cerebral cortex following destruction of noradrenergic neurons (Minneman et al 1979d) has subsequently been confirmed by U'Prichard et al (1980). However, methodological considerations raise doubts about the validity of this study. One of the two drugs used by U'Prichard et al (1980) to distinguish between β_1- and β_2-receptors was salbutamol. This drug has been shown to be a selective β_2-agonist in physiological studies and in studies of adenylate cyclase activity; however, salbutamol is an agonist at β_2-receptors and a potent antagonist at β_1-receptors. It shows no selectivity between β_1- and β_2-receptors in binding studies performed with either ^{3}H-DHA or ^{125}I-IHYP (Rugg et al 1978, Minneman et al 1979a,c). The apparent selectivity of salbutamol in the study of U'Prichard et al (1980) may have been due to the fact that the binding assays were carried out in the absence of GTP and sodium (see above). Furthermore, the affinity of salbutamol was twenty-five-fold higher for β_1-receptors in the cortex than for β_1-receptors in the cerebellum. This discrepancy raises further doubts as to the validity of this study.

SUBSENSITIVITY   Administration of pharmacologic agents that lead to an increase in neurotransmitter availability and therefore in receptor occupancy have been associated with decreased responsiveness in the brain (Vetulani & Sulser 1975, Wolfe et al 1978). The decrease in isoproterenol-stimulated cyclic AMP accumulation following the chronic administration of antidepressants has been shown to correlate with a similar decrease in the density of β-adrenergic receptors (Banerjee et al 1977, Wolfe et al 1978). These changes have been observed following the administration of drugs that inhibit norepinephrine reuptake, such as desmethylimipramine (DMI); drugs that inhibit monoamine oxidase activity, such as pargyline; and drugs that have been reported to increase catecholamine-stimulated release of norepinephrine, such as iprindole (Hendley 1978). The effect of the tricyclic antidepressant DMI on the densities of β_1- and β_2-adrenergic receptors in rat cerebral cortex has been investigated (Minneman et al 1979d). The administration of DMI to adult rats for ten days resulted in a marked decrease in the density of β_1-receptors in the cerebral cortex, with no effect on the density of β_2-receptors (Minneman et al 1979d).

Since neurally active drugs such as 6-hydroxydopamine and DMI affect only β_1-receptors in the cerebral cortex, it seems reasonable to conclude that β_1-receptors are the neuronally innervated β-receptor in this tissue. The endogenous input to the β_2-receptors in the cerebral cortex and the cellular localization of these receptors remain open to question.

EFFECTS OF STRESS   A variety of stresses are known to alter levels of neurotransmitters in the CNS (see Usdin et al 1976). Decreases in the concentration of NE (Bliss et al 1968, Kobayashi et al 1976) and increases in NE turnover (Gordon et al 1966, Thierry et al 1968) following various types of acute and chronic stress have been reported. Alterations in CNS activity accompanying stress may also lead to changes in the density of either or both subtypes of β-adrenergic receptor. Since catecholamines are not only neurotransmitters but are also adrenal hormones released in large quantities during stress (Kvetnansky & Mikulaj 1970, Popper et al 1977), it is possible that stress could affect β-adrenergic receptors in the brain that are outside of the blood brain barrier. For example, high blood levels of catecholamines accompanying stress may lead to a decrease in the density of β-receptors located on the peripheral side of the cerebral vasculature. The effects of stress on the density of β-receptors in rats that were immersed in water and exposed to a temperature of 4° C for 1½ hr daily for one week were determined. This particular form of intermittent stress has been shown to decrease NE levels in the rat heart by 40% after a single stress (Nelson & Molinoff 1976). No changes in the density of either β_1- or β_2-receptors were observed, however, in the cortex or caudate of either acutely or chronically stressed animals compared to controls (R. N. Pittman, K. P. Minneman, and P. B. Molinoff, unpublished observations).

EFFECTS OF ADRENALECTOMY   As discussed above, β_2-adrenergic receptors are much more sensitive to EPI than to NE. An attractive possibility was that cerebral β_2-receptors might be associated with cerebral blood vessels on the peripheral side of the blood brain barrier. If this were the case, the endogenous input to these receptors could be circulating EPI released from the adrenal. Since the density of β-receptors in a tissue seems to be regulated by the amount of the endogenous input to the receptors (Sporn et al 1976, Wolfe et al 1978), it seemed possible that removal of the adrenal might selectively affect the densities of β_1- or β_2-receptors in a given brain area. No change was observed, however, in the density of either β_1- or β_2-receptors in the cerebral cortex or cerebellum or in the pharmacological specificity of these receptors following adrenalectomy (R. N. Pittman, K. P. Minneman, and P. B. Molinoff, unpublished observations). These experiments suggest that if β-receptors are associated with brain capillaries, then (a) the receptors are located on the "brain side" rather than on the "blood side" of capillaries and are therefore not affected by circulating adrenal catecholamines, (b) the density of these receptors is not regulated by circulating catecholamines, or (c) the β-receptors associated with capillaries do not constitute a significant proportion of the total number of β_1-receptors present in the cerebral cortex or cerebellum.

β-ADRENERGIC RECEPTORS ON BRAIN CAPILLARIES   A number of investigators have examined the possibility that β-adrenergic receptors may be associated with blood vessels in the brain. These receptors could be involved in the control of cerebral blood flow. Most non-cardiac peripheral blood vessels contain β_2-receptors mediating vasodilation (Furchgott 1972). It was therefore surprising to find that in the brain, studies of the pharmacological characteristics of receptors mediating catecholamine-induced vasodilation suggest that these receptors are of the β_1-subtype (Edvisson & Owman 1974, Sercombe et al 1977). The results indicate that there are substantial effects on blood flow in the caudate nucleus, whole brain, and middle cerebral pial artery following stimulation of β_1-adrenergic receptors.

Several recent studies have attempted to study directly the β-adrenergic receptors on cerebral blood vessels (Herbst et al 1979, Nathanson & Glaser 1979). Pial and/or cerebral microvessels have been isolated and catecholamine-stimulated adenylate cyclase activity has been measured. Unfortunately, insufficient pharmacological information was included to permit classification of these responses as being mediated by either β_1- or β_2-adrenergic receptors.

Friedman & Davis (1980) have examined the characteristics of the binding of ^{3}H-DHA to porcine pial membranes; they have shown that the β-receptors in this tissue are mostly of the β_2-subtype, although some β_1-receptors were also present. These results are difficult to interpret with regard to the receptors on blood vessels because whole pia was used and no attempt was made to fractionate or purify blood vessels.

To determine whether β-receptors associated with capillaries in the central nervous system comprise a significant proportion of the total number of β_1- or β_2-receptors, the density of β-adrenergic receptors in capillary preparations from cerebral corticies of 30-day-old rats was determined. A modification of the capillary isolation technique of Goldstein et al (1975) was used. Histological examination of the purified tissue preparation revealed an almost homogeneous population of capillaries. Results from three separate capillary preparations consisting of cortices from 73 rats indicated that the density of receptors in this purified tissue fraction was approximately one-sixth of that observed in unpurified homogenates of cerebral cortex. Due to the low density of receptors, these preparations were not examined for the subtype of β-receptor present.

Similar results have been obtained by Peroutka et al (1980). These authors measured specific DHA binding to microvessels purified from bovine cerebral cortex and found that this tissue preparation contains a low density of β-adrenergic receptors. Comparison of the affinities of EPI and NE for the β-receptors in this preparation suggested that they are of the β_1-subtype (Peroutka et al 1980).

Cerebellum

The noradrenergic input to the cerebellum arises from the locus coeruleus, and the NE-containing presynaptic terminals of these neurons have been shown to synapse primarily in the anatomically defined Purkinje and molecular layers in the cerebellum (Bloom et al 1972). Little or no catecholamine histofluorescence is observed in the granule cell layer. It has been shown using eletrophysiological techniques that the spontaneous or gluta-mate-induced firing of Purkinje cells is substantially inhibited by the ionto-phoretic application of β-adrenergic agonists (Siggins et al 1969). Evidence has been presented that the actions of β-agonists on Purkinje cells are mediated through increased levels of cyclic AMP (Hoffer et al 1971); how-ever, the uncertain nature of drug concentrations following microionto-phoresis has made it difficult to determine whether these receptors are β_1- or β_2-adrenergic receptors.

Although most regions of the central nervous system contain mainly β_1-receptors, the β-adrenergic receptors in the cerebellum are predomi-nantly of the β_2-subtype (U'Prichard et al 1978, Minneman et al 1979b). In the cerebellum, as in most other brain areas, the major adrenergic neurotransmitter is NE. Since β_2-receptors have a relatively low affinity for NE, it seems unlikely that NE acts as the natural agonist at these receptors. On the other hand, some β_1-receptors are found in the cerebellum and, depending on the age of the animal, they comprise from 18% (10-day-old rats) to 2% (180-day-old rats) of the total number of β-adrenergic receptors (Pittman et al 1980a; see below). In view of the findings of Minneman et al (1979b), who showed that the β-adrenergic receptors in rat cerebral cortex that are controlled by the noradrenergic input are of the β_1-type, it seemed possible that the small population of β_1-receptors in the cerebellum might be the neuronally innervated receptors in this tissue.

One approach that has been used to obtain information on specific cell populations in the cerebellum has been to subject neonatal rats during the first two weeks of life to intermittent X-irradiation focused on the cerebel-lum (Altman & Anderson 1973). Neonatal X-irradiation results in a 70% loss in cerebellar mass in adult rats. This procedure destroys late maturing cells such as the granule, basket, and stellate interneurons but spares the early maturing Purkinje neurons. Purkinje cells from X-irradiated rats have been examined both histologically and electrophysiologically. They appear to have properties similar to those of Purkinje cells from untreated rats (Woodward et al 1974). Measurement of the total number of β-adrenergic receptors per cerebellum in 6 or 12-week-old animals subjected to neonatal X-irradiation revealed an 80% decrease in the total number of β-adrenergic receptors per cerebellum (Minneman et al 1980b). Analysis of receptor subtypes showed that this loss was entirely due to a decrease in the number of β_2-adrenergic receptors. The number of β_1-receptors was not different from that in control animals (Minneman et al 1980b).

These results suggest that the β_1-receptors in the cerebellum, although comprising only 3 to 5% of the total number of β-receptors in the cerebel-lum of mature rats, may be the receptors on the Purkinje neurons that receive the noradrenergic input to this tissue. Although there are very few β_1-receptors in an adult rat cerebellum (about 0.8 fmol/mg protein or 18 fmol/cerebellum) the number of Purkinje cells is also quite low (about 300,000 cells/cerebellum; Armstrong & Schild 1970). If all the β_1-receptors in the cerebellum were on Purkinje cells, there would be approximately 36,000 receptors/cell. This is a plausible estimate in view of the large size of these cells. To further clarify the cellular location of β-adrenergic receptor subtypes in the cerebellum, various treatments affecting the catecholaminergic input to the cerebellum were employed. Administration of 6-hydroxydopamine to neonatal rats results in an initial decrease in cerebellar NE content and a subsequent rebound to supernormal levels in adult rats (Jonsson et al 1974). This increased NE content was associated with a 15 to 20% decrease in the density of cerebellar β-adrenergic receptors (Harden et al 1979; K. P. Minneman, B. B. Wolfe, and P. B. Molinoff, unpublished observations). Since at this age most (95%) of the β-receptors in the cerebellum are of the β_2-subtype, this effect must involve a decrease in β_2-receptors (although β_1-receptors may also be affected). These experiments demonstrate that the density of β_2-receptors is decreased by treatments that lead to sprouting of catecholaminergic neurons. The low density of β_1-receptors in adult rat cerebellum has made it difficult to determine whether or not these receptors are subject to a similar form of control.

The effect of decreases in the noradrenergic input in the cerebellum has also been examined (Wolfe et al 1981). Destruction of cerebellar noradrenergic nerve terminals following the administration of 6-OHDA to adult rats results in the loss of 95% of the NE from the cerebellum. This treatment caused a three-fold increase in the density of cerebellar β_1-receptors, but no change in the density of cerebellar β_2-receptors (Wolfe et al 1980). Similarly, chronic treatment with the β-adrenergic receptor antagonist propranolol caused a three-fold increase in the density of β_1-receptors in the cerebellum, with no change in the density of β_2-receptors (Wolfe et al 1980). These experiments suggest that β_1-receptors receive a tonic noradrenergic input in the cerebellum but that β_2-receptors do not.

In summary, the effect of neonatal X-irradiation on cerebellar β-receptors is consistent with the hypothesis that β_1-receptors are on cerebellar Purkinje cells. Pharmacological studies indicate that cerebellar β_1-receptors probably receive a tonic noradrenergic input, while the number of β_2-receptors (and possibly β_1-receptors) are decreased by pharmacological manipulations that lead to sprouting of noradrenergic neurons.

Caudate

The density of receptors in the rat corpus striatum, which contains little or no measurable EPI or NE, is similar to the density of β-adrenergic receptors in the cerebral cortex, which contains a dense noradrenergic innvervation (Alexander et al 1975, Sporn & Molinoff 1976). The possible function of receptors in a tissue lacking a relevant endogenous input remains an intriguing question.

Several studies that have attempted to address this question have recently appeared. One approach has been to examine the effect of administering neurotoxins. Direct microinjection of kainic acid, a rigid analogue of glutamate, into the rat caudate leads to the destruction of neuronal cell bodies and dendrites in the vicinity of the injection site, while seeming to spare non-neuronal elements and neuronal processes that arise from cell bodies not located in the immediate vicinity of the injection site (Coyle & Schwarcz 1976, McGeer & McGeer 1976). This technique has been used by a number of investigators to destroy intrinsic striatal neurons. Measurement of specific receptors or receptor-linked cyclic AMP systems following kainic acid treatment can thus provide information as to whether these receptors are located on intrinsic striatal neurons.

The effects of kainic acid lesions have been examined on both β-adrenergic receptor-linked cyclic AMP production in slices of rat caudate and on β-adrenergic receptor density measured by radioligand binding techniques (Minneman et al 1978, Zahniser et al 1979, Nahorski et al 1979b). Destruction of intrinsic striatal neurons by kainic acid does not decrease, and in fact potentiates, β-adrenergic receptor-mediated cyclic AMP production in striatal slices (Minneman et al 1978, Zahniser et al 1979). In one study the density of β-receptors determined by radioligand binding assays did not change following the intrastriatal administration of kainic acid (Zahniser et al 1979). In this study, the effect of the lesion was documented by the occurrence of marked decreases in the density of binding sites for ^{3}H-spiroperidol and ^{3}H-quinuclidinylbenzoate and by decreases in the activities of phosphodiesterase, glutamic acid decarboxylase, and choline acetyltransferase (Zahniser et al 1979). In another study (Nahorski et al 1979b), a 23% decrease in the number of β-receptors was observed in the caudate following kainic acid administration. Under certain conditions, however, kainic acid can cause a nonselective lesion. In the experiments where kainic acid administration led to a loss of β-adrenergic receptors, biochemical controls were not performed to validate the success of the lesion or to rule out the possibility of nonspecific damage. It is therefore possible that the observed decrease in the density of β-receptors was due to generalized tissue damage. Nahorski et al (1979b) reported that the decrease in the density of β-adrenergic receptors was exclusively due to a decrease in the number of β_1-adrenergic receptors; however, the authors used only a single concentration of the selective drug, atenolol, to estimate the proportion of β_1- and β_2-receptors. The validity of this analysis is therefore open to question.

The results of Zahniser et al (1979) suggest that neither β_1- nor β_2-receptors are located on intrinsic neurons in the caudate. As discussed above, physiological evidence suggests that some β_1-receptors are associated with blood vessels in this tissue. β-adrenergic receptors have also been shown to exist on glial tumor cells (Clark & Perkins 1971, Gilman & Nirenberg 1971). The functional role of β_1- or β_2-receptors on neurons, blood vessels, and various types of glia in the caudate is a subject for future investigation.

DEVELOPMENT OF β-ADRENERGIC RECEPTORS AND EFFECTS OF AGING

The development of β-adrenergic receptors has been studied in the rat heart, cerebral cortex, and cerebellum, and in the mouse heart (Harden et al 1977a; Chen et al 1979, Pittman et al 1980a). The results suggest that the development of the β-adrenergic receptor permits the expression of iso-proterenol-stimulated adenylate cyclase activity in the cerebral cortex (Harden et al 1977a), and leads to an isoproterenol-stimulated chronotropic response in the heart (Chen et al 1979). In the rat cerebral cortex, adenylate cyclase activity was present prior to the appearance of measurable levels of β-adrenergic receptors, but cyclic AMP accumulation was not affected by catecholamines (Harden et al 1977a). The presence or absence of a normal presynaptic input did not influence the time of appearance of adrenergic β-receptors or of receptor-stimulated cyclic AMP accumulation (Perkins 1975, Harden et al 1977b), although it had major effects on receptor density and on maximal levels of catecholamine-stimulated cyclic AMP accumulation (Kalisker et al 1973, Perkins 1975, Harden et al 1977b).

The ontogeny of β_1- and β_2-receptors has been studied in rat cerebral cortex and cerebellum (Pittman et al 1980a). The cortex was shown to contain primarily β_1-receptors and the cerebellum primarily β_2-receptors throughout development. In both brain areas, β_1-receptors reached their maximum density earlier in development than did β_2-receptors. Neither population of receptors appeared to be uniquely correlated with the on-togeny of specific cellular constituents, although the development of β_2-receptors seemed to more closely parallel gliogenesis and vascularization than neurogenesis. The early development of β_1-receptors was consistent with their being located primarily on neuronal elements. β_1- and β_2-receptors in the same brain area followed different developmental patterns, which suggests that they may be associated with different cellular elements responsive to different developmental cues.

The ontogeny of β_1-receptors was particularly interesting in the cerebellum. In this tissue the total number of β_1-receptors per cerebellum increased up to day 28 and then decreased to approximately 25% of this number in three-month-old animals (Figure 4). The number of β_1-receptors then began increasing again, and in 14-month-old rats there were 3½ times more β_1-receptors than were present in 3 to 6-month-old animals (Figure 4; Pittman et al 1980b).

The 350% increase in the number of cerebellar β_1-receptors in aging animals was unexpected. Decreases in the density of β-adrenergic receptors have in fact been shown to occur in various brain areas of aging animals, including the cerebellum (Greenberg & Weiss 1978, Maggi et al 1979, Weiss et al 1979). It is important to note that because β_1-receptors constitute only a small fraction of the total population of β-receptors in the cerebellum, a decrease in the density of β_2-receptors obscured the 350% increase in the density of β_1-adrenergic receptors (Figure 4; Pittman et al 1980b). β_1-adrenergic receptors in the cerebellum are thus far the only population of receptors in the central nervous system to show an increase during aging; however, it seems likely that as techniques for assaying receptors become more sophisticated, more subtle changes will be observed.

CONCLUSIONS

The pioneering work of Lands and his colleagues (1967a,b) represented the first clear evidence that subtypes of β-adrenergic receptors exist. More recently, evidence has accumulated that suggests that more than one sub-

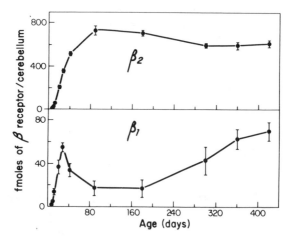

Figure 4   Development of β_1- and β_2-receptors in rat cerebellum. Each point represents the mean ± sem of 4 to 25 independent determinations. The results are expressed as fmoles of receptor subtype per cerebellum. IHYP was used to determine the density of β-receptors (Sporn & Molinoff 1976) and the percentage of β_1- and β_2-receptors was determined as described by Minneman et al (1979b).

type of receptor exists for many of the established and putative transmitters functioning in the central and peripheral nervous systems. Our understanding of receptor subtypes is based primarily on investigations of the effects of various pharmacological agents on physiological parameters. Some differences in pharmacological specificity have been reported. For the most part these differences appear to be due to the difficulty of controlling drug access, metabolism, and disposition; however, the evidence is generally consistent with the view that there are two major classes of β-adrenergic receptor corresponding to the subtypes called β_1 and β_2 by Lands et al (1967a,b). Progress in the field has been accelerated by the potential clinical usefulness of drugs selective for β_1- or β_2-adrenergic receptors.

Over the past several years the availability of radioactively labeled compounds suitable for carrying out direct in vitro studies of β-adrenergic receptors has provided additional impetus to this research. A significant advance came with the development of quantitative methods for analyzing data from tissues that contain both β_1- and β_2-adrenergic receptors. These assays have made it possible to determine the regional distribution of β-adrenergic receptor subtypes and have permitted studies of the development of receptor subtypes and of the factors involved in regulating the densities of β_1- and β_2-adrenergic receptors. It should now be possible to determine accurately the cellular location of specific β-adrenergic receptor subtypes and their functional role in given populations of cells.

The specific cellular location of β-adrenergic receptor subtypes represents a major unanswered question. Since most homogeneous tissues contain only β_1- or β_2-receptors, the observed heterogeneity of β-adrenergic receptors in tissues such as cerebral cortex and other brain regions may reflect the presence of a variety of cell types in these tissues. However, β_1- and β_2-adrenergic receptors coexist in the right atrium of cat heart and both subtypes can influence heart rate. It is possible that both subtypes of receptors are, in some cases, associated with the same cell.

Studies of β-adrenergic receptor subtypes have been markedly facilitated by the availability of a large number of selective agonists and antagonists. The pharmaceutical industry has devoted considerable effort to the development of drugs selective for these two receptor subtypes. Most of this effort has gone into developing β_1-selective antagonists that are useful in the treatment of angina pectoris and cardiac arrythmias, and β_2-selective agonists that are useful for the treatment of asthma. As future agents are developed it will become increasingly feasible to select a drug for a desired effect and to avoid complications due to stimulation or blockade of other receptor subtypes. On the other hand, the evidence obtained to date suggests that a β_1-receptor is the same in the cerebral cortex as it is in cardiac atria or ventricles, while a β_2-receptor in the liver is the same as a

β_2-receptor in the cerebellum or in skeletal muscle. This conclusion, if substantiated in future studies, will obviously limit the flexibility that can be achieved with selective therapies.

Finally, the existence of independently regulated subtypes of β-adrenergic receptor makes it imperative to assess separately the effects of a given physiological or pharmacological manipulation on each subtype of receptor. The available methods for studying the subtypes of β-adrenergic receptor must be used with caution to avoid potential artifacts. The ability to directly quantitate β-adrenergic receptor subtypes should help increase our understanding of the functional role of these receptors in the intact organism.

ACKNOWLEDGMENTS

Ms. Candace Plesha and Ms. Evelyn Hutt provided excellent secretarial and bibliographic assistance. We are also grateful to Drs. Norman Weiner, Barry Wolfe, and Nancy Zahniser for their critical review of the manuscript. During the time that this chapter was written, the authors were supported by grants from the USPHS (AA 03537 and NS 09199, 15756, and 13289).

Literature Cited

Aarons, R. D., Nies, A. S., Gal, J., Hegstrand, L. R., Molinoff, P. B. 1980. Elevation of β-adrenergic receptor density in human lymphocytes following propranolol administration. *J. Clin. Invest.* 65:949–57.

Ablad, B., Borg, K. O., Carlsson, E., Ek, L., Johnsson, G., Malmfors, T., Regardh, C. G. 1975. A survey of the pharmacological properties of metoprolol in animals and man. *Acta Pharmacol. Toxicol.* 36: Suppl 5, pp. 3

Ablad, B., Carlsson, E., Ek, L. 1973. Pharmacological studies of two new cardioselective adrenergic β-receptor antagonists. *Life Sci.* 12:107–19

Ahlquist, R. P. 1948. A study of the adrenotropic receptors. *Amer. J. Physiol.* 153:586–600

Ahlquist, R. P. 1976. Adrenergic beta-blocking agents. *Prog. Drug Res.* 20:27–42

Alexander, R. W., Davis, J. N., Lefkowitz, R. J. 1975. Direct identification and characterisation of β-adrenergic receptors in rat brain. *Nature* 258:437–39

Altman, J., Anderson, W. J. 1973. Experimental reorganization of the cerebellar cortex. II. Effects of elimination of most microneurons with prolonged X-irradiation started at four days. *J. Comp. Neurol.* 149:123–52

Apperley, G. H., Daly, M. J., Levy, G. P. 1976. Selectivity of β-adrenoceptor agonists and antagonists on bronchial, skeletal, vascular and cardiac muscle in the anaesthetized cat. *Brit. J. Pharmacol.* 57:235–46

Ariens, E. J., Simonis, A. M. 1976. Receptors and receptor mechanisms. In *Beta-adrenoceptor blocking agents,* ed. P. R. Saxena, R. P. Forsyth, pp. 3–27. Amsterdam: North Holland

Armstrong, D. H., Schild, R. F. 1970. A quantitative study of the Purkinje cells in the cerebellum of the albino rat. *J. Comp. Neurol.* 139:449–56

Arnold, A., McAuliff, J. P. 1968. Guinea pig adipose tissue responsiveness to catecholamines. *Experientia* 24:436

Arnold, A., McAuliff, J. P. 1969. Correlation of calorigenesis with other β-1 receptor mediated responses to catecholamines. *Arch. Int. Pharmacodyn. Ther.* 179: 381–87

Arnold, A., McAuliff, J. P., Colella, D. F., O'Connor, W. V., Brown, T. G. Jr. 1968. The β-2 receptor mediated glycogenolytic responses to catecholamines in the dog. *Arch. Int. Pharmacodyn. Ther.* 176:451–57

Arnold, A., Selberis, W. H. 1968. Activities of catecholamines on the rat muscle glycogenolytic (β-2) receptor. *Experientia* 24:1010–11

Atlas, D., Steer, M. L., Levitzki, A. 1976. Affinity label for β-adrenergic receptor in turkey erythrocytes. *Proc. Natl. Acad. Sci. USA* 73:1921–25

Banerjee, S. P., Kung, L. S., Riggi, S. J., Chanda, S. K. 1977. Development of β-adrenergic receptor subsensitivity by antidepressants. *Nature* 268:455–56

Barnett, D. B., Rugg, E. L., Nahorski, S. R. 1978. Direct evidence for two types of β-adrenoceptor binding site in lung tissue. *Nature* 273:166–68

Barrett, A. M. 1977. The pharmacology of atenolol. *Postgrad. Med. J.* 53: Suppl. 3, pp. 58–64

Barrett, A. M., Carter, J., Fitzgerald, J. D., Hull, R., Le Count, D. 1973. A new type of cardioselective adrenoceptive blocking drug. *Br. J. Pharmacol.* 48: 340P

Bergman, J., Persson, H., Wetterlin, K. 1969. Two new groups of selective stimulants of adrenergic β-receptors. *Experientia* 25:899–901

Berthelsen, S., Pettinger, W. A. 1977. A functional basis for classification of α-adrenergic receptors. *Life Sci.* 21:595–606

Blinks, J. R. 1967. Evaluation of the cardiac effects of several beta-adrenergic blocking agents. *Ann. NY Acad. Sci.* 139:673–85

Bliss, E. L., Ailion, J., Zwanziger, J. 1968. Metabolism of norepinephrine, serotonin and dopamine in rat brain with stress. *J. Pharmacol. Exp. Ther.* 164:122–34

Bloom, F. E., Hoffer, B. J., Siggins, G. R. 1972. Norepinephrine mediated cerebellar synapses: A model system for neuropsychopharmacology. *Biol. Psychiatr.* 4:157–77

Boissier, J. R., Advenier, C., Giudicelli, J. F., Viars, P. 1971. Studies on the nature of the bronchial β-adrenoceptors. *Eur. J. Pharmacol.* 15:101–9

Bonnelli, J. 1978. Demonstration of two different types of β₁-receptors in man. *Int. J. Clin. Pharmacol.* 16:313–19

Bowman, W. C., Rodger, I. W. 1972. Actions of the sympathomimetic bronchodilator, rimiterol (R798) on the cardiovascular, respiratory and skeletal muscle systems of the anaesthetized cat. *Br. J. Pharmacol.* 45:574–83

Bristow, M., Sherrod, T. R., Green, R. D. 1970. Analysis of beta receptor drug interactions in isolated rabbit atrium, aorta, stomach and trachea. *J. Pharmacol. Exp. Ther.* 171:52–61

Brittain, R. T., Farmer, J. B., Jack, D., Martin, L. E., Simpson, W. T. 1968. α-((t-butylamino)-methyl)-4-hydroxy-m-xylene-α¹-α³-diol (AH 3365): A selective β-adrenergic stimulant. *Nature* 219: 862–63

Brittain, R. T., Jack, D., Ritchie, A. C. 1970. Recent β-adrenoreceptor stimulants. *Adv. Drug Res.* 5:197–253

Buckner, C. K., Patil, P. N. 1971. Steric aspects of adrenergic drugs. XVI. Beta adrenergic receptors of guinea pig atria and trachea. *J. Pharmacol. Exp. Ther.* 176:634–49

Buckner, C. K., Saini, R. K. 1975. On the use of functional antagonism to estimate dissociation constants for beta-adrenergic receptor agonists in isolated guinea pig trachea. *J. Pharm. Exp. Ther.* 194:565–74

Burges, R. A., Blackburn, K. J. 1972. Adenyl cyclase and the differentiation of β-adreneroceptors. *Nature New Biol.* 235:249–50

Burns, J. J., Colville, K. I., Lindsay, L. A., Salvador, R. A. 1964. Blockade of some metabolic effects of catecholamines by N-isopropyl methoxamine (B.W. 61–43). *J. Pharmacol. Exp. Ther.* 144: 163–71

Burns, J. J., Lemberger, L. 1965. N-tertiary-butylmethoxamine, a specific antagonist of the metabolic actions of epinephrine. *Fed. Proc.* 24:298

Burns, J. J., Salvador, R. A., Lemberger, L. 1967. Metabolic blockade by methoxamine and its analogs. *Ann. NY Acad. Sci.* 139:833–40

Bylund, D. B., Snyder, S. H. 1976. Beta adrenergic receptor binding in membrane preparations from mammalian brain. *Mol. Pharmacol.* 12:568–80

Carlsson, E., Ablad, B., Brandstrom, A., Carlsson, B. 1972. Differentiated blockade of the chronotropic effects of various adrenergic stimuli in the cat heart. *Life Sci.* 11:953–58

Carlsson, E., Dahlof, C. G., Hedberg, A., Persson, H., Tangstrand, B. 1977. Differentiation of cardiac chronotropic and intropic effects of β-adrenoceptor agonists. *Naunyn Schmiedeberg Arch Pharmacol* 300:101–5

Carlsson, E., Hedberg, A. 1977. Are cardiac effects of noradrenaline and adrenaline mediated by different β-adrenoceptors. *Acta Physiol. Scand. Suppl.* 44:47

Chen, F. M., Yamamura, H. I., Roeske, W. R. 1979. Ontogeny of mammalian myocardial β-adrenergic receptors. *Eur. J. Pharmacol.* 58:255–64

Cheng, Y. C., Prusoff, W. H. 1973. Relationship between the inhibition constant (Ki) and the concentration of inhibitor which causes 50 percent inhibition (I 50) of an enzymatic reaction. *Biochem. Pharmacol.* 22:3099–3108

Clark, R. B., Perkins, J. P. 1971. Regulation of adenosine 3'5' cyclic monophosphate concentration in cultured human astrocytoma cells by catecholamines and histamine. *Proc. Natl. Acad. Sci. USA* 68:2757–60

Clark, R. B., Su, Y.-F., Ortmann, R., Cubeddux L., Johnson, G. L., Perkins, J. P. 1975. Factors influencing the effect of hormones on the accumulation of cyclic AMP in cultured human astrocytoma cells. *Metabolism* 24:343–58

Coleman, A. J., Paterson, D. S., Somerville, A. R. 1979a. The β-adrenergic receptor of rat corpus luteum membranes. *Biochem. Pharmacol.* 28:1003–10

Coleman, A. J., Paterson, D. S., Somerville, A. R. 1979b. Factors controlling the selectivity of β-blocking drugs. *Biochem. Pharmacol.* 28:1011–13

Coyle, J. T., Schwarcz, R. 1976. Lesion of striatal neurons with kainic acid provides a model for Huntington's chorea. *Nature* 263:244–46

Cullum, V. A., Farmer, J. B., Jack, D., Levy, G. P. 1969. Salbutamol: A new selective β-adrenoceptive receptor stimulant. *Brit. J. Pharmacol.* 35:141–51

Dreyer, A. C., Offermeier, J. 1975. Indications for the existence of two types of cardiac β-adrenergic receptor. *Pharmacol. Res. Comm.* 7:151–61

Duell, E. A. 1980. Identification of a beta₂-adrenergic receptor in mammalian epidermis. *Biochem. Pharmacol.* 29:97–101

Dungan, K. W., Cho, Y. W., Gomoll, A. W., Aviado, D. M., Lish, P. M. 1968. Pharmacologic potency and selectivity of new bronchodilator agent: soterenol (MJ 1992). *J. Pharmacol. Exp. Ther.* 164:290–301

Dunlop, D., Shanks, R. G. 1968. Selective blockade of adrenoceptive beta-receptors in the heart. *Brit. J. Pharmacol.* 32:201–18

Edvinsson, L., Owman, C. 1974. Pharmacological characterization of adrenergic alpha and beta receptors mediating the vasomotor responses of cerebral arteries in vitro. *Circ. Res.* 35:835–49

Erez, M., Weinstock, M., Cohen, S., Shtacher, G. 1975. Potential probe for isolation of the β-adrenoceptor, chloropractolol. *Nature* 255:635–36

Farmer, J. B., Kennedy, J., Levy, G. P., Marshall, R. J. 1970a. A comparison of the β-adrenoceptor stimulant properties of isoprenaline, with those of orciprenaline, salbutamol, soterenol and trimetoquinol on isolated atria and trachea of the guinea pig. *J. Pharm. Pharmac.* 22:61–63

Farmer, J. B., Levy, G. P., Marshall, R. J. 1970b. A comparison of the β-adrenoceptor stimulant properties of salbutamol, orciprenaline, and soterenol with those of isoprenaline. *J. Pharm. Pharmac.* 22:945–47

Friedman, A. H., Davis, J. N. 1980. Identification and characterization of adrenergic receptors and catecholamine stimulated adenylate cyclase in hog pial membranes. *Brain Res.* 183:89–102

Furchgott, R. F. 1967. The pharmacological differentiation of adrenergic receptors. *Ann. NY Acad. Sci.* 139:553–70

Furchgott, R. F. 1970. Pharmacological characteristics of adrenergic receptors. *Fed. Proc.* 29:1352–61

Furchgott, R. F. 1972. The classification of adrenoceptors (adrenergic receptors). An evaluation from the standpoint of receptor theory. In *Handbook of Experimental Pharmacology*, Vol. 33, *Catecholamines* ed. H. Blaschko, E. Muscholl, pp. 283–335. Berlin: Springer-Verlag

Furchgott, R. F. 1975. Postsynaptic adrenergic receptor mechanisms in vascular smooth muscle. In *Vascular Neuroeffector Mechanisms, 2nd Int. Symp. Odense, 1976*, pp. 131–42. Basel: Karger

Furchgott, R. F. 1976. The use of β-haloalkylamines in the differentiation of receptors and in the determination of dissociation constants of receptor-agonist complexes. *Adv. Drug Res.* 3:21–55

Furchgott, R. F., Wakade, T. D., Sorace, R. A., Stollak, J. S. 1975. Occurrence of both β₁ and β₂ receptors in guinea pig tracheal smooth muscle, and variation of the β₁:β₂ ratio in different animals. *Fed. Proc.* 34:794

Gibson, R. E., Rzeszotarski, W. J., Komai, T., Reba, R. C., Eckelman, W. C. 1979. Evaluation of beta-adrenoceptor antagonist affinity and cardioselectivity by radioligand receptor assay. *J. Pharm. Exp. Ther.* 209:153–61

Gilman, A. G., Nirenberg, M. 1971. Effect of catecholamines on the adenosine 3'5'cyclic monophosphate concentrations of clonal satellite cells of neurons. *Proc. Natl. Acad. Sci. USA* 68:2165–68

Glaubiger, G., Lefkowitz, R. J. 1977. Elevated beta-adrenergic receptor number after chronic propranolol treatment. *Biochem. Biophys. Rec. Comm.* 78: 720–25

Goldstein, G. W., Wolinski, J. S., Csejtey, J., Diamond, I. 1975. Isolation of metabolically active capillaries from rat brain. *J. Neurochem.* 25:715–17

Gordon, R., Spector, S., Sjoerdsma, A., Udenfriend, S. 1966. Increased synthesis of norepinephrine and epinephrine in the intact rat during exercise and exposure to cold. *J. Pharm. Exp. Ther.* 153:440–47

Greenberg, L. H., Weiss, B. 1978. β-Adrenergic receptors in aged rat brain: Reduced number and capacity of pineal gland to develop supersensitivity. *Science* 201:61–63

Griffin, J. P., Turner, P. 1971. Preliminary studies of a new bronchodilator (WG 253) in man. *J. Clin. Pharmac.* 11: 280–87

Gwee, M. C. E., Nott, M. W., Raper, C., Rodger, I. W. 1972. Pharmacological actions of a new β-adrenoceptor agonist, MJ-9184-1 in anaesthetized cats. *Br. J. Pharmacol.* 46:375–85

Hancock, A. A., DeLean, A. L., Lefkowitz, R. J. 1979. Quantitative resolution of beta-adrenergic receptor subtypes by selective ligand binding: Application of a computerized model fitting technique. *Mol. Pharmacol.* 16:1–9

Harden, T. K., Mailman, R. B., Mueller, R. A., Breese, G. R. 1979. Noradrenergic hyperinnervation reduces the density of β-adrenergic receptors in rat cerebellum. *Brain Res.* 166:194–98

Harden, T. K., Wolfe, B. B., Sporn, J. R., Perkins, J. P., Molinoff, P. B. 1977a. Ontogeny of β-adrenergic receptors in rat cerebral cortex. *Brain Res.* 125:99–108

Harden, T. K., Wolfe, B. B., Sporn, J. R., Poulos, B. K., Molinoff, P. B. 1977b. Effects of 6-hydroxydopamine on the development of the beta adrenergic receptor/adenylate cyclase system in rat cerebral cortex. *J. Pharm. Exp. Ther.* 203:132–43

Harms, H. H. 1976. Stereochemical aspects of β-adrenoceptor antagonist-receptor interaction in adipocytes. Differentiation of β-adrenoceptors in human and rat adipocyte. *Life Sci.* 19:1447–52

Harms, H. H., Zaagsma, J., de Vente, J. 1977. Differentiation of β-adrenoceptors in right atrium, diaphragm and adipose tissue of the rat, using stereoisomers of propranolol, alprenolol, nifenalol and practolol. *Life Sci.* 21:123–28

Hedberg, A., Minneman, K. P., Molinoff, P. B. 1980. Differential distribution of beta-1 and beta-2 adrenergic receptors in cat and guinea pig heart. *J. Pharm. Exp. Ther.* 212:503–8

Hegstrand, L. R., Minneman, K. P., Molinoff, P. B. 1979. Multiple effects of guanosine 5'-triphosphate on β-adrenergic receptors and adenylate cyclase in rat heart, lung and brain. *J. Pharm. Exp. Ther.* 210:215–21

Hendley, E. D. 1978. Iprindole is a potent enhancer of spontaneous and KCl-induced efflux of norepinephrine from rat brain slices. *Soc. Neurosci. Abstr.* 4:494

Herbst, T. J., Raichle, M. E., Ferrendelli, J. A. 1979. β-Adrenergic regulation of adenosine 3',5'-monophosphate concentration in brain microvessels. *Science* 204:330–32

Hess, A. 1979. Visualization of β-adrenergic receptor sites with fluorescent beta adrenergic blocker probes—or autofluorescent granules? *Brain Res.* 160:533–38

Hoffer, B. J., Siggins, G. R., Oliver, A. P., Bloom, F. E. 1971. Cyclic AMP mediation of norepinephrine inhibition in rat cerebellar cortex: A unique class of synaptic responses. *Ann. NY Acad. Sci.* 185:531–49

Hofstee, B. H. J. 1952. On the evaluation of the constants Vm and Km in enzyme reactions. *Science* 116:329–31

Ikegami, K. 1977. Modulation of adenosine 3'5'monophosphate contents of rat peritoneal macrophages mediated by $β_2$-adrenergic receptors. *Biochem. Pharmacol.* 26:1813–16

Imbs, J. L., Miesch, F., Schwartz, J., Velly, J., Leclerc, G., Mann, A., Wermuth, C. G. 1977. A potent new $β_2$-adrenoceptor blocking agent. *Br. J. Pharmacol.* 60: 357–62

Johnson, D. E., Sellinger, O. Z. 1971. Protein synthesis in neurons and glial cells of the developing rat brain: An in vivo study. *J. Neurochem.* 18:1445–60

Jonsson, G., Pycock, P., Fuxe, K., Sachs, C. 1974. Changes in the development of central noradrenergic neurons following neonatal administration of 6-hydroxydopamine. *J. Neurochem.* 22:419–26

Joó, F., Karnushina, I. 1973. A procedure for the isolation of capillaries from rat brain. *Cytobios* 8:41–48

Kalisker, A., Rutledge, C. O., Perkins, J. P. 1973. Effect of nerve degeneration by 6-hydroxydopamine on catecholamine-stimulated adenosine 3',5'-monophosphate forming in rat cerebral cortex. *Mol. Pharmacol.* 9:619–29

Kaumann, A. J. 1968. Supersensitivity to exogenous and endogenously released norepinephrine after COMT inhibition and cocaine. *Fed. Proc.* 27:711

Kobayashi, R. M., Palkovits, M., Kizer, J. S., Jacobowitz, D. M., Kopin, I. 1976. Selective alterations of catecholamines and tyrosine hydroxylase activity in the hypothalamus following acute and chronic stress. In *Catecholamine and Stress,* ed. E. Usdin, R. Kvetnansky, I. Kopin, pp. 29–38. New York: Pergamon

Kvetnansky, R., Mikulaj, L. 1970. Adrenal and urinary catecholamines in rats during adaptation to repeated immobilization stress. *Endocrinology* 87:738–43

Lands, A. M., Arnold, A., McAuliff, J. P., Luduena, F. P., Brown, T. G. 1967a. Differentiation of receptor systems activated by sympathomimetic amines. *Nature* 214:597–98

Lands, A. M., Luduena, F. P., Buzzo, H. J. 1967b. Differentiation of receptors responsive to isoproterenol. *Life Sci.* 6:2241–49

Lands, A. M., Luduena, F. P., Buzzo, H. J. 1969. Adrenotrophic β-receptors in the frog and chicken. *Life Sci.* 8:373–82

Langer, S. Z. 1974. Presynaptic regulation of catecholamine release. *Biochem. Pharmacol.* 23:1793–1800

Lefkowitz, R. J. 1975. Heterogeneity of adenylate cyclase coupled β-adrenergic receptor. *Biochem. Pharmacol.* 24:583–90

Lefkowitz, R. J., Mullikin, D., Caron, M. G. 1976. Regulation of β-adrenergic receptors by guanyl-5'-yl-imidodiphosphate and other purine nucleotides. *J. Biol. Chem.* 251:4686–92

Letts, L. G., Easson, P. A., Temple, D. M., Lap, B. V., Lim, C. H., Orr, A. J., Pasaribu, S. J., Quessy, S. N., Williams, L. R. 1977. The β-adrenergic activity of some monosubstituted phenethanolamines. *Arch. Int. Pharmacodyn. Ther.* 230:42–52

Levy, B. 1964. Alterations of adrenergic responses by N-isopropylmethoxamine. *J. Pharmacol. Exp. Ther.* 146:129–38

Levy, B. 1966a. The adrenergic blocking activity of N-tert-butylmethoxamine (butoxamine). *J. Pharmacol. Exp. Ther.* 151:413–22

Levy, B. 1966b. Dimethyl isopropylmethoxamine: A selective beta receptor blocking agent. *Br. J. Pharmacol. Chemother.* 27:277–85

Levy, B. 1967. A comparison of the adrenergic receptor blocking properties of 1-(4'-methylphenyl)-2-isopropylamino-propanol HCl and propranolol. *J. Pharmacol. Exp. Ther.* 156:452–62

Levy, B., Wilkenfeld, B. E. 1969. An analysis of selective β-receptor blockade. *Eur. J. Pharmacol.* 5:227–34

Levy, B., Wilkenfeld, B. E. 1970. Selective interactions with beta adrenergic receptors. *Fed. Proc.* 29:1362–64

Loakpradit, T., Lockwood, R. 1977. Differentiation of metabolic adrenoceptors. *Br. J. Pharmacol.* 59:135–40

Lucas, M., Bockaert, J. 1977. Use of (−)-³H-dihydroalprenolol to study beta-adrenergic receptor-adenylate cyclase coupling in C6 glioma cells. Role of 5'-guanylylimidodiphosphate. *Mol. Pharmacol.* 13:314–29

Lumley, P., Broadley, K. J. 1975. Differential blockade of guinea pig atrial rate and force processes to (−)-noradrenaline by practolol—an uptake phenomenon. *Eur. J. Pharmacol.* 34:207–17

Lumley, P., Broadley, K. J. 1977. Evidence from agonist and antagonist studies to suggest that the β₁-adrenoceptors subserving the positive inotropic and chronotropic responses of the heart do not belong to two separate subgroups. *J. Pharm. Pharmacol.* 29:598–604

Maggi, A., Schmidt, M. J., Ghetti, B., Enna, S. J. 1979. Effect of aging on neurotransmitter receptor binding in rat and human brain. *Life Sci.* 24:367–74

Maguire, M. E., Ross, E. M., Gilman, A. G. 1977. β-Adrenergic receptor: Ligand binding properties and the interaction with adenylyl cyclase. *Adv. Cyclic Nucl. Res.* 8:1–83

Maguire, M. E., Van Arsdale, P. M., Gilman, A. G. 1976. An agonist-specific effect of guanine nucleotides on binding to the beta-adrenergic receptor. *Mol. Pharmacol.* 12:335–39

Mayer, S. E. 1972. Effects of adrenergic agonists and antagonists on adenylate cyclase activity of dog heart and liver. *J. Pharmacol. Exp. Ther.* 181:116–24

McGeer, E. G., McGeer, P. L. 1976. Duplication of biochemical changes of Huntington's chorea by intrastriatal injections of glutamic and kainic acid. *Nature* 263:517–19

Melamed, E., Lahav, M., Atlas, D. 1976. Direct localization of β-adrenoceptor sites in rat cerebellum by a new fluorescent analogue of propranolol. *Nature* 261:420–22

Minneman, K. P., Dibner, M. D., Wolfe, B. B., Molinoff, P. B. 1979d. β₁ and β₂-adrenergic receptors in rat cerebral cortex are independently regulated. *Science* 204:866–68

Minneman, K. P., Hegstrand, L. R., Molinoff, P. B. 1979a. The pharmacological specificity of beta-1 and beta-2-adrenergic receptors in rat heart and lung in vitro. *Mol. Pharmacol.* 16:21–33

Minneman, K. P., Hegstrand, L. R., Molinoff, P. B. 1979b. Simultaneous determination of beta-1 and beta-2-adrenergic receptors in tissues containing both receptor subtypes. *Mol. Pharmacol.* 16:34–46

Minneman, K. P., Hedberg, A., Molinoff, P. B. 1979c. Comparison of beta adrenergic receptor subtypes in mammalian tissues. *J. Pharm. Exp. Ther.* 211:502–8

Minneman, K. P., Molinoff, P. B. 1979. Catecholamine receptors in the central nervous system: Multiple subtypes. In *Catecholamines: Frontiers in Basic and Clinical Research,* ed. E. Usdin, I. J. Kopin, J. Barchas, pp. 468–73. New York: Pergammon

Minneman, K. P., Molinoff, P. B. 1980. Classification of β-adrenergic receptor subtypes. *Biochem. Pharmacol.* 29:1317–23

Minneman, K. P., Pittman, R. N., Wolfe, B. B., Yeh, H., Woodward, D. J., Molinoff, P. B. 1980b. Selective survival of β₁-adrenergic receptors in rat cerebellum after neonatal X-irradiation. *Brain Res.* In press

Minneman, K. P., Quik, M., Emson, P. C. 1978. Receptor-linked cyclic AMP systems in rat neostriatum: Differential localization revealed by kainic acid injection. *Brain Res.* 151:507–21

Minneman, K. P., Weiland, G. A., Molinoff, P. B. 1980a. A comparison of the β-adrenergic receptor of the turkey erythrocyte with mammalian Beta₁ and Beta₂ receptors. *Mol. Pharmacol.* 17:1–7

Murad, F. 1973. Beta-blockade of epinephrine-induced cyclic AMP formation in heart, liver, fat and trachea. *Biochim. Biophys. Acta.* 304:181–87

Nahorski, S. R. 1978. Heterogeneity of cerebral β-adrenoceptor binding sites in various vertebrate species. *Eur. J. Pharmacol.* 51:199–209

Nahorski, S. R., Barnett, D. B., Howlett, D. R., Rugg, E. L. 1979a. Pharmacological characteristics of beta-adrenoceptor binding sites in intact and sympathectomized rat spleen. *Naunyn Schmidedebergs Arch. Pharmacol.* 307:227–33

Nahorski, S. R., Howlett, D. R., Redgrave, P. 1979b. Loss of β-adrenoceptor binding sites in rat striatum following kainic acid lesions. *Eur. J. Pharmacol.* 60:249–52

Nathanson, J. A., Glaser, G. H. 1979. Identification of β-adrenergic-sensitive adenylate cyclase in intracranial blood vessels. *Nature* 278:567–69

Nelson, D. L., Molinoff, P. B. 1976. Differential effects of nerve impulses on adrenergic storage vesicles in rat heart. *J. Pharmacol. Exp. Ther.* 198:112–22

O'Donnell, S. R. 1970. A selective β-adrenoceptor stimulant (Th 1165a) related to orciprenaline. *Eur. J. Pharmacol.* 12:35–43

O'Donnell, S. R. 1972. An examination of some β-adrenoceptor stimulants for selectivity using the isolated trachea and atria of the guinea pig. *Eur. J. Pharmacol.* 19:371–79

O'Donnell, S. R., Wanstall, J. C. 1976. The contribution of extraneuronal uptake to the trachea-blood vessel selectivity of β-adrenoceptor stimulants in vitro in guinea pig. *Br. J. Pharmacol.* 57:369–73

O'Donnell, S. R., Wanstall, J. C. 1978. Evidence that the efficacy (intrinsic activity) of fenoterol is higher than that of salbutamol on β-adrenoceptors in guinea pig trachea. *Eur. J. Pharmacol.* 47:333–40

O'Donnell, S. R., Wanstall, J. C. 1979. The importance of choice of agonist in studies designed to predict β₂ : β₁ adrenoceptor selectivity of antagonists from pA2 values on guinea pig trachea and atria. *Naunyn Schmiedebergs Arch. Pharmacol.* 308:183–90

Onaya, T., Akasu, F., Takazawa, K., Hashizume, K. 1978. Evidence for activation by β₂-adrenergic receptors of adenosine 3'5'monophosphate formation in Ehrlich ascites tumor cells. *Endocrinology* 103:1122–27

Palacios, J. M., Kuhar, M. J. 1980. Beta-adrenergic receptor localization by light microscopic autoradiography. *Science* 208:1378–80

Patil, P. N., Patel, D. G., Krell, R. D. 1971. Steric aspects of adrenergic drugs. XV. Use of isomeric activity ratio as a criterion to differentiate adrenergic receptors. *J. Pharmacol. Exp. Ther.* 176:622–33

Perkins, J. P. 1975. Regulation of the responsiveness of cells to catecholamines: Variable expression of the components of the second messenger system. *Cyclic Nucleotides in Disease,* ed. B. Weiss, pp. 351–75. Baltimore: Univ. Park Press

Peroutka, S. J., Moskowitz, M. A., Reinhard, J. F. Jr., Snyder, S. H. 1980. Neurotransmitter receptor binding in bovine cerebral microvessels. *Science* 208: 610–12

Persson, H., Olsson, T. 1970. Some pharmacological properties of terbutaline (INN), 1-(3,5-dihydroxyphenyl)-2-(t-butylamino)-ethanol. A new sympathomimetic β-receptor stimulating agent. *Acta Med. Scand.* 512:11–19

Petrack, B., Czernik, A. J. 1976. Inhibition of isoproterenol activation of adenylate cyclase by metoprolol, oxprenolol and the para isomer of oxrenolol. *Mol. Pharmacol.* 12:203–7

Pittman, R. N., Minneman, K. P., Molinoff, P. B. 1980a. Ontogeny of β_1 and β_2-adrenergic receptors in rat cerebellum and cerebral cortex. *Brain Res.* 188:357–68

Pittman, R. N., Minneman, K. P., Molinoff, P. B. 1980b. Alterations in β_1 and β_2-adrenergic receptor density in the cerebellum of aging rats. *J. Neurochem.* 35:273–75

Popper, C. W., Chiueh, C. C., Kopin, I. J. 1977. Plasma catecholamine concentrations in unanesthetized rats during sleep, wakefulness, immobilization and after decapitation. *J. Pharmacol. Exp. Ther.* 202:144–48

Rodbell, M. 1980. The role of hormone receptors and GTP-regulatory proteins in membrane transduction. *Nature* 284:17–22

Rugg, E. L., Barnett, D. B., Nahorski, S. R. 1978. Coexistence of beta$_1$ and beta$_2$ adrenoceptors in mammalian lung: Evidence from direct binding studies. *Mol. Pharmacol.* 14:996–1005

Salvador, R. A., Colville, K. I., April, S. A., Burns, J. J. 1964. Inhibition of lipid mobilization by N-isopropyl methoxamine. *J. Pharm. Exp. Ther.* 144:172–80

Schild, H. O. 1949. pAx and competitive drug antagonism. *Br. J. Pharmacol.* 4: 277–80

Sellinger, O. Z., Legrand, J., Clos, J., Ohlsson, W. G. 1974. Unequal patterns of development of succinate dehydrogenase and acetylcholinesterase in Purkinje cell bodies and granule cells isolated in bulk from the cerebellar cortex of the immature rat. *J. Neurochem.* 23:1137–44

Sercombe, R., Aubineau, P., Edvinsson, L., Mamo, H., Owman, C., Seylaz, J. 1977. Pharmacological evidence in vitro and in vivo for functional beta$_1$ receptors in the cerebral circulation. *Pflügers Arch.* 368:241–44

Sidman, R. L., Green, M. C., Appel, S. H. 1965. *Catalog of the Neurological Mutants of the Mouse.* Cambridge, Mass.: Harvard Univ. Press

Siggins, G. R., Hoffer, B. J., Bloom, F. E. 1969. Cyclic adenosine monophosphate: Possible mediator for norepinephrine effects on cerebellar Purkinje cells. *Science* 165:1018–20

Sporn, J. R., Harden, T. K., Wolfe, B. B., Molinoff, P. B. 1976. β-Adrenergic receptor involvement in 6-hydroxydopamine-induced supersensitivity in rat cerebral cortex. *Science* 194:624–26

Sporn, J. R., Molinoff, P. B. 1976. β-Adrenergic receptors in rat brain. *J. Cyclic Nucl. Res.* 2:149–61

Strosberg, A. M. 1976. The cardiovascular pharmacology and hemodynamic activity of tazolol, a selective myocardial β-stimulant. *Arch. Int. Pharmacodyn. Ther.* 222:200–15

Thierry, A. M., Javoy, F., Glowinski, J., Kety, S. S. 1968. Effects of stress on the metabolism of norepinephrine, dopamine and serotonin in the central nervous system of the rat. I. Modifications of norepinephrine turnover. *J. Pharm. Exp. Ther.* 163:163–71

Thoenen, H., Tranzer, J. P. 1968. Chemical sympathectomy by selective destruction of adrenergic nerve endings with 6-hydroxydopamine. *Naunyn Schmiedebergs Arch. Pharmakol. Exp. Pathol.* 261:271–88

Turnheim, K., Kraupp, O. 1971. Pulmonary and systemic circulatory effects and β-adrenergic selectivity of hexoprenaline, salbutamol, oxyfedrine and isoproterenol. *Eur. J. Pharmacol.* 15: 231–39

Tuttle, R. R., Mills, J. 1975. Development of a new catecholamine to selectively increase cardiac contractility. *Circ. Res.* 36:185–96

U'Prichard, D. C., Bylund, D. B., Snyder, S. H. 1978. ($\pm$)-(^{3}H)-Epinephrine and (–)-(^{3}H)Dihydroalprenolol binding to β_1- and β_2-noradrenergic receptors in brain, heart and lung membranes. *J. Biol. Chem.* 253:5090–5102

U'Prichard, D. C., Reisine, T. D., Yamamura, S., Mason, S. T., Fibiger, H. C., Ehlert, F., Yamamura, H. I. 1980. Differential supersensitivity of β-receptor subtypes in rat cortex and cerebellum after central noradrenergic denervation. *Life Sci.* 26:355–64

Usdin, E., Kvetnansky, R., Kopin, I. J., eds. 1976. *Catecholamines and Stress,* pp. 631. New York: Pergammon

Van den Brink, F. G. 1973. The mode of functional interaction. II. Experimental verification of a new model: The antagonism of β-adrenoceptor stimulants and other agonists. *Eur. J. Pharmacol.* 22: 279–86

Vaughan Williams, E. M., Bagwell, E. E., Singh, B. N. 1973. Cardiospecificity of beta receptor blockade. *Cardiovasc. Res.* 7:226–40

Vauquelin, G., Lacombe, M. L., Guellaen, G., Strosberg, A. D., Hanoune, J. 1976. Tazolol (1-isopropylamino-3-(2- thiazoloxy)-2-propanol) as a β-adrenergic blocker. *Biochem. Pharmacol.* 25:2605–8

Vetulani, J., Sulser, F. 1975. Action of various antidepressant treatments reduces reactivity of noradrenergic cyclic AMP-generating system in limbic forebrain. *Nature* 257:495–96

Vlietstra, R. E., Blinks, J. R. 1976. Heterogeneity of cardiac beta-adrenoceptors. *Fed. Proc.* 35:210

Wardell, J. R. Jr., Colella, D. F., Shetzline, A., Fowler, P. J. 1974. Studies on carbuterol (SK and F 40383-A) a new selective bronchodilator agent. *J. Pharm. Exp. Ther.* 189:167–84

Wasserman, M. A., Levy, B. 1972. Selective beta-adrenergic receptor blockage in the rat. *J. Pharm. Exp. Ther.* 182:256–63

Weiland, G. A., Minneman, K. P., Molinoff, P. B. 1979. Fundamental difference between the molecular interactions of agonist and antagonist with the β-adrenergic receptor. *Nature* 281:114–17

Weiland, G. A., Minneman, K. P., Molinoff, P. B. 1980. Thermodynamics of agonist and antagonist interactions with mammalian β-adrenergic receptors. *Mol. Pharmacol.* In press

Weinstock, M., Schechter, Y., Erez, M., Shtacher, G. 1974. Changes in β-adrenoceptors blocking activity produced by chemical modifications in the practolol molecule. *Eur. J. Pharmacol.* 26:191–97

Weiss, B., Greenberg, L., Cantor, E. 1979. Age-related alterations in the ability of the β-adrenergic receptor-adenylate cyclase complex of brain to develop denervation supersensitivity. *Fed. Proc.* 38:1915–21

Williams, L. T., Lefkowitz, R. J. 1978. *Receptor Binding Studies in Adrenergic Pharmacology.* New York: Raven

Wolfe, B. B., Harden, T. K., Sporn, J. R., Molinoff, P. B. 1978. Presynaptic modulation of beta adrenergic receptors in cerebral cortex after treatment with antidepressants. *J. Pharmacol. Exp. Ther.* 207:446–57

Wolfe, B. B., Minneman, K. P., Dibner, M. D., Molinoff, P. B. 1980. Regulation of β-adrenergic subtypes in rat cerebellum, cerebral cortex and caudate. *Soc. Neurosci. Abstr.* 6:257

Woodward, D. J., Hoffer, B. J., Altman, J. 1974. Physiological and pharmacological properties of Purkinje cells in rat cerebellum degranulated by postnatal X-irradiation. *J. Neurobiol.* 5:283–304

Yabuuchi, Y., Yamashita, S., Tei, S. S. 1977. Pharmacological studies of OPC-2009, a newly synthesized selective beta adrenoceptor stimulant, in the bronchiomotor and cardiovascular system of the anaesthetized dog. *J. Pharmacol. Exp. Ther.* 202:326–36

Yoshizaki, S., Tanimura, K., Tamada, S., Yabuuchi, Y., Nakagawa, K. 1976. Sympathomimetic amines having a carbostyril nucleus. *J. Med. Chem.* 19: 1138–42

Zahniser, N. R., Minneman, K. P., Molinoff, P. B. 1979. Persistence of β-adrenergic receptors in rat striatum following kainic acid administration. *Brain Res.* 178:589–95

Ann. Rev. Neurosci. 1981. 4:463–503

THE USE OF CONTROL SYSTEMS ANALYSIS IN THE NEUROPHYSIOLOGY OF EYE MOVEMENTS

◆11560

D. A. Robinson

Department of Ophthalmology, The Johns Hopkins University,
Baltimore, Maryland 21205

INTRODUCTION

When we ask how the brain works, the question is perceived quite differently by people working at the many levels of the nervous system. Certainly it is necessary to know how the hardware of the nervous system works—how transmitters affect membranes, how myelin is formed, how action potentials are conducted—but we all recognize that the solutions to these problems are not an end in themselves. They are the means that will enable us to understand how the brain processes information, which, we assume, is the substrate for behavior. If our knowledge of the connections between nerve cells and the signals that they carry were adequate, many theoreticians would be at work using the techniques of information theory, signal theory, and automatic control theory to explain the bases of such things as perception, recognition, and memory because these mathematical techniques are the tools of those who must deal with information-processing systems, of which the brain is by far and away the most complex and powerful. The fact that such theoreticians are remarkable for their scarcity simply reminds us that our knowledge of signals and connections in neural networks is, in most cases, so fragmented that such progress is impossible.

Our knowledge of signals and connections in the oculomotor system is still at a stage where much more information is needed, but, thanks to recent advances in tracing nerve cell processes and recording from cells in alert animals, coupled with certain simplifying features of the eye movement

463

control system, enough data are available that one can at least begin to use quantitative, analytic methods for organizing these data into circuits and systems to explain how oculomotor control is effected. Thus, one begins to see in the literature increasingly complex models of oculomotor organization and more sophisticated methods of analyzing their responses and comparing them to experimental results. In short, control systems analysis is being used more and more in the study of the oculomotor system. This review attempts to examine this rather new phenomenon and illustrate how the use of analytical techniques is employed in interpreting the wealth of new data coming from laboratories, and in guiding the course and nature of future research.

There are a number of levels at which systems analysis can be applied in the oculomotor system. The device being controlled—the eyeball and its muscles—must be described in a format suitable for systems analysis. Next, the oculomotor signals that flow about in the brainstem and cerebellum can be described with emphasis on their quantification. At higher levels of organization one may consider simple neural circuits that do rather basic signal conditioning before sending commands on to the motoneurons, and more complex circuits that go beyond the data and require additional hypothetical connections. Such proposals challenge the experimenter by providing possible explanations for a good deal of observed behavior and indicating further experiments to test the hypotheses. Finally, at the highest levels, models have been proposed to explain the overall behavior of one or several entire oculomotor subsystems. These models, however, are often filled with black boxes that are usually of little interest to the neurophysiologist since they do not suggest methods of experimental testing at the level of neural circuits. Before reviewing these levels of signal processing, however, it is interesting to examine those features of the oculomotor system that have permitted such extraordinary progress in recent years.

UNIQUE FEATURES OF THE OCULOMOTOR SYSTEM

One of the features of eye movement control that facilitates its study is the simplicity of the organization of the eyeball and its muscles (Robinson 1978). Because of the following simplifying features, it is easy to relate the discharge rate of motoneurons and other, more central, neurons directly to the motion of the eye:

1. The eyeball may be considered to rotate around a fixed point so there is only one "joint" in the system.
2. There are only two muscles to rotate the eye in any one plane.

3. The muscles are straight with parallel fibers so that the force of each fiber is applied directly to the globe.
4. The tendons wrap around the globe so that the moment arm of the muscles does not depend on eye position.
5. The muscles are reciprocally innervated and usually do not cocontract.
6. Most importantly, the eyeball is not used to apply forces to external loads, as are most other muscles; so much of the circuitry, such as the stretch reflex, required by other motor systems to deal with a wide variety of changing loads is absent in the oculomotor system.

Conceptually the oculomotor system is simple because we are able to understand what it does and why it does it. In afoveate, lateral-eyed animals there are three major oculomotor subsystems: (a) the vestibulo-ocular reflex, (b) the optokinetic system, and (c) the saccadic (or quick-phase) system. The purpose of the first two is to prevent images from moving on the retina when an animal's head (or body) turns. The vestibulo-ocular reflex senses head velocity by means of the semicircular canals and causes the eyes to move in the opposite direction, at the same speed, so that the line of sight remains constant in the visual environment. This reflex, described in detail below, enables animals to move and see at the same time. So, it is not surprising to find that the reflex is common to all vertebrates, in essentially the same form, and is even found, with modifications, in invertebrates. Because of the dynamic properties of the canals, this reflex works best at intermediate and high frequencies (0.1 to 7.0 Hz) but not at low frequencies (below 0.01 Hz). To supplement this reflex at low frequencies, the optokinetic system evolved. This system, also described in detail below, uses image slip on the retina as an error signal in a negative feedback scheme to move the eyes so as to lessen the motion of images on the retina. It is designed, not to duplicate the vestibulo-ocular reflex at high frequencies, but to complement it at the low frequency range where the canals do not operate correctly. Together these two systems allow an animal to turn slowly or rapidly, in a transient or sustained manner, while maintaining clear, stable vision by rotating the eyes in such a way that the images of the visual environment remain relatively stable on the retina.

The purpose of the saccadic system, on the other hand, is to reorient the eyes quickly in space. Since vision during saccades is poor, this system has specialized in making such eye movements very rapid to minimize the time during which vision is lost. In afoveate animals, such as rabbits and goldfish, the rapid eye movements occur as part of a coupled, programmed, eye-head reorientation. Frontal-eyed, foveate animals, such as cats, monkeys, and humans, have extended the saccadic system so that the rapid movement can also put the image of a specific target of interest onto the fovea and they are able to make these eye movements without an associated head movement. These animals have also developed a vergence system designed to put the images of targets at various distances on the fovea of each eye simultaneously for binocular vision, and, especially in primates, they have developed a smooth pursuit system to track a moving target with smooth eye movements and keep its image relatively stationary on the fovea.

The objectives of these five major subsystems (pursuit, vergence, saccadic, optokinetic, and the vestibulo-ocular reflex) seem fairly obvious and the manner in which they achieve these objectives is so stereotyped that the function of each system can be specified mathematically. Thus, what the neural networks do is known; one is therefore free to concentrate on how they do it. This is not true for most other motor control systems, where one usually does not even know what a neural circuit is trying to accomplish, let alone how it might achieve it. Understanding what a neural system is trying to achieve is a powerful advantage in any sort of neurophysiology, one that is often not appreciated, and without which the application of systems analysis is impossible.

Experimental methods also play an important part in the recent increase in our knowledge of eye movement control. Through the work of Evarts (1968), it became possible to record from single cells in the central nervous systems of alert, behaving animals, and this technique began to be applied to the oculomotor system in the late 1960s. Because the entire oculomotor system is contained in the cranial vault, all of its circuits became accessible to exploration with microelectrodes, and such investigations, which have been going on now for over ten years, have given us a rich supply of new data. This happy situation is not yet possible for the study of the control of limb movements because recordings within the cranial vault from such structures as the motor cortex, basal ganglia, and cerebellum describe the behavior of neurons that appear to be rather distantly related to events in the spinal cord, and mechanical instability has so far prevented extensive recordings from the complex, signal-processing circuits in the spinal cords of behaving animals.

In summary, a number of features—technical, functional, and conceptual —combine in the oculomotor system to permit the gathering of large amounts of interpretable data. One is thus in a position to get on with the business of interpreting these data. Since the eye movement control system is just that—a control system—it is natural to explain its workings in the language developed over the last fifty years by those who design, describe, and analyze control systems. Thus, interpreting the data means drawing the wiring diagram and specifying the signal processing in some format, such as transfer functions.

THE OCULOMOTOR PLANT

Physiological Observations

When it became possible to record from single nerve cells in alert monkeys, the motoneuron became the obvious first target of the oculomotor neurophysiologist. It was found first that the discharge rate of these cells depended upon eye position (Fuchs & Luschei 1970, Robinson 1970, Schiller 1970, Henn & Cohen 1973). For an abducens motoneuron, for example, the discharge rate was higher the farther the monkey looked ipsilaterally; in the opposite direction, the discharge rate decreased and often became zero at some contralateral eye position. When the animal made saccades, motoneurons usually burst at high rates in association with those movements that were in the pulling direction of the muscle and were inhibited during saccades in the opposite direction. Henn & Cohen (1973) divided the motoneurons they observed into four categories: (*a*) tonic cells, whose firing rates were modulated with changes in eye position but not eye velocity; (*b*) purely phasic cells, which were very active during saccades, but not with variations in eye position; (*c*) predominantly phasic cells; and (*d*) predominantly tonic cells. The activity of the latter two types was partly phasic and partly tonic in different proportions.

To proceed from this qualitative description to a description suitable for mathematical analysis, one must describe the behavior of a motoneuron in terms of its instantaneous discharge rate, $R_m(t)$, in spikes sec^{-1} and relate it to instantaneous eye position, $E(t)$, in degrees, measured in the plane of action of the muscle being considered. Independent studies by Fuchs & Luschei (1970) by Robinson (1970), and by Schiller (1970) found that the behavior of all ocular motoneurons could be described by the equation

$$R_m = R_o + kE + r\frac{dE}{dt}.\tag{1.}$$

When the monkey looks straight ahead, where E is defined as zero, and fixates so that eye velocity, dE/dt is also zero, the motoneuron discharges at a constant rate, R_o, with a typical value of 100 spikes/sec. If the monkey fixates (so that dE/dt remains zero) at some angle E in the pulling direction of the muscle, called the on-direction, the discharge rate increases by the amount, kE. In the opposite, or off-direction, the rate decreases by kE. This change in rate represents the change in muscle force required to oppose the elastic elements in the orbit and hold the eye in its new position. If the eye is also in motion, an extra force, proportional to velocity, is required to overcome the viscous impedances in the muscles and orbit. This force is represented by the term, $r(dE/dt)$. The behavior of the typical motoneuron may be found by substituting into Eq. 1 values for R_o, in spikes sec^{-1}, k, in (spikes sec^{-1})deg^{-1}, and r, in (spikes sec^{-1})(deg sec^{-1})$^{-1}$, that are the means of a large population of cells observed in many laboratories:

$$R_m = 100 + 4\ E + 0.95\frac{dE}{dt}.$$
2.

Thus, if the monkey fixates 30 deg in the on-direction, the typical motoneuron fires at 220 spikes sec^{-1}. If the monkey fixates at 25 deg in the opposite direction, the discharge rate is zero. If the eye passes through zero position at a velocity of 100 deg sec^{-1} in the on-direction, the rate will be 195 spikes sec^{-1}.

The Transfer Function

There are large differences from cell to cell in the values of the parameters R_o, k, and r, which probably are related to the different types of muscle fibers found in eye muscles. The physiologist is usually drawn to examine these differences, hoping to find them sufficiently large to justify subdivisions and classifications. To understand how the eye behaves during the operation of some oculomotor subsystem, however, it is necessary to regard the eyeball and its muscles simply as a device, or physical plant, to be controlled, and one need only be able to predict how it will respond to any signal that reaches the motoneurons. For this purpose one makes the assumption that the activity of the entire motoneuron pool can be approximated by the behavior of the typical motoneuron described by Eq. 2, thereby emphasizing the similarities rather than the differences in the motoneuron pool. It is simply impractical, in actual simulation, to represent the plant with an equation for each motor unit. The approximation represented in Eq. 2 is reasonable since there are no qualitative differences in behavior from cell to cell and the distributions of R_o, k, and r over the population are broad and flat, thus indicating that motor units cannot even be usefully divided into quantitatively different subgroups. Of course, the total force on the eye depends on the number of fibers recruited into activity and the strength of each particular fiber as well as the discharge rate, but Eq. 1 and 2 make no pretense at describing internal forces: they describe only the input and output behavior. If one specifies a given eye motion, Eq. 2 tells one what most motoneurons are doing. If one took the population distributions of R_o, k, and r into account, one would know what all the motoneurons were doing, but we are essentially saying that this amount of detail is unnecessary. What is most important is that if any signal reaches the motoneurons, we can predict, with fair accuracy from Eq. 2, what eye movement, $E(t)$, it will produce.

The practice of allowing the typical neuron to represent the activity of a pool of neurons is common in neurophysiology in general (although the pitfalls of doing this are obvious) and this is true of the oculomotor system. As we see later, there exist groups of cells, such as burst neurons, vestibular neurons, and gaze Purkinje cells, in which cells differ from one another only quantitatively, and it seems reasonable to put these cells together and regard them as forming a pool carrying a single signal—that carried by the average cell. The fact that many oculomotor signals appear to be carried in this way, at least at the premotor levels of the brain stem and cerebellum, constitutes another powerful advantage in studying the oculomotor systems. At these levels one need not worry about complex, spatial interactions between neighboring cells, as one must in, say, the inner and outer plexiform layers of the retina; one need only deal with groups of similar cells, between which flow rather simple analogue signals coded in firing rate. The practice, however, of representing the information carried in a neural pool by the signal of a typical cell does raise several issues. There is always the possibility that some target nucleus may receive fibers from only a special fraction of cells in a given nucleus and that fraction may carry a signal quantitatively rather different from the typical cell. Also, when the typical cell is driven to silence, it is no longer representative, since other, atypical cells are still transmitting a signal. In fact, the typical cell will not reflect the nonlinearities of a system with much accuracy. Thus, one must use the typical signal with caution.

Given these precautions, Eq. 2 may be said to describe the signal carried by the ocular motor nuclei. It is, however, in the form of a differential equation. The fact that it is only a linear, first-order differential equation is another, fortuitous advantage of the oculomotor system, but this form is not the most convenient one for describing how a system will respond to a variety of presynaptic signals arriving at the nucleus. While there are many ways to describe such behavior, the one that has been in most common use for the last 40 years is that of frequency analysis. If one delivers a sinusoidal stimulus of frequency s to a system that is approximately linear, the response will be also a sinusoid of frequency s. The ratio of the amplitude of the response sinusoid to the stimulus sinusoid is called the *gain* of the system. There will also be a phase shift between the two signals. The ratio between the response and stimulus can be conveniently represented by a complex number, $G(s)$, which contains both the gain and phase information. The complex gain depends on frequency and is called the *transfer function* of the system. Its form indicates how the system will deal with stimuli of all possible frequencies. Nonperiodic stimuli are made up of sums of component sinusoids so that even for such stimuli the transfer function indicates how the system will alter the frequency components of the stimulus to produce those of the response. The system may be thought of as a device that operates on its input signal to produce a different output signal.

The method of finding the transfer function of a system from its differential equation involves the use of Laplace transforms. Such mathematical manipulations are the subject of many engineering textbooks and are beyond the scope of this review, but the system at hand—the oculomotor plant—can serve as an illustrative example. Since we are mainly interested in the modulation of the discharge rate, R_m, in Eq. 1 around that in the primary position, R_o, it is convenient in Eq. 1 to replace $(R_m - R_o)$ by the modulation ΔR_m, which results in

$$\Delta R_m = kE + r\frac{dE}{dt}. \qquad 3.$$

In these terms the transfer function $G(s)$ for the oculomotor plant has the form

$$G(s) = \frac{E(s)}{\Delta R_m(s)} = \frac{1}{sT_e + 1}. \qquad 4.$$

The most important parameter in Eq. 3 is not so much k and r themselves but their ratio, r/k, which is the time constant, T_e. This parameter describes how rapidly the eye will respond to changes in the central command. If, for example, the innervation ΔR_m suddenly changes from one value to another—called a *step command*—the eye will respond with an exponential movement with the time constant T_e. In the amount of time T_e, the eye will have traveled to within 37% (e^{-1}) of its final displacement. From the values of r and k given in Eq. 2, the value of T_e is 0.24 sec.

If the input is sinusoidal at a high frequency (s large), the transfer function is approximately $(sT_e)^{-1}$. This is the equation of an integrator: the output lags the input by 90° and for every increase in frequency of, say, a factor of ten, the output will decrease by the same factor. At a very low frequency (s small), the gain is constant (at 1.0) and the output will follow the input exactly. The ratio between the two should actually be k^{-1}, or 0.25. But a liberty has been taken in Eq. 4 by adjusting the scale factor so that the gain is 1.0 in the low frequency range, which is the range that contains the signal components of most eye movement commands. The reason for this choice is that little is known about amplification factors within the nervous system—that is, the ratio between the modulation of the typical cell in a neuron pool and that of a typical cell in a presynaptic pool—and one can avoid this problem by describing the rate modulations in terms of the physical variables they represent, in this case eye position, rather than in spikes sec^{-1}. In fact, the modulation of one motoneuron pool is not the total drive to the eye; ΔR_m should represent the difference in drive between the motoneuron pools of two antagonistic muscles, and it is simplest to express this drive in terms of the steady-state eye deviation it produces. The boundary between the high and low-frequency behavior in Eq. 4 occurs when, in the denominator, sT_e changes from being less than 1.0 to greater than 1.0. This occurs when the frequency equals $(2\pi T_e)^{-1}$. In the present case, this frequency is 0.66 Hz.

Before incorporating the transfer function of Eq. 4 into models of the oculomotor system, it was necessary to demonstrate that it describes the plant during all types of eye movements, to exclude possibilities that certain types of eye movements, such as vergence, might be made by a special subset of muscle fibers (Keller & Robinson 1972). Other studies showed that Eq. 4 also describes the plant for vestibularly induced eye movements (Skavenski & Robinson 1973) and that there was no stretch reflex that could possibly influence the nature or parameters of the transfer function (Keller & Robinson 1971). Thus, the oculomotor plant processes all signals alike, according to the transfer function in Eq. 4, regardless of the type of eye movement required. Keller (1973) found that closer inspection revealed a small relationship between R_m and eye acceleration, which becomes apparent when the eye changes velocity abruptly, as in a saccade. Thus, a better differential equation to describe the plant for high-frequency signals is

$$R_m = 100 + 4E + 0.95\,\dot{E} + 0.015\,\ddot{E}, \qquad\qquad 5.$$

where $\dot{E}$ and $\ddot{E}$ denote the first and second time derivatives of E. This equation may also be rewritten in the form of a transfer function:

$$\frac{E(s)}{\Delta R_m(s)} = \frac{e^{-s\tau}}{(sT_{e_1} + 1)(sT_{e_2} + 1)} \qquad\qquad 6.$$

where T_{e_1} is similar in value to T_e in Eq. 4 and T_{e_2} is a second, smaller time constant with a value of about 16 msec. The term containing T_{e_2} causes the eye to respond even more poorly to input signals that contain frequency components above 10 Hz. The term in the numerator is the Laplace representation of the latency or pure delay, τ, (about 8 msec, e.g. Fuchs & Luschei 1970) between changes in neuronal activity and changes in eye position. Eq. 6 describes the eye movement a bit more accurately when the input changes quickly, but, for the purposes of analyzing the behavior of some proposed model of an oculomotor subsystem, one need only choose this transfer function if its complexity is warranted by the nature of the input signals considered and the accuracy of simulation desired.

There have, of course, been reductionist attempts to relate the observed behavior described by Eq. 4 or 6 to the mechanical elements of the globe and eye muscles (Clark & Stark 1974, Collins 1975; for a review, see Robinson 1980), but, if one is to analyze one or several entire oculomotor subsystems, the emphasis must be on obtaining as simple a description of the plant as possible, commensurate with observed behavior. Thus, all details of motoneuron size, conduction velocities, muscle fiber types, muscle force-velocity and length-tension relationships are subordinated, or placed inside a "black box," and, at least at one level, it is sufficient if we can say what the plant does in response to any stimulus (the transfer function of the black box), if not how it does it. Consequently, we may regard the study of the oculomotor plant as complete at this level and pass on to a consideration of the signals that impinge upon it.

OCULOMOTOR SIGNALS

When it became possible to record from neurons in the brain stems of alert animals (usually rhesus monkeys), many researchers began to explore the oculomotor system, and the 1970s saw a burgeoning of such investigations. In these studies the behavior of cells could be related to eye movements, but one did not know the anatomical connections of the cells so one could only guess where the signals one observed came from and where they went. Very recently it has become possible to answer these questions (e.g. Yoshida et al 1979). It is possible, but difficult, to record intracellularly from cells or axons in alert animals to discover the signal the cell or axon carries during normal eye movements and then inject a tracer that fills the cells' processes. Such methods will undoubtedly produce important results in the 1980s, but as of this writing we are left with a variety of cell types in the cerebellum and brain stem, characterized by the signals they carry, and one can only try to guess how the cell groups might be interconnected. It should be stated at once that a workable arrangement has yet to be found and it would seem that important parts of the circuit are still undiscovered.

Nevertheless, it is interesting to look at the collection of signals observed so far to at least appreciate the sorts of problems that oculomotor physiologists currently face. As one might guess from Eq. 1, because the motoneurons receive their signals from premotor neurons, the signals observed on the latter often consist, in part, of various components proportional to eye position and eye velocity. The only general rule that has emerged so far is that the eye position components (the E signal) is independent of which subsystem moved the eye. Thus, if the eye went from fixation at 5 deg left to 10 deg right, a central neuron would change its discharge rate (if it carried an eye position component) from one value to another regardless of whether the movement were a saccade, a pursuit, vestibular, or optokinetic movement. On the other hand, the velocity components can depend very much on which system commanded the movement. For example, a neuron might participate vigorously (in a manner related to eye velocity) if the movement were a pursuit movement, but not if it were a saccade, and vice versa. This behavior suggests that the various occulomotor subsystems generate their own eye velocity commands, by visual or vestibular afferent signals according to the purpose of each subsystem, and they are then added together at the input of some element that converts this sum into a single eye position command.

Fortunately, the signal components seen on most central oculomotor neurons are analogues of various physical variables coded in discharge rates. The variables are such as: E, eye position in some plane of interest (usually horizontal); $\dot{E}$, eye velocity; $\dot{E}_p$, eye velocity commanded by the pursuit system; $\dot{E}_r$, eye velocity commanded by rapid eye movement systems (saccades and quick phases of nystagmus); H, head position in space; $\dot{H}$, head velocity; G, eye position in space; $\dot{G}$, eye velocity in space; and $\dot{e}$, the velocity of image motion on the retina. The following signals are those most commonly observed on oculomotor pathways.

Burst-Tonic Cells

Burst-tonic cells are found in a variety of locations, such as the vestibular nucleus (Keller & Daniels 1975), prepositus nucleus (Lopez-Barneo et al 1979), and the interstitial nucleus of Cajal (Büttner et al 1977, King & Fuchs 1977). The term "tonic" has been used, somewhat unfortunately, to denote a discharge rate component proportional to eye position, E, just as in Eq. 1. The term "burst" comes from the vigorous discharge that occurs during saccades in a certain, preferred, direction. The actual discharge rate, however, can be approximated by

$$R_{bt} = R_o + kE + r_p\dot{E}_p - r_v\dot{H} + r_r\dot{E}_r. \qquad 7.$$

R_o, in this as in other equations, is the discharge rate when the eye and head are stationary and the eye looks straight ahead. The term kE indicates, as in Eq. 1, that should the monkey fixate in the on- or off-direction, the neuron will increase or decrease its discharge rate. If, for example, the monkey looked 30 deg in the on-direction and k were 2.5 (spikes sec^{-1})(deg sec^{-1})$^{-1}$, the rate increase would be 75 (spikes sec^{-1}). If at any gaze angle the eyes were also moving in pursuit, the rate would change by the amount $r_p\dot{E}_p$. If, however, the eyes were moving at the same velocity during the execution of the vestibulo-ocular reflex, the extra discharge rate, $-r_v\dot{H}$, might be different even though eye velocity, $\dot{E}$, was equal to $-\dot{H}$, because the coefficients, r_p and r_v, describing the sensitivities to pursuit and vestibularily induced eye velocities, can be different (e.g. Keller & Daniels 1975). Similarly, the term $r_r\dot{E}_r$ describes the additional modulation of the discharge rate related to eye velocity during rapid eye movements such as saccades and quick phases.

The coefficients r_p, r_v, and r_r are, in general, not equal except in the special case of the motoneuron. In that case, since the eye will either be making a pursuit ($\dot{E}_p$), a vestibular ($-\dot{H}$), or a rapid ($\dot{E}_r$) eye movement, and r_p, r_v, and r_r all have the same value, all the velocity terms can be replaced by a single term $r\dot{E}$, as in eq. 1. The signal components of burst-tonic cells illustrate rather well that the individual oculomotor subsystems generate their own eye velocity commands but that they are combined before integration to produce a single eye position component. Burst-tonic cells that are not motoneurons appear to carry all of the components of the latter's signal and are probably close to the final output of the oculomotor system. It seems reasonable to suppose that many of them are a source of input to the motoneurons. These are, however, by no means the total source of such input since it is known that other cells (e.g. tvp cells, burst cells; see below) also project directly to motoneurons and provide important parts of their signal.

Primary Vestibular Afferents

The vestibulo-ocular reflex is a very important reflex common to all vertebrates. It allows animals to see and move at the same time by rotating the eyes backward in the head, when the head moves, so that the visual axes remain stationary in space. More concisely, the reflex makes eye velocity in the head, $\dot{E}$, approximately equal to $-\dot{H}$ so that $\dot{G}$ (their sum), which is eye velocity in space, is kept close to zero. The signal $\dot{H}$ is obtained from the semicircular canals. In the squirrel and rhesus monkey, within the frequency range of 0.03 to 3.0 Hz, the signal sent by the canals to the vestibular nucleus, encoded in the discharge rate R_{v_1} of the typical primary afferent fiber, is approximately

$$R_{v_1} = 90 + 0.4\ \dot{H} \qquad\qquad 8.$$

(Goldberg & Fernandez 1971, Miles & Braitman 1980). When the head is still, the resting discharge rate is 90 spikes sec^{-1}. If the head should start to turn at, say, 100 deg sec^{-1} in the plane of the canal, the rate will increase or decrease by 40 spikes sec^{-1}, depending on which direction the cupula of the canal deflects the haircells of the crista. Note that while head acceleration, $\ddot{H}$, is the raw stimulus that causes the endolymph to move, the hydraulics of the canal (i.e. the very large viscous resistance to endolymph flow) cause the cupula deflection, and so R_{v_1}, to reflect head velocity. From the standpoint of signal processing, the canals integrate: they are stimulated by head acceleration but produce an output signal proportional to its integral, namely head velocity.

At very low and high frequencies, the canal behavior departs from Eq. 8. The major cause of this departure is the elasticity of the cupula that causes it to return slowly to its resting position after the start of a rotation of constant velocity. This causes R_{v_1} also to return exponentially to its resting rate of 90 spikes sec^{-1} with a time constant, T_c, of about 4 sec in cat (Melvill Jones & Milsum 1971) and 5.7 sec in squirrel monkey (Fernandez & Goldberg 1971). Although one could continue to describe this behavior with differential equations, the equations would be cumbersome, especially when one considers, as we shall, two additional departures from Eq. 8, one at low and one at high frequencies. Such equations give less insight to those familiar with mathematical representations than the transfer function, which can reveal how the canals operate on input signals and produce output signals for either sinusoidal or transient stimuli. Consequently, the canal behavior is best described by the transfer function, which relates its input, $\dot{H}$, to its output, the change, ΔR_{v_1}, of the afferent discharge rate from the resting rate:

$$\frac{\Delta R_{v_1}(s)}{\dot{H}(s)} = \frac{sT_c}{(sT_c + 1)}. \qquad\qquad 9.$$

Just as with Eq. 4, the scale factor in Eq. 9 has been adjusted so that the gain is 1.0 in the high frequency range (when sT_c is larger than 1.0) and ΔR_{v_1} is no longer measured in spikes sec^{-1} but in terms (deg sec^{-1}) of the head velocity that causes it. At lower frequencies (when sT_c is less than 1.0, which is below 0.03 Hz if T_c is 6 sec), Eq. 9 indicates that the gain is approximately sT_c. This operator describes differentiation; the gain decreases in direct proportion to a decrease in frequency and the phase shift approaches a 90° lead—the canal output now reflects head acceleration rather than velocity. The step response of Eq. 9 describes the slow return of the discharge rate to its resting level during per- or post-rotatory stimulation.

Equation 9 describes the major dynamic behavior of the canals—that they transduce head velocity over most of the spectrum of head movements but fail to do so at rather low frequencies—and may be used for many purposes in simulating systems involving the canals, such as the vestibulo-ocular reflex. For other situations involving very high frequencies (brief transients) or low frequencies (rotations of long duration), a more complete description is

$$\frac{\Delta R_{v_1}(s)}{\dot{H}(s)} = \frac{sT_c}{(sT_c + 1)} \frac{sT_a}{(sT_a + 1)}(sT_z + 1) \qquad 10.$$

(Fernandez & Goldberg 1971). The term containing T_a describes peripheral adaptation with a time constant, T_a, of 80 sec. The term containing the time constant, T_z, which has the value 0.49 sec, describes a high-frequency, phase lead term. For purposes of simulation, one may use the canal signal predicted by either Eq. 9 or 10, depending on the stimulus being considered and the accuracy required.

Second-Order Vestibular Neurons

Primary vestibular afferents may be the only neurons in the brain to carry a purely vestibular signal (i.e. Eq. 9) since all second-order cells observed so far in the vestibular nuclei of alert animals carry other signals as well, which converge on these cells from central sources. The most prevalent signal is associated with the optokinetic system. There exists a visual pathway (at least in subprimates) from the retina to the nucleus of the optic tract (Collewijn 1975, Hoffmann & Schoppmann 1975) and thence through the nucleus reticularis tegmenti pontis to the vestibular nuclei (Precht & Strata 1980). This pathway allows cells in the vestibular nucleus to be driven by optokinetic stimuli (Dichgans et al 1973, Henn et al 1974, Waespe & Henn 1977). It is shown below that this signal is proportional to head velocity as determined by the visual system, so it may be denoted by $\dot{H}_{ok}$. If one uses the simpler canal model of Eq. 9, the signal ΔR_{v_2}, which is proportional to the discharge rate modulation of many second-order cells, can be written

$$\Delta R_{v_2} = \frac{sT_c}{(sT_c + 1)}\dot{H} + \dot{H}_{ok}. \qquad 11.$$

Experiments reveal that the signal $\dot{H}_{ok}$ created by rotation of an animal in the light is approximately equal to $[1/(sT_c + 1]\dot{H}$. In response to a sudden rotation at a constant velocity, this signal rises slowly (with time constant T_c) to a sustained level. This signal, then, just complements the canal signal, which, in this instance, rises instantly and then falls slowly back to zero. These signals are illustrated in Figure 2, where the nature of $\dot{H}_{ok}$ is discussed in more detail. When this signal is substituted into Eq. 11 for $\dot{H}_{ok}$, one has

$$\Delta R_{v_2} = \frac{sT_c}{(sT_c + 1)}\dot{H} + \frac{1}{(sT_c + 1)}\dot{H} = \dot{H}. \qquad 12.$$

This equation shows how, when vision is available during head rotation, the transient or high-frequency response of the canals (first term) is supplemented by the sustained or low frequency optokinetic response (second term) so that the total signal carried by those vestibular neurons is proportional to head velocity at all frequencies from the lowest (including zero) to the highest. In connection with Figure 2, below, it is shown that $\dot{H}_{ok}$ does not depend entirely on vision, but can affect the vestibulo-ocular reflex even in the dark.

Tonic Cells

Tonic cells are found in the reticular formation in the region of the abducens nucleus (Keller 1974) and the prepositus nucleus (Lopez-Barneo et al 1979). These cells carry a signal component proportional to eye position but do not burst during saccades, which may suggest that they do not carry any eye velocity signals at all. Closer inspection, however, shows that some tonic cells do have an additional modulation during pursuit ($\dot{E}_p$) and vestibularly induced movements ($-\dot{H}$). Thus, their discharge rate, R_t, may be described by

$$R_t = R_o + kE + r_p E_p - r_v H. \qquad 13.$$

This equation is similar to Eq. 7 for burst-tonic cells, except that the burst term, $r_r \dot{E}_r$, is missing. For some cells, however, r_p and r_v are zero so that such cells do reflect only eye position with no eye velocity signal components. In the reticular formation these cells are evidently small and hard to hold with a microelectrode and so have not been adequately studied. This is unfortunate since they may represent the output of an important element called the neural integrator, to be described subsequently.

Burst Cells

There are cells scattered in the pontine and mesencephalic reticular formations that discharge vigorously during saccades or quick phases with components in some particular direction and are otherwise silent. There are a variety of such cells (Luschei & Fuchs 1972). Long-lead burst cells discharge well in advance of an impending saccade and their discharge rate is usually poorly correlated with any specific aspect of the eye movement, such as saccade size or eye velocity. On the other hand, medium-lead burst neurons discharge at rates that are clearly related to rapid eye velocity in a certain direction (Keller 1974, Cohen & Henn 1972, van Gisbergen et al 1981). Thus, the behavior of their discharge rate, R_r, may be roughly approximated by

$$R_r = r_r \dot{E}_r. \hspace{4cm} 14.$$

These cells do not discharge during other types of movements, such as pursuit or fixation. Closer inspection of instantaneous discharge rate (van Gisbergen et al 1981) confirms that R_r is also related to eye acceleration, as Eq. 5 would suggest, and is also influenced by either a nonlinearity or a nonstationarity in the plant. The evidence to date indicates that medium-lead burst neurons contact motoneurons monosynaptically (Igusa et al 1980), cause a burst in motoneurons (and burst-tonic cells), and through them create saccadic eye movements. Since these burst neurons discharge at about 1000 spikes sec^{-1} during large saccades, when eye velocity is near 1000 deg sec^{-1} (in the monkey), r_r has a value of roughly 1.0 in that animal.

Pause Cells

There are cells in the reticular formation, especially clustered near the midline at the level of the abducens nerve rootlets (Keller 1974, Raybourn & Keller 1977), that fire at a fairly constant rate but pause during all rapid eye movements. Thus, their discharge, R_p, might be expressed,

$$R_p = R_o - r_{ps} \, |\dot{E}_r|. \hspace{3cm} 15.$$

The absolute value sign around E_r indicates that inhibition occurs for saccades in any direction, a characteristic that causes these cells also to be called omnidirection pause cells. Of course, as in all these equations, R cannot be negative. It is currently thought that pause cells inhibit burst cells and that the former must be turned off to allow the latter to create saccades (Keller 1977, King & Fuchs 1977).

Tonic-Vestibular-Pause Cells

A subset of cells in the vestibular nucleus send their axons via the medial longitudinal fasciculus (mlf) to the oculomotor nucleus to complete the vertical vestibulo-ocular reflex. The discharge rate, R_{tvp}, of the typical cell is,

$$R_{tvp} = 130 + 2.5 \, E + 0.47 \, \dot{E}_p - 0.98 \, \dot{H} - |\dot{E}_r| \hspace{2cm} 16.$$

(King et al 1976, Pola & Robinson 1978). This equation came from recordings from the fibers of these cells in the mlf, but, subsequently, many similar cells have been observed in the vestibular nucleus (Lisberger & Miles 1980) with activity related to horizontal as well as vertical movements. The term "tonic" refers, as usual, to the eye position signal, $2.5 \, E$, "vestibular" refers to the signal component, $-0.98\ \dot{H}$, and the last term indicates that the cell pauses during all rapid eye movements; hence the name tonic-vestibular-pause, or tvp. Most, if not all, of these cells are also second-order vestibular neurons and so constitute the middle portion of the three-neuron arc: the backbone of the vestibulo-ocular reflex. The surprising feature about the behavior described by Eq. 16 is that eye movement signals ($2.5\ E$, $0.47\ \dot{E}_p$, and $-|E_r|$) emerge from some central structure to converge on these second-order vestibular cells, thereby starting the process immediately, in the vestibulo-ocular reflex, of converting the canal signal (Eq. 8) to the motor signal (Eq. 2).

Gaze-Velocity Purkinje Cells

There is a class of Purkinje cells in the monkey flocculus that discharges in relation to eye velocity in space, $\dot{G}$, (Miles & Fuller 1975, Lisberger & Fuchs 1978). The discharge rate, R_{gPc}, of the typical cell is

$$R_{gPc} = 79 + 0.9\ (\dot{H} + \dot{E}) = 79 + 0.9\ \dot{G}. \qquad 17.$$

When the monkey makes pursuit movements with the head still ($\dot{H}$ zero), the rate modulates in proportion to eye velocity, $\dot{E}$. When the monkey is rotated but cancels its vestibulo-ocular reflex by fixating a target rotating with it ($\dot{E}$ zero), the rate modulates with head velocity, $\dot{H}$. During head rotation in the dark, when the vestibulo-ocular reflex causes $\dot{E}$ to be about $-0.9\ \dot{H}$, the modulation falls almost to zero. Since the sum of $\dot{H}$ and $\dot{E}$ is $\dot{G}$, the activity of the cell reflects eye velocity in space.

Retinal Image Slip

Several oculomotor subsystems are designed to prevent images from slipping about on the retina due to self motion or the motion in space of visual targets, presumably to improve vision. There are cells in the retinas of animals, such as rabbit and cat, called direction-selective cells, that respond to retinal image slip (e.g. Oyster et al 1972). These cells discharge most vigorously when images slip across the retina in one particular direction and the discharge rate is then a function (usually nonlinear) of the slip velocity, $\dot{e}$. Thus, the cells carry information of both the direction and speed of the retinal slip. A rather oversimplified, linear, description of the discharge rate, R_{ds}, of these cells in their direction of sensitivity is

$$R_{ds} = R_o + a\dot{e} = R_o + a(\dot{W} - \dot{G}), \qquad 18.$$

where a is some constant of proportionality. The second half of Eq. 18 simply expresses the fact that the velocity with which images move across the retina ($\dot{e}$) is the difference between the velocity of visual objects in space and the velocity of the eye in space or gaze ($\dot{G}$). The optokinetic system is mainly concerned with the motion of the entire visual environment relative to the observer so that the velocity of the visual objects in space in that case is the velocity of the seen world, $\dot{W}$. In nature, the entire seen world never moves en bloc so that $\dot{W}$ is normally zero and retinal slip, $\dot{e}$, is created by motion of the eye in space ($\dot{G}$). For a subject inside an optokinetic drum, $\dot{W}$ is the drum velocity. The pursuit system of foveate animals is concerned with motion of objects moving within the visual environment. In that case, $\dot{e}$ refers to the retinal motion of the image of a particular, selected target to be tracked and $\dot{W}$ should be replaced by $\dot{T}$, the velocity of that target in space. For the optokinetic system, the signal $\dot{e}$ is relayed from the retina to the nucleus of the optic tract in cats and rabbits. In primates, the retina apparently does not produce an $\dot{e}$ signal and the role of cortical and subcortical visual pathways in generating the optokinetic $\dot{e}$ signal is not yet clear. In such foveate animals, however, the striate cortex appears to be essential in generating the $\dot{e}$ signal for pursuit.

A Synthesis

The problem that remains is to propose neural circuits containing cells that carry the above signals and that also explain the overall organization of eye movements. Such proposals will become the working hypotheses to be shaped by subsequent anatomical and physiological findings until the circuits of the oculomotor system are correctly understood. In some oculomotor subsystems, considered below, this hope is close to realization; in others, much more study, theoretical as well as experimental, is needed. For example, many of the intermediary signals are still missing, as evidenced by the continual discovery of new oculomotor nuclei and the discovery of cells carrying new combinations of signal components in well-known nuclei (e.g. the vestibular nuclei), as well as the more newly discovered nuclei. Many of the cell groups described above (e.g. tonic cells) have not been studied in sufficient detail to allow quantitative relationships to be established with other cell groups. Some of the signal components described above are rather nonlinear and have been approximated as linear only for simplicity of discussion. Consequently there is still much guess-work in proposing which cell groups drive which other cell groups.

There are a few connections that seem likely. It is probable, for example, that burst-tonic cells receive much of their input from burst cells and tonic cells, although the appropriate anatomical connections are not yet established. On the other hand, there are puzzles: Purkinje cells in the flocculus, on anatomical evidence, are generally believed to inhibit cells in the vestibular nucleus; yet, in the monkey, no cells in the latter nucleus have been observed, so far, that appear capable of receiving the signal carried by gaze-velocity Purkinje cells described by Eq. 17. There are many similar problems and new data will be required to solve them. Of course, as one works backwards from the motoneuron in the visual-oculomotor subsystems, one soon reaches the interface between the motor and the sensory systems. In some cases, as in the optokinetic system, the action of the visual system is easy to describe, as in Eq. 18, and its link to the oculomotor system is easy to guess (see below). In the saccadic system, on the other hand, the signal given to the brain-stem saccadic circuits comes from a process of visual pattern recognition and cognitive target selection that is not understood at all. Furthermore, the specification of target locations is coded retinoptically. That is, in both the visual cortex and superior colliculus, the location of a visual target, with respect to the fovea, is indicated by which cells are active in these structures, according to the well-established retinoptic maps. Yet the final saccadic command, represented by the activity of burst neurons (Eq. 14) is a temporally-coded signal (intensity of discharge rate for a desired duration). How the transformation takes place between retinoptic specifications of target location by the position of a cell within a population, regardless of the exact nature of its discharge rate, to the specification of saccade size by the time course of the discharge rate of a cell, regardless of its exact location in a pool of similar neurons, is one of the major problems in understanding the oculomotor system. Consequently, one can only expect to proceed centrally just so far with the type of analogue signals listed above before coming to a point where the coding of signals changes and becomes more complicated.

A discussion of the details of the problems one encounters in trying to fit all of the above signals into a hypothetical network that describes the flow of the signal processing in the premotor circuits is beyond the scope of this review. Equations 1 through 18 are presented only to represent the state of our current knowledge of signals in the oculomotor system and to emphasize that the recent progress in oculomotor physiology has brought us close to one of the final goals of neurophysiology—understanding how the nervous system processes signals to produce behavior.

SIMPLE PREMOTOR CIRCUITS

If we are not yet certain how the cells described above are interconnected to perform their tasks, we at least know what those tasks are. Consequently, one can model parts of the oculomotor system at a higher level of organization in which relatively simple neural networks are represented by a transfer function that describes what must be done, although we do not yet understand just how it is done. One reason for proceeding in this way is to make it very clear, in the unambiguous language of mathematics, just what signal processing must be done by the neural networks so that one can then propose hypothetical networks that can be tested experimentally. Another reason is to permit the oculomotor theorist to use such premotor circuits as building blocks in efforts to model larger sections of the system, including one or more entire oculomotor subsystem. Such considerations are best illustrated by example.

The Vestibulo-Ocular Reflex

The vestibulo-ocular reflex, shown in Figure 1, is a good example of modeling at this level. As described earlier, this reflex causes the eyes to rotate in the direction opposite to a head rotation so that the direction of gaze in space is kept constant and the location of images of the visual environment on the retina are not perturbed by head motion. The dynamic behavior of this reflex has been well studied and the transfer function that relates eye position in the head, E, to head position in space, H, is

$$\frac{E(s)}{H(s)} = -g \frac{sT_{vor}}{(sT_{vor} + 1)} \frac{sT_a}{(sT_a + 1)}. \qquad 19.$$

The term containing T_a represents the peripheral adaptation of the canals already encountered in Eq. 10. The term containing T_{vor} is related to the behavior of the cupula and is similar to the term involving T_c in eq. 10. When a subject is rotated in the dark at a constant velocity, slow-phase eye velocity decreases exponentially with the time constant T_{vor}. The difference between T_c and T_{vor} will be explained shortly. At frequencies for which sT_a and sT_{vor} are both larger than 1.0 (above about 0.03 Hz), Eq. 19 has a value close to $-g$: the gain of the reflex (the minus sign simply indicates that eye motion is opposite to head motion). If g were 1.0, the eye movements would be perfectly compensatory. In cat and monkey, g is about 0.9. In man, it is around 0.6 during mental arithmetic, but rises to 0.95, even in the dark, if the subject tries to use the reflex by looking at imagined targets (Barr et al 1976). The gain, g, is also under some form of adaptive control (not shown), in which the cerebellum plays some part, to calibrate the gain, g, just after birth and maintain it during growth, disease, and aging (Ito et al 1974, Robinson 1976).

Since the behavior of the canals is known (Eq. 10), the behavior of the plant is known (Eq. 6), and the overall behavior is known (Eq. 19), one can deduce what must occur in the central pathways of the reflex and this is illustrated in Figure 1. The plant transfer function (Eq. 6) is shown on the right in Figure 1. The canal transfer function, on the left, has been modified

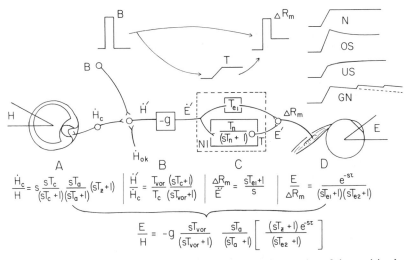

Figure 1 A model of the vestibulo-ocular reflex. The transfer function of the semicircular canals (*left*) and eyeball (*right*) are shown just below as transformations *A* and *D*. Transformation *B* describes how the time constant of the cupula, T_c, is replaced by the larger time constant, T_{vor}, of the vestibulo-ocular reflex. Transformation *C* describes the neural integrator (NI) in parallel with a direct signal path which effectively cancels the main lag of the plant (T_{e_1}). The bottom equation is the product of these four transformations and the overall gain factor $-g$. The terms in the square brackets affect behavior at high frequencies (greater than 3 Hz) and their effects approximately cancel out. Figures at the top illustrate how a neural pulse from burst neurons (*B*) is integrated to produce a step on tonic cells (T) so that the motoneurons (ΔR_m) can transmit a pulse-step waveform to create a saccade. At upper left are shown a normal saccade (N) and three types of abnormal saccades commonly seen in the clinic.

by abandoning the notation of discharge rate, ΔR_{v_1} in Eq. 10 and denoting the canal output as an internal signal, $\dot{H}_c$, reflecting head velocity, as reported by the canals. Also, head position, *H,* is used as the input instead of head velocity, $\dot{H}$ (or, in operator notation, sH). As Figure 1 indicates, there are two major signal transformations that occur in the central pathways. The first step (transformation *B,* Figure 1) creates an apparent increase in the cupula time constant. It is known, in the cat, that the cupula time constant, T_c, is about 4 sec (Melvill Jones & Milsum 1971), but the main time constant of the entire reflex, T_{vor}, is about 15 sec (Robinson 1976). A similar situation is seen in the monkey: T_c is about 6 sec (Fernandez & Goldberg 1971), but T_{vor} is around 16 sec (Waespe & Henn 1977, Buettner et al 1978). Waespe & Henn (1977) also discovered that the transformation of the major system time constant from T_c to T_{vor} occurs directly on second-order vestibular neurons so it must be effected by a signal that converges on those neurons from some central source. The signal is indicated as $\dot{H}_{ok}$ in Figure 1 because, as is argued below, there is good evidence that this signal is associated with the optokinetic system when vision is available, as well as with the vestibular system, and it is the same signal that appears in Eq. 11. The effect of the signal $\dot{H}_{ok}$, in the dark, is to create the transfer function marked B in Figure 1. When this function is multiplied by the transfer function for the canals (marked A in Figure 1), the terms containing T_c cancel out so that the resultant contains only the parameter T_{vor}. Consequently, the behavior of the overall reflex does not reflect the canal time constant, T_c, but a larger time constant, T_{vor}, created by central signal processing. Thus, transformation B simply describes what must be done to account for the experimental observations. How this transformation might actually be accomplished by neural circuits is described when the optokinetic system is discussed. In any event, the result of the transformation by $\dot{H}_{ok}$ is to create an improved representation of head velocity, $\dot{H}'$, which, after multiplication by $-g$, becomes an eye movement command, $\dot{E}'$.

The second, and more important, transformation of the central pathways creates the motoneuron signal (Eq. 1) from the eye-velocity command, $\dot{E}'$. The eye-velocity component in Eq. 1 indicates that there must be a direct projection of the canal signal, $\dot{E}'$, to the motoneurons. The origin of the eye position signal in Eq. 1 requires more explanation. The signal, in the dark, must obviously be created from the eye velocity signal, the only one available, which can only be done by integrating it—that is simply a restatement of the experimental observations—but how and where the integration is done is not known. Tonic cells (T, Figure 1; Eq. 13) mainly carry an eye position signal and could represent the output of the neural integrator (NI, Figure 1). They are located in the paramedian pontine reticular formation and this region is known to be vital for eye movements based on the results of lesions (Goebel et al 1971). The cerebellum is also necessary for correct operation of the integrator (Carpenter 1972, Robinson 1974), particularly the flocculus (Zee et al 1978). It would appear that the integrator is formed by some neural circuit involving links between the vestibulocerebellum and the pontine reticular formation. Various neural models have been proposed for the integrator (e.g. Kamath & Keller 1976) but they remain speculative.

The transfer function of the integrator shown in Figure 1 is that of a leaky integrator with a time constant T_n. The transfer function of a perfect integrator is $1/s$ (T_n infinitely large). Such an integrator would store the integral of a transient input signal and produce a constant output indefinitely in the absence of any new input. A leaky integrator would not hold a signal indefinitely; its output in the absence of any new input would slowly return to zero, exponentially, with the time constant T_n. In the normal situation, T_n is about 25 sec (Becker & Klein 1973), which is so large that for most practical purposes the integrator may be regarded as essentially perfect. If we make this assumption, the parallel combination of direct and integrator path in Figure 1 has the transfer function,

$$\frac{\Delta R_{\mathrm{m}}}{\dot{E}'} = T_{\mathrm{e_1}} + \frac{1}{s} = \frac{sT_{\mathrm{e_1}} + 1}{s} \tag{20.}$$

and this is shown as transformation C in Figure 1. To produce the correct ratio between the $\dot{E}'$ and E' signals at the motoneuron, the gain of the direct path must be $T_{\mathrm{e_1}}$. In terms of transfer functions, this gain causes the numerator in Eq. 20 to cancel the term in the denominator of the plant transfer function which contains its major time constant, $T_{\mathrm{e_1}}$. In this way the sluggishness of the plant, which would otherwise fail to respond appropriately to any oculomotor signals in the frequency range above about 0.7 Hz, is compensated so it does not interfere with the vestibulo-ocular reflex.

When all the transfer functions, A to D in Figure 1, are multiplied together, the overall function, shown at the bottom, results. This equation differs from the experimentally determined behavior (Eq. 19) by the terms in the square brackets, all of which affect performance at high frequencies (above about 3 Hz). Presumably these terms more or less cancel out since observations indicate that up to 7 to 8 Hz, which seems to be the upper limit for naturally occurring head movements, the gain is relatively independent of frequency and the phase shift is small. Thus, the descriptions of the signal processing done by the various sensory, motor, and central parts of the reflex in Figure 1 appear to be correct even though, in the two central transformations, we do not know the exact neural circuitry. It must be kept in mind that Figure 1 only describes what is done centrally, not how it is done. It shows, for example, the $\dot{E}'$ and E' signals arriving at the motor nucleus by separate pathways, yet Eq. 16 shows that the middle leg of the three-neuron arc—from second-order vestibular neurons to the motoneurons—already carries an eye position signal ($2.5\ E$) in addition to the vestibular eye-velocity signal ($-0.98\ \dot{H}$). Presumably the integrator sends part of the E signal back to the vestibular nucleus to join the eye velocity command, in addition to a direct projection to the motoneurons (Pola & Robinson 1978). Thus, Figure 1 does not propose an actual neural wiring diagram. No doubt future research will delineate the actual circuit, but even then there will be some utility in the diagram in Figure 1 since it presents an equivalent circuit that indicates most clearly what the real circuit does. Such a model of the vestibulo-ocular reflex is also useful for models of more complex eye movement systems in which this reflex plays only a part.

Saccades

Figure 1 also suggests how rapid eye movements (saccades or quick phases) are made. Although we do not understand how visual targets are selected and their coordinates sent to the lower motor machinery, it does seem possible to propose a scheme for the generation of saccades in premotor circuits close to the motoneuron. As already mentioned, burst cells, described roughly by Eq. 14, also produce an eye velocity command just as do the semicircular canals. Because of the shape of the burst, it produces a high-velocity movement of short duration rather than the slower, smoother movement of the vestibular signal. In order to translate the burst into a saccade, the motoneurons, according to Eq. 1, must receive both the velocity command (rE) directly, and its time integral, the eye position (kE), to hold the eye in its new position after the saccade. In this case, as shown by the wave forms at the top of Figure 1, the velocity command is the burst, B, which must project directly to the motoneurons. The integral of the burst is the step seen on tonic cells (T), which are presumed to project also to the motoneurons. The latter then carries the sum of the two signals, the burst-step shown by the waveform ΔR_m in Figure 1. Thus, exactly the same signal processing—a direct and an integrating pathway having the same relative gains—is needed as in the case of the vestibulo-ocular reflex and the question arises as to whether the two systems share a single integrator, as shown in Figure 1. Such an arrangement would certainly seem economical, but there are also fairly good experimental and theoretical reasons to indicate that it is correct (Robinson 1975). Separate integrators would require that there exist cells that are modulated in proportion to E for certain types of eye movements, such as quick phases, but not for others, such as slow phases. Yet, as already pointed out, all the cells described above carry a signal that reflects eye position regardless of the type of movement that carried the eye to that position. Such cells always reflect, for example, the change in eye position created by both the slow and fast phases of nystagmus. This same argument indicates that all other conjugate eye movement systems, such as the pursuit system, also share a common integrator. Optokinetic movements obviously also share this integrator since, as indicated by Eq. 11, their velocity command leaves the vestibular nucleus along with the canal signal.

Several pathological eye movements can be explained by the scheme shown in Figure 1. If the neural integrator is abnormally leaky (if T_n has decreased to, say, 2 sec), as occurs with cerebellar degeneration, the eyes of a patient will drift back from an eccentric position exponentially with the time constant T_n. To maintain eccentric fixation, the patient must make repeated eccentric saccades creating a pattern called gaze nystagmus (waveform GN, Figure 1). Occasionally the transmission of the pulse or step to the motoneurons is affected by some disease with the result that the pulse is too large or too small for the step. In that case, the pulse initially carries the eyes beyond, or causes it to fall short of, the steady-state position commanded by the step, to which the eyes then drift exponentially with the time constant T_{e_1} (Easter 1973, Bahill et al 1975b). The result is an overshoot or undershoot saccade shown by the waveforms OS and US in the upper right of Figure 1. Such waveforms do not occur normally because the size of the step, relative to the pulse, is adaptively adjusted by some cerebellar-dependent mechanism to eliminate such post-saccadic drifting movements (Optican & Robinson 1980). Over- and undershoot saccades are seen in patients only when the result of the lesion is beyond the capabilities of this repair mechanism. The saccade circuit in Figure 1 is useful, then, in that it offers a mechanistic explanation for a number of eye movement disorders frequently seen in the clinic or in lesioned animals (Zee & Robinson 1979a).

The circuit in Figure 1 is, at the moment, a useful description of some of the elementary signal processing that occurs immediately prior to the motor neurons. It must, however, be regarded as a working hypothesis until the actual neural circuits have been determined. Certainly, one of the major issues raised by Figure 1 is the neural integrator: Where is it and how does it work?

COMPLEX PREMOTOR CIRCUITS

It is the nature of neurophysiology that experimental data are difficult to obtain and usually fall far short of what one needs to explain the behavior of some neural system with any degree of certainty. Yet the desire to explain at least some aspect of neural behavior, however scant the evidence, leads the curious and frustrated investigator to try to extrapolate from the meager data available and the function, if known, of a particular system, and to propose hypotheses for how such a system might be wired together. Such a hypothesis usually consists of a specific circuit topology and the transfer functions required to produce the observed responses given the appropriate stimuli. To be useful, the hypothesis should be quantitative so that it can be tested by solving its equations—usually by computer simulation—to verify its predictions numerically. Such a mathematical hypothesis is usually called a model. The plausability of a model is related to the number of experimental observations it can simulate and the number of assumptions it requires. The main usefulness of a model is to make predictions that can be tested experimentally. If verified by sufficient testing, the model becomes an accepted theory for explaining the system's known behavioral repertoire.

In oculomotor physiology, it is interesting and useful to try to guess how the visual system may project to the premotor circuit in Figure 1 and use it to effect visually guided eye movements. It is in the invention and testing of such models that the concepts and practice of the analysis of control systems plays an important part. Despite all the transfer functions in Figure 1, the signal processing in that circuit is rather simple and there are not even any feedback loops. In visually guided movements, feedback plays a large role and models for their control utilize the concepts and analysis of feedback. Perhaps the simplest visually guided system is the optokinetic system. Recent experiments have provided enough clues to allow us to make a good guess about how this system is wired together.

Models of the Optokinetic System

The scheme in Figure 2 illustrates the general format for all models proposed for the optokinetic system and, inside the dashed lines, a specific model for its central processing. The vestibulo-ocular reflex is included because it and the optokinetic system are so intertwined in structure and function that it is not useful to consider one without the other. For present purposes, however, it may be greatly simplified from the system shown in Figure 1. It is sufficient to describe the canal dynamics by Eq. 9 and all the elements affecting the responses at very high and low frequencies may be ignored. The optokinetic system is only concerned with the velocity, not the position, of eye, head, and retinal images. Consequently, by using eye and head velocities, rather than positions, as the variables of interest, the action of the neural integrator need not be shown explicitly. The actual eye response is often nystagmoid but the quick phases, which only change eye position, may also be ignored and eye velocity may be taken as that of the slow phases. The summing junction on the right in Figure 2 expresses the fact, already indicated in Eq. 17, that eye velocity in space, $\dot{G}$, is the sum of eye velocity in the head, $\dot{E}$, and head velocity in space $\dot{H}$. The summing junction on the left indicates, as in Eq. 18, that the rate of image slip on the retina, $\dot{e}$, is the difference between the velocity of the visual environment, $\dot{W}$, (usually zero) and eye velocity in space, $\dot{G}$.

The path $\dot{G}$ indicates that the optokinetic system is a negative feedback system. Retinal slip is an error signal and the function of the feedback system is to try to keep eye velocity in space, $\dot{G}$, equal to the velocity of the visual world, $\dot{W}$, which will then minimize the error $\dot{e}$. In normal situations, of course, the visual world does not move. The optokinetic system did not evolve to track a moving visual world, but the visual world always moves relative to the head when an animal rotates in space and it is this situation with which the optokinetic system was designed to deal. As the waveforms in Figure 2 illustrate, when an animal begins to rotate at a constant velocity, the vestibulo-ocular reflex initially generates compensa-

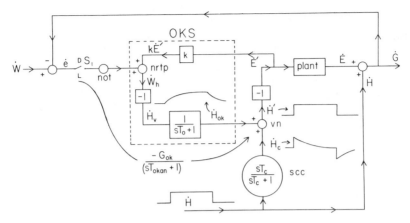

Figure 2 A model of the optokinetic system (OKS) and its connection with the vestibulo-ocular reflex. As shown by the waveforms, the optokinetic signal ($\dot{H}_{ok}$) is inserted in the vestibular nucleus (vn) and supplies the sustained activity during rotation at a constant velocity ($\dot{H}$) while the canal supplies the transient activity ($\dot{H}_c$). The general transfer function between $\dot{H}_{ok}$ and retinal slip velocity ($\dot{e}$) in the nucleus of the optic tract (not) is characterized by a gain G_{ok} and a time constant T_{okan}. The specific circuit shown in OKS utilizes a corollary discharge path ($k\dot{E}'$) from the vn to the nucleus reticularis tegmenti pontis (nrtp) and thence back to the vn, forming a positive feedback loop. Switch S_1 illustrates how going from light (L) to dark (D) opens the feedback loop $\dot{G}$.

tory eye movements but, as the cupula returns to its resting position, the canal signal $\dot{H}_c$ falls back to zero. As it does, eye velocity is no longer adequate, so a retinal slip is created that activates the optokinetic system and produces a rising signal, $\dot{H}_{ok}$, that supplements $\dot{H}_c$. Their sum, $\dot{H}'$, is thus a much better approximation to $\dot{H}$ than is $\dot{H}_c$ and eye velocity will be appropriate for the sustained, as well as the transient, portion of the rotation. It is this action that is described by Eq. 12.

The fact that the optokinetic signal appears on second-order neurons in the vestibular nucleus (vn, Figure 2) to augment the canal signal is discussed above in connection with Eq. 11. The visual input, $\dot{e}$, has been traced to the nucleus of the optic tract (not) in the pretectum in rat, cat, and rabbit (Cazin et al 1980, Hoffman & Schoppman 1975, Collewijn 1975), and this structure was shown to be essential for optokinetic responses. The situation remains unexplored in primates where there is clearly a cortical involvement in optokinetic responses (Atkinson 1979). The question has been: How does the signal $\dot{e}$ become transformed into $\dot{H}_{ok}$ and where are the neural pathways? First, one can at least characterize the nature of the $\dot{e}$–$\dot{H}_{ok}$ transformation. As indicated by the equation in Figure 2, it is largely described by a gain, G_{ok}, and a time constant, T_{okan}. These parameters have been determined experimentally. Although $\dot{W}$ is usually zero, it is convenient to study the optokinetic system in the laboratory if the animal or subject is stationary, in which case the visual environment is made to move by enclosing the subject inside a rotating drum. In this case, drum velocity, $\dot{W}$, may be considered the input, eye velocity, $\dot{E}$, the output, and the ratio $\dot{E}/\dot{W}$ as the gain—more technically, the steady-state, closed-loop gain—of the system. Experimentally, this gain is about 0.7 to 0.8 depending on drum speed and species. The gain $\dot{H}_{ok}/\dot{e}$ of the forward path (the part enclosed in the dashed lines in Figure 2) is the parameter G_{ok}, and the relationship between it and closed-loop gain, $\dot{E}/\dot{W}$, is

$$\frac{\dot{E}}{\dot{W}} = \frac{G_{ok}}{1 + G_{ok}}. \qquad\qquad 21.$$

G_{ok} must be about 3 to 5, according to this equation if eye velocity is to be 70 to 80% of drum velocity. The other feature of the optokinetic system is a lag with a time constant of T_{okan}. The value of T_{okan} may be measured by driving the system to a steady state by rotating the drum at a constant speed and then turning off the lights. This opens the feedback loop, as suggested by the switch S_1 in Figure 2, since the retina can no longer transmit $\dot{e}$ to the system. In this situation, nystagmus, called optokinetic after-nystagmus (OKAN), continues due to the activity stored in the lag element and $\dot{E}$ slowly falls back to zero. If one approximates this decline by an exponential, its time constant, T_{okan}, is about 15 to 20 sec (Cohen et al 1977).

The specific circuit shown in the box marked *OKS* in Figure 2 is only one of several models proposed for the optokinetic system. The simplest, topologically, is that the $\dot{e}$ signal projects from the not to the vn by a feed-forward path with a transfer function characterized by G_{ok} and T_{okan} as just discussed (Collewijn 1972, Schmid et al 1979). Specifically, such models do not contain internal, feedback pathways such as the one marked $k\dot{E}'$ in Figure 2. Such models deal reasonably well with responses to stimulation by optokinetic drums but fail to reflect important interactions with the vestibular system. These models cannot, for example, account for the transformation of the main vestibular time constant from T_c to T_{vor} described as transformation B in Figure 1. In contrast, the studies of Cohen et al (1977) and Raphan et al (1979) revived the old ideas of ter Braak from the 1930s (e.g. Rademaker & ter Braak 1948) that there was a common circuit—called a velocity storage element by Cohen and Raphan—that was shared by the optokinetic and vestibular systems and that carried a signal proportional to the nervous system's current estimate, $\dot{H}'$, of head velocity based on both visual and vestibular information. The signal in this element, if one likes functional interpretations, may be thought to represent the action of the inertia of the body: When the circuit is excited, it perseverates that activity, in the absence of new information, according to Newton's First Law of Motion (a body set in motion continues moving at a constant velocity unless acted upon by a force). In approximating such behavior, the circuit creates OKAN and transforms T_c into T_{vor}. This notion of a common velocity storage element would account naturally for the many similarities between the responses of the vestibulo-ocular reflex and the optokinetic system (Takemori 1974). For example, it explains why, in most circumstances, T_{okan} and T_{vor} are equal.

Two models have been proposed for this storage element. One suggests the existence of a storage element with a time constant of T_{vor} (or T_{okan}) that was fed by the primary vestibular afferent signal ($\dot{H}_c$) and sent its output ($\dot{H}_{ok}$) to the vn forming a feed-forward path in parallel with the primary afferents (Raphan et al 1979). The alternate model shown in Figure 2 uses a positive feedback loop to achieve the same result (Robinson 1977). This model is discussed in more detail because it illustrates how simple properties of feedback can be used in analyzing the behavior of a model. The rationale of the model is that the visual system desires to augment the canal signal by determining head velocity independently. To do this in Figure 2, a copy of eye velocity, $\dot{E}'$, (k is close to 1.0), is added to the velocity of the world with respect to the eye, $\dot{e}$, to recreate the velocity of the world with respect to the head, $\dot{W}_h$. This positive, internal feedback is similar to the efference copy notion of von Holst & Mittelstaedt (1950) and the corollary discharge of Sperry (1950) that has stimulated the ideas of Young (1977) in applying it to the oculomotor system. Since the visual world is presumed stationary, $-\dot{W}_h$ is taken by the nervous system to be the velocity of the head in space. This signal is denoted by $\dot{H}_v$ to indicate that it is head velocity according to the visual system. If this signal is to be used to augment the canal signal at low frequencies, its high frequency components must be removed, by the lag element $1/(sT_0 + 1)$ in Figure 2, so as not to duplicate the canal signal in the high-frequency range. For this purpose T_0 must have a value close to T_c.

Recent evidence suggests that this model might resemble the actual neural circuit. It has been discovered that the nucleus reticularis tegmenti pontis (nrtp) is a major relay station between the not and the vn in rat and cat (Cazin et al 1980, Precht & Strata 1980). Moreover, the vn projects back to the nrtp so as to form a positive feedback loop (W. Precht, unpublished observations). If one interprets the vn–nrtp projection as an eye velocity command $\dot{E}'$ rather than a canal signal, it would appear possible that the addition of $\dot{e}$ and $\dot{E}'$ to form $-\dot{H}_v$ occurs in the nrtp. There remain many aspects of the neurophysiology and anatomy of this circuit to be explored so, at the moment, one must still regard it as only a reasonable working hypothesis.

The theory of the operation of the circuit in Figure 2 can be seen by applying the feedback equation (similar to Eq. 21) to the relationship between H_{ok} and $\dot{e}$, taking into account that the feedback is positive,

$$\frac{\dot{H}_{ok}}{\dot{e}} = \frac{\dfrac{-1}{(sT_0 + 1)}}{1 - \dfrac{k}{(sT_0 + 1)}} = \frac{-\dfrac{1}{(1 - k)}}{[s\left(\dfrac{T_0}{1 - k}\right) + 1]} \overset{\Delta}{=} \frac{-G_{ok}}{sT_{okan} + 1} \; . \qquad 22.$$

Thus, G_{ok} is identified (by definition, $\overset{\Delta}{=}$) with the term $1/(1-k)$. If G_{ok} is, for example, 4.0, k would have the value 0.75. The value of T_{okan} is given by $T_0/(1-k)$. If T_0 were 4 sec (a typical value for T_c for laboratory animals such as the cat) then T_{okan} would be 16 sec. Thus, the model provides reasonable gains and time constants for the basic open- and closed-loop responses of the optokinetic system to visual stimulation. Of more interest is demonstrating the circuit's effect on the vestibulo-ocular reflex by deriving the transfer function between $\dot{H}_c$ and $\dot{E}'$ in the dark, again taking the positive feedback loop into account,

$$\frac{\dot{E}'}{\dot{H}_c} = \frac{-1}{1 - \dfrac{k}{(sT_0 + 1)}} = -\frac{1}{(1 - k)} \frac{(sT_0 + 1)}{[s\left(\dfrac{T_0}{1 - k}\right) + 1]} \overset{\Delta}{=} -\frac{T_{vor}}{T_0} \frac{(sT_0 + 1)}{(sT_{vor} + 1)} \; . \qquad 23.$$

The last step on the right defines T_{vor} as $T_0/(1-k)$ which, therefore, also has the same value as T_{okan}. If T_0 is equal to T_c, Eq. 23 describes the operator needed to transform T_c to T_{vor} illustrated by transformation B in Figure 1. Thus, even a rather simple application of feedback theory allows one to demonstrate that the neural connections in Figure 2, originally proposed to simulate optokinetic behavior, can also account for the increase in the apparent time constant of the canals seen even when vision is not available.

The scheme in Figure 2 is obviously oversimplified. There are known to be nonlinearities in the system (e.g. Collewijn 1972, 1975) that might account, for example, for the fact that during OKAN the decrease in $\dot{E}$ with time often departs markedly from an exponential waveform. There is another problem especially evident in the rabbit: When its optokinetic system is examined by opening the feedback loop by mechanically holding one eye (Collewijn 1969) or electro-optically stabilizing images on the retina (DuBois & Collewijn 1979), values for G_{ok} in the region of 50 to 100 are observed that would imply, from Eq. 21, a steady-state, closed-loop gain of 0.98 to 0.99. But the latter value is actually only about 0.8 at best, which, as indicated, requires that G_{ok} be only 5.0. The cause of this large discrepancy is unknown. In both the cat and primates, some of the optokinetic drive ($\dot{e}$) is apparently obtained by a cortical pathway (Hoffmann 1979, Atkinson 1979) about which little is known. In primates, which have a well-developed pursuit system, the question arises of where, in Figure 2, the pursuit command might be injected: Before or after the corollary discharge signal $\dot{E}'$ is fed back? Obviously more research is needed to settle these problems but the scheme in Figure 2 at least offers an interesting starting point for such studies.

The scheme in Figure 2 has also been able to provide a hypothetical explanation of a clinical entity called periodic alternating nystagmus (PAN) (Leigh et al 1981). The storage element in Figure 2 produces OKAN and prolongs the time course of per- and post-rotatory nystagmus. But after each of these phenomena, which are often called phase I of the entire nystagmus pattern, $\dot{E}$ reverses with a prolonged tail of low-velocity nystagmus called phase II, and that may be followed by yet another reversal, phase III, of even smaller velocity. To create phase II, it is generally supposed that some adaptive mechanism attempts to repair the original vestibular or optokinetic nystagmus by building up an opposing signal in the vestibular system and, in so doing, it creates an eye velocity bias in the opposite direction that is unmasked when phase I has disappeared. If such a repair mechanism is added to Figure 2, the positive feedback loop causes the system to generate damped oscillations during post-stimulus nystagmus, producing not only phase II but phase III as well. If, as is evident in several ways in PAN, such patients can no longer utilize retinal slip (the $\dot{e}$ signal) to generate following eye movements or prevent inappropriate slow eye movements, control over the parameter k may also be lost and, due to some unknown aspect of the lesion (thought to exist in the caudal-dorsal brain stem, flocculi, or both), k might drift to a value above 1.0. If that happens, the damped oscillations just described become undamped, sustained oscillations and resemble PAN with remarkable accuracy. The model also predicts the changes in the amplitude and phase of the PAN oscillations when such patients are subjected to rotatory, vestibular stimuli (Leigh et al 1981). These findings by no means validate the model but do indicate that the model in Figure 2 constitutes a hypothesis that can explain a rather large number of phenomena associated with the optokinetic system and optokinetic-vestibular interaction in both normal and pathological situations.

A Model of the Saccade Generator

Another visually guided system, which has received far more attention than the optokinetic system, is the saccadic system. When we look about, the nervous system must perceive a visual object with the peripheral retina, select it from all other objects and construct a command for the lower brain-stem circuits that will move the eye where we want it. We know very little about how any of this is done, especially target selection. In terms of brain-stem circuits, however, one can speculate on the more specific question of how the burst neurons in Figure 1 are governed so that the intensity (in spikes sec^{-1}) and duration of the burst is just correct to move the eyes by an amount appropriate to the retinal error of the selected target. A theory has been proposed for this task that uses a local feedback scheme (Zee et al 1976, Zee & Robinson 1979b, van Gisbergen et al 1981) and it is interesting to examine it in the context of this review because it certainly represents an application of control system's theory to the oculomotor system. There is, however, no question that the following hypothesis is still rather speculative and must be regarded as an interesting idea that needs further investigation. Its value, at the moment, is that it requires no unreasonable assumptions and seems to account for a large amount of normal and pathological saccadic behavior.

Until recently it has been assumed that saccades were generated in a retinotopic coordinate system. That is, if a target appeared 10 deg to the right of the fovea, the activity evoked at the retinal location would be translated, by some unspecified network, into the pulse carried by burst cells in such a manner that the burst had the correct intensity and duration to create a 10 deg saccade to the right. This proposed system would operate in a manner that was independent of initial eye position, being concerned only with changes in position. Yet it would appear that other motor systems probably use internal copies of eye position in the head and head position on the body, to create an internal representation of the location of a seen target in space to which, say, the hand is directed by a command in a body-oriented coordinate system. Most body movements must be directed by signals in such a reference frame. It may therefore be the case that the input to the saccade-generating circuit is, similarly, a signal proportional to desired eye position in the head: E_d in Figure 3. Several studies support this idea (Hallet & Lightstone 1976, Crommelinck et al 1977, Mays & Sparks 1980). The virtue of the idea is that it then becomes quite simple to construct a scheme for timing the saccadic pulse automatically by feedback. At the right in Figure 3 the neural integrator (NI), parallel feed-forward path, and plant are shown just as in Figure 1; for saccades it is best to use the plant transfer function of Eq. 6. The output of the neural integrator is an internal signal, E', proportional to instantaneous eye position. If, as shown in Figure

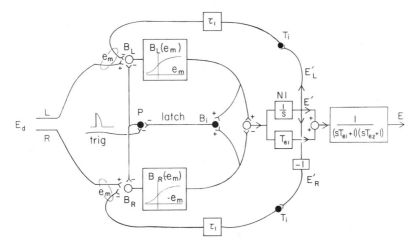

Figure 3 A model for generating saccades. An internal copy of eye position (E') from the neural integrator (NI) is hypothesized to feed back through inhibitory tonic cells (T_i) to be compared with a signal from higher centers proportional to desired eye position (E_d). The difference is motor error (e_m) which drives left and right burst cells (B_L, B_R). Pause cells (P) inhibit burst neurons. A trigger signal (trig) inhibits the pause cells to initiate a saccade. Inhibitory burst interneurons (B_i) keep pause cells off (latch) until e_m is zero, the burst is over, and the eye is on target. This model provides a hypothetical explanation for a large number of normal and abnormal saccadic behaviors.

3, this signal were compared to desired eye position, E_d, and their difference, motor error, e_m, were allowed to drive the burst cells, the eye would always be driven until E' matched E_d and e_m became zero, at which point the burst would end and the eyes would stop on target. In this way the burst amplitude and duration would automatically be always just appropriate to the desired saccade size. All that is required is an inhibitory, tonic-cell interneuron (T_i, Figure 3) to close the feedback loop.

Figure 3 shows left and right burst cells, B_L and B_R, driving the neural integrator in push-pull and being driven by separate feedback loops. The relationship between the instantaneous discharge rates B_L and B_R and motor error, e_m, is shown in the boxes in Figure 3. In the monkey this relation rises steeply as e_m, increased from zero and, for most cells, saturates around 1000 spikes sec^{-1} when e_m reaches 10 to 20 deg. It is the shape of this curve that allows the model to simulate saccades of all sizes with the correct waveform and peak velocities and durations that match experimental data. If one analyzes this feedback scheme, however, one discovers that the system is unstable. This odd situation comes about because saccades, to be useful, must be both fast and brief. The first feature requires a high gain so that even a small motor error of, say, 5 deg can cause a typical burst neuron to discharge at 700 spikes sec^{-1} and move the eye at a peak velocity of about 300 deg sec^{-1}. The second feature requires a wide bandwidth. The result is that the gain around the loop is greater than 1.0 at frequencies where the phase shift exceeds 180 deg, which, according to feedback theory, insures instability and oscillations. The neural integrator creates a constant 90° phase lag at all frequencies. Any delays in the loop, which are all lumped into τ_1, will create another 90° lag at the frequency $1/(4\tau_1)$. It would be reasonable to suppose that synaptic and recruitment delays around the loop could easily amount to 10 msec. This value for τ_1 causes a total phase shift around the loop of 180° at the frequency 25 Hz. According to theory, the system should oscillate near this frequency. In less technical terms, the system oscillates because e_m does not become zero until 10 msec after the eye has reached the target. Since the burst cells do not stop in time, the eye goes past the target before it stops. This creates an error, e_m, in the opposite direction so the contralateral burst cells are activated to bring the eye back on target. But they make the same overshoot mistake and the process continues, resulting in oscillations. The fact that the model predicts saccadic oscillations is interesting because there are several situations, normal and pathological, in which oscillations, discussed below, do occur.

Nevertheless, it seems startling to propose that nature had deliberately designed a control system to be unstable. A simple solution, however, which allows the high gain-wide bandwidth features to be retained, is to turn the circuit off when it is not in use. The pause cells (Eq. 15) seem to represent just such a mechanism. It is generally believed (and indirectly supported by anatomical studies) that pause cells inhibit burst cells so that saccades cannot occur so long as the former are active. Consequently, one might propose that saccades are initiated by turning off the pause cells. It is proposed that a trigger signal (trig, Figure 3) momentarily silences the pause cells and releases the burst cells to initiate a saccade to the position E_d. If, however, the trigger pulse disappears before the saccade is over, the pause cell would be allowed to reinhibit the burst cells and stop the saccade. To prevent this, it is proposed that an inhibitory burst interneuron exists (B_i, Figure 3) that can prevent the pause cell from firing so long as either the left or right burst cells are active. This pathway (latch, Figure 3) allows an on-going saccade to run to completion before the pause cells are released to once again disable the pulse generator.

This model has the following interesting features:

1. It produces saccades of all sizes that automatically have the correct velocity and duration.
2. It simulates the wave-shape of instantaneous burst rate for saccades of all sizes and direction.

3. It is compatible with the results of stimulating the pause cells during a saccade, which can stop the saccade momentarily in midflight (Keller 1977).
4. By decreasing the slope and amplitude of the burst-rate function [$B(e_m)$ in Figure 3], one can describe slow saccades seen in certain neurological disorders thought to affect the pontine reticular formation (Zee et al 1976).
5. If the primary saccade is over before the trigger signal is over, another small saccade in the opposite direction will occur as the system, without inhibition from the pause cells, starts to oscillate. Such movements do occur and are called dynamic overshoot. In the case of microsaccades, which have a short duration, inhibition of pause cells by the trigger signal may permit several, back-to-back, microsaccades to occur. Such microsaccadic oscillations are commonly observed in studies of human microsaccades. The model in Figure 3 mimics all these naturally occurring examples of saccadic oscillations (van Gisbergen et al 1981). If the pause cells can be kept off for many seconds, continuous saccadic oscillations occur similar to voluntary nystagmus.
6. There are patients whose abnormal eye movements can be described as an exaggeration of all the movements just mentioned in 5: very large dynamic overshoot and episodes of spontaneous oscillations called ocular flutter. Increasing the delay τ_1 and putting a lag in the latch circuit in the scheme in Figure 3 can simulate these abnormal movements (Zee & Robinson 1979b).

It has recently been demonstrated in monkeys that there is a very close relationship between instantaneous motor error $e_m(t)$ and instantaneous burst rate $B(t)$, which supports the idea that burst cells are driven by motor error e_m as indicated in Figure 3 by the relationship $B(e_m)$ (van Gisbergen et al 1981). This is the best neurophysiological evidence to date to support the hypothesis expressed in Figure 3. An obvious advantage of the hypothesis is that the feedback and latch circuits require only cell types (burst and tonic) that are already known to be present. An obvious disadvantage is that signals such as E_d have not been observed with microelectrodes and the model also fails to provide a role for other types of burst cells called long-lead burst neurons seen in the superior colliculi (Mays & Sparks 1980) and pontine reticular formation (Cohen & Henn 1972), although it is generally believed that such cells must play some role in shaping the burst delivered by those burst cells shown in Figure 1.

This model is seductive in its ability to mimic many properties of the physiology and neurophysiology, both normal and pathological, of saccades. It is especially seductive to the oculomotor neuro-ophthalmologist who must deal with such a bewildering array of eye movement disorders in the clinic that any reasonable hypothesis is better than none at all. Fortunately, many basic scientists in this field are seduced not at all, and for them the model is a challenge to be tested. For this purpose, one of the major virtues of a model should be appreciated: by its nature it is completely unambiguous. There is no way to misinterpret what is being proposed, so the testing of it can be equally unambiguous. To the oculomotor theoretician, the scheme is a challenge to produce a better model that can simulate all the phenomena listed above and more. As an example of the usefulness of the model in suggesting new experiments, it would never have occurred to van Gisbergen et al (1981) to look for, and find, a unique relationship between burst rate and motor error if the model in Figure 3, which evolved from an effort to explain clinical observations (Zee et al 1976), had not existed. In fact, constructing a model always makes one ask questions that would otherwise never have been thought of.

DISCUSSION

The examples offered in this review are intended to illustrate that the theory and practice of control systems analysis is not only useful in oculomotor neurophysiology but is rapidly becoming an essential tool. Clearly, the models in Figures 2 and 3 could not have been conceived, let alone tested, without the concepts and tools of control theory. In oculomotor physiology, we are approaching the stage of complexity where hypotheses will, of necessity, entail control systems analysis. Even if, for example, the scheme proposed in Figure 3 proves to be incorrect, the scheme that replaces it will certainly not be simpler and its conception and testing will require more, not less, systems analysis. When we start to study the interactions between subsystems such as those shown in Figures 1, 2, and 3, the dependence on quantitative analysis will increase. In fact, in visually guided systems such as the saccadic system, as we move above the level where movement commands are coded in discharge rate to those where the spatial distribution of activity within a population of cells becomes important, as in the superior colliculus, quantitative models will become more and more necessary and complex. In short, as neurophysiology grows up and addresses the main problem of how nervous tissue processes signals (or how the brain thinks), it must, in the end, come to grips with information processing and feedback regulation. It is hard to imagine how this will come about without using some form of the analytic techniques designed and utilized by those who have studied these phenomena from the time their examination was first recognized as a scientific discipline. It is now generally conceded that the facts that the nervous system is built with neurons and its effectors for movement are muscles do not constitute any reason for supposing that it should not be analyzed by theories of signal processing and feedback control, although one may readily admit that our analytical techniques must become more sophisticated to cope with higher brain functions. Clearly, if integrative neurophysiology of the mammalian brain is not to stagnate as a discipline incapable of interpreting its own data, it must progress from being descriptive to being interpretive, and it would appear that the oculomotor system is one of the first areas in which this transition is becoming clear. Thus, the question of whether control systems analysis is useful in the eye movement control system is perhaps inappropriate. The question is simply how rapidly can new data be acquired to fill in, verify, modify, and expand the systems models already being proposed.

In this review, specific models are described in some detail because there is no better way to allow the reader to judge whether the use of modeling is or is not useful in describing the neural circuits that control eye movements. Unfortunately, this practice has prevented the review of other models of oculomotor performance. Most of those models, however, describe the behavior of entire visuo-motor subsystems. Such models of eye tracking (Fender & Nye 1961, Dallos & Jones 1963, Stark et al 1962) or, in particular, saccadic tracking (Westheimer 1954, Young & Stark 1963, Robinson 1973, Wheeless et al 1966, Becker & Jürgens 1979, Bahill et al 1975a), or pursuit tracking (Robinson 1965, Yasui & Young 1975, Murphy et al 1975, Steinbach 1976, Miles 1977, Young 1977, Kowler et al 1978, Pola & Wyatt 1980) usually described the strategies of the human operator as a tracking machine, which is of interest to the psychologist and those concerned with man-machine systems. But these models had little direct impact on the neurophysiologist since they usually shed no light on how neural networks processed signals. These studies began in the early 1960s but fizzled out in the mid- to late 1970s because they could not cope with the complexities of trying to model the decision-making activities of high-order mental processes, and offered nothing testable for the electrophysiologist. They did, however, have a more subtle, long-range impact on oculomotor neurophysiology by formalizing the tasks of oculomotor subsystems and pointing out the general operations that must be done, such as integrating, amplifying, and sampling. They pointed out to us that the oculomotor control system was just that—a control system—and reminded us that there were established techniques for analyzing its behavior. It was their influence that caused the description of the canals in 1971 by Fernandez & Goldberg and of the oculomotor plant in 1970 by myself to be couched in terms of transfer functions. Those studies suggested that the qualitative, anecdotal descriptions that often characterized neurophysiological investigations of complex interactions had to be replaced by quantitative descriptions of some sort if they were to be useful in explaining behavior.

Thus, the major achievement of the behavioral models of the 1960s was to focus our attention on the need for quantitation and analysis but the more recent efforts to propose specific neural circuits, as in Figures 1 through 3, are more exciting because they address the basic issue of how neural circuits actually do process information. The specific models examined in this review are only examples of many phenomena that would benefit from modeling. How are the planes of motion sensitivities of the six semicircular canals transformed, in the vestibulo-ocular reflex, to the planes of rotation of the extraocular muscle pairs? How can long-lead burst neurons in the deep layers of the superior colliculi and in the pontine reticular formation be used to generate the burst seen on medium-lead burst cells, to challenge the model in Figure 3? Why are the velocities of saccades sometimes slowed (Morasso et al 1973), and sometimes increased (Haddad & Robinson 1977), during combined eye-head movements in various species? What constitutes appropriate pursuit and optokinetic stimuli in primates and how do they interact? Are the central commands for saccades generated in a polar or Cartesian coordinate system, or in a totally different system? These and many other questions are emminently suitable for attack by modeling.

The current modeling activity in oculomotor neurophysiology is a healthy sign because it is a measure of this discipline's vigor and growth. It marks the transition from gathering data to interpreting it. Many more data are still needed—they are the sine qua non of the models—and as they become available the use of control or systems theory will become increasingly important because, in the end, it will be the models, not the data, that will tell us how the oculomotor system works.

Literature Cited

Atkinson, J. 1979. Development of optokinetic nystagmus in the human infant and monkey infant: An analogue to development in kittens. In *Developmental Neurobiology of Vision,* ed. R. D. Freeman, 27:277–87. New York: Plenum. 446 pp.

Bahill, A. T., Bahill, K. A., Clark, M. R., Stark, L. 1975a. Closely spaced saccades. *Invest. Ophthalmol.* 14:317–20

Bahill, A. T., Clark, M. R., Stark, L. 1975b. Glissades-eye movements generated by mismatched components of the saccadic motoneuronal control signal. *Math. Biosci.* 26:303–18

Barr, C. C., Schultheis, L. W., Robinson, D. A. 1976. Voluntary, non-visual control of the human vestibuloocular reflex. *Acta Oto-Laryngol.* 81:365–75

Becker, W., Jürgens, R. 1979. An analysis of the saccadic system by means of double step stimuli. *Vision Res.* 19:967–83

Becker, W., Klein, H. M. 1973. Accuracy of saccadic eye movements and maintenance of eccentric eye positions in the dark. *Vision Res.* 13:1021–34

Buettner, U. W., Büttner, U., Henn, V. 1978. Transfer characteristics of neurons in vestibular nuclei of the alert monkey. *J. Neurophysiol.* 41:1614–28

Büttner, U., Büttner-Ennever, J. A., Henn, V. 1977. Vertical eye movement related unit activity in the rostral mesencephalic reticular formation of the alert monkey. *Brain Res.* 130:239–52

Carpenter, R. H. S. 1972. Cerebellectomy and the transfer function of the vestibulo-ocular reflex in the decerebrate cat. *Proc. R. Soc. London Ser. B* 181:353–74

Cazin, L., Precht, W., Lannou, J. 1980. Pathways mediating optokinetic responses of vestibular nucleus neurons in the rat. *Pflügers Arch.* 384:19–29

Clark, M. R., Stark, L. 1974. Control of human eye movements. I. Modelling of extraocular muscles. II. A model for the extraocular plant mechanism. III. Dynamic characteristics of the eye tracking mechanism. *Math. Biosci.* 20:191–265

Cohen, B., Henn, V. 1972. Unit activity in the pontine reticular formation associated with eye movements. *Brain Res.* 46:403–10

Cohen, B., Matsuo, V., Raphan, T. 1977. Quantitative analysis of the velocity characteristics of optokinetic nystagmus and optokinetic after-nystagmus. *J. Physiol. London* 270:321–44

Collewijn, H. 1969. Optokinetic eye movements in the rabbit: Input-output relations. *Vision Res.* 9:117–32

Collewijn, H. 1972. An analog model of the rabbit's optokinetic system. *Brain Res.* 36:71–88

Collewijn, H. 1975. Direction selective units in the rabbit's nucleus of the optic tract. *Brain Res.* 100:489–508

Collins, C. C. 1975. The human oculomotor control system. In *Basic Mechanisms of Ocular Motility and Their Clinical Implications,* ed. G. Lennerstrand, P. Bach-y-Rita, 24:145–80. Oxford: Pergamon. 584 pp.

Crommelinck, M., Guitton, D., Roucoux, A. 1977. Retinotopic versus spatial coding of saccades: Clues obtained by stimulating deep layers of cat's superior colliculus. In *Control of Gaze by Brain Stem Neurons,* ed. R. Baker, A. Berthoz, pp. 425–35. Amsterdam: Elsevier. 514 pp.

Dallos, P. J., Jones, R. W. 1963. Learning behavior of the eye fixation control system. *Inst. Electr. Electr. Eng. Trans. Automatic Controls* AC-8:218–27

Dichgans, J., Schmidt, C. L., Graf, W. 1973. Visual input improves the speedometer function of the vestibular nuclei in the goldfish. *Exp. Brain Res.* 18:319–22

Dubois, M. F. W., Collewijn, H. 1979. The optokinetic reactions of the rabbit: Relation to the visual streak. *Vision Res.* 19:9–17

Easter, S. S. Jr. 1973. A comment on the glissade. *Vision Res.* 13:881–82

Evarts, E. V. 1968. A technique for recording activity of subcortical neurons in moving animals. *Electroencephalogr. Clin. Neurophysiol.* 24:83–86

Fender, D. H., Nye, P. W. 1961. An investigation of the mechanisms of eye movement control. *Kybernetik* 1:81–88

Fernandez, C., Goldberg, J. M. 1971. Physiology of peripheral neurons innervating semicircular canals of the squirrel monkey. II. Response to sinusoidal stimulation and dynamics of peripheral vestibular system. *J. Neurophysiol.* 34:661–75

Fuchs, A. F., Luschei, E. S. 1970. Firing patterns of abducens neurons of alert monkeys in relationship to horizontal eye movement. *J. Neurophysiol.* 33:382–92

Goebel, H. H., Komatsuzaki, A., Bender, M. B., Cohen, B. 1971. Lesions of the pontine tegmentum and conjugate gaze paralysis. *Arch. Neurol.* 24:431–40

Goldberg, J. M., Fernandez, C. 1971. Physiology of peripheral neurons innervating semicircular canals of the squirrel monkey. I. Resting discharge and response to constant angular accelerations. *J. Neurophysiol.* 34:635–60

Haddad, G. M., Robinson, D. A. 1977. Cancellation of the vestibuloocular reflex during active and passive head movements in the normal cat. *Soc. Neurosci.* 3:155 (Abstr.)

Hallett, P. E., Lightstone, A. D. 1976. Saccadic eye movements towards stimuli triggered by prior saccades. *Vision Res.* 16:99–106

Henn, V., Cohen, B. 1973. Quantitative analysis of activity in eye muscle motoneurons during saccadic eye movements and positions of fixation. *J. Neurophysiol.* 36:115–26

Henn, V., Young, L. R., Finley, C. 1974. Vestibular nucleus units in alert monkeys are also influenced by moving visual fields. *Brain Res.* 71:144–49

Hoffman, K.-P. 1979. Optokinetic nystagmus and single-cell responses in the nucleus tractus opticus after early monocular deprivation in the cat. See Atkinson 1979, pp. 63–72

Hoffman, K.-P., Schoppmann, A. 1975. Retinal input to direction selective cells in the nucleus tractus opticus of the cat. *Brain Res.* 99:359–66

Igusa, Y., Sasaki, S., Shimazu, H. 1980. Excitatory premotor burst neurons in the cat pontine reticular formation related to the quick phase of vestibular nystagmus. *Brain Res.* 182:451–56

Ito, M., Shiida, T., Yagi, N., Yamamoto, M. 1974. The cerebellar modification of rabbit's horizontal vestibulo-ocular reflex induced by sustained head rotation combined with visual stimulation. *Proc. Jpn. Acad.* 50:85–89

Kamath, B. Y., Keller, E. L. 1976. A neurological integrator for the oculomotor control system. *Math Biosci.* 30:341–52

Keller, E. L. 1973. Accommodative vergence in the alert monkey. *Vision Res.* 13:1565–75

Keller, E. L. 1974. Participation of the medial pontine reticular formation in eye movement generation in monkey. *J. Neurophysiol.* 37:316–32

Keller, E. L. 1977. Control of saccadic eye movements by midline brain stem neurons. See Crommelinck et al 1977, pp. 327–36

Keller, E. L., Daniels, P. D. 1975. Oculomotor related interaction of vestibular and visual stimulation in vestibular nucleus cells in alert monkey. *Exp. Neurol.* 46:187–98

Keller, E. L., Robinson, D. A. 1971. Absence of a stretch reflex in extraocular muscles of the monkey. *J. Neurophysiol.* 34:908–19

Keller, E. L., Robinson, D. A. 1972. Abducens unit behavior in the monkey during vergence movements. *Vision Res.* 12:369–82

King, W. M., Fuchs, A. F. 1977. Neuronal activity in the mesencephalon related to vertical eye movements. See Keller 1977, pp. 319–26

King, W. M., Lisberger, S. G., Fuchs, A. F. 1976. Responses of fibers in medial longitudinal fasciculus (mlf) of alert monkeys during horizontal and vertical conjugate eye movements evoked by vestibular or visual stimuli. *J. Neurophysiol.* 39:1135–49

Kowler, E., Murphy, B. J., Steinman, R. M. 1978. Velocity matching during smooth pursuit of different targets on different backgrounds. *Vision Res.* 18:603–5

Leigh, R. J., Robinson, D. A., Zee, D. S. 1981. A quantitative hypothesis for periodic alternating nystagmus. *Proc. NY Acad. Sci.* In press

Lisberger, S. G., Fuchs, A. F. 1978. Role of primate flocculus during rapid behavioral modification of vestibuloocular reflex. I. Purkinje cell activity during visually guided horizontal smooth-pursuit eye movements and passive head rotation. *J. Neurophysiol.* 41:733–63

Lisberger, S. G., Miles, F. A. 1980. Role of primate medial vestibular nucleus in long-term adaptive plasticity of vestibuloocular reflex. *J. Neurophysiol.* 43:1725–45

Lopez-Barneo, J., Darlot, C., Berthoz, A. 1979. Functional role of the prepositus hypoglossi in the control of gaze. In *Reflex Control of Posture and Movements*, ed. R. Granit, O. Pompeiano, pp. 668–79. Amsterdam: Elsevier. 827 pp.

Luschei, E. S., Fuchs, A. F. 1972. Activity of brain stem neurons during eye movements of alert monkeys. *J. Neurophysiol.* 35:445–61

Mays, L. E., Sparks, D. L. 1980. Dissociation of visual and saccade-related responses in superior colliculus neurons. *J. Neurophysiol.* 43:207–32

Melvill Jones, G., Milsum, J. H. 1971. Frequency-response analysis of central vestibular unit activity resulting from rotational stimulation of the semicircular canals. *J. Physiol. London* 219:191–215

Miles, F. A. 1977. The primate flocculus and eye-head coordination. In *Eye Movements*, ed. B. A. Brooks, F. J. Bajandas, pp. 75–92. New York: Plenum. 223 pp.

Miles, F. A., Braitman, D. J. 1980. Longterm adaptive changes in primate vestibulo-ocular reflex. II. Electrophysiological observations on semicircular canal primary afferents. *J. Neurophysiol.* 43:1426–36

Miles, F. A., Fuller, J. H. 1975. Visual tracking and the primate flocculus. *Science* 189:1000–2

Morasso, P., Bizzi, E., Dichgans, J. 1973. Adjustment of saccade characteristics during head movements. *Exp. Brain Res.* 16:492–500

Murphy, B. J., Kowler, E., Steinman, R. M. 1975. Slow oculomotor control in the presence of moving backgrounds. *Vision Res.* 15:1263–68

Optican, L. M., Robinson, D. A. 1980. Cerebellar-dependent adaptive control of the primate saccadic system. *J. Neurophysiol.* In press

Oyster, C. W., Takahashi, E., Collewijn, H. 1972. Direction-selective retinal ganglion cells and control of optokinetic nystagmus in the rabbit. *Vision Res.* 12:183–93

Pola, J., Robinson, D. A. 1978. Oculomotor signals in the medial longitudinal fasciculus of the monkey. *J. Neurophysiol.* 41:245–59

Pola, J., Wyatt, H. J. 1980. Target position and velocity: The stimuli for smooth pursuit eye movements. *Vision Res.* 20:523–34

Precht, W., Strata, P. 1980. On the pathways mediating optokinetic responses in vestibular nuclear neurons. *Neuroscience* 5:777–87

Rademaker, G. G. J., ter Braak, J. W. G. 1948. On the central mechanism of some optic reactions. *Brain* 71:48–76

Raphan, T., Matsuo, V., Cohen, B. 1979. Velocity storage in the vestibuloocular reflex arc (VOR). *Exp. Brain Res.* 35:229–48

Raybourn, M. S., Keller, E. L. 1977. Colliculoreticular organization in primate oculomotor system. *J. Neurophysiol.* 40:861–78

Robinson, D. A. 1965. The mechanics of human smooth pursuit eye movement. *J. Physiol. London* 180:569–91

Robinson, D. A. 1970. Oculomotor unit behavior in the monkey. *J. Neurophysiol.* 33:393–404

Robinson, D. A. 1973. Models of the saccadic eye movement control system. *Kybernetik* 14:71–83

Robinson, D. A. 1974. The effect of cerebellectomy on the cat's vestibulo-ocular integrator. *Brain Res..* 71:195–207

Robinson, D. A. 1975. Oculomotor control signals. See Collins 1975, pp. 337–74

Robinson, D. A. 1976. Adaptive gain control of vestibuloocular reflex by the cerebellum. *J. Neurophysiol.* 39:954–69

Robinson, D. A. 1977. Linear addition of optokinetic and vestibular signals in the vestibular nucleus. *Exp. Brain Res.* 30:447–50

Robinson, D. A. 1978. The functional behavior of the peripheral oculomotor apparatus: A review. In *Disorders of Ocular Motility*, ed. G. Kommerell, pp. 43–61. München: Bergmann. 386 pp.

Robinson, D. A. 1980. Models of the mechanics of the orbit. In *Models of Oculomotor Behavior and Control*, ed. B. L. Zuber, W. Palm Beach, Fla: CRC Press. In press

Schiller, P. H. 1970. The discharge characteristics of single units in the oculomotor and abducens nuclei of the unanesthetized monkey. *Exp. Brain Res.* 10:347–62

Schmid, R., Zambarbieri, D., Sardi, R. 1979. A mathematical model of the optokinetic reflex. *Biol. Cybernetics* 34:215–25

Skavenski, A. A., Robinson, D. A. 1973. Role of abducens neurons in the vestibuloocular reflex. *J. Neurophysiol.* 36:724–38

Sperry, R. W. 1950. Neural basis of spontaneous optokinetic response produced by visual inversion. *J. Comp. Physiol. Psychol.* 43:482–89

Stark, L., Vossius, G., Young, L. R. 1962. Predictive control of eye tracking movements. *Trans. Inst. Radio Eng. Prof. Grp. on Human Factors in Elect.* HFE-3:52–56

Steinbach, M. J. 1976. Pursuing the perceptual rather than the retinal stimulus. *Vision Res.* 16:1371–76

Takemori, S. 1974. The similarities of optokinetic after-nystagmus to the vestibular

nystagmus. *Ann. Otol. Rhinol. Laryngol.* 83:230–38

van Gisbergen, J. A. M., Robinson, D. A., Gielen, S. 1981. A quantitative analysis of the generation of saccadic eye movements by burst neurons. *J. Neurophysiol.* In press

von Holst, E., Mittelstaedt, H. 1950. Das reafferenzprincip. *Naturwissenschaften* 37:464–76

Waespe, W., Henn, V. 1977. Neuronal activity in the vestibular nuclei of the alert monkey during vestibular and optokinetic stimulation. *Exp. Brain Res.* 27:523–38

Westheimer, G. 1954. Eye movement responses to a horizontally moving visual stimulus. *AMA Arch. Ophthalmol.* 52:932–41

Wheeless, L. L. Jr., Boynton, R. M., Cohen, G. H. 1966. Eye movement responses to step and pulse-step stimuli. *J. Opt. Soc. Am.* 56:956–60

Yasui, S., Young, L. R. 1975. Eye movements during after-image tracking under sinusoidal and random vestibular stimulation. See Collins 1975, pp. 509–13

Yoshida, K., McCrea, R. A., Berthoz, A., Vidal, P. 1979. Morphological and physiological characteristics of burst inhibitory neurons in the alert cat. *Soc. Neurosci.* 5:391 (Abstr.)

Young, L. R. 1977. Pursuit eye movements—what is being pursued? See Crommelinck et al 1977, pp. 29–36

Young, L. R., Stark, L. 1963. Variable feedback experiments testing a sampled data model for eye tracking movements. *Inst. Electr. Electr. Eng. Trans. Prof. Grp. on Human Factors in Elect.* HFE-4:38–51

Zee, D. S., Optican, L. M., Cook, J. D., Robinson, D. A., Engel, W. K. 1976. Slow saccades in spinocerebellar degeneration. *Arch. Neurol.* 33:243–51

Zee, D. S., Robinson, D. A. 1979a. Clinical applications of oculomotor models. In *Topics in Neuro-Ophthalmology*, ed. H. S. Thompson, pp. 266–85. Baltimore: Williams & Wilkins. 377 pp.

Zee, D. S., Robinson, D. A. 1979b. An hypothetical explanation of saccadic oscillations. *Ann. Neurol.* 5:405–14

Zee, D. S., Yamazaki, A., Gücer, G. 1978. Ocular motor abnormalities in trained monkeys with floccular lesions. *Soc. Neurosci.* 4:168 (Abstr.)

Ann. Rev. Neurosci. 1981. 4:505–23

CYTOSKELETAL ELEMENTS IN NEURONS

❖11561

Dennis Bray and David Gilbert[1]

MRC Cell Biophysics Unit, 26 Drury Lane, London, WC2B 5RL, England

INTRODUCTION

> Neurotubules and neurofilaments are homologous with cytoplasmic microtubules and microfilaments which are seen with the electron microscope in most plant and animal cells.

> Their subunits are transported to the growth cone of the outgrowing axon, where the subunits are added to the distal ends of the neurotubules.

These sentences from a recent edition of an influential textbook of developmental neurobiology unintentionally illustrate the need for this review. A flood of information has been obtained in the last few years on the composition and action of fibrillar proteins in vertebrate cells, much of which has implications for the nervous system. But it is difficult for a professional neuroanatomist or neurophysiologist to survey this mass of material and extricate facts that are relevant to his or her work. It is easy, as in the above example, for someone outside the field to confuse one filament with another or to draw a conclusion for which there is no evidence.

We have examined the literature on filaments, tubules, and the like and selected those findings that seem most relevant to neurobiologists. Our emphasis is molecular rather than morphological and we have concentrated on material published in the last five years. Even so we leave out several important subjects, such as the role of fibrillar proteins in neural pathology and in synaptic transmission. Access to the literature in these areas is given by two excellent previous reviews—those by Wuerker & Kirkpatrick (1972) and by Puszkin & Shook (1979), to both of which we are greatly indebted.

[1]Deceased December 11, 1979.

0147-006X/81/0301-0505$01.00

Elsewhere, for the convenience of the reader, we give one or two recent entries to the literature rather than an exhaustive bibliography. We apologize to those of our colleagues whose work is, for these reasons, not mentioned.

MORPHOLOGY

First, some definitions: the main types of fibrillar elements in their assembled form are recognized and distinguished by electron microscopy.

MICROTUBULES (ALSO CALLED NEUROTUBULES)   *Microtubules* are long straight protein tubes with an outer diameter of about 25 nm and a central core of about 10 nm. In cross-section, the microtubular wall is seen as 13 globular tubulin subunits and these comprise a simple 4 X 5 nm lattice on the microtubule surface. Microtubules are ubiquitous in eucaryotic cells and prominent in most axons and dendrites where they exist in longitudinal arrays. They are the fibrous components of eucaryotic cilia that are present on many cells of the nervous system.

MICROFILAMENTS   The term, *microfilament,* is now reserved for actin-containing filaments with a diameter of about 7 nm. These are homologous with the thin filaments of striated muscle and comprise two strings of globular subunits, about 5.5 nm in diameter, which are arranged in a helix with a pitch of about 74 nm. Microfilaments are found in virtually all animal cells in either loosely anastomosing networks or in parallel bundles (or stress fibers); the latter form has not been seen in neurons.

THICK FILAMENTS   *Thick filaments* are the myosin-containing assemblies of striated muscle and are about 12 nm in diameter. They have not been seen in neurons or other non-muscle cells.

NEUROFILAMENTS   *Neurofilaments* constitute, together with microtubules, the chief cytoskeletal elements seen in most mature nerve cells. They are about 10 nm in diameter, with irregular lateral projections in sectioned material, but with very long smooth cylinders in negative stain. They are more stable than microtubules and are probably the chief component of *neurofibrillae,* a light-microscopic term for the elaborate systems of fibrils in neurons, seen particularly with reduced silver stains. Unlike microtubules and microfilaments, these are unique to nerve cells.

GLIAL FILAMENTS   The counterpart of neurofilaments seen in glial cells, particularly fibrous astrocytes, *glial filaments* are distinct from neurofilaments in morphology, appearing slightly narrower (8 nm) and more densely packed and having a different protein composition.

INTERMEDIATE FILAMENTS   *Intermediate filament* is a generic term now widely used to describe all filaments intermediate in diameter between thick (myosin) and thin (actin) filaments. This category includes neurofilaments, glial filaments, epidermal keratin filaments, and the 10 nm filaments found in fibroblasts and in striated and smooth muscle. Each of these types has a distinct protein composition.

COLLAGEN   *Collagen* fibrils in the peripheral nervous system range between 25 nm and 55 nm in diameter, depending on the type and cell of origin. They are extracellular and therefore outside the scope of this review, except that they are sometimes so abundant as virtually to constitute an exoskeleton for the axon.

COMPOSITION

Table 1 lists 16 proteins that either form filaments or are associated with them in vertebrate cells, together with what we know of their presence or absence in nerve cells. Much of this information has been acquired in the last five years and there is no doubt that a similar table compiled a few years hence will be much longer. The reason for this rapid development is that several powerful techniques have become available for the analysis of these relatively insoluble proteins. Chief among these are acrylamide gel electrophoresis in the presence of detergent and fluorescent immunochemistry, together with such complementary methods as isoelectric focusing (O'Farrell 1975), partial proteolysis on gels (Cleveland et al 1977), and, the most recent to be applied to fibrillar proteins, antibody production by clonal selection (Kohler & Milstein 1975). The resolution and sensitivity provided by such methods is so great that even small amounts of contamination will be detectable. The limitation in many contemporary experiments becomes the preparation of a suitably pure sample—of neurofilaments, say, or synaptosomes.

Many of these proteins exist in multiple "isozymic" forms. Actin is the best documented in this respect and six different actin amino acid sequences have been identified (Vandekerckhove & Weber 1978). One form of actin (α) is found in the thin filaments of striated myofibrils, another (γ_1) in the actomyosin extracted from smooth muscle. Non-muscle tissues contain β and γ_2 in about equal proportions and this is true for neurons (Flanagan & Lin 1979). Subcellular fractions such as cytoskeletal or membrane prepa-

Table 1 Cytoskeletal proteins of neurons

Protein	Isozyme	Molecular weight[a]	Present in nervous system?
Actin	α[b] β γ[b]	42	β and γ actin are present in neurons (Flanagan & Lin 1979)
Tropomyosin	muscle[b] non-muscle[b]	35 30	Non-muscle form is present in neurons (Bretscher & Weber 1978)
α-Actinin		100	Immunofluorescent reaction in neurons (Jokusch et al 1979)
Filamin		240	Apparently absent from brain (Wallach et al 1978)
Spectrin		250	Only in erythrocytes (Hiller & Weber 1977)
Profilin		16	Abundant in brain (Carlsson et al 1977)
Myosin heavy-chain	muscle[b] non-muscle[b]	200	Brain has non-muscle type of heavy chain and light chains similar to
light-chain	muscle[b] non-muscle[b]	16–27	smooth muscle. (Kuczmarski & Rosenbaum 1979)
Troponin	C T I	18 37 23	Questionable in brain (see text)
Tubulin	α[b] β[b]	57 54	Present in neurons and other cells
MAP[c]	HMW[b,d] τ[b]	250–350 56–70	Associated with brain microtubules (Kirschner 1978)
Dynein		300–400	Similar protein in brain (Hiebsch & Murphy 1980)
Vimentin[e]		58	Schwann cells, glia, neuroblastoma (Lazarides 1980)
Desmin[f]		50	Schwann cells (Lazarides 1980)
Keratin			Only in epithelial cells (Lazarides 1980)
GFA[g] protein		50	From fibrous astrocytes (Chiu et al 1980)
Neurofilament protein		70 150 200	Abundant in axons and dendrites (Chiu et al 1980)

[a] Molecular weights are in kilodaltons.
[b] Indicates further heterogeneity.
[c] Microtubular-associated protein.
[d] High molecular weight protein.
[e] Also decamin or fibroblast intermediate filament protein.
[f] Also skeletin.
[g] Glial fibrillary acidic protein.

rations of fibroblasts (Rubenstein & Spudich 1977, Tannenbaum & Rich 1979), neurites or neuronal cell bodies (Choo & Bray 1978), or synaptosomal preparations (Kelly & Cotman 1978) contain a similar mixture of β and γ_2 actins. For at least three of these actins the differences extend to the mRNA level (Hunter & Garrels 1977) and presumably to the genetic level. Each type of muscle seems to have its own combination of myosin light and heavy chains, and embryonic and cytoplasmic myosins are different again (Whalen et al 1979, Korn 1978). The heterogeneity of tubulin and intermediate filament proteins, discussed below, is extensive.

The reason that these proteins as a class are so heterogeneous is unknown. In general it seems likely to be related to the fact that they perform a range of similar but not identical tasks within cells. Thus, the contractile machinery of smooth muscle works in a similar but not identical way to that of fast muscle; tubulin in a mitotic spindle will polymerize in response to different cytoplasmic signals and interact with different ancillary proteins to the tubulin of a dendrite. As components of larger assemblies, much of the tertiary structure (and hence amino acid sequence) will be critical. In keeping with this we find that actin is strongly conserved phylogenetically and shows greater differences between tissues than between species (Vandekerckhove & Weber 1978).

Of the dozen or so ancillary proteins of the striated myofibril, only tropomyosin has been shown by chemical tests to be present in neurons. This is the non-muscle type of tropomyosin, which is smaller in molecular size and forms paracrystals of shorter periodicity than muscle tropomyosin (Hitchcock 1977). It is not present in large amounts: The yield from brain is about 0.1% of the total protein (Bretscher & Weber 1978). Since actin comprises more than 5% of brain protein (Pardee & Bamburg 1979), only a small fraction can be complexed with tropomyosin. It may be relevant that in fibroblasts a proportion of actin filaments do not have associated tropomyosin (Lazarides 1976).

The Z-line protein α-actinin has been detected in various non-muscle cells by immunofluorescence and neurites from embryonic mouse spinal cord and dorsal root ganglion cells were stained in this way (Jockusch et al 1979). This protein is interesting because the point of attachment of microfilament bundles to the cellular membrane is formally equivalent to the Z-line. Perhaps an analog of α-actinin serves to anchor cells to the substratum or serves as the insertion point for filaments within filopodia? There is some encouragement for this view in the pattern of staining of cultured tissue cells. Fluorescent patches are observed in regions in which the cell makes close contact to the dish (Wehland et al 1979, Badley et al 1978), although they cannot be said to be confined to these regions. Earlier results suggested that α-actinin is present in the intestinal microvilli but now it seems that α-actinin is confined to the terminal web region (Geiger et al 1979).

The troponins, which are concerned with the calcium regulation of muscle contraction, have not been unambiguously identified in nervous tissue. A related protein, calmodulin, is present and is discussed below, together with the function of the neuronal cytoskeleton.

A number of proteins have been identifed in non-muscle tissues that bind to actin. These are likely to influence its state of polymerization within the cell or the state of aggregation of actin filaments. Profilin (Table 1) is a small soluble protein that forms a tight 1:1 complex with globular actin. Originally isolated from spleen, profilin is also present in blood platelets and whole brain. It seems probable that much of the unpolymerized actin in cells will be bound to profilin (Lindberg et al 1979).

A much larger protein, filamin, binds to actin filaments and causes them to aggregate. It was first isolated from smooth muscle but has now been found to be widely distributed in non-muscle tissues and to be closely related, if not identical, to the "actin-binding protein" of macrophages (Stossel & Hartwig 1976) and the "high molecular weight protein" (Schloss & Goldman 1979) of fibroblasts. A protein of similar size is present in extracts of brain but has been shown to be immunologically distinct from smooth muscle filamin (Wallach et al 1978).

Tubulin is abundant in axons and dendrites; cow and pig brain are the richest sources of tubulin for biochemical studies. In solution, tubulin exists as an $\alpha\beta$ dimer and, in the assembled microtubule, α and β alternate within the longitudinal protofilaments (Stephens & Edds 1976, Kirschner 1978). These two forms of tubulin have different amino acid sequences and are coded by separable mRNA species (Cleveland et al 1978).

Heterogeneity within the α and β tubulins has been detected. Two forms of α can be separated on hydroxyapatite columns in the presence of detergent (Forgue & Dahl 1979) and two forms of both α and β can be resolved on two-dimensional gels (Marotta et al 1979). An even greater degree of heterogeneity is revealed by one-dimensional electrofocusing, which shows a series of bands for both α and β (Nelles & Bamburg 1979). The number of tubulin bands found in brain is greater than in other tissues, such as liver or spleen, and increases with development (Gozes & Littauer 1978). It remains possible that some of the more closely related species could arise by post-translational modifications such as deamidation or glycosylation.

Of the various arms, spokes, and links that join tubules within the ciliary axoneme (Stephens & Edds 1976), only dynein has been looked for in the nervous system. The protein is of special interest because it is the active arm that produces sliding movements between adjacent microtubule doublets. If a similar protein were associated with axonal microtubules it could perhaps generate a form of transport. Preparations of brain microtubules have recently been shown to contain an ATPase that, in molecular size, affinity for microtubules, and requirements for enzyme activation, is similar to axonemal dynein (Hiebsch & Murphy 1980).

In general, any protein that forms a persistent contaminant in the purification of tubulin could be associated with tubules in the cell and modulate or supplement their activity. A number of enzymatic activities have been found in association with microtubules such as tyrosyl transferase and nucleotide diphosphate kinase, and two classes of microtubule-associated protein (MAP) are present in amounts sufficient to be detected on acrylamide gels. These two classes are the high molecular weight group known by the acronym HMW (250,000 to 350,000) and the τ group of smaller proteins (56,000 to 70,000). Both types have been shown to be distributed in patterns similar to the microtubule patterns of the cell and to be sensitive to treatment with colchicine (Connolly et al 1978). Either HMW or τ will promote the assembly of tubulin into microtubules, although this property is not unique; substances such as polylysine can also have this effect (Murphy et al 1977).

The importance of the τ proteins to microtubular function is still debated. They are lost after only a few cycles of polymerization and one τ component (the "tubulin-assembly-protein") may be related to the 68,000 constituent of neurofilaments (Runge et al 1979). The significance of one of the HMW proteins (HMW2) is more firmly established. It remains in constant proportion to tubulin through six cycles of purification (Murphy et al 1977) and in its presence tubules are seen to have small lateral projections that can link neighboring tubules (Kim et al 1979). These projections occur at regular intervals of subunit along each protofilament and appear to constitute a helix of pitch 96 nm superimposed on that of the microtubule (Amos 1977).

The composition of neurofilaments was until recently a vexing issue; neurofilaments are major components of most axons but are difficult to purify. There is no enzymatic test that can be used to identify neurofilaments (although their phosphorylation by endogenous protein kinases and their susceptibility to proteolytic attack are both distinctive features), nor are there any drugs analogous to colchicine that can provide an assay. Morphology itself is a poor criterion because several other classes of filament of similar diameter are present in nervous tissue and even partially proteolyzed filaments can retain their form (Gilbert et al 1975). Earlier preparations of vertebrate neurofilaments were consequently contaminated with other fibrillar proteins and proteolyzed to a variable extent.

These problems are less severe for the giant axons of the marine fanworm, *myxicola,* and squid, *loligo,* whose axoplasms can be easily extruded. The neurofilaments they contain were the first to be characterized biochemically. *Myxicola* neurofilaments are composed of two closely related peptides with molecular weights 160,000 and 150,000; in *loligo* the chain weights are 200,000 and 65,000 (Lasek et al 1979).

Vertebrate neurofilaments were originally identified in studies of axoplasmic transport. The slow flow of protein in guinea pig hypoglossal nerve shows a prominent triplet of proteins of molecular weights 200,000, 150,-000, and 70,000 (Hoffman & Lasek 1975). This same triplet has subsequently been found in preparations of filaments from axons in both the peripheral and central nervous systems (Schlaepfer & Freeman 1978). The differences between the components of the triplet appear to be real and not the result of proteolytic degradation during extraction (Chiu et al 1980). Concomitant with this work has been the successful resolution of other intermediate filaments (reviewed by Lazarides 1980). The intermediate filament protein from smooth muscle (Small & Sobieszek 1977), fibroblasts (Starger et al 1978), and epithelial cells (Steinert 1978) have now been identified. Immunofluorescent studies have been used to show their distribution in different cell types and this has recently been extended to neurons (Anderton et al 1980).

The disparity between the molecular weights of neurofilaments from invertebrate and vertebrate axons and the marked variation between vertebrate species (Thorpe et al 1979, Chiu et al 1980) is in marked contrast to actin and tubulin where molecular size is a strongly conserved feature. It has been suggested that this is because the component subunits of neurofilaments are fibrous rather than globular (Lasek et al 1979): the formation of a fibrous rope could then depend only on a limited region of the protein molecule and would be insensitive to variations in the amount or composition of the remainder. A clear example of such a phenomenon exists in bovine keratin where filaments can reassemble from mixtures of almost any two or three of the possible seven distinct chains of different molecular weight (Steinert 1978).

ASSEMBLY AND BREAKDOWN

The fibrillar proteins we are discussing are major components of the cell, so it is likely that they are synthesized with the bulk of neuronal proteins, i.e. on ribosomes within the cell soma (Estridge & Bunge 1978). Whether they are assembled into filaments and tubules at this point is another question. It would be perfectly possible for them to be transported down the axon before incorporation.

Studies on axoplasmic flow have not examined this directly, although on balance they seem most consistent with somal growth. Arguments in support of this include the facts that radioactively labeled protein advances as a coherent peak even after months of transport (which suggests that the radioactivity is part of a larger structure) and that continuity with the soma appears to be necessary for slow transport (McLean et al 1976). There are, morever, only minor amounts of radioactivity left behind the peak of transported protein. If protein is not deposited en route, and is indeed degraded when it arrives at the terminal (Lasek & Black 1977), then where else can it add to filaments and tubules than at the soma?

The view of a bulk flow of fibrillar material is not without its difficulties, however. In some way structural features of axonal and dendritic trees must be preserved for months or even years and maintained by the same cytoskeleton that is said to be steadily moving distally at 1 mm a day. Studies of cultured neurons show that particles on their surface—again presumably borne by the cytoskeleton—are stationary or, in some situations, move back toward the cell bodies (Koda & Partlow 1976).

It could be that actin is important here. It is transported at a rate different from tubulin and the neurofilament triplet (Hoffman & Lasek 1975), and the peak of label that contains actin seems distinctly less sharp than for the other two. This could mean that actin is continually deposited from the transported stream—perhaps into a cortical layer of microfilaments.

In a mature, non-growing nerve, any constant supply of fibrillar proteins must obviously be balanced by degradation. It was originally suggested by Weiss that this could be achieved by proteolytic degradation at the axon terminal and this has received some experimental support (Lasek & Black 1977). A protease that degrades neurofilaments is present in *myxicola* axoplasm, and breakdown can be detected within minutes of extrusion into a calcium-containing medium (Gilbert et al 1975). A similar effect has been found in squid neurofilaments (Pant et al 1979) and in rat neurofilaments (Schlaepfer & Micko 1979). It is apparent that if such a mechanism exists it will be especially important in the transition that occurs when a growing nerve meets its potential synaptic partner.

The question of the continuity of fibrillar elements within the axon has received some attention recently. The conventional view was that tubules extend in undivided continuity from soma to tip, whereas neurofilaments either branch or are in short segments. The former tenet now seems certain to be wrong. Two analyses of the number of tubules in cross-sections of axons have been described in the crayfish (Nadelhaft 1974) and in the nematode *caenorhabditis elegans* (Chalfie & Thomson 1979): Both show considerable fluctuations along the length of the axon. In the latter study, moreover, serial reconstructions showed the tubules to be in short lengths, 10 to 20 μm and apparently attached to the membrane at one end. Neurofilaments are more difficult to trace, although their numbers also change with the diameter of the axon. Microfilaments are so indistinct that it is not even possible to count them sensibly.

If one considers non-muscle cells in general, it seems that both microtubules and microfilaments are labile structures that can if necessary form and breakdown rapidly. This is clearly the case for the microtubules of the mitotic apparatus, for example, and transformations in the state of microfilaments, if not their actual depolymerization, accompany many changes in cell shape. Histochemical techniques such as immunofluorescence as well as direct biochemical assays of cell extracts show, in many situations, abundant and sometimes changing levels of unpolymerized tubulin or actin (Lindberg et al 1979, Kirschner 1978). The difficulty with such approaches is that, almost by definition, they will perturb the system they seek to study. This is why direct biochemical analysis of the properties of the purified proteins is so important. If all the factors present in the cytoplasm that influence their polymerization can be defined, then in principle we will be able to deduce their true state within the cell.

The assembly of muscle actin into filaments has been studied for many years. Indeed it is, together with bacterial flagellin, one of the classic examples of a protein polymerizing system. Globular actin may be extracted from an acetone powder of muscle at very low salt concentrations. The addition of monovalent salt (50 mM NaCl) or divalent salt (0.7 mM $MgCl_2$) then promotes a rapid formation of filamentous actin. The filaments that are formed comprise two parallel strings of subunits helically disposed with the same periodicity as in native muscle thin filaments.

The concentration of actin and the ionic conditions within the cytoplasm would be expected to produce almost complete polymerization. That it is not fully polymerized seems to be more the result of ancillary proteins than of intrinsic differences in the type of actin (Lindberg et al 1979). The protein profilin, for example, will probably sequester an appreciable amount of actin in the globular form, and the same may be true for a long list of other actin-binding proteins. The membrane-associated complexes discussed below, which have been identified by their ability to bind cytochalasin, are possible sites for the nucleation of actin filament formation in the cell.

Microtubules are larger and more complex than microfilaments and it was only comparatively recently that it was learned how to make them in the test tube (reviewed by Kirschner 1978). A preparation of tubulin dimeric subunits will form apparently normal microtubules upon warming. Very pure tubulin will do this if forced (which shows that it is competent to undergo self-assembly), but in physiological buffers and at moderate concentrations, some nucleating agent is necessary. The most likely candidates for such agents within the cell are the MAPs, which were discussed above. Both in neuroblastoma (Seeds & Maccioni 1978) and in developing rat brain (Fellous et al 1976) the growth of neuronal extensions, with its concomitant increase in polymerized tubulin, is accompanied by the accumulation of soluble proteins that are functionally equivalent to MAPs.

The elongation of tubules in the test tube occurs in a polar fashion. If short lengths of pre-formed tubule are labeled with radioactivity or by the attachment of particles, then new subunits add predominantly, but not exclusively, to one end (Bergon & Borisy 1980). Since axonal extension entails the elongation of tubules, it is logical to conclude that in the axon, too, there is likely to be preferred growth. But—with reference to the quotation at the beginning of this review—unless one knows the polarity of the microtubules in the axon, this conclusion is of limited value. One way in which this polarity may be determined is by the addition of dynein. This large protein adds to microtubules in a stereospecific manner and, therefore, could provide an index of polarity comparable to the decoration of actin filaments by muscle myosin (Haimo et al 1979).

A fascinating phenomenon that was shown to exist first for actin (Wegner 1976) and later for tubulin (Margolis & Wilson 1979) is that of head-to-tail polymerization. Polymer and subunits at equilibrium in the test tube constitute a dynamic system in which new subunit protein is continually added at one end of the filament or tubule and lost from the other. The cycle is accompanied by the dephosphorylation of the nucleotide associated with the subunit—ATP for actin, GTP for tubulin—and these cofactors, in a sense, provide the necessary energy. Colchicine and other antimitotic drugs inhibit the assembly of tubules and lead to their eventual dissolution. They appear to form a complex with the free tubulin dimer and to interfere with its ability to polymerize, although the details are the subject of debate (Margolis & Wilson 1977, Sternlicht & Ringel 1979). Although treadmilling of subunits may not occur as such, the differences between the ends, produced by the hydrolysis of ATP on GTP, could still be important. These differences could confine the growth of a filament or tubule to certain organizing centers and suppress it elsewhere (Kirschner 1980).

With regard to neurofilaments, the identification of their component proteins is so recent that there has been little opportunity to find out much about their assembly in the cell. In common with other intermediate filaments, neurofilaments were previously thought of as insoluble structures since they remain behind after actin, tubulin, myosin, and the rest have been extracted; however, both mammalian neurofilaments (Schlaepfer 1978) and the filaments of baby hamster kidney cells (Zackroff & Goldman 1979) can be dissociated in solutions of low ionic strength and the latter will reassemble on return to isotonic conditions. It should now be possible to define the conditions and accessory proteins (if any) that are necessary for polymerization and to answer a number of intriguing questions, such as whether neurofilament protein will form mixed co-polymers with desmin, vimentin, or cytokeratin.

FUNCTION

Fibrillar proteins are generally supposed to be concerned with the maintenance of form and movements of the cell and the same is presumably true for those within neurons. Indeed, in these respects neurons have special requirements. The extremely asymmetric axons, sometimes meters long, must sustain repeated stresses as the animal moves. Such huge lengths of axon call for efficient and perhaps novel forms of intracellular transport. The details of nerve cell shape—the position and size of branches and synapses—are vital to the function of the nervous system and must have a structural basis.

Direct evidence for a skeletal role is hard to obtain. Gentle lysis of fibroblastic cells leaves behind a shell of "insoluble" protein that preserves the original form of the cell (Brown et al 1976) and the same is true of cultured neurons (Marchisio et al 1978). This shell is enriched in the fibrillar components, which therefore are disposed in the shape of the cell. But which components are bearing stress and resisting the tendency of the living cell to adopt a spherical form is not shown in this way.

Tubules have been implicated in this regard because of the response of cultured neurons to antimitotic agents. Colchicine or demecolcine (Colcemid) causes neurites to retract and eventually to be absorbed into the cell body (Daniels 1972). This response is one that takes several hours to be complete, however, and it is accompanied by much cellular activity, such as the formation of lateral filopodia and flattened lamellipodial regions (Bray et al 1978). A better indication of mechanical function was devised for smooth muscle cells (Cooke & Fay 1972). When such cells were stretched mechanically it was found that the tubules and filaments rearranged with the cytoplasm. The component that moved to the central region and hence appeared to bear the tension was the 10 nm filament. A similar experiment has not been reported for nerve cells but the neurofilaments from the giant axons of *myxicola* have been shown to be mechanically robust structures with an ability comparable to that of wool to sustain tension (Gilbert 1975).

One of the most important functions of the neuronal cytoskeleton must surely be that of the transport of materials. Everything we know of the movements of vertebrate cells points to the involvement of fibrillar proteins and the axon has a full complement of these: The problem is to determine which elements are responsible. Despite the accumulation of a great deal of data on the rates and nature of materials carried, it cannot be said that this movement is produced by actin or by an interaction with microtubules. To do so, in the last analysis, will require a detailed correlation of ultrastructure with movement and the reconstitution of a functional system from purified components. Rigorous evidence of this kind presently exists only for the contraction of striated muscle.

It may be, indeed, that there is no single mechanism. The movement of large organelles such as mitochondria, which travel in a rapid saltatory fashion within the axon, probably has requirements different from the flow of neurosecretory materials and is certainly distinct from the slow movement of tubules and filaments (see the recent review by Schwartz 1979).

It is a widely held belief that microtubules are involved in fast transport. This conclusion is based on the appearance of cross-links between mitochondria or synaptic vesicles and microtubules (Smith et al 1975), and on the sensitivity of the fast transport of protein to mitotic inhibitors (Hanson & Edström 1978). Recent evidence however, seems to argue against this interpretation. Some years ago it was observed that the blockade of transport was not synchronous with the disappearance of tubules (Byers 1974), and the same has now been found to be true in axons treated with solutions containing high levels of calcium ions. This treatment is said to cause the almost complete disintegration of axonal microtubules without impairing fast transport—which is, moreover, still sensitive to colchicine (Brady et al 1980).

The second possibility is that fast axoplasmic transport is produced by actin and myosin. These two proteins are present in neurons and we know that they are able to interact with each other to produce movement. Although many agents are available that can bind to actin and myosin in the test tube and may be used in principle to interfere with their activities—agents such as deoxyribonuclease I (Hitchcock et al 1976), phalloidin (Wehland et al 1977), and myosin antibody (Mabuchi & Okhuno, 1977)—most of these will not cross the neuronal plasma membrane. A way around this difficulty is to inject the agent into the cell under pressure (Graessmann et al 1980). A preliminary use of this technique has already indicated that filamentous actin is involved in fast axoplasmic transport (Isenberg et al 1980) and other applications of this kind are likely to follow.

A group of drugs that do not have to be injected is the cytochalasins. This is a family of fungal secondary metabolites that rapidly paralyze many kinds of movement, including the division of cells, the migration of fibroblasts, and morphogenetic contractions in sheets of cells. It was found some time ago that the application of cytochalasin B to the nerve cord of crayfish inhibits both the slow transport (1mm/day) and a fast transport (10mm/day) of radioactively labeled protein (Fernandez & Samson 1974). The mechanism of action of the cytochalasins, for long a controversial subject, has now been clarified. The effects that cytochalasins have on the transport of glucose into the cell have been separated from those on motility, and a derivative, dihydrocytochalasin B, has been synthesized that acts on cell movements alone (Atlas & Lin 1978). This derivative has been used to isolate from erythrocytes and brain a binding complex that contains a short oligomeric fragment of actin in association with other proteins. Further analysis reveals that the binding site of dihydrocytochalasin B is actually the end of the actin filament and that binding results in the inhibition of actin polymerization (Flanagan & Lin 1980, Brown & Spudich 1979). In view of this, the observation that cytochalasin B inhibits axon flow is certainly a suggestion that actin is involved, although the experiment should now be repeated with the more specifically acting dihydro-derivative and over a shorter time scale.

Cytochalasin B has a dramatic and immediate effect on the movement of the growth cone of developing neurites (Yamada et al 1971): Movement immediately ceases and within minutes changes in the shape of the growth cone are seen. The long filopodia become flaccid and wilt and are eventually withdrawn into the cell. These effects are reversible and movements resume soon after removal of the drug. If, as seems probable, dihydrocytochalasin B will produce the same result, it will imply that filopodial formation involves a sudden polymerization of actin within the growth cone.

Filopodia may also have a contractile function in the normal growth of the axon. In culture, growth cones are sometimes seen to "palpate" neighboring objects or cells (Wessells et al 1980). The microspikes of settling fibroblasts can retrieve small particles from the culture surface (Albrecht-Buehler & Goldman 1976). This contraction is presumably responsible for the tension developed in cultured neurites—a tension that influences the course of migration of the growth cone (Bray 1979). In addition it seems possible that the neuronal process itself might have some contractile abilities. A matrix of filaments has been seen under the membrane of squid axon (Metuzals & Tasaki 1978).

Filopodia are in several ways similar to the microvilli present on epithelial cells in the intestine and elsewhere. They have a similar diameter of about $0.1\,\mu m$, and their principle structural component is an aggregate of microfilaments that extends into their base (Letourneau 1979). The similarity ends there, however, and in other respects the long (sometimes 30 μm or more), actively moving, transient extensions of growth cones with their untidy ultrastructure belong to a class of cellular extension different from the short ($1\,\mu m$), static, and precisely made microvilli. Extrapolation from the relatively well-understood biochemistry of the latter is therefore to be made with caution.

Even when the elemental mechanisms of movements in nerve cells have been identified it will still be necessary to understand how they are controlled and regulated: how materials are transported to the right place at the right time; how endocytosis and exocytosis occur in response to external signals. It seems likely that this physiological aspect of the cell will be controlled by ionic fluxes and, in particular, by changes in the level of calcium ions, as it is in the well-understood example of striated muscle. Here the sensitivity to calcium is conferred by a complex of three proteins —troponins C, I, and A—that is attached at regular intervals along the thin filaments, between adjacent tropomyosin molecules. Calcium ions released from the sarcoplasmic reticulum as part of the chain of responses to neuronal stimulation bind to troponin C and cause it and its associated proteins to prevent (perhaps by steric hindrance) the myosin heads from combining with the actin filaments.

It is logical to ask whether a similar mechanism exists in association with non-muscle actomyosin. Proteins of a size similar to troponin have been found in tissues such as brain and at least one of them has an action similar to troponin C. But the situation is confused by the existence of a different control mechanism employing a protein very like troponin C. Here the calcium exerts its primary action effect, not on actin, but on myosin. The myosin is "switched on"—made competent to undergo actin-activation of its ATPase—by the phosphorylation of one of its light chains. This phosphorylation is achieved by a specific protein kinase that is made sensitive to physiological changes in calcium concentration by combination with a second protein: a small (17,000) heat stable protein known as calmodulin. The kinase is inactive by itself and functions only when it is combined with a complex of calcium and calmodulin. This system was found first in smooth muscle and more recently has been found in platelets, fibroblasts, and brain (Dabrowska & Hartshorne 1978, Cheung 1980). Phosphorylation of myosin may affect its state of aggregation as well as its ATPase activity (Scholey et al 1980).

The presence of this alternative system obscures the search for an actin-linked regulation because calmodulin is so closely related to troponin C. The two proteins have a 70% homology in their amino acid sequences, they bind calcium ions to a similar extent, and they will even exchange functions to a degree. Criteria more rigorous than have yet been applied are needed to establish that troponin C and not calmodulin is present in non-muscle tissues (Van Eldick & Watterson 1979). The absence of cysteine and the presence of trimethyllysine are distinctive features of calmodulin; the detection of troponin I and A in non-muscle tissue would be strong evidence for an actin-linked regulation.

The influence of calmodulin extends beyond myosin. In its active form, bound to calcium, it affects—in well-documented interactions with purified proteins—cellular levels of cAMP, calcium, NADP, prostaglandin, and glycogen; there are suggestions that it influences processes such as synaptic release and microtubule assembly. As the mediator of so many of the effects of calcium it provides, especially since it is a protein and therefore more easily identified and located, a unique opportunity to study the coordination of the neuronal cytoskeleton and its association with the rest of the cell.

ACKNOWLEDGMENTS

Thanks to Q. L. Choo, P. A. M. Eagles, and M. Jacobs for their help.

Literature Cited

Albrecht-Buehler, G., Goldman, R. D. 1976. Microspike mediated particle transport towards the cell body during early spreading of 3T3 cells. *Exp. Cell Res.* 97:329–39

Amos, L. A. 1977. Arrangement of high molecular weight associated proteins on purified mammalian brain microtubules. *J. Cell Biol.* 72:642–54

Anderton, B. H., Thorpe, R., Cohen, J., Selvendran, S., Woodhams, P. 1980. Specific neuronal localization by immunofluorescence of 10 nm filament polypeptides. *J. Neurocytol.* In press

Atlas, S. J., Lin, S. 1978. Dihydrocytochalasin B: Biological effects and binding to 3T3 cells. *J. Cell Biol.* 76:360–70

Badley, R. A., Lloyd, C. W., Woods, A., Carruthers, L., Allcock, C., Rees, D. A. 1978. Mechanisms of cellular adhesion III preparation and preliminary characterization of adhesions. *Exp. Cell Res.* 117:231–44

Bergen, L. G., Borisy, G. G. 1980. Head-to-tail polymerization of microtubules in vitro. *J. Cell Biol.* 84:141–50

Brady, S. T., Crothers, S. D., Nosal, C., McClure, W. O. 1980. Fast axoplasmic transport in the presence of high Ca^{2+}: Evidence that microtubules are not required. *Proc. Natl. Acad. Sci. USA.* 77:5909–13

Bray, D. 1979. Mechanical tension produced by nerve cells in tissue culture. *J. Cell Sci.* 37:391–410

Bray, D., Thomas, C., Shaw, G. 1978. Growth cone formation in cultures of sensory neurons. *Proc. Natl. Acad. Sci. USA* 75:5226–29

Bretscher, A., Weber, K. 1978. Tropomyosin from brain contains two polypeptide chains of slightly different molecular weights. *FEBS Lett.* 85:145–48

Brown, S., Levinson, W., Spudich, J. A. 1976. Cytoskeletal elements of chicken embryo fibroblasts revealed by detergent extraction. *J. Supramol. Struct.* 5:119–30

Brown, S. S., Spudich, J. A. 1979. Cytochalasin inhibits the rate of elongation of actin filament fragments. *J. Cell Biol.* 83:657–62

Byers, M. R. 1974. Structural correlates of rapid axonal transport. Evidence that microtubules may not be involved. *Brain Res.* 75:97–113

Carlsson, L., Nyström, L.-E., Sundkvist, I., Markey, F., Lindberg, U. 1977. Actin polymerizability is influenced by profilin, a low molecular weight protein in non-muscle cells. *J. Mol. Biol.* 115:465–83

Chalfie, M., Thomson, J. N. 1979. Organisation of neuronal microtubules in the nematode *C. elegans. J. Cell Biol.* 32:278–89

Cheung, W. Y. 1980. Calmodulin plays a pivotal role in cellular regulation. *Science* 207:19–27

Chiu, F.-C., Korey B., Norton, W. T. 1980. Intermediate filaments from bovine rat and human CNS: Mapping analysis of the major proteins. *J. Neurochem.* 34:1149–59

Choo, Q. L., Bray, D. 1978. Two forms of neuronal actin. *J. Neurochem.* 31:217–24

Cleveland, D. W., Fischer, S. G., Kirschner, M. W., Laemmli, U. K. 1977. Peptide mapping by limited proteolysis in sodium dodecyl sulfate and analysis by gel electrophoresis. *J. Biol. Chem.* 252:1102–6

Cleveland, D. W., Kirschner, M. W., Cowan, N. J. 1978. Isolation of separate mRNAs for α and β tubulin and characterization of the corresponding in vitro translation product. *Cell* 15: 1021–31

Connolly, J. A., Kalnins, V. I., Cleveland, D. W., Kirschner, M. W. 1978. Intracellular localization of the high molecular weight microtubule accessory protein by indirect immunofluorescence. *J. Cell. Biol.* 76:781–86

Cooke, P. H., Fay, F. S. 1972. Correlation between fiber length ultrastructure and the length-tension relationship of mammalian smooth muscle. *J. Cell Biol.* 52:105–16

Dabrowska, R., Hartshorne, D. J. 1978. A Ca²⁺ and modulator dependent myosin light chains kinase from non-muscle cells. *Biochem. Biophys. Res. Commun.* 85:1352–59

Daniels, M. P. 1972. Colchicine inhibition of nerve fiber formation in vitro. *J. Cell Biol.* 53:164–76

Estridge, M., Bunge, R. 1978. Compositional analysis of growing axons from rat sympathetic neurons. *J. Cell Biol.* 79: 138–55

Fellous, A., Francon, J., Lennon, A. M., Nunez, J. 1976. Initiation of neurotubulin polymerization and brain development. *FEBS Lett.* 64:400–3

Fernandez, H. L., Samson, F. E. 1974. Axoplasmic transport: Differential inhibition by cytochalasin. *J. Neurobiol.* 4:201–6

Flanagan, M. D., Lin, S. 1979. Comparative studies on the characteristic properties of two forms of brain actin separable by isoelectric focussing. *J. Neurochem.* 32:1037–46

Flanagan, M. D., Lin, S. 1980. Cytochalasins block actin filament elongation by binding to high-affinity sites associated with F-actin. *J. Biol. Chem.* 255:835–38

Forgue, S. T., Dahl, J. L. 1979. Rat brain tubulin: Subunit heterogeneity and phosphorylation. *J. Neurochem.* 32: 1015–25

Geiger, B., Tokuyasu, K. T., Singer, S. J. G. 1979. Immunocytochemical localization of α-actinin in intestinal epithelial cells. *Proc. Natl. Acad. Sci. USA* 76:2833–37

Gilbert, D. S. 1975. Axoplasm architecture physical properties as seen in the myxicola giant axon. *J. Physiol.* 253:257–309

Gilbert, D. S., Newby, B. J., Anderton, B. H. 1975. Neurofilament disguise, destruction and discipline. *Nature* 256:586–89

Gozes, I., Littauer, U. Z. 1978. Tubulin microheterogeneity increases with rat brain maturation. *Nature* 276:411–13

Graessmann, A., Graessmann, M., Mueller, C. 1980. Microinjection of early SV40 DNA fragments and T antigen. *Meth. Enzymol.* 65:816–25

Haimo, L. T., Telzer, B. R., Rosenbaum, J. L. 1979. Dynein binds to and crossbridges cytoplasmic microtubules. *Proc. Natl. Acad. Sci. USA* 76:5759–63

Hanson, M., Edström, A. 1978. Mitosis inhibitors and axonal transport. *Int. Rev. Cytol. Suppl.* 7:373–402

Hiebsch, R. R., Murphy, D. B. 1980. Evidence for a dynein-like adenosine triphosphatase on cytoplasmic microtubules. *J. Cell Biol.* In press

Hiller, G., Weber, K. 1977. Spectrin is absent in various tissue culture cells. *Nature* 266:181

Hitchcock, S. E. 1977. Regulation of motility in nonmuscle cells. *J. Cell Biol.* 74:1–15

Hitchcock, S. E., Carlsson, L., Lindberg, U. 1976. Depolymerisation of F-actin by deoxyribonuclease I. *Cell* 7:531–42

Hoffman, P. N., Lasek, R. J. 1975. The slow component of axonal transport. Identification of major structural polypeptides of the axon and their generality among mammalian neurons. *J. Cell Biol.* 66:351–66

Hunter, T., Garrels, J. I. 1977. Characterization of the mRNAs for α-, β and γ actin. *Cell* 12:767–81

Isenberg, G., Schubert, P., Kreutzberg, G. W. 1980. Experimental approach to test the role of actin in axonal transport. *Brain Res.* 194:588–93

Jacobson, M. 1978. *Developmental Neurobiology*, pp. 121, 146. New York: Plenum. 2nd ed.

Jockusch, H., Jockusch, B. M., Burger, M. M. 1979. Nerve fibers in culture and their interactions with non-neural cells visualized by immunofluorescence. *J. Cell Biol.* 80:629–41

Kelly, P. T., Cotman, C. W. 1978. Synaptic proteins. Characterization of tubulin and actin and identification of a distinct postsynaptic density polypeptide. *J. Cell Biol.* 79:173–83

Kim, H., Binder, L. I., Rosenbaum, J. L. 1979. The periodic association of MAP-2 with brain microtubules in vitro *J. Cell Biol.* 80:266–76

Kirschner, M. W. 1978. Microtubule assembly and nucleation. *Int. Rev. Cytol.* 54:1–71

Kirschner, M. W. 1980. Implications of treadmilling for the stability and polarity of actin and tubulin polymers in vivo. *J. Cell Biol.* 86:330–34

Koda, Y., Partlow, L. M. 1976. Membrane marker movement on sympathetic axons in tissue culture. *J. Neurobiol.* 7:157–72

Kohler, G., Milstein, C. 1975. Continuous cultures of fused cells secreting antibody of predefined specificity. *Nature* 256:495–97

Korn, E. D. 1978. Biochemistry of actomyosin-dependent cell motility (a review). *Proc. Natl. Acad. Sci. USA* 75:588–99

Kuczmarski, E. R., Rosenbaum, J. L. 1979. Chick brain actin and myosin isolation and characterization. *J. Cell Biol.* 80:341–53

Lasek, R. J., Black, M. M. 1977. How do axons stop growing? Some clues from the metabolism of the proteins in the slow component of axonal transport. *Dev. Neuro.* 2:161–69

Lasek, R. J., Krishnan, N., Kaiserman-Abramof, I. R. 1979. Identification of subunit proteins of 10 nm microfilaments isolated from axoplasms of squid and *myxicola* giant axons. *J. Cell Biol.* 82:336–46

Lazarides, E. 1976. Two general classes of cytoplasmic actin filaments in tissue culture-cells. The role of tropomyosin. *J. Supramol. Struct.* 5:531–63

Lazarides, E. 1980. Intermediate filaments as mechanical integrators of cellular space. *Nature* 283:249–56

Letourneau, P. C. 1979. Cell-substratum adhesion of neurite growth cones and its role in neurite elongation. *Exp. Cell Res.* 124:127–38

Lindberg, U., Carlsson, L., Markey, F., Nyström, L.-E. 1979. The unpolymerized form of actin in non-muscle cells. *Methods Achiev. Exp. Pathol.* 3:143–70

Mabuchi I., Okhuno, M. 1977. The effect of myosin antibody on the division of starfish blastomeres. *J. Cell Biol.* 74:251–63

Marchisio, P. C., Osborn, M., Weber, K. 1978. The intracellular organisation of actin and tubulin in cultured C-1300 mouse neuroblastoma cells. *J. Neurocytol.* 7:571–82

Margolis, R. L., Wilson, L. 1977. Addition of colchicine-tubulin complex to microtubule ends: The mechanism of substoichiometric colchicine poisoning. *Proc. Natl. Acad. Sci. USA* 74:3466–70

Margolis, R. L., Wilson, L. 1979. Regulation of the microtubule steady state in vitro by ATP. *Cell* 18:673–79

Marotta, C. A., Strocchi, P., Gilbert, J. M. 1979. Biosynthesis of heterogeneous forms of mammalian brain tubulin subunits by multiple messenger RNAs. *J. Neurochem.* 33:231–40

McLean, W. G., Frizell, M., Sjöstrand, J. 1976. Slow axonal transport of labelled proteins in sensory fibers of rabbit vagus nerve. *J. Neurochem.* 26:1213–16

Metuzals, J., Tasaki, I. 1978. Subaxolemmal filamentous network in the giant nerve fiber of the squid (*loligo lealei L.*) and its possible role in excitability. *J. Cell Biol.* 78:597–621

Murphy, D. B., Valee, R. B., Borisy, G. G. 1977. Identity and polymerization-stimulatory activity of the nontubulin proteins associated with microtubules. *Biochemistry* 16:2598–2605

Nadelhaft, I. 1974. Microtubule densities and total numbers in selected axons of the crayfish abdominal nerve cord. *J. Neurocytol.* 3:73–86

Nelles, L. P., Bamburg, J. R. 1979. Comparative peptide mapping and isoelectric focussing of isolated subunits from chick embryo brain tubulin. *J. Neurochem.* 32:477–89

O'Farrell, P. H. 1975. High resolution two-dimensional electrophoresis of proteins. *J. Biol. Chem.* 250:4007–21

Pant, H. C., Terakawa, S., Gainer, H. 1979. A calcium activated protease in squid axoplasm. *J. Neurochem.* 32:99–102

Pardee, J. D., Bamburg, J. R. 1979. Actin from embryonic chick brain. Isolation in high yield and comparison of biochemical properties with chicken muscle actin. *Biochem.* 18:2245–52

Puszkin, S., Schook, W. 1979. The role of cytoskeleton in neuron activity. *Methods Achiev. Exp. Pathol.* 9:87–111

Rubenstein, P. A., Spudich, J. A. 1977. Actin microheterogeneity in chick embryo fibroblasts. *Proc. Natl. Acad. Sci. USA* 74:120–23

Runge, M. S., Detrich, H. W., Williams, R. C. 1979. Identification of the major 68000 dalton protein of microtubule preparations as a 10 nm filament protein and its effect on microtubule assembly in vitro. *Biochemistry* 18:1689–98

Schlaepfer, W. W. 1978. Observations on the disassembly of isolated mammalian neurofilaments. *J. Cell Biol.* 76:50–56

Schlaepfer, W. W., Freeman, L. A. 1978. Neurofilament proteins of rat peripheral nerve and spinal cord. *J. Cell Biol.* 78:653–62

Schlaepfer, W. W., Micko, S. 1979. Calcium-dependent alterations of neurofilament proteins of rat peripheral nerve. *J. Neurochem.* 32:211–19

Schloss, J. A., Goldman, R. D. 1979. Isolation of a high molecular weight actin-binding protein from baby hamster kidney (BHK-21) cells. *Proc. Natl. Acad. Sci. USA* 76:4484–88

Scholey, J., Taylor, K., Kendrick-Jones, J. 1980. Regulation of nonmuscle myosin assembly by calmodulin-dependent light chain kinase. *Nature* 287:233–35

Schwartz, J. H. 1979. Axonal transport: Components, mechanisms and specificity. *Ann. Rev. Neurosci.* 2:467–504

Seeds, N. W., Maccioni, R. B. 1978. Proteins from morphologically differentiated neuroblastoma cells promote tubulin polymerization. *J. Cell Biol.* 76:547–55

Small, J. V., Sobieszek, A. 1977. Studies on the function and composition of the 10 nm filament of vertebrate smooth muscle. *J. Cell Sci.* 23:243–68

Smith, D. S., Jarlfors, U., Cameron, B. F. 1975. Morphological evidence for the participation of microtubules in axonal transport. *Ann. NY Acad. Sci.* 253:472–506

Starger, J. M., Brown, W. E., Goldman, A. E., Goldman, R. D. 1978. Biochemical and immunological analysis of rapidly purified 10 nm filaments from baby hamster kidney (BHK-21) cells. *J. Cell Biol.* 78:93–109

Steinert, P. M. 1978. Structure of the three-chain unit of bovine epidermal Keratin filament. *J. Mol. Biol.* 123:49–70

Stephens, R. S., Edds, K. T. 1976. Microtubules: Structure, chemistry and function. *Physiol. Rev.* 56:709–77

Sternlicht, H., Ringel, I. 1979. Colchicine inhibition of microtubule assembly via copolymer formation. *J. Biol. Chem.* 254:10540–50

Stossel, T. P., Hartwig, J. H. 1976. Interactions of actin, myosin and a new actin-binding protein of rabbit pulmonary macrophages II role in cytoplasmic movement and phagocytosis. *J. Biol. Chem.* 68:602–19

Tannenbaum, J., Rich, A. 1979. An isoelectric focussing study of plasma membrane actin. *Anal. Biochem.* 95:236–44

Thorpe, R., Delacourte, A., Ayers, M., Bullock, C., Anderton, B. H. 1979. The polypeptides of isolated brain 10nm filaments and their association with polymerized tubulin. *Biochem. J.* 181:275–84

Vandekerckhove, J., Weber, K. 1978. Mammalian cytoplasmic actins are the products of at least two genes and differ in primary structure in at least 25 identified positions from skeletal muscle actins. *Proc. Natl. Acad. Sci. USA* 75:1106–10

Van Eldik, J., Watterson, D. M. 1979. Characterization of a calcium-modulated protein from transformed chicken fibroblasts. *J. Biol. Chem.* 254:10250–55

Wallach, D., Davies, P. J. A., Pastan, I. 1978. Purification of mammalian filamin. Similarity to high molecular weight actin-binding protein in macrophages, platelets, fibroblasts and other tissues. *J. Biol. Chem.* 253:3328–35

Wegner, A. 1976. Head to tail polymerization of actin. *J. Mol. Biol.* 108:139–50

Wehland, J., Osborn, M., Weber, K. 1977. Phalloidin-induced actin polymerisation in the cytoplasm of cultured cells interferes with cell locomotion and growth. *Proc. Natl. Acad. Sci. USA* 74:5613–17

Wehland, J., Osborn, M., Weber, K. 1979. Cell to substratum contacts in living cells: A direct correlation between interference-reflexion and indirect immunofluorescence microscopy using antibodies against actin and α-actinin. *J. Cell Sci.* 37:257–73

Wessells, N. K., Letourneau, P. C., Nuttall, P. C., Luduena-Anderson, M. A., Geiduschek, J. M. 1980. Responses to cell contacts between growth cones, neurites, and ganglionic non-neuronal cells. *J. Neurocytol.* In press

Whalen, R. G., Schwartz, K., Bouveret, P., Sell, S. M., Gros, F. 1979. Contractile protein isozymes in muscle development: Identification of an embryonic form of myosin heavy chain. *Proc. Natl. Acad. Sci. USA* 76:5197–5201

Wuerker, R. B., Kirkpatrick, J. B. 1972. Neuronal microtubules, neurofilaments and microfilaments. *Int. Rev. Cytol.* 33:45–75

Yamada, K. M., Spooner, B. S., Wessells, N. K. 1971. Ultrastructure and function of growth cones and axons of cultured nerve cells. *J. Cell Biol.* 49:614–35

Zackroff, R. V., Goldman, R. D. 1979. In vitro assembly of intermediate filaments from baby hamster kidney (BHK-21) cells. *Proc. Natl. Acad. Sci. USA* 76:6226–30

AUTHOR INDEX

(Names appearing in capital letters indicate authors of chapters in this volume.)

SUBJECT INDEX

A

Acetylcholine (ACh), 30
  axon-Schwann cell
    interaction and, 153
  calcium current in cardiac
    muscle and, 112
  dorsal root ganglion cell
    action potential and,
    113
  enteric, 232–40
  release of, 234–35
  Schwann cell
    hyperpolarization and,
    153
Acetylcholine receptors
  (AChRs)
  clustering of
    polypeptide factor and, 63
  cross-linking of
    myasthenic IgG and,
    204–5
  developing muscle fiber and,
    48
  endocytosis of, 205–7
  half-life of, 51
  ion channel of, 52
  myasthenia gravis and,
    195–200
  myasthenic antibodies and,
    203–10
  myasthenic IgG and, 203–7
  postsynaptic, 49–50
  synaptogenesis and, 51
  synthesis and incorporation
    of
    myasthenic
      immunoglobulin and,
      208
  d-tubocurarine and, 52
Acetylcholinesterase (AChE)
  deposition of, 53–55, 62–63
  electrical activity and, 54
  end-plate specific, 52–53
Acheta
  nervous system development
    in, 169–70
Actin
  amino acid sequences of, 507
  axoplasmic transport and,
    517
  filament aggregation of
    filamin and, 510
  head-to-tail polymerization
    of, 514
  muscle
    assembly into filaments,
    514
α-Actinin
  in non-muscle cells, 509–10

Actinopterygians
  evolution of, 302
Action potentials
  dorsal root ganglion, 113
  positive-going, 70
  tetrabutylammonium-
    induced, 71–72
  tetraethylammonium-
    induced, 71–72
  tetrodotoxin and, 238
Actomyosin
  in smooth muscle, 507
Adenylate cyclase activity
  adrenergic receptors and,
    430–31
  inhibition of, 434–35
Adrenalectomy
  adrenergic receptor density
    and, 446
Adrenaline
  calcium current modulation
    and, 116
  calcium currents in cardiac
    muscle and, 111
Adrenergic blockers
  impotence and, 407
Adrenergic nerves
  submucosal and myenteric
    plexuses and, 240
Adrenergic receptors, 419–54
  adenylate cyclase activity
    and, 430–31
  aging and, 451–52
  agonist interactions with
    guanosine 5'-triphosphate
    and, 438
  on brain capillaries, 446–47
  broken cell preparations and,
    430
  cardiac and broncho-selective
    drug development and,
    422–26
  in cat and guinea pig heart,
    441
  catecholamines and, 419
  in the caudate, 449–51
  in the cerebellum, 447–49
  in the cerebral cortex,
    444–47
  classification of, 420
  density of
    adrenalectomy and, 446
    drugs and, 445
  drug affinity and selectivity
    and, 423–28
  heterogeneous drug binding
    sites and, 436–38
  localization of, 441–51
    cellular, 442–44
    in mammalian tissues, 440

  in non-mammalian tissues,
    440–41
  physiological function of,
    421–22
  radioligand binding assays
    and, 431–36
  in rat brain, 441–42
  stereoselectivity of, 428
  subtypes of, 428–41
    classification of, 438–39
    coexistence of, 429–30, 439
Aequorin
  measurement of changes in
    calcium and, 88
Affective disorders
  insomnia and, 383–86
Aging
  adrenergic receptors and,
    451–52
Agnathans
  evolution of, 302
  forebrain of, 308
  telencephalic variation in,
    309–11
Agonists
  adrenergic receptor
    interactions with
    guanosine 5'-triphosphate
    and, 438
  β_1-selective, 422–23, 434
  β_2-selective, 424, 435
Alcohol
  delayed sleep phase
    syndrome and, 402
  sedative effects of, 387
Alcoholism
  insomnia and, 387–89
Alveolar hypoventilation
  syndrome
  central, 389–90
Alzheimer's disease
  hypersomnia and, 399
Amino acids
  gap junction permeability
    and, 44
γ-Aminobutyric acid (GABA)
  calcium current modulation
    and, 116
  dentate gyrus inhibition and,
    357
  dorsal root ganglion cell
    action potential and, 113
4-Aminopyridine
  calcium currents and, 75
  myasthenia gravis and, 215
  potassium channel blocking
    and, 149–50
Amphetamines
  narcolepsy and, 397

542

CUMULATIVE INDEXES

CONTRIBUTING AUTHORS, VOLUMES 1–4

CHAPTER TITLES, VOLUMES 1–4

556   CHAPTER TITLES